Communications in Computer and Information Science 891

Commenced Publication in 2007
Founding and Former Series Editors:
Phoebe Chen, Alfredo Cuzzocrea, Xiaoyong Du, Orhun Kara, Ting Liu,
Krishna M. Sivalingam, Dominik Ślęzak, Takashi Washio, Xiaokang Yang,
and Junsong Yuan

Editorial Board Members

Simone Diniz Junqueira Barbosa ⓘ
    Pontifical Catholic University of Rio de Janeiro (PUC-Rio),
    Rio de Janeiro, Brazil
Joaquim Filipe ⓘ
    Polytechnic Institute of Setúbal, Setúbal, Portugal
Ashish Ghosh
    Indian Statistical Institute, Kolkata, India
Igor Kotenko ⓘ
    St. Petersburg Institute for Informatics and Automation of the Russian
    Academy of Sciences, St. Petersburg, Russia
Lizhu Zhou
    Tsinghua University, Beijing, China

More information about this series at http://www.springer.com/series/7899

Lucio Grandinetti · Seyedeh Leili Mirtaheri ·
Reza Shahbazian (Eds.)

High-Performance Computing and Big Data Analysis

Second International Congress, TopHPC 2019
Tehran, Iran, April 23–25, 2019
Revised Selected Papers

 Springer

Editors
Lucio Grandinetti
University of Calabria
Rende, Italy

Seyedeh Leili Mirtaheri
Kharazmi University
Tehran, Iran

Reza Shahbazian
University of Calabria
Rende, Italy

ISSN 1865-0929   ISSN 1865-0937   (electronic)
Communications in Computer and Information Science
ISBN 978-3-030-33494-9   ISBN 978-3-030-33495-6   (eBook)
https://doi.org/10.1007/978-3-030-33495-6

This Springer imprint is published by the registered company Springer Nature Switzerland AG
The registered company address is: Gewerbestrasse 11, 6330 Cham, Switzerland

Preface

Recent technological developments have caused rapid growth in network data size for applications including intelligence transportation, mobile sensing, Internet of Things (IoT), M2M communications, and online social media – all bringing people into the era of big data. These network data hold much valuable information that could significantly improve performance. However, big data processing requires a vast amount of computing resources, not available in traditional data analytics infrastructures.

High-performance computing (HPC) and big data require almost diverse approaches. However, it is not possible to invest in both fields, simultaneously. Therefore, HPC techniques have been widely agreed upon as a promising solution to facilitate big data processing, keeping in mind that tremendous research challenges such as the scalability of computing performance for high velocity; high variety and high volume big data; learning with massive-scale datasets; multi-core, GPU, and hybrid distributed environments; and unstructured data processing still exist as open research issues.

TopHPC is a joint international congress between Europe and Asia including top level international organizers, Program Committee members, and keynote speakers. This event attracts many researchers focused on advances in computing methods and data analytics.

The call for papers for the TopHPC 2019 conference attracted a number of high-quality submissions. During a rigorous review process, each paper was reviewed by at least four experts. In addition, we invited ten keynote speakers listed below:

- Prof. Andrei Raigorodskii (Lomonosov Moscow State University): "Big data Projects at Phystech-School of Applied Mathematics and Computer Science"
- Prof. Benjamin W. Wah (University of Hong Kong): "Using Dominance to Harness the Complexity of Big Data Applications"
- Dr. Ali Hakim Javadi (Behsazan Farda Holding): "Digital Transformation, Bank 4.0 and HPC"
- Prof. Nahid Emad (University of Versailles): "High Performance Data Analytics and Some Applications"
- Prof. Vladimir Voevodin (Lomonosov Moscow State University): "HPC Education at Lomonosov Moscow State University: Current State, Challenges and Trends"
- Dr. Happy Marumo Sithole (Centre for High Performance Computing at the CSIR): "The Impact of High-Performance Computing in Large Scale Science Projects"
- Prof. Erwin Laure (PDC Center for High Performance Computing): "Programming Heterogeneous Systems at Exascale"
- Dr. Hossein Asadi (Sharif University): "Cost-Effective Storage Architectures for Virtualized Platforms"
- Dr. Shahram Moein (Iran Telecommunication Research Center): "Big Data: A Main Research Topic in ICT Research Institute"
- Dr. Ali Asghar Ansari (Ministry of Health): "Big Data in Healthcare"

- Dr. Parinaz Ameri (OranIT Company): "Databases Throughout and Beyond Big data Hype"
- Prof. Mikhail Petrovskiy (Lomonosov Moscow State University): "Machine Learning and Big Data Technologies for User Behavior Monitoring and Analysis"
- Dr. Masoud Mazloum (University of Amsterdam): "All Science will Become Data Science"
- Prof. Karl Svozil (Technische Universität Wien): "Quantum Clouds as Pathways to Non-(hyper)-Classicality"

We also invited 15 workshops presenters from academic and industrial experts with the following titles:

- Pawel Plaszczak (Consultant and Specialist in Data Processing and Data Science): "Data Engineering for Data Scientists"
- Parinaz Ameri (Co-Founder and Consultant of OranIT GmbH): "Intro to Machine Learning for non-Data Scientists"
- Vahid Amiri (Data Engineer at RCDAT – Research Center of Developing Advanced Technologies): "Big Data Enterprise Architecture"
- Mohammad-Shahram Moin, Mehdi Hosseinpour, and Hamze Sadeghizadeh (ITRC): "How To Govern Data: A Multi-level Perspective"
- Ehsan Aryanian, Dr. Ayoub Mohammadian, Sara Mortezagholi, and Mitra Karami (ITRC): "Legal Requirements for Data-Driven Businesses"
- Amin Sadeghi (Ostadkar Co-founder and Assistant Professor in Tehran University): "Cryptocurrency and Its Uses in Business"
- A. Abedinzade and M. Anbari (Mohaymen Company): "Real-time Processing Systems for Big Data at High Velocity Streams: Based on Distributed Computing Technology"
- Amirreza Tajally (Industrial Engineering, University of Tehran): "Big Data Analytics in Industries, Energy & Electricity Engineering"
- Arash Attari (Petroleum Engineer – Expert in Oil and Gas Industry – Itech Accelerator): "Industry 4.0 in Oil and Gas and Energy Industry"
- Masoud Mazloom (Postdoc Researcher with the University of Amsterdam): "Machine Learning Is the Next Big Thing in Computing"
- Amin Nezarat (Assistant Professor, Payame Noor University): "An Introduction to Data Science Programming with MPI"
- Yaas Arghavani (Industrial Engineering Company): "Big Data in Banking: Many Challenges, More Opportunities"
- Amir Nouri (ISC Co.): "Neshan System – LI in Banking"
- Shohreh Tabatabaei (ISC Co.): "Graph Analysis of Banking Transactions"

We also had a panel discussion titled "HPC and Big Data/AI-Convergence or Co-existence: where are we heading?" The panelists were Prof. Vladimir Voevodin, Prof. Benjamin Wah, Prof. Nahid Emad, Dr. Masoud Mazloom, Dr. Happy Sithole, Prof. Shahin Rouhani, and the panel's chair was Prof. Erwin Laure. The main issues presented at this panel were:

- We hear a lot of talk about the convergence of HPC and Big Data/AI – do we have enough application pull for this?

- While hardware is getting more similar in principle, there are important differences, like low precision for AI – will this hamper the convergence?
- Will we see one unified hardware or more "modular" systems with different "islands." If the latter, why do we bother about convergence?
- Currently we have vastly different software stacks that are often non-interoperable but people start to learn from each other – will we see full convergence or rather co-existence?
- Looking into the glass ball – what will be the next Big Thing?

We would like to individually express our gratitude to all authors for choosing TopHPC 2019 – each did their best in writing, revising, and presenting their papers. We appreciate the invaluable support of the TopHPC 2019 Program Committee and thank reviewers for their efforts and contributions in reviewing the papers.

July 2019

Lucio Grandinetti
Seyedeh Leili Mirtaheri
Reza Shahbazian

Organization

General Chair

Lucio Grandinetti   University of Calabria, Italy

Program Committee Chairs

Seyedeh Leili Mirtaheri   Kharazmi University, Iran
Lucio Grandinetti   University of Calabria, Italy

Steering Committee

Ali Ahmadi	Khaje Nasir Toosi University, Iran
Parvin Dadandish	Islamic Azad University at West Tehran, Iran
Mohammad Eshghi	Shahid Beheshti University, Iran
Lucio Grandinetti	University of Calabria, Italy
Mohammad Khansari	University of Tehran, Iran
Ehsan Mahmoudpour	University of Tehran, Iran
Seyedeh Leili Mirtaheri	Kharazmi University, Iran
Seyed Ahmad Moatamedi	Amirkabir University, Iran
Shahin Rouhani	Sharif University of Technology, Iran
Seyed Hossein Sajjadi Nayeri	Soft Tech Development Council, Iran
Hamid Sarbazi–Azad	Sharif University of Technology, Iran
Reza Shahbazian	University of Calabria, Italy
Jamshid Shanbezadeh	Kharazmi University, Iran
Mohsen Sharifi	Iran University of Science and Technology, Iran

Program Committee

Hamid Reza Arabnia	University of Georgia, USA
Mohammad Kazem Akbari	Amirkabir University, Iran
Reda Alhajj	University of Calgary, Canada
Hossein Asadi	Sharif University of Technology, Iran
Seyed Amir Asghari	Kharazmi University, Iran
Mahmud Ashrafizadeh	Isfahan University of Technology, Iran
Kambiz Badie	ITRC, University of Tehran, Iran
Keivan Borna	Kharazmi University, Iran
Rajkumar Buyya	Manjrasoft, University of Melbourne, Australia
Cristian S. Claude	University of Auckland, New Zealand
Giacinto Donvito	Tier-2 Computing Centre at INFN-Bari, Italy
Omid Mahdi Ebadati	Kharazmi University, Iran

Kimmo Koski	IT Center for Science, Finland
Erwin Laure	KTH Stockholm, Sweden
Pejman Lotfi Kamran	IPM, Iran
Masoud Mazloom	University of Amsterdam, The Netherland
Abolfazl Mirzazadeh	Kharazmi University, Iran
Mohammad Shahram Moin	ICT Research Institute, Iran
Ali Movaghar	Sharif University of Technology, Iran
Manish Parashar	Rutgers University, USA
Valerio Pascucci	University of Utah, USA
Mir Mohsen Pedram	Kharazmi University, Iran
Mikhail Petrovskiy	Lomonosov Moscow State University, Russia
Andrei Raigorodskii	Lomonosov Moscow State University, Russia
Sugam Sharma	Iowa State University, USA
Happy Sithole	University of Limpopo, South Africa
Leonid Sokolinsky	South Ural State University, Russia
Thomas Sterling	Indiana University, USA
Karl Svozil	Technische Universität Wien, Austria
Domenico Talia	University of Calabria, Italy
Alfredo Tirado-Ramos	University of Texas, USA
Hadi Veisi	University of Tehran, Iran
Vladimir Voevodin	Lomonosov Moscow State University, Russia
Benjamin W. Wah	University of Hong Kong, Hong Kong, China
Cheng-Zhong Xu	Chinese Academy of Sciences, China
Amy Yuexuan Wang	The University of Hong Kong, Hong Kong, China

Contents

Deep Learning

Big Data Analytics

Internet of Things

Data Mining, Neural Network and Genetic Algorithms

Performance Issues and Quantum Computing

Deep Learning

Providing RS Participation for Geo-Distributed Data Centers Using Deep Learning-Based Power Prediction

Somayyeh Taheri[1,2], Maziar Goudarzi[1(✉)], and Osamu Yoshie[2]

[1] Department of Computer Engineering, Sharif University of Technology,
Tehran, Iran
s.taheri@fuji.waseda.jp, goudarzi@sharif.edu
[2] Graduate School of Information, Production and System,
Waseda University, Shinjuku, Japan
yoshie@waseda.jp

Abstract. Nowadays, geo-distributed Data Centers (DCs) are very common, because of providing more energy efficiency, higher system availability as well as flexibility. In a geo-distributed cloud, each local DC responds to the specific portion of the incoming load which distributed based on different Geographically Load Balancing (GLB) policies.

As a large yet flexible power consumer, the local DC has a great impact on the local power grid. From this point of view, a local DC is a good candidate to participate in the emerging power market such as Regulation Service (RS) opportunity, that brings monetary benefits both for the DC as well as the grid. However, a fruitful collaboration requires the DC to have the capability of forecasting its future power consumption. While, given the different GLB policies, the amount of delivered load toward each local DC is a function of the whole system's conditions, rather than the local situation. Thereby, the problem of RS participation for local DCs in a geo-distributed cloud is challenging. Motivated by this fact, this paper benefits from deep learning to predict the local DCs' power consumption. We consider two main GLB policies, including Power-aware as well as Cost-aware, to acquire training data and construct a prediction model accordingly. Afterward, we leverage the prediction results to provide the opportunity of RS participation for geo-distributed DCs. Results show that the proposed approach reduces the energy cost by 22% on average in compared with well-known GLB policies.

Keywords: Deep learning prediction · Emerging power market · Geo-distributed data center · Regulation service

1 Introduction

Cloud computing is among the fastest markets which is considered as the next-generation computing paradigm because of its advantages in term of availability,

© Springer Nature Switzerland AG 2019
L. Grandinetti et al. (Eds.): TopHPC 2019, CCIS 891, pp. 3–17, 2019.
https://doi.org/10.1007/978-3-030-33495-6_1

flexibility, security as well as cost. The all-over functionality of the cloud which offers to work anywhere leads to the widespread broadcasting of cloud users, around the world. Hereupon, DCs, as the underlining infrastructure of the cloud, are imposed to be distributed over the long geographically distance, near to cloud users. As far as, replacing the centralized cloud with a geo-distributed ones is the subject of most current efforts by the cloud providers.

Although a distributed DC is much smaller than the centralized one, it is undoubtedly large enough to be considered as a big power consumer and hence, widely affects the local power grid. Thus, it can be counted as a good participant in the new power market.

Recently, the power market has provided Regulation Service (RS) to match supply and demand, while simultaneously providing economic benefits to consumers and the network. RS is the market-based compensation service for customers who have the ability to adjust their electricity demand with the electricity generation. The cooperation results affect both of the power grid as well as participants' energy cost, effectively. However, for fruitful collaboration, participants need to be aware of their future power requirement and also have the ability of real-time power management. Obviously, as a large yet flexible consumer, DCs are excellent candidates for participating in the RS. While, considering the electricity cost as a big portion of a DC expense, RS participation brings the significant beneficiary for it. Nevertheless, as a small part of a larger image, forecasting the future power requirement of a local DCs in a geo-distributed cloud is challenging.

For more insight, please note that in a geo-distributed cloud, the incoming load is dispatched toward DCs based on the specific GLB policy; Obviously, according to the wide considerations of the cloud. Taking advantages of condition variations among local DCs, GLB has the promise of significant monetary saving. However, the local DCs' power consumption presents a wide range of variations. In other words, power predicting of DCs is a complex problem which has a wide range of effective input variables. Motivated by this fact, we suggest using deep learning as a well-known and powerful technique to predict the local DCs' power consumption.

In this paper, we consider two main GLB policies including *Power-aware* as well as *Cost-aware* to acquire training data and construct the power prediction model accordingly. Leveraging the Convolutional Neural Network (CNN) for power prediction, we provide the RS participation opportunity for geo-distributed DCs and reduce the electricity cost of GLB, effectively.

2   Back Ground and Related Work

2.1   Regulation Service and DC Participation

In an RS contract, the customer declares its average power consumption ($\overline{P}$) and power adjustment capacity (R) in advance. The Independent System Operator (ISO) also declares the power price (Π^P) as well as the adjustment capacity (RS) credit (Π^R). Then, the electricity billing is calculated based on the average power consumption ($\Pi^P \overline{P}$) and received the credit based on the adjustment capacity ($\Pi^R R$). Undoubtedly, the credit does not come for free. For this, at the beginning of each billing period, the ISO clarifies the target power consumption (P_{tar}) through an RS signal ($z(t)$) as (1):

$$P_{tar} = \overline{P} + Rz(t). \tag{1}$$

Where $-1 < z(t) < 1$ [11]. Then, for a fulfilled participation, the customer has to effectively control its actual power consumption (P_{act}) so as to track the target power. Otherwise, it faces paying penalty based on the power mismatches magnitude.

Providing the RS participation opportunity for DC has been widely studied in recent researches. Proposing an integrated system of DC and plug-in electric vehicles [19], developing a coupled DC with battery storage [14], utilizing available energy storage in a DC [16], as well as benefiting from green DC [4], are among the proposed approaches. Chen et al. demonstrate the DC capabilities to participate in the emerging power markets [7]. Authors propose a dynamic server power regulation approach, which dynamically manages the DC's power consumption to track the RS signal, while satisfies the Quality-of-Service(QoS) constraint. Authors continue their work by modeling the QoS in DC and propose a new approach of processing rate modulation aimed at optimizing the RS tracking error as well as QoS violation penalty [8]. Regarding the server power proportionality, Ghasemi-Gol et al. suggest managing server resources to control the DC power consumption, subject to the Service-Level Agreement (SLA) [12].

The RS participation is being addressed in different studies; However, none of them consider geo-distributed DCs. To our best knowledge, this work is the first that pays attention to the RS participation for local DCs in a geo-distributed cloud.

2.2   DC Power Prediction

Regarding the recent advances in deep learning, a few studies have focused on utilizing them to predict the DC power requirement. Authors in [18] use historical power data to predict the future. They consider two different time scale for data acquisition and accordingly propose two different power prediction models that are based on the linear recursive Auto Encoder (AE). They reinforce the proposed models with an additional layer to correct the prediction results with respect to the prediction error analysis. [6] considers the weather condition as a major effective factor on the DC power consumption. Authors analyze the correlation between different features of weather and the DC power consumption, and extract more effective features to build a linear regression-based power prediction, accordingly. Hsu et al. [17] consider the power modeling based on a wide range of input variables and propose a prediction framework using the self-aware computing. The proposed framework dynamically updates the effective training variables according to the analysis of variables interaction as well as the prediction error monitoring. [13] uses the servers power modeling to estimate the DC power consumption. Authors utilize historical data for CPU as well as memory usage and apply the neural network to predict future values. Afterward, they estimate the DC required power based on their power modeling and explore the opportunity of RS participation in the DC.

To the best of our knowledge, the existing studies neither pay attention to the power prediction in the geo-distributed DC nor use it to provide RS participation opportunity, and this work is the first on this topic. In summary, our contributions are as follows:

- Given the benefits and challenges of RS participation for geo-distributed DCs, we utilize deep learning to construct a power prediction model for local DCs.
- Regarding the correlation between the GLB policy and the local DC power consumption, we acquire the training data through the formulation and solving two well-known GLB policies.
- Instead of using historical power data, we calculate the DCs' power consumption based on the accurate model of load-dependent PUE as well as server power consumption model.
- We consider a set of input variables for our prediction model wide enough to reach acceptable accuracy, yet narrow enough to prevent unnecessary complexity.

3   System Model

3.1   Geo-Distributed Cloud Structure

Cloud computing architecture comprises the cloud components, which are loosely coupled. At its most basic, the cloud architecture is classified into two sections including front-end which is the user part of the cloud, and back-end, which is all of the resources required to process the incoming load.

In a geo-distributed cloud, all requests reach the front-end, and then different portions are dispatched toward DCs based on different GLB policies. Each DC includes number of servers, power unit, cooling system as well as network infrastructure. To the best of pervious knowledge as well as our observations, most of the time, 25–40% of servers in a DC are idle [9]. Thus, to manage the number of on-off server switching, we divide the total number of servers in a DC into active and reserved set. We consider the power utility comprises the power grid and solar power and apply a hybrid model of a cooling system consists of an air-side economizer for free cooling as well as a chiller-based air conditioner. Figure 1 depicts an overview of our DC structure.

The amount of power consumption in a DC depends on its Power Usage Efficiency (PUE), which is defined as the ratio of the total power consumption to the power consumed by the IT equipment. Regarding the DC structure in this work, we consider the PUE as a load-dependent factor which models the amount of power consumed by the hybrid cooling system and is a function of the IT power consumption and time as (2):

$$PUE_{IT,t} = \frac{P_{IT,t} + Cooling_{IT,t}}{P_{IT,t}}. \tag{2}$$

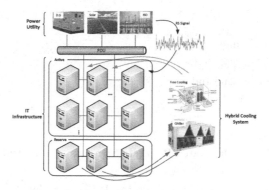

Fig. 1. Overview of DC structure.

4   Methodology and Formulation

In this section, we present the DC power modeling and then formulate the RS participation in the local DCs.

4.1   DC Power Modeling

Table 1 lists the parameters and their meanings.

We consider a geo-distributed cloud with $U = \{u_1, u_2, \ldots u_{|U|}\}$ front-ends, which dispatches the incoming loads $N_t = \{n_{1,t}, n_{2,t}, \ldots n_{|U|,t}\}$ toward local DCs; where $n_{i,t}$ denotes the number of originated requests from front-end i at time slot t. The proposed model has $| L |$ local DCs, each of which including contemporary servers as $S = \{s_1, s_2, \ldots s_{|L|}\}$. Every time slot is set to one billing period. Thus, in all equation in the rest of paper, we eliminate time index for simplicity; However, the concept of time implies with all parameters.

The average power consumption of DC j over time slot t, and without any power management method is as (3):

$$P_{DC_j} = PUE_{IT_j} P_{IT_j} - Solar_j. \tag{3}$$

Where $Solar_j$ denotes the available solar power of DC j and P_{IT_j} represents the IT power consumption, defined as (4):

$$P_{IT_j} = y_j(u_{Reqj}(p_{activej} - p_{idlej}) + p_{idlej}). \tag{4}$$

Where u_{Reqj} is the utilization that each request imposes to the servers of DC j and y_j represents the number of required servers in DC j and can be written as (5):

$$y_j = \sum_{i=1}^{|U|} X_{i,j}. \tag{5}$$

Table 1. Parameters and meaning

Parameter	Meaning
t	Time Slot Index
i	Front-End Index
j	Local DC Index
{U}	Set of Front-End
{L}	Set of Local DC
{N}	Set of Originated Requests from Front-Ends (time slot t)
{S}	Set of DCs' Servers
$X_{i,j}$	Request from Front-End i to DC j (time slot t)
y_j	DC j Number of Required Servers (time slot t)
PUE_{IT_j}	DC j Power Usage Efficiency (time slot t)
p_{active_j}	DCj's Server Active Power
p_{idle_j}	DCj's Server Idle Power
u_{Req_j}	Request Utilization on Server of DCj
P_{IT_j}	DC j IT Power Consumption (time slot t)
$\hat{P}_{DC_j}$	Prediction of DC j Average Power Consumption (time slot t)
P_{DC_j}	DC j Average Power Consumption (time slot t)
P_{tar_j}	DC j Target Power Consumption (time slot t)
P_{act_j}	DC j Actual Power Consumption (time slot t)
R_j	DC j Power Adjustment Capacity (time slot t)
Π_j^P	DC j Electricity Price (time slot t)
Π_j^R	DC j RS Credit (time slot t)
$Pentaly_j$	DC j Energy Mismatch Penalty (time slot t)
$EnergyCost$	Total Energy (Electricity) Cost of the Cloud
$Power$	Total Power Consumption of the Cloud
e_j	DC j Cost of Server on-off Switchin
τ	Time Slot Duration
$z_j(t)$	Dynamic RS Signal Declared for DC j
ϵ_j	DC j Prediction Error (time slot t)

4.2   RS Participation Formulation

Let $\hat{P}_{DC_j}$ denotes the power prediction of DC j. Note that $\hat{P}_{DC_j}$ is the prediction of P_{DC_j} and these two values are not exactly the same, because of inevitable prediction inaccuracy. We consider $\epsilon_j = |P_{DC_j} - \hat{P}_{DC_j}|$ as the power prediction absolute error in DC j over time slot t.

As already mentioned, at the beginning of each billing period, DC j declares $\hat{P}_{DC_j}$ to the ISO; Then, ISO sets the target power consumption of the DC j, that is $P_{tar_j} = \hat{P}_{DC_j} + R_j z_j(t)$. During the billing period, DC j modulates its power consumption (ΔP_{DC_j}) through various power management methods. As a result, the actual power consumption of DC j would be $P_{act_j} = P_{DC_j} + \Delta P_{DC_j}$. Finally, the total energy cost of the DC j based on the RS contract is calculated as 6 [1]:

$$EnergyCost_j = \Pi_j^P P_{act_j} - [\Pi_j^R R_j - \Pi_j^R R_j Pentaly_j(\epsilon_j + |\Delta P_{DC_j} - R_j z_j(t)|)]. \tag{6}$$

Where $Penalty_j$ is the penalty coefficient that the DC j must pay for mismatches between the target power and the actual power consumption.

Energy Cost is our objective function which want to optimize through an RS participation. The problem has certain constraints on $\hat{P}_{DC_j}$ as well as P_{DC_j}. Please note that $|\Delta P_{DC_j} - R_j z_j(t)|$ depends on the power management techniques in DCs, which is out of the scop of this paper. Thus, our objective is a function of the power prediction error.

4.3   Prediction Model

The prediction procedure includes several steps as the historical data acquisition, original data pretreatment for noise elimination, training data generation, and prediction model construction.

We acquire historical data from the local databases and formulate two basic GLB policies including Power-aware as well as Cost-aware to generate the training data.

Training Data Generation and GLB Formulation: We generate the training data by the off-line data, which is referred to as "batch learning". Batch algorithms are divided into two categories based on their training data generation policy including off-policy and on-policy. The training data generation policy in off-policy algorithms is generally independent of the policy applied for predicted data. While on-policy directly uses the predicted values as the control values concurrently. In this work, we propose an off-policy control algorithm which is based on different formulations of the GLB problem, as bellow:

Cost-Aware GLB Problem: The goal of this policy is to distribute the incoming load among local DCs in such a way that the total electricity cost is minimized. The problem statement and formulation are as follows:

Problem State: In a distributed cloud with $| L |$ DCs, each with a temporal PUE variation (PUE_{IT_j}) and electricity price variation (Π_j^P), distribute the incoming load, aimed at minimizing the overall energy cost ($EnergyCost$):

$$Minimize: \quad EnergyCost = \sum_{j=1}^{|L|}[(P_{DC_j} + e_j(|y_{j,t} - y_{j,t-1}|))\Pi_j^P \tau]. \tag{7}$$

We consider e_j as the on-off switching penalty in the DC j to limit the sudden and large fluctuation in the number of server switching over the consecutive time slots. We set e_j to the 20% of the active power as well as idle power of server in DC j, for on and off switching, respectively. Constraints of the above problem are as follows:

$$\forall i \in U, j \in L : \sum_{j=1}^{|L|} X_{i,j} = n_i. \tag{8}$$

That is, every incoming load would be served with one and only one DC.

$$\forall j \in L : 0 <= y_j <= s_j. \tag{9}$$

That is, the total required servers in DC j would be less than or equal to the total servers number.

$$\forall i \in U, j \in L : y_j >= \sum_{i=1}^{|U|} X_{i,j}. \tag{10}$$

That is, the number of active servers in DC j would be sufficient.

Power-Aware GLB Problem: This policy is aimed at minimizing the total power consumption of serving the incoming load through local DCs and is formulated as (11):

$$Minimize: \quad Power = \sum_{j=1}^{|L|} ((P_{DC_j} + e_j(|y_{j,t} - y_{j,t-1}|)). \tag{11}$$

S.t: (8), (9), (10).

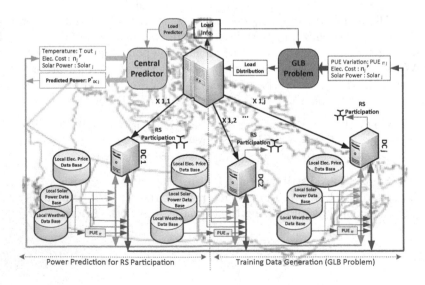

Fig. 2. Overall system structure. (Color figure online)

Prediction Model Construction: Figure 2 shows the overall structure of our simulated cloud as well as the separation of the problem steps. Blue lines represent the training data generation procedure; While green ones show the prediction procedure.

We consider two levels of prediction including the local predictor for incoming load prediction as well as the central predictor for DCs' power prediction.

To predict the incoming load, we utilized Seasonal Autoregressive Integrated Moving Average (SARIMA) or Seasonal ARIMA, which is an extension of ARIMA. SARIMA explicitly supports univariate time series data with a seasonal component and is formed by including additional seasonal terms in the ARIMA model.

For power prediction through the central predictor, we leverage the Convolutional Neural Network-based (CNN-based) prediction model. CNN is a type of deep neural network that has the ability of reading sequences of input data and automatically extracting the features. In this paper, we adopt the feed forward back propagation CNN to construct the power prediction model. Our prediction model is a one-step multi-variate problem which include three layers as: input layer, hidden layer and output layer. We consider the input layer with 13 neurons (including the input variables) and the hidden layer with 25 neurons, respectively. The input layer is fed with single time slot data and each output is the next time slot data ($\hat{P}_{DC_j, t+1}$). We use Rectified Linear Unit (ReLU) as the activation function which defined as:

$$f(x) = \frac{e^x}{1 + e^x}. \tag{12}$$

We initialized the interconnected neurons' weights with the uniform distribution and updated them in response to the back propagated errors, by adopting adam optimization during the training process. Adam optimization is a method for efficient stochastic optimization that only requires first-order gradients with little memory requirement. For more learning efficiency, we also shared the same weight and bias among every neuron in the same layer. We considered Root Mean Squared Error (RMSE) as an evaluation metric. Table 2 summarizes the input variables of our predictor.

Please note that the computation complexity of the prediction model is low since (i) the number of neurons in the hidden layer is small and(ii) the frequency of he prediction execution is low.

Table 2. Input variables of prediction model.

Category	Input variable	Repository
Temperature	DCs' Outdoor Temp.	Local Data Base
Solar Power	DCs' Solar Power	Local Data Base
Cost*	DCs' Electricity Price	Local Data Base
Work Load	Incoming Load	Local Predictor

*Only for Cost-aware based prediction.

Table 3. Cloud specification.

Location	Server type	Number	P_{active}	P_{idle}
Manitoba	Intel E5-2699	1500	0.529	0.102
Quebec	Intel X5570	1000	0.352	0.153
Minnesota	Intel E5506	2000	0.419	0.146
Ontario	AMD EPYC 7601	1500	0.483	0.138

5   Performance Evaluation

5.1   Experimental Setup

We considered a cloud with 4 geo-distributed DCs located at Manitoba, Quebec, Ontario, Minnesota namely DC1, DC2, DC3 and DC4, respectively. The characteristics of our DCs including the server type and specification as well as the total number of servers is shown in Table 3. We obtained the available solar power data using the PVWatts calculator [2]. In the hybrid cooling system, we modeled the air-side economizer's power consumption as a cubic function of IT equipment power consumption [15] and used the Coefficient of Performance (CoP) model of a typical water-chilled cooling system [5], for the chiller-based section. Finally, the cooling power consumption of the hybrid system would be calculated as (13):

$$Cooling_{IT} = \begin{cases} \alpha P_{IT}^3 & P_{IT} < P_{ITRef} \\ \alpha P_{ITRef}^3 + \frac{(P_{IT} - P_{ITRef})}{CoP} & \text{otherwise.} \end{cases} \tag{13}$$

where α is the outside-inside temperature difference coefficient, CoP is the chiller Coefficient of Performance, and P_{ITRef} is the optimal cooling power capacity between the air-side economizer and the chiller, that is, $\sqrt{\frac{1}{3\alpha CoP}}$.

We employed Google cluster workload through a period of two weeks from May 1–14 as our workload model [3]. The incoming load arrival rate over every time slot was extracted and normalized with respect to the total server number in our cloud. We used the BARON solver through the General Algebraic Modeling System (GAMS) 24.6 to solve the GLB problems.

For electricity pricing, we used the Real-Time (RT) energy pricing model in which the electricity price declared at the beginning of each billing period is valid until the end of it. We acquired the electricity pricing data from [1] for each location.

We considered each billing period as one hour and set Π^R to 30–50% of Π^P and *Penalty* as a random variable between 0.1 and 0.5. The RS signal ($z(t)$) is set up as a random variable with a uniform distribution between -0.2 and 0.2 [10].

5.2   Algorithm in Comparison

We considered five different algorithms and compared their total electricity cost:

- Baseline: The basic model of our system which dispatches the incoming loads toward the DCs based on the available computing capacity in them.
- Cost-aware: that distributes the incoming load based on the Cost-aware GLB.
- Power-aware: that distributes the incoming load based on the Power-aware GLB.
- Cost-Aware based RS participation (CARS): that uses the Cost-aware GLB results to generate the training data and constructs the power prediction model accordingly. This approach provides the RS participation opportunity for DCs. The total energy cost is calculated based on the (6).

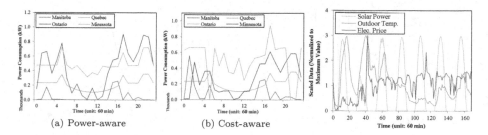

Fig. 3. Local DCs' power consumption variation on My 5.

(a) Power-aware   (b) Cost-aware

Fig. 4. Variation of effective factors in DC2. (Color figure online)

- Power-Aware based RS participation (PARS): that applies the Power-aware GLB results to generate the training data and provides the opportunity of RS participation for the DC. The total energy cost is calculated based on the (6).

5.3 Simulation Results

System Evaluation. We begin our evaluation by considering the DCs' power variation over the consecutive time slots as a function of GLB policy, as shown in Fig. 3.

As the fig. shows, basically, the DC's power consumption is influenced by the GLB policy. In addition, given a specific GLB policy, the DC's power consumption does not indicate a clear pattern of variations over the consecutive time slots. That is due to a wide range of factors influencing the results of GLB. Figure 4 depicts the variation of the effective factors in DC2 during May 1–7.

Figure 4 clearly represents variations in local conditions over the consecutive time slots, which results in changes in GLB results. In addition, as already mentioned, not only local conditions but the whole system's conditions affect the power consumption of a DC. For more insight, please let to compare the (brown) power consumption of a DC with its available solar power, under the Cost-aware GLB. Figure 5 depicts this for DC2 and DC4 during May 1–2.

As we expect and Fig. 5 shows, so far as (green) solar power is available, the DC consumes less utility power. However, while solar power varies periodically, the DC's power consumption does not indicate the same pattern and is different for different days as well as among different DCs.

All of the above confirms our argument about a wide range of effective factors on DCs' power consumption and the high complexity of the power prediction problem in a geo-distributed cloud.

GLB Problem Analysis. Along with considering the existing computing capacity, one of the other important issues during GLB is the number of server switching in DCs. We considered the on-off switching penalty in our proposed

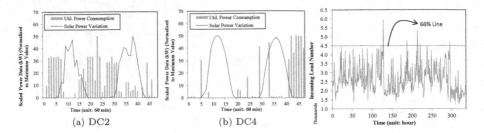

Fig. 5. Comparison of power consumption and solar power during My 1–2. (Color figure online)

Fig. 6. Incoming load variation during the simulation days.

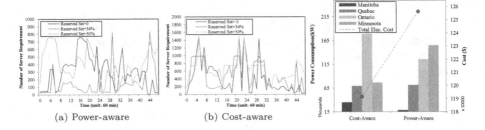

Fig. 7. Number of sever requirement in DC1 as a function of the reserved set size.

Fig. 8. Comparison of power consumption and electricity cost.

GLB formulation (Eqs. (7), (11)). In addition, since most of the times, the existing computing capacity is much higher than needed, we divided the existing servers into active set as well as reserved set. Figure 6 shows the incoming load variation over the simulation days.

Regarding the Fig. 6, we consider the size of the reserved set in each DC to 34% of the total available server. Note that we have the chance to be aware of DCs' power requirement at the begging of each time slot, which means sufficient time to change the server status. Variation in the number of server requirement as a function of size of the reserved set in DC1 and within May 1-2 is shown in Fig. 7.

The fig. shows that server reservation policy results in a lower and smoother server switching frequency over the consecutive time slots. However, the more restriction on the size of the reserved set causes larger fluctuations. In addition, the satisfactory operation of this approach could be influenced by the GLB policy.

Finally, we represent the comparison between different GLB policies in term of power consumption as well as total electricity cost in Fig. 8. As we expect, Cost-aware GLB reduces the electricity cost, while the total power consumption is roughly the same. In addition, the load distribution in different GLB policies is quite different.

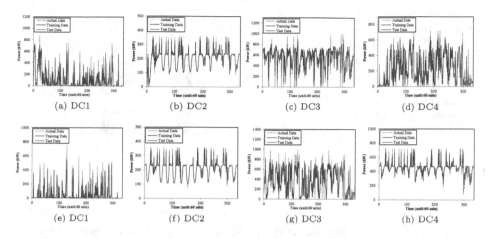

Fig. 9. Prediction model accuracy (a)–(d) Cost-aware based. (e)–(h) Power-aware based.

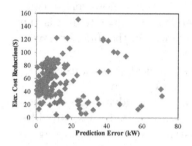

Fig. 10. Electricity cost reduction in DC2 as a function of ϵ_2.

Fig. 11. Electricity cost reduction in DC3 as a function of R_3 to $\hat{P}_{DC_3}$ ratio and Π_3^R.

Prediction Model Evaluation. In this section, we evaluate the accuracy of the prediction model. As mentioned before, after feeding the prediction model with training data, the DCs' power consumption could be predicted (training procedure). Then, some test data is applied to verify the prediction accuracy. We used 80% of the total data for training and the rest was used as test data. Figure 9 compares the accuracy of our prediction model during the training and test procedure. The evaluation metric of RMSE for Cost-aware based prediction as well as Power-aware based prediction reached 6.86% and 4.62% respectively.

RS Participation Monetary Saving. Figure 10 represents the electricity cost reduction of DC2 as a function of prediction error (ϵ_2).

As the fig. shows, our prediction model successfully manages the prediction error which leads to more cost saving. Nevertheless, the other given inputs also affect our objective function as Eq. (6). Figure 11 depicts the electricity cost

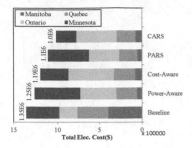
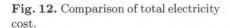

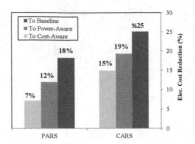

Fig. 12. Comparison of total electricity cost.

Fig. 13. Electricity cost saving percentage.

reduction of DC3 as a function of the ratio of R to the predicted power ($\hat{P}_{DC_3}$) as well as Π_3^R.

The fig. clearly indicates that as R increases, the effect of Π^R increases. Thus, a higher R with higher Π^R results in more cost saving. However, since R represents the flexibility in the DC power management, a higher R means the more effective capability of power management, deep within the different layers of the DC infrastructures.

Figure 12 compares the total electricity cost of simulated algorithms. While, Fig. 13 depicts the percentage of electricity cost saving by them.

Regarding the final results in Figs. 12 and 13, taking advantages of deep learning to predict the power consumption of geo-distributed DCs, provides new opportunities for RS participation in geo-distributed DCs. According to the obtained results, our proposed approach reduces the electricity cost of a geo-distributed cloud by 22% on average, which is undoubtedly a significant monetary saving.

6   Conclusion and Future Work

In this paper, we studied the benefits of RS participation in geo-distributed DCs. We leveraged deep learning to build a power prediction model and formulated two well-known GLB policies to acquire training data for a feed forward, back propagation CNN-based model. Our experimental results reaches 22% electricity costs reduction. In our future work, we plan to investigate other deep learning techniques (LSTM, RNN, etc.) and compare their results. We are also going to consider a wider range of input variables. Other direction includes leveraging deep learning for power management in DCs.

References

1. Independent Electricity System Operator (2018). http://www.ieso.ca/Power-Data/Price-Overview/Time-of-Use-Rates
2. PVWatts Calculator (2018). http://pvwatts.nrel.gov/pvwatts

3. Google Cluster Data (2018). https://github.com/google/cluster-data
4. Arnone, D., Barberi, A., La Cascia, D.: Smart grid integrated green DCs as ancillary service providers. In: International Conference on Clean Electrical Power (2015)
5. Pakbaznia, E., Ghasemazar, M., Pedram, M.: Temperature-aware dynamic resource provisioning in a power-optimized DC. In: Conference on Design, Automation and Test (2010)
6. Smpokos, G., Elshatshat, M.A., et al.: On the energy consumption forecasting of DCs based on weather conditions: remote sensing and machine learning approach. In: International Symposium on Communication Systems, Networks and Digital Signal Processing (2018)
7. Chen, H., Hankendi, C., et al.: Dynamic server power capping for enabling DC participation in power markets. In: International Conference on Computer-Aided Design (2013)
8. Chen, H., Zhang, B., et al.: DC optimal RS provision with explicit modeling of quality of service dynamics. In: IEEE Conference on Decision and Control (2015)
9. Barroso, L.A., Clidaras, J., Hölzle, U.: The DC as a Computer: An Introduction to the Design of Warehouse-Scale Machines, 2nd edn. Morgan and Claypool Publishers (2013)
10. Market-Based Regulation (2018). http://www.pjm.com/~/media/markets-ops/ancillary/regulation-market-concepts-benefits-factor-calculation.ashx
11. Caramanis, M., Paschalidis, I.C., et al.: Provision of RS by flexible distributed loads. In: 51st IEEE Conference on Decision and Control (CDC), pp. 3694–3700 (2012)
12. Ghasemi-Gol, M., Wang, Y., Pedram, M.: An optimization framework for DCs to minimize electric bill under day-ahead dynamic energy prices while providing RS. In: International Green Computing Conference (2014)
13. Liu, N., Lin, X., Wang, Y.: DC power management for RS using neural network-based power prediction. In: 18th International Symposium on Quality Electronic Design (ISQED) (2017)
14. Guruprasad, R., Murali, P., et al.: Coupling a small battery with a DC for frequency regulation. In: IEEE Power and Energy Society General Meeting (2017)
15. Zhou, R., Wang, Z., McReynolds, A.: Optimization and control of cooling micro grids for DCs. In: Thermal and Thermomechanical Phenomena in Electronic Systems (2012)
16. Li, S., Brocanelli, M., et al.: Integrated power management of DCs and electric vehicles for energy and regulation market participation. IEEE Trans. Smart Grid 5(5), 2283–2294 (2014)
17. Hsu, Y.F., Matsuda, K., Matsuoka, M.: Self-aware workload forecasting in DC power prediction. In: International Symposium on Cluster, Cloud and Grid Computing (2018)
18. Li, Y., Wen, Y., Zhang, J.: Learning-based power prediction for DC operations via deep neural networks. In: 5th International Workshop on Energy Efficient DCs (2016)
19. Shi, Y., Xu, B., et al.: Leveraging energy storage to optimize DC electricity cost in emerging power markets. In: 7th International Conference on Future Energy Systems (2016)

Event Detection in Twitter Big Data by Virtual Backbone Deep Learning

Zahra Rezaei[1], Hossein Ebrahimpour Komleh[1(✉)], and Behnaz Eslami[2]

[1] Department of Computer and Electrical Engineering,
University of Kashan, Kashan, Iran
z.rezaei2010@gmail.com, ebrahimpour@kashanu.ac.ir
[2] Department of Computer Engineering, Science and Research Branch,
Islamic Azad University, Tehran, Iran
behnazeslami30@gmail.com

Abstract. In addition to knowledge enhancement, recreation and providing chat, development of Social Network Sites leads to big data, such data can be of great value, as it shows the tendency of the members based on geographical zone, language and culture. The data can also be useful for content oriented planning. In addition, special events of society can be discovered and classified using these data. In some cases, the events have previously existed in society and are considered to be repetitive, like flood in Indian or Typhoon in Florida state, and sometimes the events are unprecedented in which cases, the new event is classified under a new class. The high cost for computations associated with event detection in real time is considered as a major challenge encountered in this context. In the present paper, a model is presented based on deep learning. In the first phase, the first class is trained based on labeled data, then unlabeled data are introduced to the model in a flow manner, and are classified into current classes based on the model through which they have been trained. The data which are higher than a specified threshold are classified into a new class, and if they are lower than the threshold, they are classified as temporary event. At the end, the effectiveness of the model will be evaluated through an available corpus as a benchmark data set. A significant improvement is shown in recall and precision over five state-of-the-art baselines.

Keywords: Deep learning · Classification · Big data · Topic modeling · Virtual backbone

1 Introduction

An event on social networking sites is an interesting event in the real world that leads to discussion of events by users and is immediately commented upon. Sometimes it takes a lot of time for events that users can comment on. Like our definition of an event, [1] defines an occurrence as an update that changes the volume of data and defines its issues over a specific time period. This event

© Springer Nature Switzerland AG 2019
L. Grandinetti et al. (Eds.): TopHPC 2019, CCIS 891, pp. 18–31, 2019.
https://doi.org/10.1007/978-3-030-33495-6_2

is identified with time and subject. And usually relates to issues like people and places. In discussing Topic Detection and Tracking (**TDT**), an event is "something that comes with all the requirements at a specific time and place and the inevitable consequences" defined. [2] Dredze and Osborne found that the latest news reports are faster than social networking sites. In addition, the event detection efficiency in the tweeter is fixed event detection. Atefeh and Kheirich [3] presented some of the event detection methods based on three main categories: the first type of event identified by an event identification system (e.g., identified and unknown events), the second, the identification task (for example, Identifying the new event and detecting the old event) and the final event detection methods (such as observed and invisible and combined methods). In the present paper, the original model is firstly trained in terms of intrusive data in offline mode, then stream data is introduced into the model and classified into current models. If some of the data is produced at high frequency, they are lulling as incidents, then the same class is used to model the subject in event detection. If the topics are the same and are classified in the current classes and the property level is higher than the threshold, then the topics are classified as the event class; however, if the level of the events of the temporary events is low to the current class, then the topics are new class titles are categorized. Such categories provide a daily categorization of views that lead to the identification of particular events. The purpose of classifying all events, in addition to detecting an event, is the discovery of knowledge of the problem. The purpose of this article is to provide a wide range of data coverage in different classes to identify the event. The remainder of this article is as follows: In the second section, previous studies and important approaches to detecting the event are presented. In the third and fourth sections, the proposed model and its architecture are discussed. Then, in the next section, the results of the implementation of the model and in Sect. 4, the results of the data analysis are presented.

2   Related Work

Different event detection methods have been classified using temporal based approaches, Topic modeling based approaches and Incremental clustering based approaches for a detailed discussion on the related literature in the field.

Twitter Live Detection Framework (**TLDF**) Gaglio et al. [4] employs the Soft Frequent Pattern Mining (**SFPM**) algorithm Petkos et al. [5] Term interestingness based methods in order to detect related topics in a broad macro event from the Twitter data stream. A dynamic temporal window size is used by Twitter Monitor Mathiou-dakis and Koudas [6], TLDF so that it can detect events considering their co-occurrences in real time. The TLDF can adjust its event detection performance according to the real volume of tweets related to an event through the dynamic temporal window. The term selection method in the modified SFPM extracts top-k terms from the tweets in the current time window in order to reduce the number of terms which should be studied. A value according to which each term is weighted is identified based on two factors. The first factor is the term which has been identified as entitled entities like persons and locations by the NER Boom et al. [7] and its tf-idf score; the second factor is the maximum proportion of the similarity of appearance for a term in the current time window as well as the reference volume of the tweets which have been collected randomly. The underlying assumption of the Topic modeling based approaches is that some hidden topics always can be seen in the tweets that are processed. Tweets are demonstrated as a combination of topics, and each topic has a probability distribution in the terms that are included in those tweets. The most used probabilistic topic model has been Latent Dirichlet Allocation (**LDA**) Blei et al. [8] in which the topic distribution has a Dirichlet prior. Extracting good topics from the limited context is a challenge that should be solved because of to the limit on the length of a tweet. In addition, the topic modeling based approaches usually consider too much computational price as an effective issue in a streaming setting, and are not that effective in managing the events that are described in parallel Aiello et al. [9].

According to Stilo and Velardi [10], the LDA-based methods can only work offline, since the temporal dimension of the events are usually not considered in clustering approaches, Hasan et al. [11] Twitter News+, employed an incremental clustering based approach which could solve the problem of event detection from the Twitter data stream with a low cost solution to detect both major and minor events from the Twitter data stream. In Hasan et al. [11,12] TwitterNews+ which is an event detection system that includes specialized reversed guides and an incremental clustering approach to detect both major and minor newsworthy events in real time with a low cost, the most expensive operations in the Search Module and the Event Cluster Module algorithms of Twitter News+ have a computational complexity of **O(1)** with parallel processing. Various filters which are applied after the generation of candidate events, together make a computational cost of **O(n2)**, in which n is the number of tweets in an event cluster.

3   Proposed Methods

Observed classification is widely employed in model detection and computer intelligence, and it needs a proper number of labels in order to properly train classification model. By the term *proper number* it is meant that the labeled data must be at a level which they can display the data structure properly. The number of labeled data in factual application is low and reaching the labels is costly, though. Accordingly, we should know how to use unlabeled valuable data to train clustering. There are different methods to solve the problem of label shortage and learning through low number of labels such as self-training, co-training and generative models.

In self-training method which is the common technique in semi-observed learning, the labeled data are used for classification, and then classifier is employed for clustering of unlabeled data. Unlabeled data which are classified with a high assurance are added to labeled data cumulatively. This process continues to the point of convergence. If the labeled data cannot show the general

structure of the data, they face problems, since the primary training of classifier
with labeled data leads to weak results for unlabeled data. Recently, an enhance
version of self-training technique called Help training has been presented. The
main idea is to use a generative model to reselect part of unlabeled data to help
for the training process of the main classifier. The results show that this method
yields to better results compared with self-training method. Still, this method,
due to the generative model which uses labeled data to choose unlabeled subset,
has not solved the basic problem of self-training method. Accordingly, when the
labeled data cannot display the total amount of real data, help training model
is not very helpful. In order to discuss existing events, semi supervised model
twitter is also used. One useful way to understand the reaction of the users to
what is happening, is the ability to check the tweets that discuss a special event.
Accordingly, in this paper, a semi-supervised method is proposed which monitors related tweets, considering a particular event, in order to make a timeline.
In domain vocabulary and Knowledge Base (**KB**) are used to extract related
sub events. In addition, the graph theory was employed to identify relationships
between the sub events and NE mentions. This latter approach is used to the
soccer domain. The main concept of the present paper is to use unlabeled, valuable data in the learning process of classifier. In this paper a framework has been
presented for semi-observed classification in which classification and clustering
have been combined. In this method, Semi Supervised has been combined with
self training process for better learning of classifier. The main advantage of this
method is using labeled and unlabeled data to show the real structure of data
atmosphere through clustering to compensate the limitation of labeled data. The
novelty of this paper are as follows: considering the fact that unlabeled data can
include essential information about data atmosphere, in this paper clustering
methods have been used for revealing the data atmosphere structure in order
to enhance the learning process of the classifier. The labeled data for instruct-
ing the clustering process are used through semi-observed clustering. The new
unlabeled data are used not only for updating the classifier, but also for better
instruction of clustering methods. The importance of social media is that each
user is a potential writer. The language of the writings is closer to reality more
than any other linguistic rule. This special type of data mishmash provides the
opportunity for the enhancement of event detection using models which have
been developed from big data analysis. Through time, it has been shown that
Tweeter is an important data source to plan for objections and to be aware of
crimes, and it is useful in detection of various events. A system which can extract
such data from the vast amount of data in tweeter, and can give a presentation of
imminent events such as sport games and administrations, is of great value. The
systems which are at work for analysis of event in weblogs may fail in processing
of short tweets, and cannot extract rich data. Currently, the existing works either
ignore the structural aspect of Tweeter data or employ traditional methods like
NLP to extract this structure. The characteristic of the proposed method is feature learning for recovering this structure. It's very difficult to find a relationship
explicit and implicit relation among tweets, locating spare data in tweet also to

build metadata from tweets, etc., Considering the fact that most tweets, due to their non-imperative nature cannot be analyzed, the preprocessing of data and making the structure of events and the mentioned limitations in tweets, still can extract some information through relation between these tweets and other tweets in related fields. Therefore, the primary classification helps keeping the structure of tweets. The stream of dig data from social media usually has the following features: unofficial language in using comments and 140-characters limitation for each comment, leads to summarizing the content of message and sometimes block message delivery, for example using pls instead of please, using colloquial language as well as using hashtag. In addition, the tweeter data are so much and sometimes related documents with an event cannot be retrieved easily.

The points which should be taken into account regarding tweets: The information which are shared in social media, especially the information which has argumentative form (like weblogs and tweets), contain rich data. In these social media, the information has noise, unusual spelling and misspelling; therefore, the next stage is for removing noise with the following feature: Those tweets containing hashtag, Tweets with similar content and different user names and with the same temporal label, which are considered as multi-user accounts, Tweets with identical user name and content, are considered as repetitive tweets, Tweets with identical user name and content, but different time are considered as spams, which cause NLP methods can not extract data easily. Figure 1, illustrates the overall architecture of the offline and online phases.

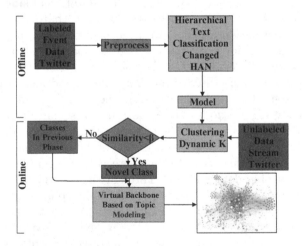

Fig. 1. Architecture of the proposed model

3.1   Data Collection Phase

Table 1, shows input datasets. In this paper we used a collection which includes data for 30 different Twitter datasets associated with real world events [13].

The datasets were collected between 2012 and 2016, always using the streaming API with a set of keywords that we list below. Based on the twitter datasets, a number of 70000 comments related to 2016 were extracted Twitter event dataset comments, In addition, these datasets were used to compare comments with Twitter which has a wide variety of point of views. In addition, these datasets are used to evaluate the method for each category of 1000 tweets which were in English was extracted. The total number of tweets was 70000 from among which 63000 were chosen for the training phase and 7000 were chosen for the test phase in a 10-fold manner.

Table 1. Tweets distribution on 2016

Title	From	To
Brexit	2/24/2016 12:54	4/13/2016 6:43
Brussels-airport-explosion	3/22/2016 8:28	3/22/2016 13:54
hijacked-plane-Cyprus	3/29/2016 7:16	3/29/2016 10:38
irish-ge16	2/3/2016 11:05	2/15/2016 23:56
Lahore blast	3/27/2016 15:33	3/27/2016 17:03
Panama papers	4/3/2016 18:58	4/4/2016 6:10
Sismoecuador	4/17/2016 16:36	4/18/2016 0:21

3.2   Classification Phase

Classification of the stream of social media data can be observed as an instance of stream data in which a group of classes are permanently growing and evolving. The methods which are used for data classification include HNN [14] when word2vec is generated, this file is used for later steps of study. Hierarchical Attention Network (**HAN**) has two distinguishing features: (**I**) a hierarchical structure that shows the hierarchical structure of documents; and (**II**) two stages of attention mechanisms which are used at the word and sentence-level, and which enable it to perform differentially to more and less important content when building the document representation. Furthermore, the HAN network includes quite a few parts including; a word sequence encoder, a word level attention layer, a sentence encoder and a sentence level attention layer. HAN assumed that considering sentence and documents structure in modeling, leads to better representation of document structure in the model architecture. The attention mechanism in neural networks is a copy of the visual attention mechanism in human being, meaning that the human visual attention can concentrate on a specific point of an image with *high resolution* while perceiving it's around in *low resolution* and then s/he can adjust the focal point over time. The attention mechanism in NLP provides the model with the capability to learn what to attend to according to the input text and what it has produced so far. In fact, it does not encode the full source text into a fixed length vector like standard RNN and LSTM. The main difference of this method is that our proposed approach uses context in order to detect when a certain sequence of tokens is linked rather than easily filter various mixtures if signs in a text. Our method has significantly improved the previous approaches. All words do not cooperate in presentation of a sentence similarly. Accordingly, through the attention mechanism, we detect the words which are important in understanding of the text, and then we juxtapose these meaning rich words in the form of a sentence vector. In order to detect sentences which are clues for classification of a text, the importance of each sentence is identified through a vector and using a context vectors us. Then, considering n sentences in m documents a stronger probability of multi label classification of the text can be achieved, that is how much of a text present a specific topic and how much of that show another topic. The purpose of this is finding web based texts which are comprised of various classes, and sometimes within a text, for example an economical text, political issues are raised. The level of dependence of text on each class is important. The number of attention modules in each document differs based on the length of the document. For example, in short documents which include one sentence, this section is deactivated, and in long documents in which the attention module of sentence level between two classes reaches lower levels than threshold. Attention module of document level is used. Then, based on the size of $(\frac{n}{2} + 1)$ of the previous and next sentences, sentences with low dependency are investigated using RNN short term memory. Details of the proposed method are presented in the following section. Additionally the most important HAN's method parameters are listed in Table 2.

Table 2. HAN hyper parameters

Optimization	Loss function	Learning rate	Embedding_dim	Pad_seq_Len
Adam	Sigmoid	0.5	100	150

3.3 Dynamic Topic Modeling Phase

Non-negative Matrix Factorization (**NMF**) Lee and Seung [15] which is a linear algebraic model introduces high-dimensional vectors into a low dimensional image. Similar to Principal Component Analysis (**PCA**) Wold et al. [16], NMF takes the fact into consideration that the vectors are non-negative.

The two matrices of **W** and **H**, would be obtained through original matrix **A**, in which **A** = **WH**.

Also, NMF has an inborn clustering property and **W** and **H** represent the following information:

A (Document-Word Matrix): input that shows which words appear in which documents.

W (Basis Vectors): the topics (clusters) are elicited from the documents.
H (Coefficient Matrix): the membership weights for the topics in each document.
W and **H** are calculated by optimizing an objective function (like the *EM algorithm*), and updating both **W** and **H** iteratively until they are converged.

In this way, Table 3 shows the configuration of the NMF method for extracting topics by evaluating number of topics dynamically, based on the **TC-W2V** coherence measure.

Table 3. NMF topic modeling configuration

Initializations:
Step1- Applying Term Weighting with TF-IDF // Create TF-IDF-normalized document-term matrix as a pickle (**.pkl**) file

Step2- <Finding the **best k-value** for number of topic dynamically>
    Create the Topic Models by pre-specifying an initial range of
    "sensible" values:
    decomposition NMF init = "**nndsvd**" // better for sparseness
    Build a Word Embedding model from all documents in the input file
    Selecting the Number of Topics by implement a simple version of the
    TC-W2V coherence measure
    Number_of_Topics = Select max value as a number of topics

Step3- <Extracting Topics>
    Examine the Final Model
    Number of Topics: {"**Number_of_Topics**"}
Run to extracting Topics

3.4 Clustering and Virtual Backbone Phase

In Clustering and virtual backbone phase, in order to model relationships between terms in the NE contexts, an event graph is made. To divide the graph into subgraphs, a graph theory which will be considered as event candidates is applied. Tweets are about the same events usually having a few similar keywords. On the contrary, tweets which are not related to different events usually have different keywords. This phenomenon is shown by stronger links between nodes which are about the same event in the event graphs. Differently stated, when the edges connect terms from tweets about similar events their weights are higher compared with the time when edges connect terms from tweets about different events. The goal of dividing the edges is to recognize such edges that, if removed, will divide the large graph **G** into sub graphs. Sometimes, the obtained graphs at a definite time window are disconnected, that is all the nodes in the graph are not linked to each other. The generated graph is processed first by analyzing the link between different nodes; therefore, those nodes that are connected with few other nodes are deleted from the graph. At this stage, the primary event graph is categorized into a set of sub graphs that include keywords which are highly related. In our detection approach, it is assumed that events from various sub graphs are not related to each other. Thus, in the event detection sub module, each sub graph is processed separately from others. If you consider that each input sample (the tweets) is introduced to the model for clustering, and are allocated to more than one class according to the classification model, the following definitions need to be presented regarding backbone. Independent set (**IS**) Griggs et al. [17] is considered as a subset of cluster sample or graphs which have no two neighboring members. An independent set with the highest possible number of members in a graph is known as Maximal Independent Set (**MIS**) Johnson et al. [18] Differently stated, adding every cluster sample result in the loss of the independence feature of an independent set. Thus, each sample that is not a member of MIS is at least next to the sample of MIS's cluster which is presented in Fig. 2.

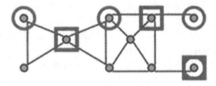

Fig. 2. Maximal independent Set-Johnson et al. [18]

Another subset of cluster sample is the Dominating set (**DS**) which is like **V**; every sample belongs to V-DS and it is adjacent to at least one of the nodes in DS. Accordingly, it can be said that MIS itself is a DS but not vice versa. DS is usually used to relate sample to the cluster heads in clustering of twitter. Finding a dominating set which has the least size or MDS is a NP-Hard problem. An example of DS is Connected Dominating Set (**CDS**) which is a dominating set the inductive sub graphs of which are connected. A CDS with the minimum possible cluster in the model is called Minimum Connected Dominating Set (**MCDS**) Cheng et al. [19] a virtual backbone can be considered. As with definition, CDS is a subset of network nodes in a way that every node belongs to CDS subset or at least is adjacent to a node in CDS. As this structure is connected, it can be employed to start a virtual backbone to find the best cluster in model. Considering the fact that the proposed classification model is multi label classification, and each input sample belongs to more than one degree, this virtual backbone is created using the model, and is gradually converged to one of the clusters based on the results of belonging degree clustering. Another important point is addition of a new class. Sometimes the current discussions do not belong to one of the present classes, and due to low level of similarity to the current clusters, it is needed to make a new cluster. Because of the variability and the dynamic nature, sometimes the clusters of new class are created. Creating event in tweeter has conceptual difference with trend. Usually an event is created in tweeter, and sometimes due to high frequency of re-tweeting, become a trend. For event determination the topic modeling phase helps in this model.

3.5 Event Detection Phase

An event is an incidence which occurs either at a place or time, or occurs in different places during several days like infectious diseases. An event is detected using a group of expressions the frequency of which is increased intensively during the analysis period, one or several times. As hashtags provide a general view on topics in tweets, one of the methods for this paper is using a method for outstanding topics. The first phase is normalizing and classifying hashtags, so that they can be used as the event index. Event detection algorithm introduced in Table 4, which shows the novelty of the method for creating the virtual backbone based on the connected dominating set.

4 Result

In this paper we introduced sink node which changes dynamically. In fact, the sink node has many edges as hot topics which represents events and eventually assigned in each window with a high probability to one or more events. In this paper, the relationship between the topics is calculated based on the correlation matrix.

Considering the fact that the number of topics in each cluster is different, event detection in limited range is very difficult (The number of topics is also considered to be dynamic, a cluster may have two or several important topics). With the change of each topic compared with the previous topics, if a significant change occurs in the distribution of data in each cluster, then the cluster is defined as a new event. Figures 3 and 4, sequentially show the number of effective topics in each event in the stream data and graph visualization of effective topics.

A new topic model TopicVB is provided so that the Correlated Topic Modeling is available for phrase level topics. More contextual information of phrases is used by TopicVB; and the phrases are placed in dynamic virtual backbone; therefore, at phrase level TopicVB can present at phrase level. The results of experiments show that the correlated topic modeling on phrases is more useful for interpretation of the fundamental themes of a corpus. The efficiency of TopicVB is optimized in future, so that it can be fit $\beta\rho$ will be trained. Words in the same document and their component words within semantically coherent links form the contextual information of phrases. The component words within semantically coherent links were modeled in the previous subsection. As with words, the phrases and words in a same document d have the same topic parameter ηd, which is a K-dimension vector derived from a Gaussian Distribution N (μ, Σ). Σ which models the correlation between topics is the covariance matrix.

Table 4. Event detection by virtual backbone and deep learning

Door list is hot topic that event class label
Dee list is relation hot topic
Algorithm: **EDVB DEEP**
1: Input: Stochastic Graph $G<V,E>$, Tweet Stream with name "t", Insert real time.
to graph ,V is event classes and create by hot topic of Twitter in t time slice.
2: Output: Connected Dominating set (called backbone formation) multi label classification
Begin Algorithm
Step1-<construct virtual backbone phase in stream tweets>after classification of offline mode
Let SN be Sink Node;
Let VS be the class node of multi label classification for any event node; Start from VS to create a backbone for multi label classification; Vi be selected node *ith*;
Door_list_labeled = Door_list_determined= VS ;
Dee_list = Neighbors (VS) ;
Door_list_labeled = Door_list_labeled ∪ Vi ;
Dee_list = Dee_list ∪ Neighbors (Vi) ;
Step2-<Learning phase>
begin of while
While (Sink node did not Dominator yet) do:
Select one if the elements Door_list_labeled;
If (Ci>Cavg) // $corr_{ij}^{p} = \dfrac{\Sigma_{ij}(p)}{\sqrt{\Sigma_{ii}(p)\Sigma_{jj}(p)}}$
Create edge in topic and increase weight of edge;
else panelize selected action update Action Vector;
Door_list_ determined = Door_list_labeled ∪ Vi ; // based max probability Action Vector
Dee_list = Dee_list ∪ Neighbors (Vi) ;
C= C+1 ;
E[c] = Wi ;
end of while
if (Vi is Dee and hasn't neighbor of Door) **then**
Vi will be one of elements Door_list_determined and isn't correlated hot topic;
End of Algorithm

When we want to gain the correlation for $\beta\rho$, we should note that it cannot be gained from Σ, which was learned in the first stage, this is due to the fact that Σ consists the impact from word topics. So, with $W^{(p)}$ and $\beta\rho$ in hand, the variation inference will be used again to learn $\Sigma^{(p)}$ Accordingly, the following equation can be used to compute the correlation matrix. $corr_{ij}^{p} = \dfrac{\Sigma_{ij}(p)}{\sqrt{\Sigma_{ii}(p)\Sigma_{jj}(p)}}$

In Table 5, based on the hyper parameters of the HAN model, the parameters of the analytical measure examined and shows that by changing in HAN parameters, the evaluation criteria are improved, because at the same time, in attention mechanism to the level of the word, the sentence, the phrases are used.

Table 5. Result of classification phase

Metric	Section 1	Section 2
Batch size	64	128
Learning rate	0.001	0.001
Accuracy	0.9775	0.9785
Kappa	0.9738	0.975
Recall	0.9775	0.9785
Precision	0.9777	0.9786
F1-Score	0.9775	0.9785
95% CI	(0.9738257769727767, 0.9809115773206379)	(0.9749016079476367, 0.9818346597857032)

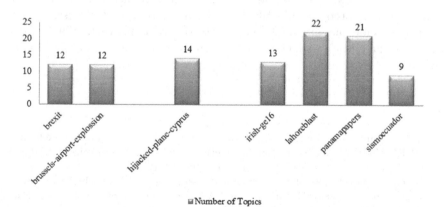

■ Number of Topics

Fig. 3. Dynamic value of topic's number

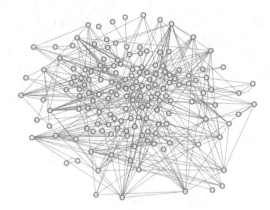

Fig. 4. Graph visulization of extracted topics from 7 categories

5   Conclusion

In the present paper, a semi-observational framework is proposed for detection of data events of tweeter. The modified classification model of HAN is used for classification. 70000 tweets are introduced to the model to make the model and then, stream data are introduced to the classification phase at the same time span without any labels, and they either make a new cluster or are subsumed under previous clusters. In this phase, the classification phase is shaped based on virtual backbone, in a way that the stream data can belong to one or some clusters with different grades. This grade of belongingness enhances or reduces as time goes on. When the virtual backbone is shaped, topic modelling with dynamic k with stream data is used, and the events of each cluster which have been more commented are identified as the main topic. At this stage, topic relations between topics are identified for detection of main events. Finally, based on the time span, the event clusters are shaped and important topics are also investigated.

References

1. Dou, W., Wang, X., Ribarsky, W., Zhou, M.: Event detection in social media data. In: IEEE VisWeek Workshop on Interactive Visual Text Analytics-Task Driven Analytics of Social Media Content, pp. 971–980 (2012)
2. Osborne, M., Dredze, M.: Facebook, Twitter and Google Plus for breaking news: is there a winner?. In: Eighth International AAAI Conference on Weblogs and Social Media (ICWSM), Michigan, USA (2014)
3. Atefeh, F., Khreich, W.: A survey of techniques for event detection in Twitter. J. Comput. Intell. **1**(31), 132–164 (2015)
4. Gaglio, S., Re, G.L., Morana, M.: A framework for real-time Twitter data analysis. J. Comput. Commun. **7**, 236–242 (2016)

5. Petkos, G., Papadopoulos, S., Aiello, L., Skraba, R., Kompatsiaris, Y.: A soft frequent pattern mining approach for textual topic detection. In: Proceedings of the 4th International Conference on Web Intelligence Mining and Semantics (WIMS 2014), Greece, pp. 25–40 (2014)
6. Mathioudakis, M., Koudas, N.: Twittermonitor: trend detection over the twitter stream. In: Proceedings of the 2010 ACM SIGMOD International Conference on Management of Data, USA, pp. 1155–1158 (2010)
7. Boom, C.D., Canneyt, S.V., Dhoedt, B.: Semantics-driven event clustering in Twitter feeds. Making Sense Microposts **1395**, 2–9 (2015)
8. Blei, D.M., Ng, A.Y., Jordan, M.I.: Latent Dirichlet allocation. J. Mach. Learn. Res. **3**, 993–1022 (2003)
9. Aiello, L.M., et al.: Sensing trending topics in Twitter. J. IEEE Trans. Multimed. **15**(6), 1268–1282 (2013)
10. Stilo, G., Velardi, P.: A survey on real-time event detection from the Twitter data stream. J. Data Min. Knowl. Discov. **30**(2), 372–402 (2016)
11. Hasan, M., Orgun, M.A., Schwitter, R.: Efficient temporal mining of micro-blog texts and its application to event discovery. J. Inf. Sci. **44**(4), 443–463 (2018)
12. Hasan, M., Orgun, M.A., Schwitter, R.: Real-time event detection from the Twitter data stream using the TwitterNews+ Framework. J. Inf. Process. Manag. **56**(3), 1146–1165 (2018)
13. Arkaitz, Z.: A longitudinal assessment of the persistence of Twitter datasets. J. Assoc. Inf. Sci. Technol. **69**(8), 974–984 (2018)
14. Yang, Z., Yang, D., Dyer, C., He, X., Smola, A., Hovy, E.: Hierarchical attention networks for document classification. In: Proceedings of the 2016 Conference of the North American Chapter of the Association for Computational Linguistics: Human Language Technology, pp. 1480–1489 (2016)
15. Lee, D.D., Seung, H.S.: Algorithms for non-negative matrix factorization. In: Advances in Neural Information Processing Systems, pp. 556–562 (2001)
16. Wold, S., Esbensen, K., Geladi, P.: Principal component analysis. J. Chemometr. Intell. Lab. Syst. **2**, 37–52 (1987)
17. Griggs, J.R., Grinstead, C.M., Guichard, D.R.: The number of maximal independent sets in a connected graph. J. Discret. Math. **68**, 211–220 (1988)
18. Johnson, D.S., Yannakakis, M., Papadimitriou, C.H.: On generating all maximal independent sets. J. Inf. Process. Lett. **27**(3), 119–123 (1988)
19. Cheng, X., Huang, X., Li, D., Wu, W., Du, D.Z.: A polynomial-time approximation scheme for the minimum-connected dominating set in ad hoc wireless networks. J. Netw.: Int. J. **42**(4), 202–208 (2003)

Privacy Preserved Decentralized Deep Learning: A Blockchain Based Solution for Secure AI-Driven Enterprise

Amin Fadaeddini[1], Babak Majidi[1(✉)], and Mohammad Eshghi[2]

[1] Department of Computer Engineering, Khatam University, Tehran, Iran
{m.fadaodini,b.majidi}@khatam.ac.ir
[2] Department of Computer Engineering, Shahid Beheshti University,
Tehran, Iran

Abstract. Deep learning and Blockchain attracted the attention of both the research community and the industry. In the financial enterprise by using the Blockchain technology, financial transactions could be performed in shorter periods and with higher transparency and security. In Blockchain ecosystem, there is no need for having a central reliable authority to regulate and control the system. In Blockchain many entities which cannot trust each other in normal conditions can join together to achieve a mutual goal. Deep learning algorithms are currently the best solution for many machine learning applications and provide high accuracy models for robotics, computer vision, smart cities and other AI-driven enterprise. However, availability of more data can boost the performance of deep models considerably. In this paper, a secure decentralized deep learning framework for big data analytics on Blockchain for AI-driven enterprise is proposed. The proposed framework uses the Stellar Blockchain infrastructure for secure decentralized training of the deep models. A Deep Learning Coin (DLC) is used for Blockchain compensation. The security of the proposed framework incentivizes people and organizations to share their valuable data for training the deep neural models while the privacy of their data is preserved.

Keywords: Deep learning · Stellar framework · Privacy preserving · Blockchain · Big data

1 Introduction

The promise of deep learning models is to find the complex and nonlinear patterns in the datasets. This promise has been proved by numerous results in the domains like object recognition, synthesizing samples, image retrieval and many other applications [1–6]. However, in many of these applications security and privacy is a vital issue. These aforementioned advances in deep learning are strongly dependent on presence of big data resources. Without this immense sources of data, trained deep models will fail to achieve acceptable accuracy and performances. Unfortunately, this valuable information cannot be accessed without considerable cost and efforts. Many of these valuable data are in possession of the large companies and only are available with high

© Springer Nature Switzerland AG 2019
L. Grandinetti et al. (Eds.): TopHPC 2019, CCIS 891, pp. 32–40, 2019.
https://doi.org/10.1007/978-3-030-33495-6_3

cost. This cost makes it almost impossible for fledgling startups and researchers to train high accuracy deep models.

Along with strict regulations posed by legislatures and administrations which will restrict sharing of end user's data, there is also privacy concerns for ordinary individuals to share their personal data due to possibility of the leakage of these private data. Using Blockchain, many of these impediments can be mitigated. In this new decentralized ecosystem everyone can process the data locally and only share the deeply learned parameters. Blockchain provides a solution for registering all kinds of assets. Along with its application for production of the cryptocurrencies, the Blockchain has the ability to tokenize every kind of assets. Machine learning and deep learning can use this fact. We can build an infrastructure on Blockchain to share a deeply learned models between multiple parties. Parties does not obliged to share their private data to mistrustful authorities. Instead they train the shared model on their data locally and at the end of the training they only share the learned parameters. The proposed model has the following characteristics:

1. Every computing partner can leave the network at any stage of training while taking the proportionate rewards,
2. The malicious partner will be punished by paying more Deep Learning Coin (DLC) (currency of the proposed model) in order to cooperate in future training,
3. Deterring the malicious activities by implementation of Know Your Customer (KYC),
4. Using the Stellar native assets to incentivizing the computing partners.

In Sect. 2, we represent some of the past efforts in the literature towards collaborative training. In Sect. 3, we explain the ecosystem of Blockchain and list potential applications in the industry. In Sect. 4, we show how Blockchain and deep learning can join together to increase the privacy. Then we propose to use stellar Blockchain network for decentralized learning and explain our proposed architecture. Finally, Sect. 5 concludes the paper.

2   Related Works

Models with high accuracy is the goal of almost every researcher around the world. To bypass privacy concerns and inviting individuals to invest in the model training process by sharing their data, a series of methods are proposed. One of the early steps towards collaborative learning was the federated deep learning. In this design one repository known as parameter server updates the model parameters by receiving gradients from the parties that are involved. DistBelief [7] by Google is an example of this implementation. While in this design it is still possible for malicious actors to infer some private information from gradients features, Deep chain designed by Weng et al. [8] uses a collaborative encryption and decryption of gradients to curb the disclosure of private data by malicious parties. Since federated deep learning has many problems new models have been designed to tackle this bottleneck. For instance Mendis et al. [9] leveraged the infrastructure of Ethereum to devise a decentralized deep learning mechanism to share the data with more security. Addair [10] compared the rate of errors in the model and overhead in both centralized and decentralized architecture and have founded that it is a tradeoff between the use of a shared parameter server or a decentralized one.

3   Blockchain Ecosystem

The reports of data breaches increase every year and prominent companies are responsible for about half of the data breaches. Blockchain has been devised to increase the privacy and transparency among anonymous participants. In Blockchain there is a Distributed Ledger Technology (DLT) that contains the history of transactions and operations. After the set of data transactions, i.e., blocks in Bitcoin network, added to the ledger, they become utterly immutable and clear to every person. Cryptography science plays a pivotal role to achieve anonymity and immutability in Blockchain. When specific quantity of data aggregates to form a block, a hashing function which is mostly SHA-256 is deployed to anonymize the data and to make it impossible to alter. This process is with reference to previous block and will result in forming a chain of blocks. To be more elaborative, because of this chain of blocks if a malicious party at current ledger of M decides to alter an information in ledger N, this malicious party must compute the consensus algorithm from ledger N up to ledger M. This is theoretically impossible because as the chain of blocks are in progress, the hostile node either have to compute an infeasible quantum or have to attain consent of other majority nodes on M-N ledgers. The means that enable parties to update their information about the system and progress the ledger is called consensus. Generally, Blockchain networks classified based on the algorithm that they use for consensus. Some networks prefer to judge based on processing power scale known as proof of work and some of them prefer participant's credit known as proof of stake.

The most important concern about Blockchain is the increasing volume of data. As the time passes the infrastructure to store this big data becomes more important. Until the hostile parties do not hold majority of the network, i.e., more than 51%, we can say that the data is authentic. In settings like a decentralized deep learning model if one company which have a great deal of computing power and owns more than 51% of the training power it can break the procedure of training.

3.1   Industrial Utilization

With manifestation of Blockchain technology many areas of industry along with the financial sector become concerned with this trending technology. The primary advantage of using Blockchain is to provide more security, which is an imperative topic to cover in business models. Below we summarize potential deployments of Blockchain in the Information Technology (IT) industry.

Financial Services: Financial businesses and the transfer of money was the first application of Blockchain in industry. Bitcoin proposed by Nakamoto [11] was the first secure decentralized cryptocurrency that uses the Blockchain to prevent tampering and double spending among users. After Bitcoin other altcoins have been created, e.g.,

Ether, XLM and etc. In recent years, some attempts started to use machine learning and artificial intelligence to improve the Blockchain systems by detecting the hostile nodes. i.e., preventing double spending. Also the faster consensus can be achieved using artificial intelligence.

Healthcare: Blockchain can help patients to share their sensitive data without apprehension that their information leaks. Blockchain provides a solution for patients to track the status of their data at any time. In the situation that the record of a disease is rare, scarcity of the related data halts the research progress due to the mentioned privacy issues. Kuo et al. [12] design a cross-institutional healthcare predictive model using Blockchain to help researchers and patients safely share the records.

Internet of Things and Robotics: As the volume of data increases every year the concern of losing and tampering of this data becomes more challenging. In the era that connected machines are pervasive, malicious attack to this data can put human lives in fatal danger. Also internet of things and connected sensors deployed to provide data for monitoring purposes. The wrong data that injected by potential attackers can mislead the Decision Support Systems (DSS). As a solution for this problem, Blockchain can ensure manufacturers and consumers that their data is legitimate. As another application smart contracts can be used to purchase new components in a situation where the old component either have deprecated or caused heavy damages due to accidents.

4 Privacy Preserved Decentralized Deep Learning

As pointed out in Sect. 3, privacy preserving in training deep models is of crucial impact. This means that individuals and companies who hold sizeable and useful source of data, show reluctance to give their private data for research purposes. In literature, privacy preserving means that agents do not need to share their private and sensitive data. This concept is called privacy preserving. In order to solve this problem the federated learning has been investigated by researchers [13]. The federated learning, i.e., collaborative learning is also deals with decentralized computation and sharing of processing power among certain participants with a portion of data which could be their local private data. In federated deep learning there is a central repository for the model. This model is updated based on the parameters that parties sent periodically.

4.1 Architecture of the Proposed Framework

Most of the available federated deep learning models is a black box that store the model in one central server. Although, this is acceptable in many cases that participants want to only help the training by sharing their private data, sometimes this design is agnostic to the concept of distributed computation while all participants like to have a copy of the model. In Blockchain, it is possible that we train the model in a way that the model is distributed between multiple participants. This concept defines the DLT.

After the emergence of Bitcoin and Blockchain, variety of Blockchain frameworks with different characteristics has been proposed. One of these improved characteristics is the ability to tokenize any abstracts and real world's assets. Ethereum [14] and Stellar [15] are two of these frameworks. In our proposed architecture we leverage the stellar framework to build our distributed deep learning model. A participant who contributes to the training of a model must have reasonable motivation for doing his job. Thus, including an implementation of incentivization is important in distributed deep learning. In the proposed implementation we define an asset called DLC as the base currency of the model. Beside this native asset, all model initiators must issue assets that represent their own training model. Because of the embedded order book in the stellar framework known as Stellar Decentralized EXchange (Stellar DEX), it is possible for individuals to make an offer based on their balance of token. In the proposed mechanism we treat models as a token that issued by the model initiator.

As depicted in Table 1, the proposed model has four classes of assets:

1. The Deep Learning Model (DLM) that is issued by the model initiator and represents the model in the order book.
2. Verified Learned Model (VLM) which is also issued by the model initiator and is distributed to the validators that's been designated to validate the authenticity of the trained model.
3. Deep Learning Coin (DLC) is the asset that computing partners must pay to cooperate in training procedures. Volunteers can buy this coin for a low price. The primary reason behind using this asset is to compensate the hostile participant to pay more DLC in future cooperation.
4. Stellar native asset (XLM) that been paid to computing participants.

Table 1. List of assets used in the proposed model.

Asset code	Issuer
DLM	Model initiator
DLC	Issued by the proposed model
VLM	Model initiator
XLM	Stellar native asset

Suppose that A as a model initiator decides to train a model. Because of the role that the big data plays in machine learning, A decides to cooperate with $p \in N$ computing partners that have valuable data. A first announces the planned model as an offer (sell XLM $\rightarrow$ buy DLM). Number of this offer is multiplied by the number of nodes that A wants to cooperate with, which in this example is p. It is better that A have a diverse list of offers based on the number of DLC and XLM. This will enable the participants to leave the network at any stage of training while taking the proportionate rewards. Beside the aforementioned settings, A must also specifies a quorum of validators for the

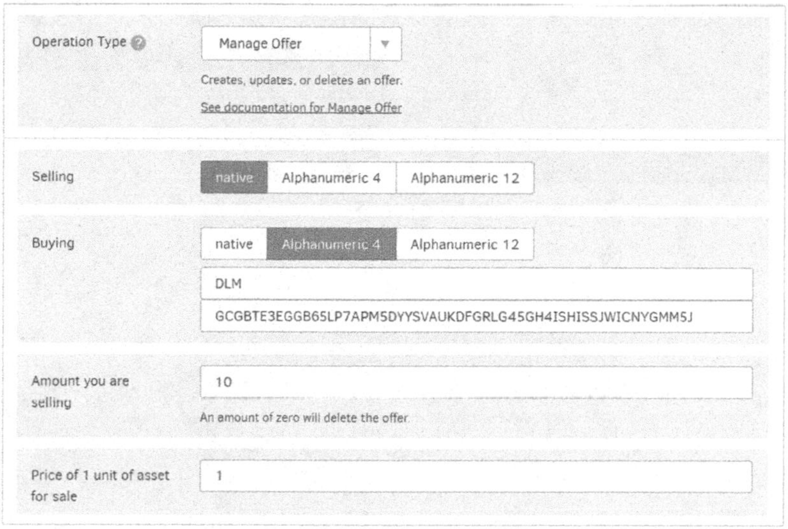

Fig. 1. Depiction of required arguments for submitting the offer operation.

deep model and the accurate details about the environment of the model. This information along with other offers are available in the dashboard interface. The volunteers thus first examine the dashboard interface of the framework to find the model that they like to cooperate with. Figure 1 shows an example of submitting an offer operation.

Let's denote the volunteer computing partner as B. B first creates an account by generating a pair of private and public keys in ED25519 cryptography algorithm and download the predefined model from dashboard to start the training procedure on his local private data. When B finished the training in any epoch it will submit an offer to buy VLM for specific amount of DLC (sell DLC $\rightarrow$ buy VLM). In this transaction computing partner must upload the model checkpoints to interplanetary file system (IPFS) and insert the hash checksum in memo field of transaction. If the trained model has an acceptable accuracy, validator of the model submits two transactions. First submit a direct payment of VLM to B, and secondly submit an offer to transfer $\frac{1}{epoch}$ DLM over VLM (sell DLM $\rightarrow$ buy VLM). The range of these accuracies is determined using voting mechanism. When the computing participants decide to cash out the XLM he set another offer to buy XLM for VLM (sell VLM $\rightarrow$ buy XLM). The rule of cross asset payment in Stellar enable the participant to have the corresponding amount of XLM in his account. Figure 2 is an example of offers in the order book that are waited to be traded and Fig. 3 represents the proposed model in detail.

DLM	VLM	Price
5	27	0.09000
9	107	0.08861
2	38	0.08949
4	5	0.08825
7	8	0.08819

(a)

Price	XLM	DLM
	50	4
0.08137	73	5
0.08002	499	54
0.08000	8	9
0.07989	2	1

(b)

VLM	XLM	Price
5	50	0.08696
1	21	0.08652
9	109	0.08645
8	97	0.08496
5	56	0.08462

(c)

Fig. 2. List of offers in the order book where: (a) the offers submitted by model validators that will be used for completing computing partners offers through cross assets payments, (b) the offers submitted by model initiator, (c) the offers submitted by computing partners. (Offers highlighted in blue is one chain of offers that were matched.) (Color figure online)

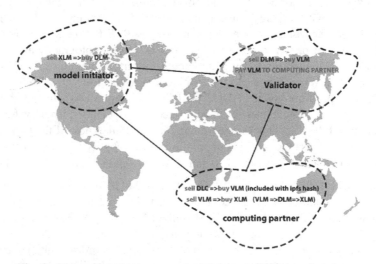

Fig. 3. The proposed privacy preserved decentralized deep learning.

It is essential to note that some part of this mechanism like many proposed distributed learning methods are still performing off-chain. For example like [9] the model learned by computing partners will be saved on IPFS. The result of this saving is a hash checksum that refers to the data. This hash checksum can now be saved on-chain for more privacy and immutability.

4.2   Controlling the Authenticity of Trained Models

There are many ways that some computing participants can harm the quality of model. For example, one computing participant can stop the process of training too early or can train the model on completely irrelevant data set. In order to deterring these malicious activities, the proposed distributed framework must have a plan to prevent these activities. In the proposed framework we control this by two means. First, we issue an asset called DLC. The primary purpose of issuing this asset is to control the authenticity of the computing partner's model. Any volunteer that wants to be involve in training of a specific model, must pay some DLC from his account. If the number of malicious activities by an account increases, in the future cooperation it will be forced to pay more DLC. Secondly, we have an implementation of Know Your Customer (KYC) in our framework to further control the authentication. Required information about the computing partners are recorded on chain.

5   Conclusion

In this paper, a secure decentralized deep learning framework leveraging big data on Blockchain for AI-driven enterprise is proposed. The proposed framework uses the Stellar Blockchain infrastructure for secure decentralized training of the deep models. A Deep Learning Coin (DLC) is used for Blockchain compensation. The security of the proposed framework incentivizes people and organizations to share their valuable data for training the deep neural models and ensures that the privacy of their data is preserved.

References

1. Li, G., Yu, Y.: Contrast-oriented deep neural networks for salient object detection. IEEE Trans. Neural Netw. Learn. Syst. **29**(12), 6038–6051 (2018)
2. Anbari, S., Majidi, B., Movaghar, A.: 3D modeling of urban environment for efficient renewable energy production in the smart city. In: 2019 7th Iranian Joint Congress on Fuzzy and Intelligent Systems (CFIS), pp. 1–4 (2019)
3. Kumar, K., Shrimankar, D.D.: F-DES: fast and deep event summarization. IEEE Trans. Multimed. **20**(2), 323–334 (2018)
4. Fadaeddini, A., Majidi, B., Eshghi, M.: Generative adversarial networks for procedural synthesis of original textures in video games. In: 2nd Digital Games Research Conference (2018)
5. Safari, G., Majidi, B., Khanzadi, P., Manzuri, M.T.: Cross-platform e-management for smart care facilities using deep interpretation of patient surveillance videos. In: 2018 25th National and 3rd International Iranian Conference on Biomedical Engineering (ICBME), pp. 1–6 (2018)
6. Fadaeddini, A., Eshghi, M., Majidi, B.: A deep residual neural network for low altitude remote sensing image classification. In: 2018 6th Iranian Joint Congress on Fuzzy and Intelligent Systems (CFIS), pp. 43–46 (2018)
7. Dean, J., et al.: Large scale distributed deep networks. In: Advances in Neural Information Processing Systems, pp. 1223–1231 (2012)

8. Weng, J., Weng, J., Zhang, J., Li, M., Zhang, Y., Luo, W.: Deepchain: auditable and privacy-preserving deep learning with blockchain-based incentive. Cryptology ePrint Archive, Report 2018/679 (2018)
9. Mendis, G.J., Sabounchi, M., Wei, J., Roche, R.: Blockchain as a service: an autonomous, privacy preserving, decentralized architecture for deep learning. arXiv preprint arXiv:1807.02515 (2018)
10. Addair, T.: Decentralized and distributed machine learning model training with actors
11. Nakamoto, S.: Bitcoin: a peer-to-peer electronic cash system (2008)
12. Kuo, T.-T., Ohno-Machado, L.: ModelChain: decentralized privacy-preserving healthcare predictive modeling framework on private blockchain networks. arXiv preprint arXiv:1802.01746 (2018)
13. Konečný, J., McMahan, H.B., Yu, F.X., Richtárik, P., Suresh, A.T., Bacon, D.: Federated learning: strategies for improving communication efficiency. arXiv preprint arXiv:1610.05492 (2016)
14. https://www.ethereum.org/
15. Mazieres, D.: The stellar consensus protocol: a federated model for internet-level consensus. Stellar Development Foundation (2015)

Unsupervised Deep Learning for Conflict Resolution in Big Data Analysis

Zeinab Nakhaei[1(✉)] and Ali Ahmadi[2,3]

[1] Department of Computer Engineering, Science and Research Branch,
Islamic Azad University, Tehran, Iran
zeinab.nakhaei@srbiau.ac.ir
[2] School of Computer Science, Institute for Research in Fundamental
Sciences (IPM), Tehran, Iran
ahmadi@kntu.ac.ir
[3] Faculty of Computer Engineering, K.N. Toosi University of Technology,
Tehran, Iran

Abstract. One of the important challenges in big data era is veracity. It means that the data sources are of widely differing qualities, with significant differences in the coverage, accuracy and timeliness of data provided. Therefore, different and conflicting descriptions for a single entity may be found. The aim of conflict resolution in the context of data integration is to identify the true values from among different and conflicting claims about a single entity, provided by different data sources. Most data fusion methods for resolving conflicts between objects are based on two estimated parameters: the truthfulness of data and the trustworthiness of sources. The relations between objects are however an additional source of information that can be used in conflict resolution. The main challenge in these methods is deciding about the existence of relation between entities. Previous methods focus only on apparent characteristics of entities and use a similarity function for drawing inference about relations. In this paper, we propose a method that the relationship between the entities can be estimated by mapping the data to an embedding space and creating enhanced and more informative features and then clustering it. Then the true values are determined by defining a confidence score based on the distance between data and center of clusters. We evaluated this approach on three categories of data, synthetic, simulated and real data. The evaluation results showed that our proposed approach outperforms existing conflict resolution techniques, especially where there are few reliable sources.

Keywords: Data fusion · Conflict resolution · Big data integration · Relation assessment · Unsupervised deep learning

1 Introduction

Big data integration differs from traditional data integration in many aspects including volume, velocity, variety and veracity. In this paper we focus on veracity aspect of big data analysis. Veracity means the data sources are of widely differing qualities, with significant differences in the coverage, accuracy and timeliness of data provided [1]. In other words, there are multiple sources describing the same real-world entity, providing

© Springer Nature Switzerland AG 2019
L. Grandinetti et al. (Eds.): TopHPC 2019, CCIS 891, pp. 41–52, 2019.
https://doi.org/10.1007/978-3-030-33495-6_4

different values for the same attribute of the entity, thereby causing the so-called *conflict* at the value level. The aim of *conflict resolution* is combination of values that describe a similar entity from the real world leading to one value which is closer to the real world. This process is considered as a *data fusion* process [2].

In a conflict resolution problem, there are usually many claims about entities provided by a number of different sources. Almost all works on data fusion employ a basic heuristic for resolving conflicts, *"The reliable source provides true value"*. In theory, we need a number of labeled examples proportional to the number of sources in order to find the reliability of the sources. But the number of sources may be in the scale of thousands or millions and the training data is very limited. So almost all of conflict resolution methods are conducted in an unsupervised manner in the sense that the ground truth is only used in the evaluation step. These methods consider the reliability of sources as a main quality factor that is unknown apriori. So, these methods typically estimate this factor using expectation maximization (EM) and ERM approach [3]. It is considerable that at the higher level of data abstraction, it is not necessary knowing about low level data parameters such as reliability of sources and basically we deal with high level information like *relation* between objects.

As described in the survey carried out by Li et al. [4], most data fusion methods assume that objects are independent, whereas in reality objects may have relations between them and may affect each other. For example, two people who are classmates at university are likely to have the same level of education. Or two books published by a certain publisher may cover the same or similar subject matter. There are few studies that use relation as an additional information for increasing precision of data fusion. These studies are grouped into two categories: the methods that assume the relations between entities are provided as an input of the problem. And the methods that consider similarity between values of feature of entities as relationships. It means that, they define a similarity function and the probability of the correctness of the values between the related entities is considered the same.

Our proposed approach, in contrast to previous works, finds relations in the semantic space instead of feature space, which means that we consider semantic relations between objects besides the apparent similarities. First, the entities are transformed to the deep embedding space and then clustered in this new space. For training of embedding network we define a loss function that can be used for our proposed unsupervised deep learning. The intuition behind our loss function is that related entities in the new embedding space are close together. So the embedding space that contains dense clusters with more distance from each other, is a good representation of entities. Finally, all claims are mapped to the embedding space and confidence score is calculated for each one and then true value is selected.

In summary, this paper makes the following contributions to knowledge in this area:

- We proposed new high level data fusion method for conflict resolution in big data integration.
- We employed deep embedding network for finding latent semantic space for entities and define loss function that can be used for unsupervised deep learning.
- We use clustering in embedding space for finding relations that are not predefined.

- We evaluate our approach by three category of data set: synthetic, simulated and real dataset and show improvement of the fusion performance, particularly in unreliable environments.

The rest of this paper is organized as follows. In Sect. 2, we review the existing data fusion techniques in the literature and separate them into low level and high level data fusion approaches. In Sect. 3, we define the problems around conflict resolution and explain why we use relations to solve them. Section 4 describes our proposed approach in more details including the stages and required formulations. Finally, the results of our experiments are analyzed in Sect. 5.

2 Related Works

The first study that precisely defines the data fusion in the context of data integration and expresses the goals of data fusion is provided by Bleiholder and Neumann in [5]. In this survey, different strategies for dealing with value conflicts are categorized into three groups, conflict ignoring, conflict avoidance, and conflict resolution. Although in this review, the fusion methods are very simple for resolving the conflicts, like voting and averaging, but since then, more efficient and intelligent methods are proposed, which are also referred as truth discovery (some of important methods are reviewed and compared by Dong et al. in [6]). In the following, data fusion techniques for conflict resolution are reviewed in two groups, low level and high level data fusion.

The truthfulness of data and the trustworthiness of sources are two main factors of data quality that in the context of data integration are estimated by low level data fusion methods. One of the inspiring methods in this field is proposed by Yin et al. naming TrustFinder [7]. TruthFinder adopts the Bayesian analysis to infer the trustworthiness of sources and the probabilities of a value being true. Using graphical model for modeling the parameters of a conflict resolution problem is proposed first time by Zhao et al. [8]. This method uses a Bayesian network to model the relationship between data correctness and source accuracy and uses EM to obtain the solution. Recently, SLiMFast is proposed by Rekadsinas et al. [3] as a discriminative model that also enables considering other features of data sources (e.g., update date, number of citations, etc.) for fusion; in presence of sufficient labeled data SLiMFast uses empirical risk minimization (ERM). In low level data fusion category, some methods like in Li et al. [9], use optimization approach.

In the following, we review some papers that use relation between entities in the fusion process. These relations are partially addressed by Meng et al. [10]. However, this work is based on the key assumption that a correlation graph already exists. In this paper, the optimization framework is applied to extract true information where resources are mobile users' reports about a specific entity such as weather temperature (a crowd sensing technique). The intuitions behind the proposed method are that truths should be closed to the observations given by the reliable users, and correlated entities should have similar true values. Ge et al. in [11] and [12] apply conflict resolution in the problem of the diagnosis of correct or false comments that are produced by users about a particular topic. In the first method, matrix decomposition and in the second one deep belief networks are used for finding latent common feature between users.

Almost all of these methods consider the reliability of sources as a main quality factor that is unknown apriori. So, these methods must estimate this factor using EM or ERM approach

3 Problem Definition

Problems of conflict resolution generally involved dealing with often conflicting claims about an entity. The task of a conflict resolver (or truth finder) is to determine the correctness of each claim. Depending on the level of data abstraction, the conflict resolution problem engages with several concepts, including *entity (or object), attribute, data source, claim,* and *truth value.* The main approach used in most current methods is based on estimating the reliability of data sources. As mentioned earlier, two heuristics are used in this approach: first, that the claim provided by a reliable source is likely to be correct; and second, that a source that provides true value will be reliable. For example, if a source S_1 is assessed as the source with higher reliability, then the value provided by S_1 is returned as the correct value. According to evaluations in [13], at higher levels of fusion, the addition of another concept, called the relation, causes to increase the fusion accuracy. An example is given below to illustrate this:

Example 1: Suppose the entity about which there are claims from multiple sources is a symbol of company in the stock. This entity is described by a range of attributes such as open price, last price, volume, and so on. The values of these features are available through the disseminating sites of the stock exchange. Each of these values is considered as a claim, and the purpose of fusion is to resolve possible inconsistencies among these values. Table 1 shows part of the dataset.

Table 1. Part of stock dataset and the output of fusion

Source	Symbol	Change	Last price	Open price	Volume
Input					
bloomberg	vocs	2.03%	31.23	30.45	119,929
nasdaq-com	vocs	2.03%	$31.23	$30.45	119,237
google-finance	vocs	−2.03%	31.23	30.45	119,237
bloomberg	pfcb	3.36%	41.59	40.37	513,983
nasdaq-com	pfcb	3.35%	$41.59	$40.37	510,801
google-finance	sial	−1.17%	74.24	73.54	673,614
bloomberg	sial	1.17%	74.24	73.54	674,663
nasdaq-com	sial	1.17%	$74.24	$73.54	673,614
Output					
	vocs	2.03%	$31.23	$30.45	119,237
	pfcb	3.35%	$41.59	$40.37	510,801
	sial	1.17%	$74.24	$73.54	673,614

According to the table, we see that the values obtained for some of the features are inconsistent. In high-level fusion, relationships between entities are considered. So two main challenges are as follows:

1- How to find relationships while the correct values are inaccessible and we must use unsupervised learning?
2- How to use relationships to find the correct value?

4 Proposed Method

The core of our approach is *relation assessment*. There are two main challenges for relation assessment in the context of conflict resolution. First, there is no predefined relation type in dataset in contrast to other applications such as link prediction [14] or knowledge graph completion [15] that relation types are defined and exist. Second, a conflict resolution problem due to huge volume of data is an unsupervised problem inherently, in contrast of relation learning that is a supervised problem with a labeled training data set include objects and relations.

For the second problem, we apply clustering, which consists of dividing the dataset into clusters of similar examples. For the first problem, our proposed method finds relations automatically using this heuristic: *"entities that belong to the same cluster are related"*. So the number of clusters determines the number of relations without knowing the type of relationship. But applying clustering method to the feature space is not sufficient. Because clustering in the feature space focuses only on apparent similarities like in previous approaches, while in big data, the dimension of feature space is very huge and we want to extract more complex and semantic relations between entities. To achieve this aim, proposed method explores new representation for entities by finding fewer but more meaningful features. Recently, deep learning has drawn much attention because it's highly nonlinear architecture can help to learn powerful feature representation [16]. Thus, to take advantage of it, we propose a deep embedding network which maps entities to new embedding space such a way that entities form dense and separated clusters. In the new space, entities that are far from centroids are more untrustworthy than other entities that are close to centroids.

4.1 Entity Representation

In this section, we present a general neural network framework for mapping entities to the embedding network such that related entities get closer to each other in the embedding space. Each claim about one entity corresponds to a high-dimensional feature space vector, denoted by $x_c = [a_1, a_2, \ldots, a_d]^t$, which a_i is i^{th} feature and d is the number of features. Let $X = \{x_i\}_{i=1,2,\ldots,n}$ denotes the set of input data. The weights of each layer denoted by $W^j = \{w_i\}_{i=1,2,\ldots,h}$. And for the last layer h = k, k ≪ d. So the output of network is a k-dimensional space and the network defines a transformation $f(.) = \mathbb{R}^d \rightarrow \mathbb{R}^k$ which transforms an input x to a d-dimensional representation $f(x)$.

$$f(x) = W^j \phi\left(W^{j-1}\ldots\phi\left(W^1 x + b^1\right) + b^j\right) \tag{1}$$

where $\phi(.)$ is an activation function and b is a bias term for each layer.

After transforming input data to the embedding space, we use a clustering method like k-means for utilizing objective function.

4.2 Loss Function

We define a loss function inspired by [17] with two competing terms to achieve this objective: a term to penalize large distances between embeddings in the same cluster, and the other one to penalize small distances between embeddings in different clusters.

$$L_1 = \frac{1}{C}\sum_{i=1}^{C}\frac{1}{N_c}\sum_{j=1}^{N_c}\left\|f(x_j) - \mu_i\right\|^2 \tag{2}$$

where C and N_c are the number of clusters and the number of members belonging to cluster c, respectively.

$$L_2 = \frac{1}{C(C-1)}\sum_{CA=1}^{C}\sum_{CB=1,\neq CA}^{C} -\left\|\mu_{CA} - \mu_{CB}\right\|^2 \tag{3}$$

And finally loss function is defined as

$$L = L_1 + \alpha L_2 \tag{4}$$

4.3 Confidence Score

We define confidence score to map each claim about a specific entity to a real number in $\mathbb{R}$. So that the larger number represents the greater probability of the correctness. In our proposed method, this function is defined in terms of distance from the centers of the clusters, which is given as follows:

$$confidence\_score(x_i) = 1/D(f(x_i), \mu), \tag{5}$$

$$D(f(x_i), \mu) = \frac{1 + \left\|f(x_i) - \mu_c\right\|^2}{\sum_{c'} 1 + \left\|f(x_i) - \mu_{c'}\right\|^2}$$

where μ_c is the centroid of cluster that $f(x_i)$ belongs to. $f(x_i)$ is defied in relation (1). That is deep feature vector of entity x_i. By applying the confidence score function on each claim, we can rank the claimed values of each entity in terms of the output of confidence function, and then, select a value that has a lower rank as the correct value and the fusion output.

Figure 1 shows schematic illustration of our proposed method for finding relations and resolving conflicts. On the left side, data in the feature space are illustrated and data with incorrect values are displayed by red color. After passing data to the network and creating embedding representation for each entity (deep feature vector), the related entities in the embedding space are closer to each other and the wrong data goes away from the center of the cluster. On the other words, confidence score for wrong entities becomes high.

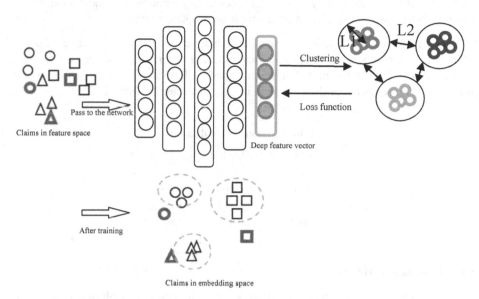

Fig. 1. Schematic illustration of the proposed fusion method (Color figure online)

5   Experimental Results

To evaluate the performance of our proposed approach in conflict resolution problem, we have done some experiments on synthetic and real datasets.

5.1   Datasets

We have applied three types of *synthetic*, *simulated* and *real* dataset:

Synthetic Data. This dataset are produced to further explore and evaluate the proposed method in terms of the number of noisy and context features of the data. This dataset is made of objects with 10 real valued features and 10 classes. The value of each feature is randomly selected from a specific interval of real numbers. This interval is different for each feature. To simulate the relation between the entities, we have followed specific rules while generating the data in each class. In this way, for example, when the value of the attribute a_1 is in range r_a, the value of attribute a_2 should be in range r_b, in which r_a and r_b are intervals in the real numbers. After generating this dataset, it is regarded as ground truth and based on it, we generate a dataset consisting of multiple conflicting sources by injecting different levels of noise into the original data. A parameter γ indicates the percentage of noisy data. In this way, we can control the reliability degree of each source.

Simulated Data. We used simulated data from a real dataset, Adults[1] dataset. This dataset is real dataset in terms of entities and features, but the sources are simulated. We select 5000 entities from the dataset and this dataset is considered as ground truth. We generate a data set consisting of multiple conflicting sources by injecting different levels of noise into the original data as the input to our program.

Real Data. *Stock* dataset is a popular data fusion dataset [8] where sources provide information on multiple 16 stock-attributes including open price, change and volume, for July 2011. We use the data that are provided by NASDAQ.com to obtain ground truth data. The statistics of datasets used in this paper are in Table 2.

Table 2. Statistics of datasets

	Synthetic data	Adult dataset	Stock dataset
#features	10	9	16
#entities	2000	5000	1000
#sources	5	5	55
#claims	10000	25000	54307

5.2 Evaluation Metric

As previously mentioned in Introduction section, the conflict resolution is an unsupervised process and only a limited set of data is considered as ground truth. This dataset is used to compare output results and to evaluate the fusion method. In this paper, we used accuracy metric, which shows the percentage of output values similar to those of the ground truth set.

$$accuracy = \frac{1}{n_g} \sum_{i=1}^{n_g} 1\{g_i = f_i\} \tag{6}$$

where n_g is the number of entities in ground truth, g_i is the entity in ground truth set and f_i is the fusion output. The basis of all the previous proposed methods is voting. Therefore, the performance of the proposed method is compared with the voting.

5.3 Setup

For all datasets, we set a five layers network with d-20-50-20-e dimensions. where d is the dimension of input data and e is the dimension of embedding space. All layers are fully connected. According to what is recommended for new networks [18], we use a rectified linear unit or ReLU [19] as activation function for each layer. For layer three

---

[1] https://archive.ics.uci.edu/ml/datasets/Adult?ref=datanews.io.

with most hidden units we apply dropout with $\mu = 50$ as a probability of units that must be multiplying by zero. ASGD[2] is applied for optimization.

For clustering, k-means algorithm with varying number of clusters is applied. The parameters of our methods are as follow:

- k: the number of clusters
- e: dimension of embedding space
- loss function parameters: $\alpha = 1$ and learning rate $\eta = 0.001$.

5.4   Result and Discussion

The basis of all previous methods is voting. Therefore, the performance of the proposed method is compared with the voting. We design some experiments and then compare the results of the proposed method with other methods in the literature. Each experiment is repeated 5 times.

- **Clustering in feature space vs latent semantic space** - in this experiment, we show that finding embedding space and enriched features has a great effect in finding better relationships between entities. We apply the clustering in the feature space of the problem and the embedding space and then measure the fusion accuracy.
- **Effect of the parameters** - in Sect. 5.3 we defined parameters of our approach. It should be noted that for each dataset, the training dataset is different from the evaluation dataset.

Table 3 shows the accuracy of proposed method with two approaches of clustering in feature space and clustering in embedding space, compared with results of voting method. According to Table 3, when data is mapped to the embedding space and then clustered, the accuracy of fusion is increased. Clustering in the feature space yields to better accuracy than of the voting method. These results show that using relations between objects can improve the accuracy of a fusion method.

Table 3. Comparison of accuracy in proposed method and voting method

Dataset	Synthetic	Adults	Adults	Adults	Adults	Stock
Average reliability of sources (m)	m = 0.7	m = 0.7	m = 0.6	m = 0.55	m = 0.5	–
Voting method	0.7	0.7	0.6	0.55	0.5	0.74
Proposed method (clustering in feature space)	0.95	0.93	0.89	0.79	0.77	0.78
Proposed method (clustering in embedding space)	**0.97**	**0.94**	**0.92**	**0.85**	**0.81**	**0.80**

---

[2] Averaged Stochastic Gradient Descent.

From Fig. 2, it can be seen that our approach is converged in epoch 6 for Adult dataset but it needs more epochs for Stock dataset.

We have tested different dimension for embedding space by changing the number of output layer nodes in the network, e = {2, 3, 4, 5, 6, 7, 8, 9, 10, 11, 12, 13}. The accuracy of network for Adult dataset with average noise $\gamma = 0.4$ is graphed in Fig. 3.

From Fig. 3, we can see that the best accuracy rate of network is 0.9 ± 0.2, which is related to e = 12, 13.

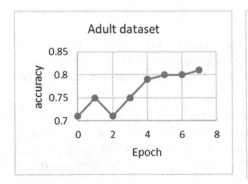
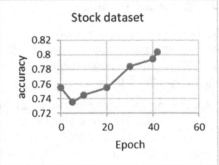

Fig. 2. Convergence of proposed method in Adult (left) and Stock (right) dataset.

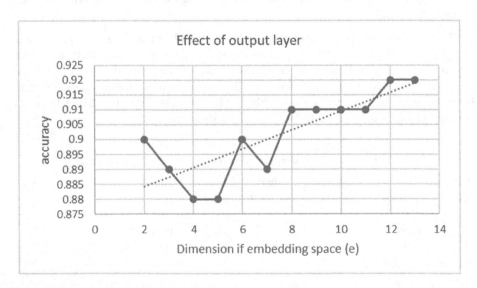

Fig. 3. Effect of parameter e (embedding space dimension) on accuracy of fusion in Adult dataset $(\gamma = 0.4)$

Figure 4 demonstrates the impact of number of clusters on the accuracy. We can see that when the number of nodes is not high, we can improve the accuracy by increasing the number of clusters.

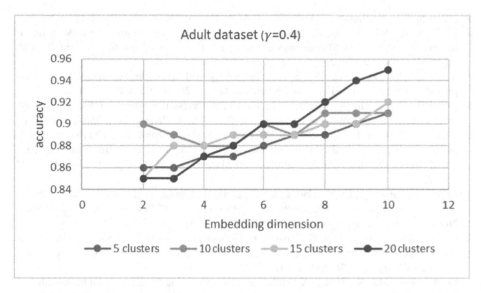

Fig. 4. Effect of number of clusters on accuracy of fusion in Adult dataset $(\gamma = 0.4)$

6 Conclusion

In this paper, a method for data fusion is proposed in order to resolve the conflicts in the context of big data integration. The methods proposed so far mostly operate at low level data fusion and are designed based on the estimation of source reliability. The proposed method here, alternately is characterized by high-level fusion because it estimates the relationships between entities and uses the information in the relationships to determine the truth value.

This method has three main stages: 1 - Finding embedding vector for entities by deep network; 2 - Clustering the data in the new embedding space; 3 - Identifying the true value using the confidence score function. An evaluation of the results shows that our proposed approach outperforms existing conflict resolution techniques, especially where there are few reliable sources.

References

1. Dong, X.L., Srivastava, D.: Big data integration. In: 2013 IEEE 29th International Conference on Data Engineering (ICDE), pp. 1245–1248. IEEE (2013)
2. Dong, X.L., Naumann, F.: Data fusion: resolving data conflicts for integration. In: Proceedings of the VLDB Endowment, no. 2, pp. 1654–1655. VLDB Endowment (2009)
3. Rekatsinas, T., Joglekar, M., Garcia-Molina, H., Parameswaran, A., Ré, C.: SLiMFast: guaranteed results for data fusion and source reliability. In: Proceedings of the 2017 ACM International Conference on Management of Data, pp. 1399–1414. ACM (2017)
4. Li, Y., et al.: A survey on truth discovery. ACM SIGKDD Explor. Newsl. **17**(2), 1–16 (2016)

5. Bleiholder, J., Naumann, F.: Data fusion. ACM Comput. Surv. (CSUR) **41**(1) (2009)
6. Li, X., Dong, X.L., Lyons, K., Meng, W., Srivastava, D.: Truth finding on the deep web: is the problem solved? In: Proceedings of the VLDB Endowment, vol. 6, no. 2, pp. 97–108. VLDB Endowment (2012)
7. Yin, X., Han, J., Philip, S.Y.: Truth discovery with multiple conflicting information providers on the web. IEEE Trans. Knowl. Data Eng. **20**(6), 796–808 (2008)
8. Zhao, B., Rubinstein, B., Gemmell, J., Han, J.: A Bayesian approach to discovering truth from conflicting sources for data integration. Proc. VLDB Endow. **5**(6), 550–561 (2012)
9. Li, Q., Li, Y., Gao, J., Zhao, B., Fan, W., Han, J.: Resolving conflicts in heterogeneous data by truth discovery and source reliability estimation. In: Proceedings of the 2014 ACM SIGMOD International Conference on Management of Data, pp. 1187–1198. ACM (2014)
10. Meng, C., Jiang, W., Li, Y., Gao, J., Su, L., Ding, H., Cheng, Y.: Truth discovery on crowd sensing of correlated entities. In: Proceedings of the 13th ACM Conference on Embedded Networked Sensor Systems, pp. 169–182. ACM (2015)
11. Ge, L., Gao, J., Yu, X., Fan, W., Zhang, A.: Estimating local information trustworthiness via multi-source joint matrix factorization. In: 2012 IEEE 12th International Conference on Data Mining (ICDM), pp. 876–881. IEEE (2012)
12. Ge, L., Gao, J., Li, X., Zhang, A.: Multi-source deep learning for information trustworthiness estimation. In: Proceedings of the 19th ACM SIGKDD International Conference on Knowledge Discovery and Data Mining, 11 August, pp. 766–774. ACM (2013)
13. Nakhaei, Z., Ahmadi, A.: Toward high level data fusion for conflict resolution. In: 2017 International Conference on Machine Learning and Cybernetics (ICMLC), vol. 1, pp. 91–97. IEEE (2017)
14. Lin, Y., Liu, Z., Sun, M., Liu, Y., Zhu, X.: Learning entity and relation embeddings for knowledge graph completion. In: AAAI, vol. 15, pp. 2181–2187 (2015)
15. Nickel, M., Murphy, K., Tresp, V., Gabrilovich, E.: A review of relational machine learning for knowledge graphs. Proc. IEEE **104**, 11–33 (2016)
16. Xie, J., Girshick, R., Farhadi, A.: Unsupervised deep embedding for clustering analysis. In: International Conference on Machine Learning, pp. 478–487 (2016)
17. De Brabandere, B., Neven, D., Van Gool, L.: Semantic instance segmentation with a discriminative loss function. arXiv preprint arXiv:1708.02551 (2017)
18. Goodfellow, I., Bengio, Y., Courville, A., Bengio, Y.: Deep Learning, vol. 1. MIT Press, Cambridge (2016)
19. Glorot, X., Bordes, A., Bengio, Y.: Deep sparse rectifier neural networks. In: Proceedings of the Fourteenth International Conference on Artificial Intelligence and Statistics, pp. 315–323 (2011)

Smart Recommendation System Based on Understanding User Behavior with Deep Learning

Reza Mahdavi[(⊠)] and Afsaneh Hasanjani Roshan

Department of Applied Mathematics, Iran University of Science and Technology,
Tehran, Iran
reza.mahdav@gmail.com,
Hasanjaniroshan.afsaneh@gmail.com

Abstract. Consumer behavior is one of the most important issues that has been discussed in recent decades. Organizations always want to understand how consumer make decisions so that they can use it to design their products and services. Having a correct understanding of the consumers and the consumption process has many advantages. These advantages include helping managers make decisions, providing a cognitive basis through consumer analysis, helping legislators and regulators legislate on the purchase and sale of goods and services, and ultimately helping consumers make better decisions. Here is a solution for recommending goods based on the users' past behavior over deep learning. The architecture expressed for deep learning is trained by users' past behavioral data. Amazon data was studied and the results indicated that the proposed method has a much higher accuracy than similar methods.

Keywords: Recommendation systems · Deep learning · Users' behavioral past

1 Introduction

The rapid and growing spread of information provided on the global Internet network has faced users with numerous and notable problems regarding the resources and information they need, and it is possible that without proper guidance, users make mistakes in making right decisions or choosing the goods and services they need, which will have many consequences, including dissatisfaction, discouraging users and customers from the websites on the Internet. Hence, there is a need for tools and systems to help users choose the right information they need. In recent years, to meet these needs, recommendation systems have been proposed and developed, and there are a variety of different algorithms, articles and scientific texts in this field.

Meanwhile, the creation and expansion of social networks, trust networks, and the existence of a variety of relationships among the users of these networks have opened a new horizons to researchers and developers of recommendation systems so by utilizing the social sciences and psychological sciences dominant in these networks, and in

© Springer Nature Switzerland AG 2019
L. Grandinetti et al. (Eds.): TopHPC 2019, CCIS 891, pp. 53–67, 2019.
https://doi.org/10.1007/978-3-030-33495-6_5

particular the existence of a trust relationship among users, they can introduce a new generation of recommendation systems called "trust-based recommendation systems". These systems are able to provide a greater percentage of users with the right answers, and their results are more accurate.

Here, we have tried to give suggestions of items to the users considering their behavioral past. Here, the relationship between users and items is created graphically, and according to the graphic theorem, there are suggestions that exactly depend on the behavioral background of the users, hence the users will respond positively to the suggestions with a much greater desire. In the proposed approach, deep learning has been used that leads to high accuracy of the proposed method.

2 Recommendation Systems

Recommendation systems try to suggest the most suitable items (data, information, goods, etc.) by analyzing the behavior of their users. This system will help its user get closer to their target faster within a massive amount of information. Some consider recommendation systems same as a group refinement.

Recommendation systems are systems that help users find and select the items they want. It is natural that these systems are not able to suggest without having adequate and right information about the users and their intended items (such as movies, music, books, etc.); therefore, one of their most basic goals is to collect various information in relation to tastes of users and items available in the system.

In addition to the implicit and explicit information, there are some systems that use personal information of users. For example, the age, gender, and nationality of the users can be a good source for knowing the user and making suggestion to him. This kind of information is called Demographic Information that a group of recommendations systems are based on this information. With the advent of web 2 and the expansion of social networks in recent years, researchers have found another information source to improve the quality of the suggestions, which is the same information on social networks, and based on this, many research work has been done in this field [1, 2].

2.1 Principles of Recommendation Systems

There are some things that need to be addressed and in the design and implementation process of the system they should be considered in order to establish an efficient recommendation system. They are as follows:

1. The type of data in the system context
2. The filtering algorithm used
   a. Collaborative Filtering
   b. Content-based Filtering
   c. Social-based Filtering

    d. Knowledge-based Filtering
    e. Context-aware Filtering
    f. Hybrid Filtering
3. Selected model for the system
4. The technique used to suggest
5. Expected scalability of the system
6. Optimal system performance
7. Quality of the presentable results

Undoubtedly, the most basic component in the recommendation systems is the algorithm and its filtering solution. In the following, we will look at the most important solutions used in this area:

- Collaborative Filtering
- (KNN) K Nearest Neighbors
- Demographic Filtering
- Content-based Filtering
- Social-based Filtering
- Context-aware Filtering
- Location-aware Recommendation Systems
- Knowledge-based Filtering
- bio-inspired

According to official statistics provided by the Amazon book sales website, 35% of the website's sales are due to the existence of a recommendation system and the provision of appropriate suggestions to book enthusiasts [3]. Also, the Netflix website has set a prize of over $1 million for researchers to improve its accuracy by 10% for its recommendation system, Cinematic. Both examples show the importance of recommendation systems and their role in business. In following, we mention some of the reasons for the importance of having an appropriate and efficient recommendation system on an e-commerce website:

- Helping users and visitors choose the right goods, product or service
- Accelerating the time of choosing a customer's product, such as choosing a suitable movie among the millions of videos available on the website
- Collecting valuable information about user tastes and behavior for future planning
- Attracting customers and visitors who visit the website for the first time
- Increasing user satisfaction and naturally increasing the profit of commercial investors
- Creating confidence in customer by providing other users' feedback

In contrast to the advantages mentioned, these systems have deficiencies and limitations that are referred to below:

- The impossibility of evaluating the user's profile in all aspects and not providing a fully consistent answer to their point of view

- Customer uncertainty about the comments and suggestions provided by the system
- Lack of optimal accuracy in the suggestions provided by the system
- Failure to respond in certain circumstances, especially for new users or new customers
- There are many technical problems and barriers to implementing and providing algorithms for recommendation systems, some of which are:
- The implementation of these systems is very complicated due to the numerous factors affecting the mechanized decision-making process.
- The possibility of penetrating attackers and fraudsters into such systems and creating incorrect data in the system that may change or distort the outcome of the system.
- The complexity of the existing algorithms and low speed of responsiveness of these systems, especially in providing online suggestions.

3   Related Works

In 2011, Ying Hei provided a model from combining the item-based method and the Bayesian method that could solve the problem of the rating matrix privacy to some extent. For his model, he used the content data of the item and the rating matrix [4]. In the article [5], authors have argued the need to set personal recommendation systems in learning-enhancing technology for a specific learning feature, rather than using recommendation systems for other areas. In this work, specific learning needs were defined and it was concluded that such personal recommendation systems should focus on learning objectives, learner characteristics, learner groups, ranks, learning ways, and learning strategies for better recommendations. In fact, the purpose of this article is to provide the appropriate techniques for building a personal recommendation system for lifelong learning. In this article, memory-based methods are used, such as: collaborative filtering that includes context-based techniques and hybrid techniques. In the collaborative filtering method, items used by similar users in the past are recommended. The method of content-based techniques recommends options similar to those that users preferred in the past. In the hybrid technique, both techniques are used to provide more precise recommendations. In this paper [6], the problem of finding a time expert based on meaning is presented to identify a person with specific expertise for different time periods. In this regard, this paper has proposed the development of a time modeling method for the TET issue based on the STMS method, which can provide expert ratings in different groups in an unmonitored manner. This method was inspired by a previous ACT model that was inspired by a separate document (sub-group) (without the conference effect) and developed to all conference journals from the "whole group" of the conference (the conference effect). In the paper [7] the problem of finding the expert in the community of questions and answers (CQA) is studied. The goal of finding an expert in the CQA is to find users who can provide a large number of high-quality, complete, and reliable answers. For this reason, a probabilistic model sensitive to topic is suggested for finding experts in this paper. In the CQA, there are a set of users in the community, first the topics that are interesting to users within the community are automatically found by analyzing the content of the questions they are asked or they answered, and then, based on the topics found, question-answer relationships sensitive to topic can be constructed between questioners and respondents. This, in its turn, makes it possible to calculate the outstanding score of expertise by considering both the link structure and the topic similarity between the questioners and the respondents. Accordingly, a probabilistic model is made for ranking the candidate experts with the consideration of the user's expertise and credibility. In [8], expert knowledge has been applied to build a solution recovery system and to find expert and problem-detecting. In order to change the expert knowledge and the base knowledge of a solution recovery system, the idea of developing a recovery system based on the RCBR and RBR combination method that uses CBR is proposed in this paper.

Article [9] examines a content-based collaborative recommendation system. In the content-based collaborative system, efforts are made to suggest a specific user the similar options to those that were previously interesting to him. While in the collaborative recommendation system, users with similar interests are identified, and same options are recommended to them. The method used in this article, called fab, is derived from combining these two methods, and is part of the Stanford University Digital Library Project.

Trust in recommendation systems has been discussed and evaluated in [10]. Some authors have argued that traditional emphasis on user similarity may be exaggerated. This paper proves that traditional factors play an important role in guiding recommendations, and especially the trustworthiness of users should be considered as an important point. To this end, two computational models of trust are presented and illustrated that how they can be used to combine with the standard collaborative filtering frameworks in a variety of ways. In fact, the goal in this article is to improve the collaborative filtering method.

In paper [11], Lee et al. provided a social recommendation system that could create personality product recommendations based on similarity of preference, trust, and social relationships. The Advantage of the proposed method in this paper is comprehensive evaluation of that from the resources recommended.

Recommendation systems based on the multi-meaning value theory are suggested to deal with issues of existing recommendation systems, such as the problem of cold start and change of settings.

In paper [12], Scholz et al. state that existing MAVT-based methods are not suitable for measuring the weighting of attributes to purchasing tasks that are used for common recommendation systems. Learner systems are often used by consumers who are not familiar with the features and scopes available, and are eager to save time and effort. In contrast to this background, in this paper a new approach is developed based on a product configuration process that is devoted to the features of these decision-makers.

The method outlined in this article can be considered as a promising option for supporting decision-making processes of consumers in tasks of business e-purchasing.

4 Proposed Method

In the present era, in many cases, machines have been replaced by humans, and many of the physical work done by humans in the past is carried out by machines today. Although the power of computers in storage, data recovery, and office automation is undeniable, there are still some things that man has to do himself. But in general, machine related issues include systems in which, due to the complex relationships between components, the human brain is incapable of mathematical understanding of these relationships. The human brain can over time detect system habits by observing the sequence of system behaviors and sometimes testing the result obtained by manipulating one of the system components.

In the proposed method, deep learning is used, which can be considered as a more advanced state of the multi-layered perceptron neural networks, which has much higher layers. The proposed method focuses on users' general behaviors. Behaviors obtained through binary relations between the user and the objects. Here we formulate the user behavior so that we can extract the rules from it.

In the previous section, it was stated that this method has many advantages that we need considering the type of work we want to do, and it is described in this regard in following.

The behavior of users in this proposed method is stored as tuples {a, o, t} in which:

- a: is user's action type and specifies the type of work that he has done.
- o: is the object to which the user has shown an action.
- t: is time when action a is performed on object o.
- Here a and t are unique or atomic. Therefore, for a user, one can express all actions in relation to objects as following.

$$U = \{(a_j, o_j, t_j) | j = 1, 2, \ldots, m\} \tag{1}$$

The general framework of the proposed method is shown in Fig. 1. As we can see, we divide our proposed strategy into several blocks, which are respectively as following:

- Finding raw space for features for each user
- Grouping the users' behavioral raw space obtained
- Time division in each group
- Semantic grouping of behaviors

- Reducing the semantic space obtained and dividing closer semantic groups in new groups
- Using groups created and assigning these groups to deep learning network layers
- Recommending to user based on the user's behavioral characteristics and predicting learned network (trained model)

Flowchart of the proposed method is shown in Fig. 1.

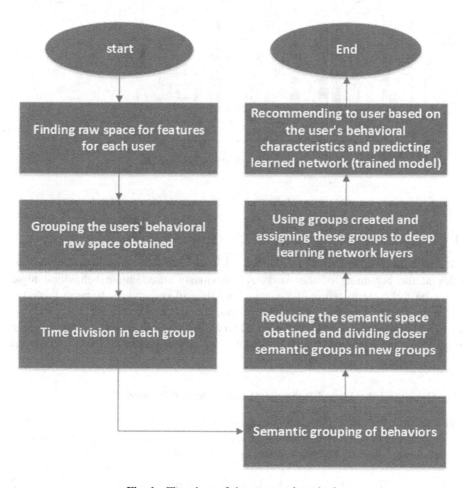

Fig. 1. Flowchart of the proposed method

In Fig. 2, you can see a general view of the deep learning network architecture used here.

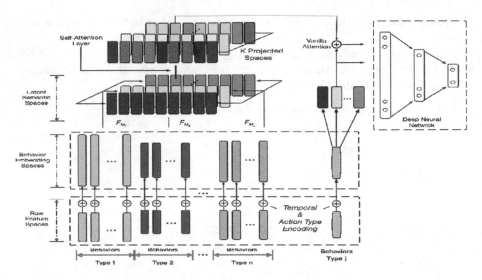

Fig. 2. Detailed Flowchart

In following, we want to explain each of these sections completely.

4.1   Finding Raw Feature Space for Each User

First, for each user, according to relation (1), we create behavioral tuples and get the tuples at the beginning of the work. As previously stated, these behavioral tuples cannot be the same in terms of behavior and time of occurrence, because these two have to be atomic. Logically, in the real world, it is also impossible for a user with one action and at a time to affect some objects, so at this stage we get these behavior tuples for all users and somehow it can be said that we create the past of these users in this way.

4.2   Grouping the Users' Behavioral Raw Space Obtained

In this section, the users' behaviors collected in the previous section and became in the form of relation (1) are divided here into various groups. The grouping performed here depends on the type of objects that is used in relation (2).

$$G = \{bg_1, bg_2, \ldots, bg_n\}bg_i \cap bg_j = \emptyset \ and \ U = \ \cup_{i=1}^{n} bg_i \qquad (2)$$

Given the relationship (2), different groups that are obtained have nothing in common, and eventually all tuples that fall from a user in different groups, if they are gathered, are equal to the user-related tuples, so all tuples are divided into different groups, and this division of tuples in different groups is such that it is divided according to the type of objects, tuples related to different objects are divided from different groups, and this mode is also possible that for a user different tuples with different times or different actions are placed in one group. Because this division depend on objects, not the action or the time of action.

4.3 Time Division in Each Group

After grouping, we need a key property to be able to differentiate between the tuples of a group and to search among them. In this step, we sum up the action time of all the tuples and, given the time diversity, we create different timing classes, for example $(1, 0], (2, 1], (3, 2], \ldots$ in this case the tuples can be divided into different classes, and according to the action, one can differentiate between the tuples in a given time class because we said that a tuple has an action time and an atomic action, and these two cannot be for two tuples of one user, so in these time classes, two behavioral tuples can be clearly distinguishable from one user.

4.4 Semantic Grouping of Behaviors

After time classification of the tuples in this section, the tuples are divided into different semantic groups. In fact, the same time classes are divided into different semantic groups, and this semantic grouping is according to the action.

4.5 Reducing the Semantic Space Obtained and Dividing Closer Semantic Groups in New Groups

In the previous section, semantic grouping was done in time classes for tuples, in this section, semantic groups that can be merged into each other are merged, and it can be somehow said that in this way, the groups get less and the semantic space of the previous step is reduced. In this case, it can be said what actions can lead to another action, that is, somehow one can obtain dependence of the tuples here, and using this information, the tuple dependence graph can be created.

4.6 Using Groups Created and Assigning These Groups to Deep Learning Network Layers

In this section, semantic groups that were optimized and their dependency were obtained are attributed to different layers of the deep learning network at this stage. So here in some way, one can say the possibility of training network is provided and this network is trained.

4.7 Recommending to User Based on the User's Behavioral Characteristics and Predicting Using the Trained Model

The network that was trained at the previous stage is used at this stage. At this stage, we use the network that was trained to recommend to users, and the objects considering the user's past behavior and similarity to the behavior of this user with previously trained users, the suggestions are made and for the objects that should be displayed to users are ranked.

5   Evaluating the Proposed Method

In this section, the proposed method is evaluated. Implementation is done in Python 2.7, and here tensorflow-Gpu is used that uses the gpu to perform computations, so we need a graphic card of over G10. The system used here has the following characteristics:

- Windows 10
- 16G Ram
- NVIDIA 900-52401-0020-000 GRID K1 16 GB Graphics Card

5.1   The Dataset Used

The dataset used in this section is related to Amazon data. We collected several subsets of Amazon product data, but here we tried to collect 5 categories. The features we used from this data include user_id, item_id, cate_id, and times_stamp. Table 1 shows the number of data used. In this section, we show the set of behaviors of user u with $(b_1, b_2, \ldots, b_n)$. Here we intend to examine k behavior of the user and identify the k + 1 behavior of the user, k can be between 1 and n − 1, in other words, we intend to examine n − 1 behavior and predict n behavior.

Table 1. Statistics of dataset used.

Dataset	(Users)	(Items)	(Cates)	(Samples)
Electro.	192403	63001	801	1689188
Clothing.	39387	2303	484	278677

Taobao Dataset: In this paper, we collect a variety of user behaviors on an e-commerce website. We used three groups of behaviors in this section, which are item behavior group, search behavior group, and coupon receiving behavior group. The behavior item group is all user activities on items such as purchases and searches made for shopping centers and so on. The features that are considered for an item are item_id, shop_id, brand_id, and category_id.

The behavior search group is considered as only a type of activity because it contains only the words used in the search. In coupon receiving behavior group, coupon_id, shop_id and coupon_type features are used. There is only one type of activity in this group. In all of this behavior, a timestamp is placed so that it can be detected when the user performs the behavior. Table 2 shows the statistics used of each category. Of course, it should be noted here that we only input users in our database that have at least three behaviors.

Table 2. The number of data used to evaluate multiple behaviors.

	Dataset-multiple behavior
(Users)	30358
(Items)	447878
(Cates)	4704
(Shops)	109665
(Brands)	49859
(Queries)	111859
(Coupons)	64388
(Records)	247313
(Avg length)	19.8

5.2 Competing Methods for Comparison with the Proposed Method

In this section, different methods are presented to be compared to the proposed method. These methods are mostly similar to proposed method but are different in architecture while learning and suggesting. Thus, these methods are as follow:

- BPR-MF: Bayesian ranking offered by Randall provides a framework for ranking based on business networks. In this solution, user-product pairs are used which, if the user chooses or orders the product, the pair between the user and that product gets the value of true and otherwise the value of false and consequently, after training the network with this method is able to predict based on these pairs.
- Bi-LSTM: In this method, the past of a user is encoded and he able to evaluate and train users' past with higher performance and speed. This method was provided by Zhang in 2014.
- Bi-LSTM+Attention: In this method, through Bi-LSTM, which only deals with the purchase relationship or user search for the product, an extension was added that can take into account the user's attention so it can provide better recommendations.
- CNN+Pooling: This method uses the CNN structure, which is able to demonstrate excellent performance using these neural networks. In this method, pooling is also used for encoding in order to reduce the volume of users' past behaviors and speed up the evaluation process.

5.3 Parameters Used in the Program

Network Schema: Here we set the size of each of the grouping features to 64. The size of the hidden layers of the neural networks is 128.

Batch size: Here, 32 is set for it. In fact, it can be said that the batch size is a module of input data that is given to the system, such as any spoon of food, here the spoon is input data which is fed to the system.

Optimizer: The optimizer used in this neural network is SGD, which is one of the best optimizers which the learning rate is considered in this optimization function.

5.4 Evaluation Criteria

Here we considered the AUC for evaluation. The AUC represents the level below the ROC chart, the higher the number for a batch, the better the performance of the batch will be. The ROC chart is a method for evaluating the performance of the batches. In fact, ROC curves are two-dimensional curves in which the DR or the True Positive Rate (TPR) on the Y axis and similarly the FAR or False Positive Rate (FPR) on the X axis are drawn up. In other words, a ROC curve shows the relative compromise between profits and costs.

The AUC of the average user in our method is calculated as relation (3) where $p_{u,i}^{\wedge}$ is the probability of predicting the action on I in the test set of user $u \in U^{Test}$ and $\delta(.)$ is the index function.

$$AUC = \frac{1}{|U^{Test}|} \sum_{u \in U^{Test}} \frac{1}{|I_u^+||I_u^-|} \sum_{i \in I_u^+} \sum_{j \in I_u^-} \delta\left(p_{u,i}^{\wedge} > p_{u,j}^{\wedge}\right) \tag{3}$$

5.5 Results Obtained from a Behavior with Only One Type

We first calculated the average AUC of the user for all data received from Amazon in Table 3. It can be seen that our proposed method is much better than its competitor methods, especially when the user's behavior is of depth, because the method we mentioned is capable of managing and evaluating users' deep behaviors.

Table 3. Calculating the AUC on the Amazon data.

	Electro.	Cloth.
BPR	0.7982	0.7061
Bi-LSTM	0.8757	0.7869
Bi+LSTM+Attention	0.8769	0.7835
CNN+Pooling	0.8804	0.7786
MyAlgorithm	0.8921	0.7905

It can also be seen in the chart of Fig. 3 that our proposed method is capable of creating the coverage with a very high speed compared with CNN-based methods on the dataset. This shows high performance and high quality of our proposed method, which is able to have a proper performance on any behavior.

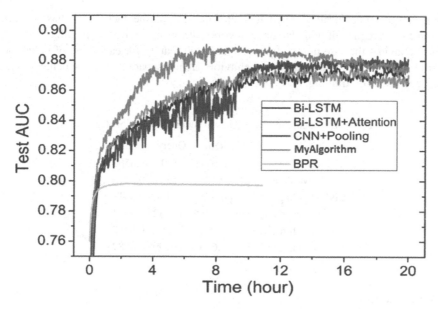

Fig. 3. AUC chart in Amazon Electro training data.

5.6 Results Obtained from Several Types of Behavior

We trained the typical model so it can predict the user's next activity based on its past behavior, which can include any kind of behavior. We considered and evaluated two sets of training and testing in this section.

One-to-One Model: We initially trained triad models (for each user, three features) with just one behavior, and we predicted one type of behavior so that it can be called item2item, coupon2coupon, and query2query we named it one2one because it was trained with one type of behavior and one type of behavior was predicted.

All-To-One Model: In the following, we trained the system other triad models of any type of behavior, that is, all kinds of selection behaviors as the user's behavioral past, and predicted a separate type of behavior, that is, outputs are proportional to the type of behavior. So it can be called all2item, all2coupon and all2query, which we named it all2one

All-To-All Model: Ultimately, we used an integrated one including a variety of behaviors to train, so that we could predict any behavior at the same time that we named this model all2all.

In the following, we evaluated these three types of models, the results of which can be seen in Table 4. According to this table, the all2one model is much better than the one2one model, because one2one model only examines one type of behavior. While in all2one, several behaviors are all examined together and eventually a prediction takes place for each behavior. One interesting result here is that all2one also has better performance than all2all, because it is not possible to use all behavioral models for training, and ultimately, predict any other behavior using this training. Because in this case the accuracy is reduced because in prediction we may require a behavior that the system has not been trained by that, therefore, due to the fact that it has not seen training in this regard, it will also have a worse performance. But in the all2one model, all the behaviors that are trained by that system ultimately for each kind of behavior a separate prediction is made, but these behaviors are not separate from the behaviors that have been used for training.

Table 4. Calculating the AUC on multiple behavior dataset.

	Item	Query	Coupon
Bi-LSTM	0.6779	0.6019	0.8500
Bi-LSTM+Attention	0.6754	0.5999	0.8413
CNN+Pooling	0.6762	0.6100	0.8611
MyAlgorithm-one2one	0.6785	0.6132	0.8601
MyAlgorithm-all2one	0.6825	0.6297	0.8725
MyAlgorithm-all2all	0.6759	0.6199	0.8587

6   Conclusion

The proposed method in this article is of a high accuracy in evaluating users' behaviors and is able to offer very useful suggestions in proportion to user behavior. This evaluation of behaviors is based on neural networks and is able to predict future behaviors of the users by examining the past behaviors of the users, and can therefore provide users with very useful suggestions using the user's future behavior. We looked evaluated various models such as one2one, all2one and all2all, and it was observed that all2one performs much better than the other models, in which different behaviors are trained and a prediction is made for each behavior.

7   Future Suggestions

Various solutions can be used here. One of the solutions that can be used well here is to get help from deep learning with different architectures, so that it can be used in different systems according to the architecture. That is, it can also function properly on mobile or low-power systems. Currently, the proposed method is implemented using the graphic card of NVidia 16G. Of course, with NVidia 10G it can also be implemented, but the method somehow can be upgraded so that it can be run using parallel and opencl processing on the graphic card of AMD.

    Another improvement that can be made to the proposed method is that are a model is constructed for the proposed method and this trained model is used for predictions, but this trained model is not able to learn new data so the data will be old after some time, while transfer learning can be used, which is new knowledge, and it can be used to add new data to the model during the work and train the model with new data, that is, during the work that the prediction is done, this new data is also use as training data after the prediction and over time, in this mode, the system can perform predictions with higher power.

Another possible solution is to improve the neural network used here by the genetic algorithm in order to be able to perform the analysis with better accuracy and speed, and thus able to perform better than our proposed method.

References

1. De Pessemier, T., Leroux, S., Vanhecke, K., Martens, L.: Combining collaborative filtering and search engine into hybrid news recommendation. Universiteit Gent (2015)
2. Ricci, F., et al. (eds.): Recommender Systems Handbook. Springer, New York (2015). https://doi.org/10.1007/978-1-4899-7637-6
3. Linden, G., Smith, B., York, J.: Amazon.com recommendations: item-to-item collaborative filtering. IEEE Internet Comput. 7(1), 76–80 (2013)
4. Yun, S.-Y., Youn, S.-D.: Recommender system based on user information. IEEE (2011)
5. Drachsler, H., Hummel, H., Koper, R.: Recommendations for learners are different: applying memory based recommender system techniques to lifelong learning (2007)
6. Halder, S., Sarkar, A.M.J., Lee, Y.-K.: Movie recommendation system based on movie swarm. In: 2012 Second International Conference Cloud and Green Computing (CGC) (2012)
7. Zhou, G., Zhao, J., He, T., Wu, W.: An empirical study of topic-sensitive probabilistic model for expert finding in question answer communities (2014)
8. Tung, Y.-H., Tseng, S-.S., Weng, J-.F., Lee, T-.P., Liao, A.Y.H., Tsai, W-.N.: A rule-based CBR approach for expert finding and problem diagnosis (2009)
9. Balabanović, M., Shoham, Y.: Fab: content-based, collaborative recommendation (1997)
10. Li, Y.M., Wu, C.T., Lai, C.Y.: A social recommender mechanism for e-commerce: combining similarity, trust, and relationship. Decis. Support Syst. 55, 740–752 (2013)
11. Lee, J.-H., Yuan, X., Kim, S.-J., Kim, Y.-H.: Toward a user-oriented recommendation system for real estate websites. Inf. Syst. 38(2), 231–243 (2013)
12. Scholz, M., Dorner, V., Schryenc, G., Benlian, A.: A configuration-based recommender system for supporting e-commerce decisions. Eur. J. Oper. Res. 259(1), 205–215 (2017)

Big Data Analytics

Genome of Human-Enabled Big Data Analytics

Mohammad Allahbakhsh[1,3(✉)], Saeed Arbabi[1],
Hamid-Reza Motahari-Nezhad[2], and Boualem Benatallah[3]

[1] University of Zabol, Zabol, Iran
{allahbakhsh,sarbabi}@uoz.ac.ir
[2] Ernst & Young AI Lab, Palo Alto, CA, USA
motahari@ieee.org
[3] The University of New South Wales, Sydney, Australia
{mallahbakhsh,boualem}@cse.unsw.edu.au

Abstract. Nowadays, we live in the big data era, the gigantic information resource that Organizations are so interested to analyze in order to know their customers better, provide better services and increase their income. Recently, after the emergence of crowdsourcing techniques, people are also involved in the process of big data analysis. Inclusion of human computing power, while helps improve the quality of the results, can raise serious challenges that needs deeper investigations. In this paper, we propose a genome for human-enabled big data analytics, to understand them better, and to study where and why people get involved in such systems. We then study the challenges that raise as the result of such an involvement and propose some future research directions.

1 Introduction

With the pervasive use of digital gadgets, sensors, systems, mobile phones, etc., we are moving towards an age of total digital footprints. Sharing opinions in social networks, searching the web, twitting, purchasing online products, participating in online polling and many other digital aspects of our lives leave behind a tremendous digital footprint. Billions of sensors embedded in cars, mobiles and other forms of devices constantly sense, generate and communicate trillions of bytes of information [1]. This gigantic generated data, which is also referred to as *big data*, is a rich source of information about how people and things behave and interact. The process of extracting this information from big data is called big data analytics and is applied using different techniques and methods [2–6], and in different context.

A key observation in data analysis, based on various machine learning and artificial intelligence technique is that humans are still involved in taking steps, and making decisions that impact the success of data analysis techniques. The examples ranges from data preparation and cleaning, selection of algorithms, tuning the parameters of the algorithms, to performance analysis and results

© Springer Nature Switzerland AG 2019
L. Grandinetti et al. (Eds.): TopHPC 2019, CCIS 891, pp. 71–83, 2019.
https://doi.org/10.1007/978-3-030-33495-6_6

interpretation. There are also another class of problems, in which machines are not yet good enough. For instance, various data cleansing, information extraction, optimization and social behaviour analysis that may benefit for having human in the loop of problem solving. People might be employed for such tasks, in combination with machines, to extract information from ordinary or big data sources [7,8]. Presence of human in big data analytics process, either as the subject of data analysis, in training systems [9] or in the data analysis process as a worker, can raise serious challenges such as security, trust, privacy, quality of outcome, etc. Addressing these issues requires fundamental understanding of different aspects of big data analytics, and how it relates to the involvement and impacts of people on the outcome of the analysis. This is what we propose in this paper.

The rest of the paper is organized as follows. In Sect. 2 we propose a framework and a genome to study big data analytics. In Sect. 3, we study how and where people may get involved in the process of big data analytics. The challenges that might raise after inclusion of people in big data analytics are studied in Sect. 4, and finally, we conclude in the Sect. 5.

2   A 5-Dimension Framework for Data Analytic Projects

Big data analytics is becoming more and more popular. A large number of big data analytics approaches have emerged so far and the number is increasing day by day. Taking advantages of this opportunity and also dealing with the challenges it arises, needs a clear understanding of the nature of big data analytics, the important aspects it might have, and the dimensions along which it is characterized. At the first glance, big data analytics systems are so diverse in terms of their architecture, leveraged techniques, data sources being used, extracted insight and so on [10–14]. All these research efforts show the complexity and diversity of big data analytics systems and, hence, the hardship of understanding and leveraging them. In this section, we try to approach the problem from a different jargon-free point of view, in which, we put all the details away and grasp the core elements of the system.

Having a deeper look at big data analytics systems, despite the aforementioned diversity and complexity, we identify a relatively small set of dimensions that characterize the systems. We use the following five key questions to identify these characterizing dimensions:

- *Why* data is being analyzed?
- *Where* does the data come from?
- *What* is the expected outcome?
- *Who* are the data analyzers?
- *How* is the analysis being accomplished?

When studying a big data analytics system, each of these five questions can have specific answers, each of which reflects a particular aspect of the system. Employing an analogy from biology, and inspired by [15], we call these answers,

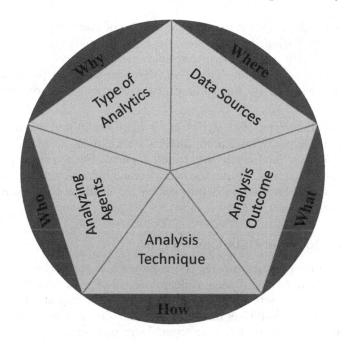

Fig. 1. Five dimension that characterize big data analytics.

which are in fact building blocks of systems, *genes*. We define a gene as a particular answer to one of those key questions. Similar to biological genes, these genes are the key characterizing aspects of big data analytics systems. Figure 1 depicts the five different dimensions along which a typical big data analytics system is characterized, according the aforementioned key questions. In the followings, we describe these questions and their corresponding answers, i.e., genes, in more details.

2.1   *Why* Data is Being Analyzed?

The first important question regarding using big data analytics is the reason of data analysis. In other words, we know that dealing with big data is challenging due to huge volume, variety and velocity of data. So, why someone should care about big data analysis. What kind of benefits one can get from data analysis? The answer to this questions is: *extracting analytics*. Extracted analytics can be from one of the following types: *descriptive*, *predictive* and *prescriptive* [16,17].

Descriptive. Descriptive analytics describe the behaviour and help users better understand and model the past behaviour of the system, or probably a community, by identifying behavioral patterns and creating management reports. More precisely, they answer *"what happened?"* and *"why did it happen?"* questions.

Predictive. Predictive analytics are the results of analyzing the current and past states and behaviour of the system, in order to predict a future behaviour or state. In contrast with descriptive analytics, predictive analytics answer *"what will happen?"* question. Therefore, they are useful in predicting future behaviours, and planning or fine tuning future directions of an organization.

Prescriptive. This type of analytics is far more useful and informative than the previous two types. Prescriptive analytics anticipate *"what will happen?"*, *"when it will happen"* and even *"why it will happen"*. Furthermore, prescriptive analytics assist analysts in finding the best decision options to take advantage of future events, and show their users the possible implications of choosing different decision options.

2.2 Where Does the Data Come From?

The second question regarding a big data analytics, is about the *source of data*. Different data sources might have different levels of quality and, hence, may impact quality of extracted analytics. Therefore, it is important to take into account quality or credibility of data sources. The question regarding the source of information can have different answers. More precisely, there exist a list of answers, acting as genes of the system, which can be used as building blocks, in terms of data source. We identify three different genes as general answers to the data source question: *humans, sensors/Devices* and *organisations*.

Human. This gene appears in the architecture of human-enabled big data analytics systems when the data is generated by human. Mass collaborative systems such as Facebook or YouTube are examples of systems in which members generate big data [18]. Based on a report, in 2016, in every minute, YouTube users share about 400 h of new video, Instagram users like more than 2.4 million photos, more than 3.5 million text messages are sent in the U.S., and so on [19]. Hence, humans are one of the main sources of big data.

Sensor/Device. Sensors are another important source of data in big data systems [18]. A sensor is a device (a thing in the IoT terminology) that senses data from the environment and sends the raw data to an analytics system without modification. RFID tags, GPS sensors, weather forecast sensors, etc., are examples of such sensors which collect trillions of bytes of information every day. According to Gartner, in 2016 there have been 6.4 billion connected things in use worldwide, and this number will reach 20.8 billion by 2020 [1]. These sensors generate trillions of bytes of data.

Organization. Organizations also can act as data sources for big data analytics systems. They almost provide structured data such as staff data, transactions records, etc. Recommendations and indices provided by stock market analysts are examples of big data items provided by organizations.

2.3   Who Are the Data Analyzers?

The next key question is about the agents who are responsible for analyzing data. Extracting information from big data needs computing power and probably intelligence. Therefore, big data analytics should be done using *machines*, *humans* or in a *hybrid* manner.

Machine. Leveraging machines for the purpose of big data analysis is the most popular data analysis method. Analytics extraction usually needs a level of machine intelligence, and hence, it might be necessary to employ artificial intelligence and machine learning techniques as well. For example, natural language processing techniques usually need artificial intelligence (AI) for analytics extraction.

Human. There might be cases in which AI is not enough to extract analytics. More precisely, there are problems which are very easy for a human to do, but very difficult or even impossible for a machine. Human intelligence tasks (HITs) such as product labelling, photo tagging or music transcription are examples of HITs [20]. In such cases, human intelligence is employed separately, or in combination with AI, to extract analytics. For instance, in Wall-Mart product classification project (http://www.walmart.com) a huge volume of data is constantly being received from various retailers from all across the globe. The received data is structured but, in most of the cases, it is incomplete. So, Wall-Mart cannot use only machines for the purpose of entity matching and resolution. The results, as tested, do not have the required level of accuracy. To overcome this problem, Wall-Mart selects some individuals with adequate expertise to refine the results produced by machine [8]. There are also research show that combining human computing and machine learning can lead to better results, compared to the cases in which one of the two techniques has been used [7, 21–23].

2.4   How the Analysis Is Being Accomplished?

The answer to the "how" question specifies the manner in which the data is analyzed. The data might be processed all at once, i.e., in *batch*, or gradually and upon arrival, that is, *interactively*.

Batch. In the systems having this gene, arrival/collection and analysis of data is performed in batch. These are mostly map-reduced based systems in which data is first collected, stored, pre-processed and then analysed [16, 24, 25].

Interactive. In contrast to batch methods, interactive methods (may also refer to as real-time, near-time or streams [16]), extract analytics in real-time. Interactive approaches aim at adopting the process of data analysis in order to match the real-time requirements of users [26, 27].

Table 1. Human-enabled big data GeneTable

Key question	Gene	Abbreviation
Why	Descriptive	D
	Predictive	P
	Prescriptive	S
Where	Human	H
	Sensor/Device	V
	Organization	O
Who	Machine	M
	Human	H
How	Batch	B
	Interactive	I
What	Tag	T
	Named Entity	N
	Fact	F
	Association	A
	Event	E

2.5   What Is the Expected Outcome?

Several research evidences approve that using big data analytics improves performance and competence of organizations by providing them with valuable information and intelligence extracted from a humongous amount of data [28]. But, what are the outcomes of big data analytics process with such a great impact? We identify five different types of outcomes: *tag, named entity, fact, associations* and *event*.

Tag. A tag, also referred to as a keyword or a term, is a label or a piece of information attached to an item, e.g., an entity, document, etc., in order to provide more information about the item. A tag, can have an important role in the process of big data analysis, e.g., to extract the sentiment of a community. The terms Google Trends (https://www.google.com/trends/) extracts from searches being done on Google search engine are examples of tags extracted form big data. Also, annotating a text, document, video, etc., with tags extracted from big data is another example of tag extraction.

Named Entity. An entity refers to an object, person or an artefact. Entities might be extracted using entity recognition techniques [29]. While entity recognition is a challenging task, recognized entities are still just noun phrases and hence ambiguous. Entities are classified into broad categories such as person, product, date, etc. Mapping an entity to a particular entity that is registered in a knowledge base is called the name entity disambiguation (NED) [29,30]. For instance, Steve Jobs, extracted from a text being analyzed is a name entity. Extracting dates, products or books from huge logs of online markets are other examples of named entity extraction.

Fact. A fact is a piece of information, extracted or inferred from data, that reveals a fact about the system or one or more components or entities of the system. Annotating an object with a new attribute or a tag which is obtained from analyzing big data is an example of such facts. Another type of facts are concepts that are extracted from data. For example, assume that a data analyzer uses a feedback received from a customer containing phrases such as "great quality", "fair price", "will recommend to friends", to infer that the customer is happy. This customer happiness is a concept/fact that is the outcome of the data analysis process.

Association. An association refers to a discovered relationship among two or more components or entities of a system. A relationship can be either directed or undirected. Also an association might have attributes. Finding co-purchase relationship between two or more products in an online market is an example of extracted associations. Computing friendship probability between two members in a social network is another example of associations extraction [31]. Another type of associations between items is having them in the same clusters, groups, etc. Being in a same itemset of products or a same collaborating group of people [32] are other examples of associations that might be brought out to the open by analyzing big data.

Event. Detecting occurrence of an event is another possible outcome of big data analytics. An extracted event reveals information such as "what has happened", "when it has happened", etc. For example, when analysing oil global prices, a drastic change, such as a revolution, in an oil producing country is an event of a very high important that should be extracted out using analysis techniques.

3   Human Involvement in Big Data Analytics

In Sect. 2, we proposed a framework introducing a genome for the big data analytics systems. We identified five dimensions along which a big data analytics system can be characterized, and we introduced genes for each of them as well. The summary of the proposed genes is depicted in the Table 1. According to the proposed gene table, people might be involved in two out of five dimensions of a big data analytics system: data sources or data analyzers. In the followings, we analyze how people are involved in big data analytics platforms.

As the sources of data, people can get involved in several ways. In online social networks or online mass collaborative systems, Such as Twitter, Facebook, YouTube, etc., people, as the primary source of data, generate tremendous amount of data in their daily life activities [19]. People also might generate

Table 2. Genome of studied big data analytics platforms

No.	Tool	Where	Who	What	How	Why
1	IBM SPSS Modeler	HVO	M-	TNFAE	BI	DP-
2	Rapid Miner's tools	HVO	M-	TNFAE	BI	DPS
3	Oracle Advanced Analytics	HVO	MH	TNFAE	BI	DPS
4	SAP Predictive Analytics	HVO	MH	TNFAE	BI	DPS
5	Microsoft Revolution Analytics	HVO	M-	TNFAE	B-	DP-
6	Tableau	HVO	M-	T- - - -	B-	D- -
7	CrowdFlower	HVO	MH	TNFAE	BI	DPS
8	Kaggle	HVO	MH	TNFAE	B-	DPS
9	WorkFusion	HVO	MH	TNFAE	BI	DPS
10	IBM Watson Analytics	HVO	M-	TNFAE	B-	DPS

data implicitly, while doing another work. For instance, they leave their digital footprints while surfing or searching the web, doing online transactions, etc.

In some platforms, people are also involved as the data curators [23]. For instance, CrowdFlower Big Data Analytics platform[1], having people in the loop, offers its users very sophisticated and advanced mechanisms to organize and streamline their big data. More precisely, in CrowdFlower, human computing power is used along with AI algorithms, to curate, organize, and optimize data.

Looking at the Table 1 again, another role of human in big data analytics platforms is contributing to the data analysis. In this role, people can act as trainers, or analysts. As trainers, people might generate data to train AI algorithms in order to create more accurate results [9,21–23,33]. As another example, in CrowdFlower, human generated and curated data is used to train customer algorithms and models.

Finally, people might be employed in big data analytics platforms as analysts. People might analyze data and extract insight individually or collaboratively. For instance, a person might summarize a text, tag a photo or write a short essay individually. In contrast, a crowd of people might be employed to extract the sentiment of a community, quality of an artifact, or contribute to a prediction task [20,32,34]. Moreover, people might contribute to an analysis task explicitly or implicitly. When a user rates a movie on IMDb[2], for instance, he contributes explicitly, but when a user searches something on the google, and google uses his search activity along with others, to extract community trends, the contribution of the user is implicit.

Researchers and practitioners have proposed several big data analytics systems, some of which have humans included [11–14,18,35]. Due to large number of related systems, it is almost impossible to list all of them. Therefore, we select only few well-known candidates, that we believe well represent the area,

---

[1] https://www.crowdflower.com/solutions/open-source-big-data-analytics.
[2] imdb.com.

and apply the aforementioned taxonomy proposed in Sect. 2, in order to identify their genomes and assist with the better understanding of the proposed genome for human-enabled big data analytics processes. A list of the selected candidate systems along with their identified genome is represented in Table 2.

4   Open Challenges

Processing, analyzing and taking useful insights out of a tremendous amounts of data, i.e., big data, needs both efficiency and accuracy. Manual methods are inefficient, while automatic approaches are inaccurate. Therefore, at least in very serious situations such as disaster responding [21], lifesaving and rescue missions [36], monitoring urban emergency events [37] and endangered species monitoring [22], inclusion of people in solving big data problems as source of data, curators or analyzers seems inevitable. Such an inclusion can raise very serious challenges, some of which are highlighted in the followings.

4.1   Quality of People

Research show that, one of the first considerations to take into account, when it comes to people contribution, is their level of *trustworthiness* [20,32,38–40]. People might provide low quality contributions to gain unfair advantages, to promote their interests and demote what they do not like [20,38]. They might act so, individually or collaboratively [34,41,42]. People may also submit low quality contributions because of lack of expertise, misunderstanding, poor task design, etc.

As explained earlier in Sect. 3, people also might be used to train AI systems to provide more accurate results [9,21–23,33]. Although, in most of the research prototypes or projects, people-trained algorithms provide better results, the real-world story sometimes is different. For instance, on March 23, 2016, Microsoft unveiled a Twitter chatbot, called Tay, as an experiment in conversational understanding. As company described, the more you chat with Tay, the smarter it gets. Tay learns from talking to people to talk more causal and playful, like a teen [43]. Unfortunately, right after Tay lunched, people started talking to it with all sorts of misogynistic and racist remarks. The result is obvious. In less than 24 h, Tay was trained to be a nasty racist bot [9]. This example shows that if the low quality people train an algorithm, the result could be predictably useless.

Availability of workers can be another challenge to deal with. On-demand nature of crowdsourcing, time differences and different incentives and motivations can lead to having no contributors for some tasks [39,44]. Even, when the tasks are performed collaboratively with human and machine, any lack of availability or delays from human side, which is likely to happen, can cause serious problems in decomposition, execution and result aggregation of the big data analytics task.

Tight people selection criteria, can guarantee, to some extent, to have high quality people on board, but it negatively impacts the number of potential workers. Relaxing people selection conditions might lead to higher number of recruited people, but makes the task vulnerable to low quality contributions, collusion or other forms of unfair behaviour. Therefore, recruiting sufficient number of right people as source of information, or as contributors to online analysis tasks, is a serious challenge that needs further investigations.

4.2   Credibility of Results

Organizations leverage big data analytics to tune their direction, plan for their future and decide on how to spend their valuable resources. So, they expect big data analytics platforms to provide them with credible and reliable insights and recommendations. An analytics platform can guarantee the credibility of its results, only when the data has been collected from reliable sources and has been analyzed by reliable analysts. Referred to the Table 1, these are exactly the places in which humans are involved in the process of big data analytics.

As we explained earlier, quality of people is a challenge, so credibility of the human contributions remains as a challenge, too. The data collected from low trust people, even if they are processed by the most accurate algorithms, cannot lead to dependable results. Moreover, even if the data is collected from reliable sources, if they are not analyzed by trustworthy people, the result would not be credible. Non-dependable results mean waste of time, efforts and resources. Over-coming these challenges needs novel approaches that take into account important factors from all sides including, human, machine, the way they are combined, etc. Coming to such techniques need more comprehensive research in the future.

4.3   Privacy Preservation

Involving people in data analysis processes can result in privacy breaches. When the source of data is a human, the way data is stored, processed and disseminated is very important, in terms of privacy preservation. Human generated data might need to be anonymized. It also might be necessary to omit some parts of data provided by people, or at least, a consent from the person is necessary for each data item is being used.

When people contribute as analyzers or curators, the story changes a little bit, but the challenge still remains. People who have access to the data can easily breach the privacy of people or organizations. Many pieces of evidence show that even de-identified data can result in serious privacy breach [45, 46]. So, preserving privacy is a challenge, particularly, in the presence of people; and it is necessary to devise privacy-aware solutions. This is not straightforward and needs deeper research and investigation.

5   Conclusion

Due to inefficiency of manual approaches and inaccuracy of automatic data analytics, sometimes inclusion of human in the process of big data analytics is unavoidable. To be able to understand and manage this inclusion, and to address challenges that raises after this inclusion, one needs to have a clear understanding of the nature of big data analytics, and to clearly know the places in which people are involved.

In this paper, we first proposed a taxonomy and genome for the big data analytics systems, and used this genome to clarify when and why people are included. We also studied some well-known big data analytics tools and platforms, and identified their genome, according to the proposed taxonomy. Then, we studied the involvement of people in big data analytics, the roles they can take and the contributions they can have. We also identified the challenges that such an inclusion can raise.

We believe that big data analytics platforms that have people onboard will emerge increasingly in near future. So, it is crucial to have trust and privacy aware analytics to be able to take the benefits of people inclusion, and at the same time, manage the drawbacks and side challenges of presence of people in such analytics.

References

1. Gartner says 8.4 billion connected "things" will be in use in 2017, up 31 percent from 2016. https://goo.gl/qnQalb. Accessed 22 Jan 2019
2. Cuzzocrea, A., Song, I.-Y., Davis, K.C.: Analytics over large-scale multidimensional data: the big data revolution!. In: Proceedings of the ACM 14th International Workshop on Data Warehousing and OLAP, pp. 101–104. ACM (2011)
3. Srinivasa, S., Bhatnagar, V.: Big Data Analytics: Proceedings of the First International Conference on Big Data Analytics BDA, pp. 24–26. Springer, Heidelberg (2012). https://doi.org/10.1007/978-3-642-35542-4
4. Manyika, J., et al.: Big data: the next frontier for innovation, competition, and productivity (2011)
5. Pike, R., Dorward, S., Griesemer, R., Quinlan, S.: Interpreting the data: parallel analysis with Sawzall. Sci. Program. **13**(4), 277–298 (2005)
6. Sakr, S., Liu, A., Batista, D.M., Alomari, M.: A survey of large scale data management approaches in cloud environments. IEEE Commun. Surv. Tutor. **13**(3), 311–336 (2011)
7. Little, G., Sun, Y.-A.: Human OCR: insights from a complex human computation process. In: Workshop on Crowdsourcing and Human Computation, Services, Studies and Platforms, ACM CHI (2011)
8. The power of crowdsourcing. https://goo.gl/KH4y4p. Accessed 14 Mar 2017
9. Twitter taught Microsoft's AI chatbot to be a racist asshole in less than a day. https://goo.gl/kJKr7X. Accessed 26 Mar 2017
10. Chen, M., Mao, S., Liu, Y.: Big data: a survey. Mob. Netw. Appl. **19**(2), 171–209 (2014)
11. Fan, J., Han, F., Liu, H.: Challenges of big data analysis. Natl. Sci. Rev. **1**(2), 293–314 (2014)
12. Chen, C.P., Zhang, C.-Y.: Data-intensive applications, challenges, techniques and technologies: a survey on big data. Inf. Sci. **275**, 314–347 (2014)
13. Lohr, S.: The age of big data (2012). http://www.nytimes.com/2012/02/12/sunday-review/big-datas-impact-in-the-world.html. Accessed 26 Mar 2017

14. Yi, X., Liu, F., Liu, J., Jin, H.: Building a network highway for big data: architecture and challenges. IEEE Netw. **28**(4), 5–13 (2014)
15. Malone, T.W., Laubacher, R., Dellarocas, C.: The collective intelligence genome. MIT Sloan Manag. Rev. **51**(3), 21 (2010)
16. Assunção, M.D., Calheiros, R.N., Bianchi, S., Netto, M.A., Buyya, R.: Big data computing and clouds: trends and future directions. J. Parallel Distrib. Comput. **79**, 3–15 (2015)
17. Davenport, T.H.: Analytics 3.0. Harv. Bus. Rev. 91(12), 64–+ (2013)
18. Lv, Z., Song, H., Basanta-Val, P., Steed, A., Jo, M.: Next-generation big data analytics: state of the art, challenges, and future research topics. IEEE Trans. Ind. Inform. **13**(4), 1891–1899 (2017)
19. Data never sleeps 4.0. https://www.domo.com/blog/2016/06/data-never-sleeps-4-0/. Accessed 14 Mar 2017
20. Daniel, F., Kucherbaev, P., Cappiello, C., Benatallah, B., Allahbakhsh, M.: Quality control in crowdsourcing: a survey of quality attributes, assessment techniques, and assurance actions. ACM Comput. Surv. (CSUR) **51**(1), 7 (2018)
21. Ofli, F., et al.: Combining human computing and machine learning to make sense of big (aerial) data for disaster response. Big Data **4**(1), 47–59 (2016)
22. Matabos, M., et al.: Expert, crowd, students or algorithm: who holds the key to deep-sea imagery 'big data' processing? Methods Ecol. Evol. **8**, 996–1004 (2017)
23. O'Leary, D.E.: Embedding AI and crowdsourcing in the big data lake. IEEE Intell. Syst. **29**(5), 70–73 (2014)
24. Moretti, C., Bulosan, J., Thain, D., Flynn, P.J.: All-pairs: an abstraction for data-intensive cloud computing. In: 2008 IEEE International Symposium on Parallel and Distributed Processing, IPDPS 2008, pp. 1–11. IEEE (2008)
25. Hashem, I.A.T., Yaqoob, I., Anuar, N.B., Mokhtar, S., Gani, A., Khan, S.U.: The rise of "big data" on cloud computing: review and open research issues. Inf. Syst. **47**, 98–115 (2015)
26. Chen, Y., Alspaugh, S., Katz, R.: Interactive analytical processing in big data systems: a cross-industry study of mapreduce workloads. Proc. VLDB Endow. **5**(12), 1802–1813 (2012)
27. Crotty, A., Galakatos, A., Zgraggen, E., Binnig, C., Kraska, T.: Vizdom: interactive analytics through pen and touch. Proc. VLDB Endow. **8**(12), 2024–2027 (2015)
28. Hurwitz, J., Nugent, A., Halper, F., Kaufman, M.: Big Data for Dummies. Wiley, Hoboken (2013)
29. Suchanek, F., Weikum, G.: Knowledge harvesting in the big-data era. In: Proceedings of the 2013 ACM SIGMOD International Conference on Management of Data, pp. 933–938. ACM (2013)
30. Gandomi, A., Haider, M.: Beyond the hype: big data concepts, methods, and analytics. Int. J. Inf. Manag. **35**(2), 137–144 (2015)
31. Leskovec, J., Huttenlocher, D., Kleinberg, J.: Predicting positive and negative links in online social networks. In: Proceedings of the 19th International Conference on World Wide Web, pp. 641–650. ACM (2010)
32. Allahbakhsh, M., Ignjatovic, A., Benatallah, B., Foo, N., Bertino, E., et al.: Representation and querying of unfair evaluations in social rating systems. Comput. Secur. **41**, 68–88 (2014)
33. Guo, K., Tang, Y., Zhang, P.: CSF: crowdsourcing semantic fusion for heterogeneous media big data in the internet of things. Inf. Fusion **37**, 77–85 (2017)
34. Allahbakhsh, M., Ignjatovic, A., Motahari-Nezhad, H.R., Benatallah, B.: Robust evaluation of products and reviewers in social rating systems. World Wide Web **18**(1), 73–109 (2015)

35. Tsai, C.-W., Lai, C.-F., Chao, H.-C., Vasilakos, A.V.: Big data analytics: a survey. J. Big Data **2**(1), 21 (2015)
36. Meier, P.: Results of the crowdsourced search for Malaysia flight 370 (2014). https://irevolutions.org/2014/03/15/results-of-the-crowdsourced-flight-370-search/. Accessed 31 Mar 2017
37. Xu, Z., et al.: Crowdsourcing based description of urban emergency events using social media big data. IEEE Trans. Cloud Comput. (2016). https://doi.org/10.1109/TCC.2016.2517638
38. Wang, G., et al.: Serf and turf: crowdturfing for fun and profit. In: Proceedings of the 21st International Conference on World Wide Web, pp. 679–688. ACM (2012)
39. Amintoosi, H., Kanhere, S.S., Allahbakhsh, M.: Trust-based privacy-aware participant selection in-social participatory sensing. J. Inf. Secur. Appl. **20**, 11–25 (2015)
40. Amintoosi, H., Kanhere, S.S.: A reputation framework for social participatory sensing systems. Mob. Netw. Appl. **19**(1), 88–100 (2014)
41. Allahbakhsh, M., Ignjatovic, A., Benatallah, B., Beheshti, S.-M.-R., Bertino, E., Foo, N.: Collusion detection in online rating systems. In: Ishikawa, Y., Li, J., Wang, W., Zhang, R., Zhang, W. (eds.) APWeb 2013. LNCS, vol. 7808, pp. 196–207. Springer, Heidelberg (2013). https://doi.org/10.1007/978-3-642-37401-2_21
42. Mukherjee, A., Liu, B., Glance, N.: Spotting fake reviewer groups in consumer reviews. In: Proceedings of the 21st International Conference on World Wide Web, pp. 191–200. ACM (2012)
43. Microsoft made a chatbot that tweets like a teen. https://goo.gl/v3uX4Y. Accessed 31 Mar 2017
44. Salehi, N., McCabe, A., Valentine, M., Bernstein, M.: Huddler: convening stable and familiar crowd teams despite unpredictable availability. In: Proceedings of the 2017 ACM Conference on Computer Supported Cooperative Work and Social Computing, CSCW 2017, pp. 1700–1713. ACM, New York (2017)
45. Zeller, T.L.: AOL executive quits after posting search data (2006). https://nyti.ms/2DHjkx0. Accessed 31 Mar 2017
46. Truta, T.M., Tsikerdekis, M., Zeadally, S.: Privacy in social networks. In: Zeadally, S., Badra, M. (eds.) Privacy in a Digital, Networked World. CCN, pp. 263–289. Springer, Cham (2015). https://doi.org/10.1007/978-3-319-08470-1_12

A Distributed Method Based on Mondrian Algorithm for Big Data Anonymization

Amin Nezarat[1]([✉]) and Khadije Yavari[2]

[1] Payame Noor University, Tehran, Iran
aminnezarat@pnu.ac.ir
[2] Yazd Branch, Islamic Azad University, Yazd, Iran
kh.yavari@gmail.com

Abstract. There exist a multitude of techniques for privacy preservation in data publishing, but these methods are mostly designed for traditional and small databases and are incapable of handling big data. The leading method of privacy preservation in data publishing is called de-identification, and one of the best-known techniques of de-identification is k-anonymity. The Mondrian algorithm is among the best and fastest algorithms developed for implementing this technique. This algorithm has been made scalable through solutions such as data partitioning for multiple workers and then running Mondrian on individual nodes, or implementing the algorithm in parallel runs. This study examined and tested a solution involving the use of k-means clustering for data partitioning and then distributing clusters among workers. The test was performed to measure the improvement over the serial method in terms of de-identification accuracy, memory usage, and runtime. The results show that the serial method is completely incapable of handling big data, but the problem can be resolved with the proposed solution.

Keywords: K-anonymity · Big data · Mondrian

1 Introduction

The past decade has seen a dramatic increase in the speed and quantity in which data is published for purposes such as assisting the research community to verify the results of data content analyses, generating the data necessary for training, and providing real data for meta-analysis.

One of the challenges in data publishing is how to protect the privacy of the individuals to whom the published data pertains, or in other words, avoid publishing any information that may reveal their personal identities. Often, there is a tradeoff between privacy preservation and usefulness of the published data, in the sense that focusing on one objective undermines the other. Hence, the question is how to ensure privacy preservation without losing data usefulness as much as possible. A new challenge in this regard is how to reach such a balance in big data and whether the existing methods are scalable enough to undertake such a task. The existing algorithms can be made scalable in two ways. The first method is to modify the algorithm for parallel execution. The second method is to partition the data, process multiple

© Springer Nature Switzerland AG 2019
L. Grandinetti et al. (Eds.): TopHPC 2019, CCIS 891, pp. 84–97, 2019.
https://doi.org/10.1007/978-3-030-33495-6_7

partitions simultaneously by multiple runs of the algorithm, and aggregate the output. The problem with the data partition approach is the scattering of data that may be related together in the context of de-identification across different partitions, which will undermine the de-identification accuracy. In other words, it is essential to distribute the data among different processing nodes such that distribution does not affect the quality of de-identification [1]. The present study evaluates the use of a data clustering approach for this data partitioning based on three hypotheses:

- Implementing the MapReduce framework on the Mondrian algorithm will reduce the big data de-identification runtime.
- Implementing the MapReduce framework on the Mondrian algorithm will reduce the big data de-identification memory usage.
- Combining the Mondrian algorithm with the k-means clustering algorithm will improve the big data de-identification accuracy.

The objective of this study is to provide a model with the MapReduce framework for faster and more accurate de-identification of big data with the Mondrian algorithm.

2   Research Background

De-identification is a traditional method for protecting confidential information. This method involves using generalization and suppression techniques to strip the data from identifiers before publication for data mining. In de-identification, the goal is to facilitate data publishing for research purposes without jeopardizing the privacy of the people involved [1].

All models developed for privacy preservation in data publishing pursue two core objectives: ensuring sufficient data usefulness and ensuring satisfactory privacy preservation. The challenge lies in the fact that these objectives are conflicting, in the sense that a model that is completely focused on privacy preservation will only let useless data to be published. In fact, absolute privacy preservation is extremely difficult if not impossible, mainly because attackers get to pick their targets with prior knowledge. One of the best and most commonly used methods of de-identification is the k-anonymity. In this method, attributes are categorized as follows:

- **Identifier attribute:** the primary identifiers of a person, which often include unique personal attributes such as name, national code, insurance number, etc.
- **Quasi identifier:** attributes that allow a person to be identified if combined with external information.
- **Sensitive attribute:** Information that is personal and confidential, such as disease type, income, etc.
- **Non-sensitive attribute:** Information that is public and can be published without privacy concerns.

The k-anonymity model was first developed by Sweeney in 1998 [2] for privacy preservation in relational data. A table satisfies k-anonymity if and only if every permutation of quasi identifier set occur at least k times in the table. This means that trying to find the record of a specific person will lead the attacker to at least k − 1 other records, which make it more difficult to extract data.

Another concept related to this discussion is the notion of equivalent classes, which refers to the set of records where quasi identifier have the same field values. Hence, after applying a k-anonymity, there should be at least k records in equivalent classes.

To meet the k-anonymity condition, the field values of the quasi identifier set fields should be modified through one of the two following methods to make sure that the table contains k indistinguishable record:

Suppression: In this method, some or even all of the field values must be replaced with * or another character to make them unidentifiable.

Generalization: In this method, the field values are replaced with a range of data. For example, if the age of an individual is 24, it must be replaced with the interval "20 < age < 30" [4].

In 2006, LeFevre introduced a top-down greedy algorithm called the Mondrian algorithm for ensuring k-anonymity in relational databases [4]. The original Mondrian algorithm is designed for numerical attributes. For categorical attributes, Mondrian needs to convert them to numerical data, which is not suitable for some applications.

Later, LeFevre developed another algorithm called the Basic Mondrian which supports both numerical and categorical attributes [5]. Basic Mondrian can use generalization hierarchies to split categorical attributes [5]. Examples of generalization hierarchies are provided in Figs. 2, 3 and 4. Over the following years, this algorithm became the basis of many works in the field of de-identification. Although many papers claim that they utilize Mondrian with generalization hierarchies, in fact, they use the original Mondrian, because Basic Mondrian does not require generalization hierarchies [6]. In general, Mondrian algorithm functions based on generalization. The workflow of Mondrian consists of the following steps:

Use of KD-Tree to split the data set into groups of k elements.

Generalization of groups so that each group has the same equivalent class.

This is equivalent to the implementation of the recursive function partition (region, k) with the following function.

Select the best dimension for the K-anonymity conditions.

Partition into two regions R1 and R2 if possible.

Return the values of the recursive functions *partition (R1, k)* and *partition (R2, k)*. Return *Null* if there was no possibility for partition.

In the Mondrian method, the idea is to consider each database as: [QID, SA], where QID denotes quasi-identifier attributes, such as age and birth date, and SA denotes sensitive attributes such as disease information. In the Mondrian method, the set of similar qid's are generalized to a single qid (Fig. 1).

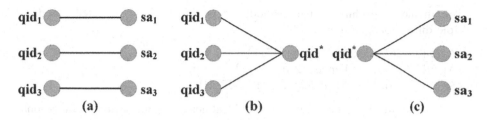

Fig. 1. Mondrian method

It can be stated that the Mondrian method has modified the "Single Dimensional" method into the "Multi-Dimensional" approach. Consider the following example. A hypothetical set of patient data is given in Table 1.

k = 2

Quasi Identifiers:

Age, Sex, Zipcode

Table 1. Patient data

Age	Sex	Zipcode	Disease
25	Male	53711	Flu
25	Female	53712	Hepatitis
26	Male	53711	Bronchitis
27	Male	53710	Broken Arm
27	Female	53712	AIDS
28	Male	53711	Hang Nail

In the Single Dimensional method, the partitions will be as follows and patient data will be de-identified as shown in Table 2.

Age: {[25–28]}
Sex: {Male, Female}
Zip: {[53710–53711], 53712}

Table 2. De-identification of patient data with the single-dimensional method

Age	Sex	Zipcode	Disease
[25–28]	Male	[53710, 53711]	Flu
[25–28]	Female	53712	Hepatitis
[25–28]	Male	[53710, 53711]	Bronchitis
[25–28]	Male	[53710, 53711]	Broken Arm
[25–28]	Female	53712	AIDS
[25–28]	Male	[53710, 53711]	Hang Nail

But in the multi-dimensional method, the partitions will be created based on multiple dimensions, as shown below:

{Age: [25–26], Sex: Male, Zip: 53711}
{Age: [25–27], Sex: Female, Zip: 53712}
{Age: [27–28], Sex: Male, Zip: [53710–53711]}

Table 3 shows the patient data after de-identification with the multi-dimensional method. In Fig. 2, the mechanisms of these methods are illustrated alongside each other.

Table 3. De-identification of patient data with the multi-dimensional method

Age	Sex	Zipcode	Disease
[25–26]	Male	53711	Flu
[25–27]	Female	53712	Hepatitis
[25–26]	Male	53711	Bronchitis
[27–28]	Male	[53710, 53711]	Broken Arm
[25–27]	Female	53712	AIDS
[27–28]	Male	[53710, 53711]	Hang Nail

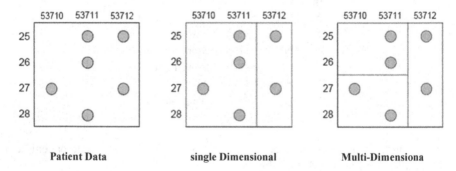

Patient Data   single Dimensional   Multi-Dimensiona

Fig. 2. Comparison of single-dimensional, multi-dimensional methods [4]

Given the importance of privacy preservation for today's big data and the scalability of the existing methods for ever-larger databases, this paper proposes the use of MapReduce framework in this application. The core idea of the proposal is to distribute the computational load of the de-identification processes over processing nodes in such a way as to ensure this distribution has no effect on the quality of de-identified data.

Zakerzadeh et al. have developed two versions of the MapReduce-based de-identification method [7]. In the first version, all data are initially considered to be of the same equivalent class and are then classified according to attribute tuples until the newly created equivalent class meets the k-anonymity conditions. After each iteration, the data file is updated based on the new equivalent class and is fed to the next iteration as input. The main issue of this version is the need to share a global file among all nodes in order to update equivalent class information after each iteration, and the fact that this global file becomes progressively larger in successive iterations [1].

The second version of the MapReduce-based de-identification method was developed to overcome this issue. In this version, instead of creating a single global file for all nodes, file fragments are generated and distributed among the nodes. In the Map phase, each node adds a unique file ID to each file fragment for the identification purposes, and in the next iteration, each node has to access only those files that are needed for Reduce operation. This removes the need to maintain a global file.

The major flaws of this technique are in the handling of multiple iterations and file management. As the number of iterations increases, system performance decreases and file management with MapReduce becomes progressively more difficult [15].

Zhang et al. introduced a new method for multidimensional de-identification of big data in the MapReduce framework [8]. This technique is also based on the Mondrian method and hence named MRMondrian. In this technique, MapReduce is used to split the large datasets into smaller ones until all partitions are broken to the size of individual computational nodes, and then the Basic Mondrian algorithm is run in parallel on all nodes [9]. These researchers also developed a tree structure called PID-tree (Partition ID indexing tree) for implementing their model.

3 Proposed Method

Given the importance of publishing big data without creating privacy risks, this paper proposes and tests a distributed approach to the implementation of the Mondrian method. At present, several open-source implementations of Mondrian algorithm are available for public use. In these implementations, the algorithm is generally executed serially. The Mondrian algorithm code used in the proposed method is one of the best and fastest currently available and was selected after repeated testing of the existing codes.

This code takes three inputs. The first input is related to the Mondrian model and specifies whether relax or strict mode will be used. In the strict mode, kd-tree partitioning is performed without allowing any node to remain on the boundary between left (lhs) and right partitions, but in the relax mode, such partitioning is allowed, which means the intersection of the two sets may not be null. The second input, a, is related to the dataset of interest. Here, the Adult dataset is used for testing. The final input is k, which is related to k-anonymity.

The dataset used for testing is the Adult dataset, which contains 32,000 records. The reason for using the Adult dataset is the frequent use of this dataset in previous articles on de-identification. The pseudocode of the proposed method is presented in algorithm 1 and the workflow is presented in Fig. 3.

Algorithm 1 Mondrian_K-Means

1: **Get parameter**
{CreateData, Number of dataset record, Iteration of k-means, Number Of cluster in k-means, Mood of Mondrian,
K Of k-Anonymity}

2: **IF** CreateData=yes **then** {DS=Create_Data(Number of dataset record)}

3: Run-kmeans(DS, Iteration of K-means, Number Of cluster in k-means)
→Cluster of k-means[] \\ Run k-mean in parallel on created data set and produce clusters

4: **For**[I=1 to Number of clusters]

5: Mondrian(Cluster of k-means[I], Mood of Mondrian, K Of k-Anonymity) →
Cluster_Anonymized_Dataset [I] \\ Run the serial Mondrian algorithm on each cluster on a worker In parallel and produce Anonymized Clusters

6: Merge_Anonymized_Dataset(Cluster_Anonymized_dataset [I=1 to number of clusters]) → Full_Anonymized_Dataset

7: **Return** Full_Anonymized_Dataset

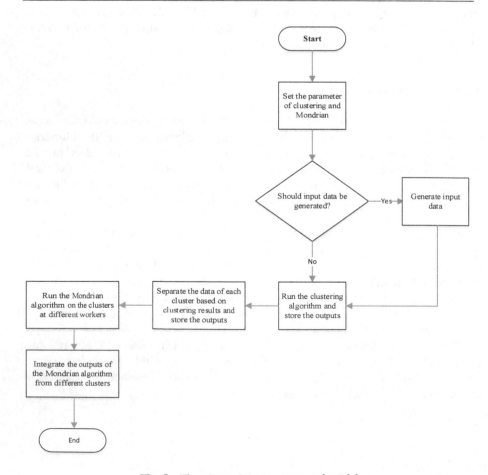

Fig. 3. Flowchart of the implemented model

Since the objective is to handle big data but the Adult dataset has only 32,000 records, first, a big data set was generated based on the Adult dataset's field values. Data continuity and accuracy were also considered in the dataset generation. For example, the fields related to education and education codes were chosen such that there would be no mismatch between these values. An input parameter was defined to stop the generation of new data, which can be used in case of need to work on the existing data. Input parameters of the algorithm are:

- Is it necessary to generate data?
- How many records are needed?
- How many clusters are needed?
- How many iterations of k-means are needed for accuracy improvement (here, this is fixed at 100).
- Whether to use Relax or Strict mode of anonymity
- The value of k for k-anonymity

Considering the size of the dataset, the k-means clustering algorithm is implemented with the spark library and is set to be executed in multiple parallel runs. Clustering is performed based on the age field. The k-means algorithm randomly selects k member (where k is the number of clusters) from among n members as cluster centers, then allocate the n-k remaining members to the nearest cluster. Once all members are allocated, this algorithm recalculates the cluster centers and redoes the allocation according to the new centers. This loop continues until the cluster centers remain constant in successive iterations. Here, the parameter k and the number of iterations should be specified. After clustering, the output is processed to get ready for the next step. This processing involves the separation of clusters, the generation of suitable files for feeding into the next step, proper naming of these files.

For cluster separation, the output is stored in a number of output files (this number depends on the volume of data). This separation is performed based on the cluster number that is given in the header of each record, which is subsequently deleted be become similar to the Adult dataset (these are implemented within the program). These clusters are then sent to Workers and each cluster is run on one Mondrian worker. The outputs of each worker are the de-identified file of the cluster, the runtime of Mondrian, and Normalized Certainty Penalty (NCP). Next, the outputs of workers are combined to create the final de-identified file as output.

The metrics for comparison are runtime, memory usage, and Normalized Certainty Penalty (NCP), which is used to measure accuracy. Expressed as a percentage, NCP can vary from 0 to 100, with 0 representing the loss of no information and 100 signifying of loss of all information [3, 10, 11].

The features that distinguish this method from other models of parallelization with data partitioning is the use of k-means clustering for efficient distribution of data between workers and also the assessment of the effect of data fragment clustering in parallelization on the accuracy and speed of de-identification.

4   Results

First, a comparison was made between the serial model and the parallel model with different numbers of clusters. For this purpose, a dataset with 5 million records was generated. In this run, the Mondrian mode was set to Relax and the parameter k of k-anonymity was set to 10. The resulting NCP is plotted in the following chart. In this chart, it is well illustrated that the serial model enjoys good accuracy, and among the parallel models, the one with four clusters has the most accurate results.

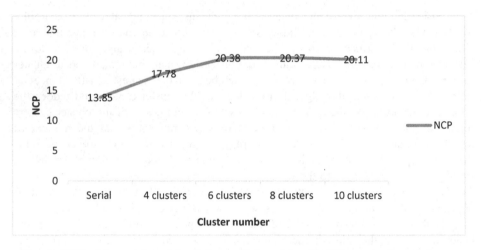

Fig. 4. NCP of the serial model and the parallel models with different number of clusters

In Fig. 5, it can be seen that the parallelization helps reduce the memory consumption and this contribution increases with the number of clusters. Note that the reported values are the percentage of memory usage.

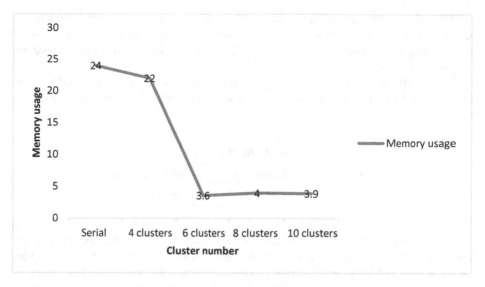

Fig. 5. Memory usage of the serial model and the parallel models with different number of clusters

Figure 6 illustrates the runtime of the tested models in seconds. The parallel model shows a significant improvement in this regard and, as can be seen, the execution speed has increased with the number of clusters involved.

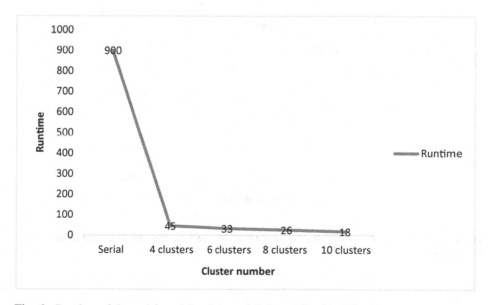

Fig. 6. Runtime of the serial model and the parallel models with different number of clusters

Next, the models were tested with a 20 GB dataset. Using the serial model, the memory usage reached 100% after 4 h, leading to a system crash. But the parallel model managed to produce an output in 3 h. The serial model was also irresponsive for 20 million records.

The parallel model was then evaluated using the parameters given in Table 4. In Fig. 7, it can be seen that as the k value of k-anonymity increases, the accuracy of de-identification decreases.

Table 4. Input settings

Number of records	Number of clusters	Mondrian mode
50,00,000	4	Relax

Figure 8 shows that as the k value of k-anonymity increases, the operation speed decreases. In this chart, the time is in minutes. Figures 9 and 10 show the similar results that were obtained with the Mondrian mode set to Strict. By comparing Figs. 7 and 9 with Figs. 8 and 10, it can be concluded that the Strict mode results in a lower speed and higher accuracy compared to the Relax mode.

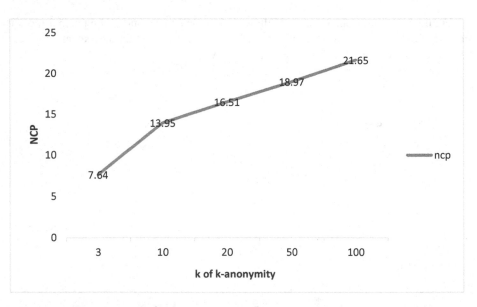

Fig. 7. Effect of the parameter k of k-anonymity on the accuracy of de-identification in the Relax mode

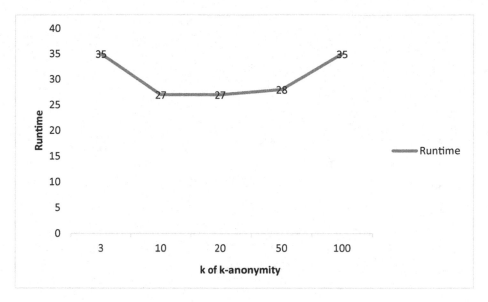

Fig. 8. Effect of the parameter k of k-anonymity on the Runtime of de-identification in the Relax mode

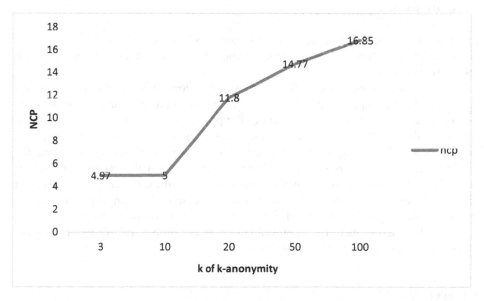

Fig. 9. Effect of the parameter k of k-anonymity on the accuracy of de-identification in the Strict mode

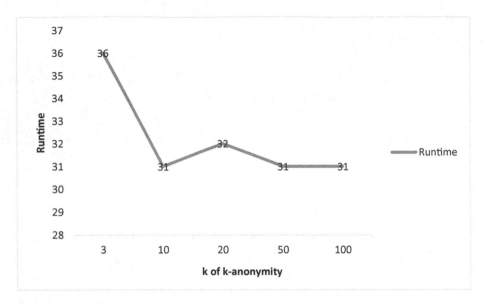

Fig. 10. Effect of the parameter k of k-anonymity on the Runtime of de-identification in the Strict mode

5 Conclusion

The results show that the serial implementation of Mondrian is incapable of handling big data, but parallelization can greatly reduce the memory usage and increase the speed of this method.

Overall, running the model in the strict mode with lower k values (for k-anonymity) and more clusters improved the accuracy of the operation as measured by NCP or the loss of data. In this paper, after examining the privacy preservation methods and their problems, the existing solutions for privacy preservation in the publishing of big data were reviewed. There are still many flaws in the privacy preservation models and the methods developed to make them scalable, which highlight the necessity of further research in this area. Future research can focus on each of many variables involved in data de-identification, including de-identification method, algorithm type and specifications, scalability solutions, the choice of the field that constitute the basis of clustering, and clustering methods, in order to develop an optimal scalable method for the de-identification of big data for publication.

References

1. Abouelmehdi, K., Beni-Hssane, A., Khaloufi, H., Saadi, M.: Big data security and privacy in healthcare: a review. Procedia Comput. Sci. **113**, 73–80 (2017)
2. Samarati, P., Sweeney, L.: Protecting privacy when disclosing information: k-anonymity and its enforcement through generalization and suppression. In: Proceedings of the IEEE Symposium on Research in Security and Privacy, pp. 384–393 (1998)

3. Sacharidis, D., Mouratidis, K., Papadias, D.: K-anonymity in the presence of external databases. IEEE Trans. Knowl. Data Eng. **22**(3), 392–403 (2010)
4. LeFevre, K., DeWitt, D.J., Ramakrishnan, R.: Mondrian multidimensional K-anonymity. In: Proceedings of the International Conference on Data Engineering, vol. 2006, p. 25 (2006)
5. LeFevre, I.K., DeWitt, D., Ramakrishnan, R.: Efficient full-domain k-anonymity. In: ACM SIGMOD International Conference on Management of Data (2005)
6. Mehmood, A., Natgunanathan, I., Xiang, Y., Hua, G., Guo, S.: Protection of big data privacy. IEEE Access **4**, 1821–1834 (2016)
7. Zakerzadeh, H., Aggarwal, C.C., Barker, K.: Privacy-preserving big data publishing. In: Proceedings of the 27th International Conference on Scientific and Statistical Database Management - SSDBM 2015, pp. 1–11 (2015)
8. Zhang, X., et al.: MRMondrian: scalable multidimensional anonymisation for big data privacy preservation. IEEE Trans. Big Data (2017)
9. Machanavajjhala, A., Kifer, D., Gehrke, J., Venkitasubramaniam, M.: L-diversity. ACM Trans. Knowl. Discov. Data **1**(1), 3 (2007)
10. Ghinita, G., Karras, P., Kalnis, P., Mamoulis, N.: Fast data anonymization with low information loss. In: Proceedings of the International Conference on Very Large Data Bases, pp. 758–769 (2007)
11. Loukides, G., Gkoulalas-Divanis, A.: Utility-preserving transaction data anonymization with low information loss. Expert Syst. Appl. **39**(10), 9764–9777 (2012)

An Ensemble Angle-Based Outlier Detection for Big Data

Raghda Al-taei and Maryam Amir Haeri[✉]

Department of Computer Engineering and Information Technology,
Amirkabir University of Technology, Tehran, Iran
{raghdafaris,haeri}@aut.ac.ir

Abstract. Outlier detection is an important problem in machine learning and it has many applications such as fraud detection, network anomaly detection, medical diagnosis and data cleaning. However, traditional outlier detection techniques tend to fail in dealing with high-dimensional data especially when they encounter with big data. By the continues generation of data, we are facing massive datasets rather than just high dimensional data, in such conditions, the methods that depend on distance fail to provide correct results due to the vast distance between points which lead to similar distance between all points. Angle based outlier detection is a method proposed for outlier detection in high dimensional spaces. However, it is very time consuming and cannot be used for big data. In this paper, we introduce an angle based outlier detection ensemble method for big data. A dimensionality reduction technique called locality sensitive hash function (LSH) is used to reduce the complexity of big data. Moreover, the method utilizes an ensemble technique to improve angle based outlier detection and then a supervised combination method such as SVM is used to aggregate the ensemble subsets. The method is investigated with several real-world datasets and the results show that the proposed method can efficiently find the outliers.

Keywords: Angle based outlier detection · Big data · Outlier detection · Locality sensitive hash function · Ensemble learning

1 Introduction

Big data can be analyzed for insights that lead to better decisions, but with a large number of data, the chance of having outlier data grows high. An outlier is a data object that deviates significantly from the normal objects as if it were generated by a different mechanism [1,2]. Outliers are different from noise data. Noise is random error or variance in a measured variable. Outliers can be divided into two categories (global outliers and the local outliers). Global outliers are data objects that significantly deviate from the other samples of the dataset. Finding global outliers are much simpler than the local outliers. On the other hand, local outliers are a data instance which its density significantly deviates

© Springer Nature Switzerland AG 2019
L. Grandinetti et al. (Eds.): TopHPC 2019, CCIS 891, pp. 98–108, 2019.
https://doi.org/10.1007/978-3-030-33495-6_8

from its neighbors. It is important to note that, in a dataset, different types of outliers may exist. Thus, we need different types of outlier detection methods [3].

Detecting outliers in big data could be very challenging [4], due to its large size (volume), the high speed of data changing according to time(complexity). The well-known methods may fail to deal with big data to detect outliers. Moreover, big data generated from different sources. Thus they contain various types of outliers. It is essential to pay attention to all types of outliers in big data. Using an ensemble learning method provide us with an opportunity to use different outlier detection strategies to find different types of outliers [5].

The concept of *ensemble model* can be constructed by integrating several learners t in-order to solve a specific problem [6]. The traditional learning methods tend to use one learning technique on the whole data which may contain unwanted data types like noise or outliers making it difficult to set parameters for these datasets, leading to uncertainty about the model that should be used. Ensemble model address this problem by gathering various models together named base-learners they can be a decision tree, neural network, and others. Ensemble method is suitable for big data because ensemble learning provides the possibility of distributed and parallel implementation. Problems such as fraud detection and intrusion detection can be solved by choosing a proper method to detect outliers [7]. Thus, in this paper, we utilize the ensemble learning for detecting the outliers.

Most of the outlier detection methods rely on calculating the distance between points such as [8–10]. However, in case of high dimensional data defining a distance measure could be challenging, because the Euclidean distance will be the same for all points due to the sparsity of data [11,12]. Kriegle et al. [13] suggested a method called angle based outlier detection which is less sensitive to distance. This method is designed for high dimensional data and works appropriate in such space. However, this method is not efficient for big data and it is quit time-consuming. The aim of this paper is to improve this method for big data.

In this paper, we consider the high dimensional big data. In this case, bagging can be considered as a good approach for outlier detection in big data where the dataset is split into parts of subsets then, on each subset an outlier detection method is performed. By bagging several versions of the outlier detection methods are use a sample of data. These samples are generated with replacement; this can help to achieve class balance through sampling without losing much information. Moreover, in this paper in order to reduce the complexity of high dimensionality a dimension reduction technique proper for big data such as locality sensitive hashing (LSH) is used. Then an ensemble bagging method is applied to partition the data set into sub-samples, and on each sample, an angle-based outlier detection is applied, this can improve the ability to detect different types of outliers that exist in the data set. To aggregate all ensemble results an SVM combination method is performed.

The organization of this paper is as follows: next section devoted to the preliminary knowledge. Section 3 explains the proposed method in detail. In Sect. 4 the ensemble angel-based outlier detection method is evaluated. Finally, Sect. 5 is the conclusion.

2   Backgrounds

In this section, some of the most well-known algorithms that inspired our work will be explained starting with the locality sensitive hash function to reduce the dimensions then explaining the SVM method that we use to aggregate the final result of each bag then, an angle-based outlier detection approach which is suitable to detect outliers in big data.

2.1   Locality Sensitive Hash Function (LSH)

The locality sensitive hash function is used as a dimension reduction technique, which works by grouping the most similar data in one bucket without the need to examine every pair (in contrast to methods such as PCA). According to some conditions Fig. 1 shows how the close points set to be in one bucket [14].

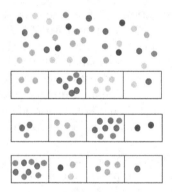

Fig. 1. Basic LSH structure

LSH assign data points into the same buckets if they are similar. LSH is a proper method to use while working with big data because it works effectively in dealing with big data. LSH indicates that whether the two points s_1 and s_2 are candidate to be similar points or not. LSH supposes that the points are located in space S and a distance measure d such as Hamming, Cosine and Euclidean. Based on the distance measure the hash functions are defined. For each distance measure, a hash function family with random parameters should be defined. The hash function family requires four parameters to be defined (d_1, d_2, p_1, p_2)-sensitive [12]. For each two points s_1 and s_2 in S in each LSH family the following two statements should be valid and based on them the two similar points belong to the same bucket with a high probability and the different points belong to the same bucket with low probability.

- If $d(s_1, s_2) < d$ then $h(s_1) = h(s_2)$ i.e. candidate (of being similar) with a p_1 probability at least.
- If $d(s_1, s_2) > d$ then $h(s_1) = h(s_2)$ i.e. candidate (of being similar) with a p_2 probability at most.

LSH Families can be used for many distance measures, here, we explain the LSH family for Euclidean distance measure. Each LSH family needs a random factor. Here the random factor is a random line in the space S. The random line is divided into buckets of size a. The hash function projects each point s_i onto the line and the bucket number is the results of hashing s_i, considering the random line. Two close points in the space S have a high chance of being projected on the same bucket [12,14]. In order to increase the accuracy of the method several random lines are chosen and each data instance is projected to these lines. Thus, when this method is utilize to reduce the dimensionality of a dataset with d dimensions, if m lines chosen randomly and the data instances project on them and hashed to the bucket numbers, a dataset with m dimensions is obtained.

2.2 Angle Base Outlier Detection Method

Traditional methods tend to perform the worse while dealing with high-dimensional data because they depend on the data distances, angle base outlier detection (ABOD) method is less sensitive to distance. Consider Fig. 2 the angle for a data point (O) that is in the cluster to any pair of points tend to differ extensively, but the angle for a point that is out of the cluster will be the same to other pairs of points. Thus, the points that have low variance are considered to be outliers.

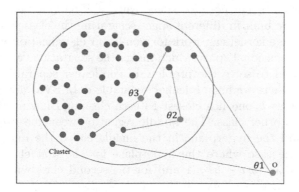

Fig. 2. ABOD for different data points

There are three types of points shown in Fig. 2, points within the cluster, border points and outliers the spectrum angles for all three types are considered the variance angle for inlier point tends to be the highest and gets a little lower for a border point but, still not so low as the angle variance of an outlier point [13]. ABOD assign a degree of 'outlierness' to all points by sorting them an outlier can be detected, assigning different degrees to different types of points is not possible with other outlier detectors. The most significant advantage of ABOD is that it does not require any additional parameter. This algorithm was based on the assumption that outliers are data point that are far from the normal data thus, the angles between inlier data are bigger than the angle between an outlier point and the other points, therefore a normal data tend to have a various different angle set between the surrounding points, while abnormal data angles set is a set of several small angles with less variance. The angle between three points can be calculated as in Eq. 1 for data points s_1, s_2 and s_3 [1,15].

$$cos(\overline{s_2} - \overline{s_1}, \overline{s_3} - \overline{s_1}) = \frac{< (\overline{s_2} - \overline{s_1}), (\overline{s_3} - \overline{s_1}) >}{\|\overline{s_2} - \overline{s_1}\|_2 . \|\overline{s_3} - \overline{s_1}\|_2} \qquad (1)$$

The ABOF of data X is:

$$ABOF(\overline{s_1}) = Var_{s_2, s_3 \in S} cos(\overline{s_2} - \overline{s_1}, \overline{s_3} - \overline{s_1}) \qquad (2)$$

However, performing the ABOD method for high dimensional big data it is very challenging. The dataset cannot be stored in the RAM. And the algorithm needs to read the dataset in many passes thus, this method is quite time-consuming. In the following, we describe that how we can use the ABOD for the high dimensional big data by utilizing the LSH method for the dimensionality reduction and the ensemble bagging for partitioning the data.

2.3   Support Vector Machine (SVM)

It is used in the case of classification and regression analysis. It can classify given data into two or more categories. By given a sample of the dataset as training data the SVM produce a leaner model that can classify new instances. The model and data points are built in different space separating the classes by a huge gap, SVM with the usage kernel can work for non-linear classification by moving the data into high dimensional space which makes the separation easier. Mainly the SVM was proposed to solve the problem of non-leaner separable classification [16]. In the case where we have a leaner separable data SVM aim to define two hyperplanes that each one lies closest to each class and having the maximum distance from the other class. The middle region between the two hyperplanes called margin and the hyperplane in the middle called the maximum margin hyperplane, in the case where the hyperplane for the first class with label 1 can be calculated by $\boldsymbol{wx} - b = 1$ and for the second class with the label -1 $\boldsymbol{wx} - b = -1$ the distance between the two hyperplanes is $\frac{2}{\|w\|}$.

3   Methodology

In this section, an outlier detection method that we proposed will be explained. First, LSH which is a suitable dimensionality reduction method that works well for big data angle base outlier detection method is explained with ABOD. After that ensemble LSH-angle base method which, ensemble method with SVM for combining the ensembles results is suggested.

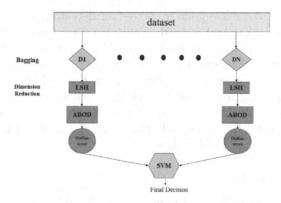

Fig. 3. Ensemble LSH-ABOD

Figure 3 shows the algorithm structure of the method that we are proposing. In the following subsections, we will describe the proposed algorithm in more details.

3.1   LSH Angle Base Outlier Detection

In the LSH angle-based outlier detection, at first, the dataset is hashed with a family of locality sensitive hash functions (for Euclidean distance) with different random parameters. Using LSH results in a low dimension data and then the angle-based method is performed over the lower dimensional data. The angle is calculated for every other pair of points and using Eq. 2 the variance of that point is measured assigning an outlier degree to all point the least variant points are outliers.

3.2   Ensemble LSH-ABOD Method

In order to reduce the time complexity of the LSH based outlier detection method and to improve the results, we suggest an ensemble method based on the LSH-angle based outlier detection. By using an ensemble, the result of a single base learner can be advance this is because ensemble allows the algorithm to be trained on different subsets of the data, which makes it familiar with more data characters in different small spaces. The ensemble helped to produce many accurate results than just using a single outlier detector, in this paper after applying locality sensitive hash functions to reduce the dataset complexity. Afterward an ensemble bagging is performed which, divide the large dataset into chunks with replacement this helps us to obtain accurate result. For each produced bag, angle base outlier detection is applied to provide each point with an outlierness score low scores indicate outliers and high scores belongs to normal data instances.

3.3   Aggregation of the Outlier Scores

Thus, in order to find a reasonable threshold for detecting outliers for each bag, a supervised learning method is used. The supervised learning method can be SVM. Moreover each date point might exist in several bags. The SVM can be trained over a minimal dataset, in order to detect the label (normal or anomaly) of the data from the outlier score. After training the SVM in each bag, all the data points of each bag are fed to the SVM, and it detects its label if a data points assigned to more than one bags a majority voting is performed for the final decision. It should be note that in order to learn the SVM the method only needs to very small set of labeled data.

3.4   Time Complexity

At first, the time complexity of the ensemble LSH-angle based method is investigated. The time complexity of the simple angle-based method when all the dataset can be stored in the RAM is $O(n_3)$, where is the size of the dataset. When it encountered with big data, the whole dataset cannot be stored in the RAM. Thus, it needs to access to the disk and read the data points (in each time we want to compute the angles between each data instance to the other ones) which need IO operations, and it can be very time to consume. Therefore, in this case, the time complexity of the angle-based method is $O(n_3 + T_{IO})$. For high dimensional big data, it can be assumed that by using the LSH the volume of data decreases significantly, and the whole dataset or significant part of it can be stored in the RAM. As in the ensemble LSH angle-based method, the dataset is divided into some bags, and the dimensions of the datasets decrease by the LSH. Thus it is possible to assume that the dataset can be stored in the RAM and the time of reading the data repetitively from the disk is saved in this case. Therefore, the time complexity of the ensemble LSH- angle based outlier detection is mostly related to the time complexity of angle-based method over each bag. In the ensemble model computing over each bag can be performed in a parallel manner. The time complexity of angle based method over each bag is $O(n_b^3)$, where n_b is the size of each bag. The final part is aggregating the results which we utilize SVM trained on a minimal training data which the training time is negligibly incomparable to the time complexity of the ABOD and the test part of SVM time complexity is $O(n)$.

4   Experimental Results

In this section, we will discuss the effectiveness and efficiency of the outlier detection method suggested in this paper, in comparison with traditional outlier detection methods such as (iforest, LOF, Loop). The method is performed on real datasets. First, the detection methods both suggested and traditional will be discussed. Second, we will represent their results on real datasets.

4.1 Evaluation of Outlier Detection Methods

Three methods such as (iforest [11], LOF [8], Loop [9]) are used to compare our outlier detection method with for the Euclidean space. By using LOF and LoOP methods local outliers can be detected in a dataset, they require distance calculation of all pair point in the dataset and save the resulted calculation in the RAM. Base on points density they tend to be accurate when normal size data is used while in the case of big data this will get more problematic due to the lack of space to save all calculations. The iforest method is based on the fact that outliers are less than normal data it constructs small trees with outlier conditions which are fewer than the normal data conditions, it can perform well in parallel computation. Principal Component Analysis (PCA) is used to reduce the dimension before applying the outlier detection methods (iforest, LOF, Loop) in order to see the effect of dimensionality reduction on them.

4.2 Parameters

For the angle base ensemble outlier detection, the parameters that we use are the percentage of data points taken from the dataset. In order to structure the bag size of bagging ensemble model which is equal to 20% of the whole dataset drown with replacement into ten bags, then, by the usage of LSH for all the dataset, the dimension of our big data is decreased to 20 dimensions.

Table 1. Real data information

Datasets	Data	Dimension	Outliers
Wbc	278	30	21
Speech	3686	400	61
satimage-2	5803	36	71
optdigits	5216	64	150
Musk	3062	166	97
Mnist	7603	100	700
arrhythmia	452	274	66
portsweepnormal	80793	39	3740
r2l	80247	39	3194
nmapnormal	78619	39	1566

4.3 Results

In order to evaluate the proposed method (Ensemble LSH-ABOD) we used the most well-known datsets. All of these real dataset are available in [17] where preprocessed by [18] that remove the unnecessary features and assign the mode instead of the missing values this will help the outlier detection methods to provide much better results with the least error, the datasets were labeled by 1 referring to an outlier and 0 referring to normal data. Table 1 Shows the datasets in details the number of normal data is higher than the number outliers. To evaluate and analyze the methods the area under the curve (AUC) is used the model is accurate if the result is close to one AUC provides a value between $[0, 1]$ the area under the curve is equal to one in perfect conditions. By calculating the ratio between the true positive rate (TPR) and false positive rate (FPR) for various threshold values $[0, 1]$. The true positive rate is $TPR = \frac{TP}{\#P}$ and false positive rate is $FPR = \frac{FP}{\#N}$, where TP is the instance that was classified correctly over the total number of the positive class and FP is the misclassified data as positive over the total number of the negative class.

The results are shown in Table 2 is the results of AUC according to deferent methods and different datasets (real datasets), in most cases provide better results compared to iforest, LOF, LoOP. For the satimage-2 dataset, our method provided better results in detecting outliers both global and local compared to the well-known method which they failed to provide better detection results. Also for Musk, Mnist, optdigits and speech datasets, we get very accurate results in which none of the (PCA-iforest, PCA-LOF, PCA-LoOp) outlier detectors perform well, the portsweepnormal dataset also our method provided better results than both type detectors. Even when the well-known methods performed better, our result was very close to them, and their difference can also be ignored. Finally, the r2l dataset provided unignorable different between the results in comparison with PCA-LOF and PCA-LoOP.

Table 2. Real dataset results

Datasets	Ensemble LSH-ABOD	Pca-iforest	Pca-LOF	Pca-LoOP
Wbc	0.590	0.6128	0.941	0.9377
Speech	0.682	0.4505	0.4202	0.4156
satimage-2	0.9284	0.6978	0.5849	0.4842
optdigits	0.771	0.4378	0.6518	0.5982
Musk	0.712	0.5345	0.3842	0.4076
Mnist	0.843	0.5328	0.6137	0.5859
arrhythmia	0.747	0.5679	0.7965	0.7964
portsweepnormal	0.831	0.8293	0.4936	0.4824
r2l	0.662	0.7577	0.5816	0.5361
nmapnormal	0.611	0.7934	0.443	0.4864

However, the result of our method over Wbc dataset shows some limitation against detecting collective outliers. A clustering method like K-means can be used at first to cluster collective outlier in one cluster then another outlier detection method can be applied to the rest of the dataset in order to detect the other types of outliers.

The results show that although the time complexity of the ensemble angle-based method is low it can provide accurate results which is comparable with the most famous (and time-consuming) outlier detection methods.

5 Conclusion

This paper suggested an angle based ensemble method for outlier detection in big data. In this method at first the dataset is partitioned into N parts to construct the ensemble learning phase, then LSH function is applied to each data bag. After that SVM aggregates the results of each bag and label the data as inlier or outlier. The time complexity of the method is $O(n_b^3)$ where n_b is the size of each bag and this is quite smaller than the time complexity of ABOD which is equal to $O(n^3)$, where n is the size of the whole dataset. The method is evaluated by real-world data sets. The results demonstrated that the method can provide better results in most cases compared with the well-known methods. However, it was limited to detecting collective outliers and it should be improved in the future studies.

References

1. Aggarwal, C.C.: Outlier analysis. In: Aggarwal, C.C. (ed.) Data Mining, pp. 237–263. Springer, Cham (2015). https://doi.org/10.1007/978-3-319-14142-8_8
2. Sreevidya, S., et al.: A survey on outlier detection methods. (IJCSIT) Int. J. Comput. Sci. Inf. Technol. 5(6), 8153–8156 (2014)
3. Carter, P.: Big data analytics: future architectures, skills and roadmaps for the CiO. IDC White Paper (2011)
4. Nasser, T., Tariq, R.: Big data challenges. J. Comput. Eng. Inf. Technol. 9307(2) (2015)
5. Dietterich, T.G.: Ensemble methods in machine learning. In: Kittler, J., Roli, F. (eds.) MCS 2000. LNCS, vol. 1857, pp. 1–15. Springer, Heidelberg (2000). https://doi.org/10.1007/3-540-45014-9_1
6. Zhou, Z.-H.: Ensemble Methods: Foundations and Algorithms. Chapman and Hall/CRC, New York (2012)
7. Zimek, A., Schubert, E., Kriegel, H.-P.: A survey on unsupervised outlier detection in high-dimensional numerical data. Stat. Anal. Data Min.: ASA Data Sci. J. 5(5), 363–387 (2012)
8. Breunig, M.M., Kriegel, H.-P., Ng, R.T., Sander, J.: LOF: identifying density-based local outliers. ACM SIGMOD Rec. 29(2), 93–104 (2000)
9. Kriegel, H.-P., Kröger, P., Schubert, E., Zimek, A.: Loop: local outlier probabilities. In: Proceedings of the 18th ACM Conference on Information and Knowledge Management, pp. 1649–1652. ACM (2009)
10. Huang, J., Zhu, Q., Yang, L., Feng, J.: A non-parameter outlier detection algorithm based on natural neighbor. Knowl.-Based Syst. 92, 71–77 (2016)
11. Liu, F.T., Ting, K.M., Zhou, Z.-H.: Isolation forest. In: 2008 Eighth IEEE International Conference on Data Mining, pp. 413–422. IEEE (2008)
12. Leskovec, J., Rajaraman, A., Ullman, J.D.: Mining of Massive Datasets. Cambridge University Press, Cambridge (2014)

13. Kriegel, H.-P., Zimek, A., et al.: Angle-based outlier detection in high-dimensional data. In: Proceedings of the 14th ACM SIGKDD International Conference on Knowledge Discovery and Data Mining, pp. 444–452. ACM (2008)
14. Hu, W., Fan, Y., Xing, J., Sun, L., Cai, Z., Maybank, S.: Deep constrained siamese hash coding network and load-balanced locality-sensitive hashing for near duplicate image detection. IEEE Trans. Image Process. **27**, 4452–4464 (2018)
15. Pang, G., Cao, L., Chen, L., Lian, D., Liu, H.: Sparse modeling-based sequential ensemble learning for effective outlier detection in high-dimensional numeric data. In: AAAI (2018)
16. Hsu, C.-W., Chang, C.-C., Lin, C.-J., et al.: A practical guide to support vector classification (2003)
17. Hettich, S., Bay, S.: The UCI KDD archive. University of California, Department of Information and Computer Science, Irvine, CA, vol. 152 (1999). http://kdd.ics.uci.edu
18. Rayana, S.: Odds library. Stony Brook University, Department of Computer Science, Stony Brook, NY (2016). http://odds.cs.stonybrook.edu

Big Data Analytics Help Prevent Adolescents Suicide: An Introduction to Mindpal

Brianna Turner[1] and Ali Eslami[2(✉)]

[1] Department of Psychology, University of Victoria, Victoria, Canada
`briannat@uvic.ca`
[2] Department of Psychiatry, University of British Columbia, Vancouver, Canada
`ali.eslami@ubc.ca`

Abstract. Suicide is cited as one of the three leading death causes on the rise, according to the centers for disease control and prevention. The field of adolescent mental health is in urgent need of better methods to lower suicide risk and psychiatric re-admission rates after discharge from inpatient psychiatry. Frequent follow-up is needed to detect fluctuations in suicide risk factors, such as sleep problems, mood disturbances and social withdrawal. Mobile Health approaches and digital phenotyping are proposed as a promising solution. In this paper, we first study the problem of suicide and its relation with data analytics especially among youths. We further present Mindpal Platform developed to provide rich, time-sensitive information about health behaviors by gathering information both in active and passive manner. Further analysis of gathered data Using the Mindpal Platform can help prevent the adolescents suicide.

Keywords: Mindpal · Data analytics · Suicide · Smartphone

1 Introduction

Identifying individuals at high risk of future mental health problems is a priority objective in psychiatry [1] with suicide prediction as arguably the "holy grail" of the field. Despite decades of active research and many advances in understanding the risk factors and correlates of suicidality, our ability to predict suicidal behavior remains no better than chance [2]. One reason for this lack of progress may be the use of infrequent assessment schedules to predict a highly dynamic process [3,4]. Recent studies show that the transition from suicidal thoughts to action often occurs in less than 4 h [3] and that suicidal crises frequently resolve within a matter of days [5].

A critical question is how we can feasibly monitor, predict, and intervene in suicide risk that evolves and resolves over minutes, hours, or days. Answering this question would significantly improve efforts to reduce suicide risk in youth, which is a top priority for health research and policy [6]. Such efforts, in turn,

© Springer Nature Switzerland AG 2019
L. Grandinetti et al. (Eds.): TopHPC 2019, CCIS 891, pp. 109–118, 2019.
https://doi.org/10.1007/978-3-030-33495-6_9

could reduce this leading cause of adolescent mortality and attenuate the annual loss of 15,000 potential years of life that result from youth suicidal behaviours [7].

This goal is addressed by innovative mobile Health (mHealth) approach known as digital phenotyping [8] as presetned in Fig. 1. We would like to identify which youth are most likely to experience suicidal thoughts or behaviors in the year following inpatient treatment and therefore, we aim to develop a self-monitoring application that with youth who are already connected with psychiatric care, in order to identify high-risk periods and support youth and their families in the moments when intervention can have the greatest impact.

Data provided by smartphones whether collected actively via user-generated surveys (ie, participatory sensing) or passively via native sensors (ie opportunistic sensing) have three important features that make them especially well-suited to monitoring youths' mental health [9]. These data have: (1) Velocity, meaning information can be sampled frequently. This is critical for understanding suicide risk. Suicidal crises are often brief and lasting only a few hours [3]. To capture dynamic risk states, we need frequent updates about a youth's social, physical, and emotional context. (2) Volume, meaning that the same individual can be sampled repeatedly, facilitating within-person comparisons to detect departures from baseline. In contrast to retrospective interviews (e.g. "how has your sleep been over the last week, compared to normal?"), digital phenotyping assesses risk states much closer to the times and events in which we are interested. This should improve the accuracy and sensitivity of our predictions about deviations from baseline risk. (3) Variety, meaning that many types of information are available. As presetned in Fig. 2, today's smartphones are equipped with numerous native sensors, which can be combined using machine-learning to make inferences about emotional, social, or physical states. Inferences are then compared against user-inputted information (i.e., data "tagging") to assess their accuracy. Our approach combines youth self-report, sensor-generated behavioral data, and parent reports, allowing predictions to be verified against various "ground-truth" indicators. Previous studies show that:

1. Depressed mood is classified with 74–96% accuracy from distance traveled and time per location based on GPS data [10].
2. Sleep problems are classified with 81–93% accuracy from accelerometer and gyroscope, device "lock" duration, and ambient light and sound levels [11].
3. Social withdrawal is classified with 83–88% accuracy from call and SMS logs, app usage, and microphone or camera activity (not content) [10].

Clinicians currently use various heuristics to assess "warning signs" for acute suicide risk. Theoretical models of suicide risk that inform such heuristics emphasize combining distal factors (e.g., demographic and historical factors) with proximal states (e.g., recent behaviors, cognitions, and emotions) to determine risk [12]. Proximal risk factors include social withdrawal, sleep problems, agitation or recklessness, mood changes, and purposelessness [13]. However, many heuristics have low predictive accuracy when assessed in a single clinical interview [14]. To date, we have lacked sufficiently sensitive and accessible means to monitor risk factors in a way that reflects their hypothesized proximal role.

Sensor data, collected in a manner that complies with health research privacy/security standards, hold tremendous promise in this regard. Digital phenotyping can thus provide rich information to inform proximal risk predictions. It has been used to track various psychiatric symptoms, including unipolar, bipolar mood, psychosis, anxiety and has excellent acceptability in people who live with mental illness.

In this paper we first study the related works followed by presenting the Minapp platform that could be used to gather data and identifying crisis based on data analytics and machine learning algorithms.

The rest of this paper is organized as follows: In Sect. 2 we present the related works. Section 3 presents the fundamentals on Mindpal platform and its data usage and analytics strategy and finally Sect. 4 concludes the paper.

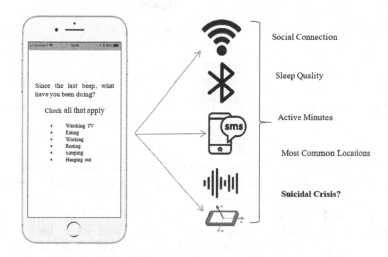

Fig. 1. Schematic of digital phenotyping to detect risk of suicidal crisis.

2 Related Works

Each year, thousands of adolescents present to hospital with suicidal thoughts or behaviors [15]. Despite efforts to promote community mental health care inpatient psychiatric admissions rose by over 60% between 2006 and 2014 [15]. While inpatient admissions can ensure youths' safety, stabilize or reduce acute symptoms, and initiate or revise outpatient care plans, many youth struggle to sustain the gains achieved during hospitalization. As such, youth who receive inpatient psychiatric treatment remain at high risk for chronic mental health problems and suicide [16]. The transition from inpatient to outpatient care poses a particular challenge. Many (40–75%) youth fail to engage with or complete the recommended outpatient treatments after they leave the hospital [17]. Within one year of discharge, nearly half of adolescent patients experience a recurrence

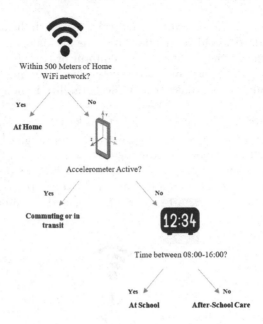

Fig. 2. Simple decision tree illustrating how sensor data can be combined to identify probable locations.

of suicidal ideation, 30–40% require another inpatient admission and 10–25% make a suicide attempt. By the time they are 18 years old, 11% of these youth will have required 3 psychiatric hospitalizations [15].

The field of adolescent mental health is in urgent need of better methods to lower suicide risk and psychiatric re-admission rates after discharge from inpatient psychiatry. Youth and caregivers want interventions that are minimally burdensome, time-sensitive, and accessible [18]. At the same time, frequent follow-up is needed to detect fluctuations in suicide risk factors, such as sleep problems, mood disturbances [19] and social withdrawal [20]. Meeting both challenges simultaneously could lower the personal and societal burdens associated with suicidal crises by giving youth and their caregivers the tools to recognize interventions are needed before a crisis escalates. mHealth approaches may address many of these challenges, because they can provide rich, time-sensitive information about youths' health behaviors and are accessible, because smartphones are owned and regularly used by over 80% of youth.

3   Mindpal Platform

Digital phenotyping uses native smartphone sensors to infer youths' emotional, social, and behavioral states, allowing ongoing monitoring in everyday life (see Fig. 1). This method will allow us to classify which youth are most at risk of suicidal crisis, and when their risk is most acute, with a minimally burdensome assessment strategy that reduces the need for user input.

In this paper we present the developed and pilot tested the MindPal App that will be used to collect sensor and survey data. Ongoing piloting with 20 community adults shows that the App was successfully installed without issue on a variety of smartphones. MindPal is an Android and iOS compatible application that allows passive collection of sensor data without user input, as well as active user surveys about youths' activities, emotions, and social contexts. The MindPal App was developed in-house by experts in cyber-security to ensure compliance with the privacy standards and to retain intellectual property rights, enhancing future scalability. MindPal is compatible with all smartphones running Android or iOS versions from 2014 and later. MindPal facilitates collection of prompted user surveys and the following sensor and summary data:

- Accelerometer, Gyroscope, & Device Motion
- Motion Activity (e.g., stationary, walking, in a vehicle)
- GPS (only distance travelled and time in location)
- Screen state (duration locked vs. in use)
- Ambient light and sound intensity
- Application usage (number, type and duration of app use)
- Call and SMS log (i.e., count, duration of incoming/outgoing calls or SMS text messages)

MindPal does not gather content related information, including: social networks credentials, activities, or content; browsing history; SMS or call content; contact names or phone numbers; image, video, or stored multimedia content, including notes, calendars, etc; sound recordings or recorded voices. MindPal has a number of features to protect the security and privacy of participants' data. First, MindPal transfers only summaries of the raw data (i.e., meta-data), in JSON format, to a secure server (see Fig. 3). Personally-identifiable data such as GPS coordinates are either aggregated (e.g., distance traveled, not raw coordinates) or anonymized using hashing and encryption so that we can extract features but cannot "read" the data. Second, data are encrypted prior to transfer and stored on secure servers. Once data are transferred to the server, temporary raw data are deleted from the device. Thus, "stewardship" of sensitive data occurs at the level of the phone. The app uses a store-and-forward system to transfer data only over Wifi, which minimizes the impact of network connectivity disruptions on data integrity and minimizes storage capacity, battery, and cellular data plan usage.

3.1   Data Collection

The modern machine-learning landscape is complex and rapidly evolving. The most effective modeling scheme will depend on the distribution of the data, making the optimal approach challenging to predict. Potential algorithms include Random Forests, which construct a multitude of decision trees at training time and output the modal class (classification) or mean prediction (regression) of the individual trees, Support Vector Machines (SVMs), which are supervised non-probabilistic binary linear classifiers that build models by representing a set of training examples as points in space so that categories are divided by as wide a gap as possible, and Gradient-Boosted tree Models (GBMs), which perform well across learning tasks. These models can accommodate linear and non-linear classification using the kernel trick, implicitly mapping inputs into high-dimensional feature space. We assume that the smartphone sensors will provide sufficiently sensitive and informative data to detect patterns indicative of baseline functioning and departures from baseline. This assumption is considered reasonable given previous studies showing that sensors can accurately classify a range of behavioral and emotional states [21] and simulations showing our sample size and study parameters are robust to noise in sensor data and missing data.

Data collection using Minpal could be performed in two phases of active and passive monitoring.

1. Active monitoring phase, during which patient and parents will respond to daily MindPal surveys regarding the patient's emotional, social, and physical well-being, and sensor data will be passively collected from the patient's smartphone;
2. Passive monitoring phase when the frequency of user surveys will be minimized, and MindPal will primarily acquire sensor data.

Active Monitoring

Patient. In active monitoring Phase, patient will complete four daily surveys (2 min, 10 questions each) to capture risk state. Surveys will be programmed so that they arrive at convenient intervals when youth are likely to have their smartphone (e.g., before school, at lunch, after school, before bed). Each survey will ask about the youth's current location (e.g., home/school/in transit, etc.), social context (e.g., with friends/with family/with classmates/alone, etc.), current emotions (e.g., hopeless, lonely, encouraged, etc.), and risk-state since the last survey (i.e., fleeting vs. persistent suicidal thoughts, desire vs. plan to act on suicidal thoughts). The Safety Monitoring Protocol will guide the research team in promptly responding to these reports if necessary. Youth will also be asked to rate their sleep quality in the first survey of the day, and, in the final survey of the day, any stressors (e.g., arguments), or positive/negative coping behaviors (e.g., support seeking, alcohol/substance use) that occurred.

MindPal will also passively collect sensor data. Participants can receive a summary of their passive data (e.g., average daily traveled distance, average daily time using social media, etc.).

Caregivers. Caregivers will complete one daily survey in the evenings (3 min, 15 items) regarding their child's activities, emotions, behaviors, treatment compliance (e.g., attendance at medical/psychiatric appointments, medication use), and overall rating of their child's risk that day. Sensor data from parents's smartphones will not be collected.

Passive Monitoring

Patient. After active monitoring phase, youth may continue to provide sensor data passively through MindPal, but surveys will reduce to one every two weeks. This will provide on-going safety monitoring and compliance reminders, while minimizing reactivity to the monitoring protocol. At the end of this phase, patients can receive a summary of their sensor data.

Caregivers. Caregivers can complete telephone surveys every two months to assess whether their child required emergency or inpatient psychiatric services. If a patient withdraws, parents can rate how easy it was for their child to use MindPal using the Single Ease Question and describe any obstacles to the protocol.

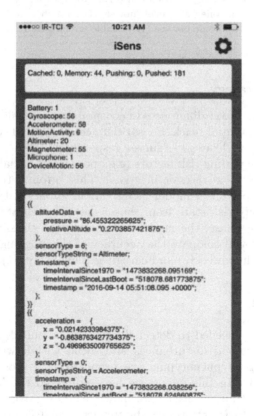

Fig. 3. Example of data as recorded and stored by MindPal App.

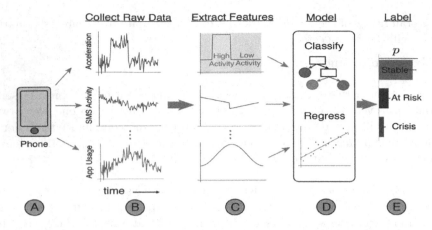

Fig. 4. Raw signals are collected from a participant's phone (A) and broadcast to a server. The raw signals (B) are processed to extract features (C), which summarize key properties of the data (e.g., applying filtering to time series data, binning data). Features are then fed into one or several models (D), which together produce a set of probabilities, r, for the outcome classes (E - Stable [green], At Risk [blue], Crisis [red]) (Color figure online)

3.2   Analytic Strategy

Sex, gender, and diagnostic differences in acceptability or feasibility will be investigated to inform potential reach. Sensor data could be used to classify patient' suicide risk. Patient and caregiver survey responses will serve to externally validate ("tag") data regarding risk factors (e.g., poor sleep, social withdrawal) and risk state (e.g., stable, at-risk, or in crisis). This "ground truth" information will then be used to train a model using various supervised machine-learning algorithms to classify risk state from sensor data as accurately as possible. In the second –passive–phase, the model will continue to classify sensor patterns and risk states. We will compare the accuracy of this ongoing classification (in Fig. 4) against youth (biweekly) and caregiver (bimonthly) reports.

4   Conclusion

Frequent follow-up is needed to detect fluctuations in suicide risk factors, such as sleep problems, mood disturbances and social withdrawal. Mobile Health approaches and digital phenotyping are proposed as a promising solution to decrease suicide rate. In this paper, we have presented the Mindpal Platform developed to gather time-sensitive data in active and passive manner. The gathered data from Mindpal Platform can be furthered analyzed to help prevent the adolescents suicide.

References

1. Eslami, A., Jahshan, C., Cadenhead, K.S.: Disorganized symptoms and executive functioning predict impaired social functioning in subjects at risk for psychosis. J. Neuropsychiatry Clin. Neurosci. **23**(4), 457–460 (2011)
2. Franklin, J.C., Ribeiro, J.D., Fox, K.R., et al.: Risk factors for suicidal thoughts and behaviors: a metaanalysis of 50 years of research. Psychol. Bull. **143**(2), 187–232 (2017)
3. Kleiman, E.M., Turner, B.J., Fedor, S., Beale, E.E., Huffman, J.C., Nock, M.K.: Examination of real-time fluctuations in suicidal ideation and its risk factors: results from two ecological momentary assessment studies. J. Abnorm. Psychol. **126**(6), 726–738 (2017)
4. Kleiman, E.M., Turner, B.J., Chapman, A.L., Nock, M.K.: Fatigue moderates the relationship between perceived stress and suicidal ideation: evidence from two high-resolution studies. J. Clin. Child Adolesc. Psychol. **47**(1), 116–130 (2017)
5. Olfson, M., et al.: Short-term suicide risk after psychiatric hospital discharge. JAMA Psychiatry **73**(11), 1119–1126 (2016)
6. Saxena, S., Funk, M.K., Chisholm, D.: Comprehensive mental health action plan 2013–2020. EMHJ-East. Mediterr. Health J. **21**(7), 461–463 (2015)
7. Navaneelan, T.: Suicide Rates: An Overview. Statistics Canada, Ottawa (2012). Catalogue no. 82–624-x. Health at a Glance
8. Onnela, J.P., Rauch, S.L.: Harnessing smartphone-based digital phenotyping to enhance behavioral and mental health. Neuropsychopharmacology **41**(7), 1691–1696 (2016)
9. Torous, J., Staples, P., Onnela, J.P.: Realizing the potential of mobile mental health: new methods for new data in psychiatry. Curr. Psychiatry Rep. **17**(8), 61 (2015)
10. Place, S., et al.: Behavioral indicators on a mobile sensing platform predict clinically validated psychiatric symptoms of mood and anxiety disorders. J. Med. Internet Res. **19**(3), e75 (2017)
11. Ben-Zeev, D., Scherer, E.A., Wang, R., Xie, H., Campbell, A.T.: Next-generation psychiatric assessment: using smartphone sensors to monitor behavior and mental health. Psychiatr. Rehabil. J. **38**(3), 218–226 (2015)
12. Shneidman, E.S.: Overview: a multidimensional approach to suicide. In: Jacobs, D.G., Brown, H.N. (eds.) Suicide: Understanding and Responding: Harvard Medical School Perspectives on Suicide, pp. 1–30 (1989)
13. Lester, D., McSwain, S., Gunn III, J.F.: A test of the validity of the IS PATH WARM warning signs for suicide. Psychol. Rep. **108**(2), 402–404 (2011)
14. Asarnow, J., McArthur, D., Hughes, J., Barbery, V., Berk, M.: Suicide attempt risk in youths: utility of the Harkavy-Asnis suicide scale for monitoring risk levels. Suicide Life Threat. Behav. **42**(6), 684–698 (2012)
15. Canadian Institute for Health Information: Care for childre and youth with mental disorders, Ottawa, Ontario (2015)
16. Chung, D.T., Ryan, C.J., Hadzi-Pavlovic, D., Singh, S.P., Stanton, C., Large, M.M.: Suicide rates after discharge from psychiatric facilities: a systematic review and meta-analysis. JAMA Psychiatry **74**(7), 694–702 (2017)
17. Timlin, U., Riala, K., Kyngäs, H.: Adherence to treatment among adolescents in a psychiatric ward. J. Clin. Nurs. **22**(9–10), 1332–1342 (2013)
18. Boulter, E., Rickwood, D.: Parents' experience of seeking help for children with mental health problems. Adv. Ment. Health **11**(2), 131–142 (2013)

19. Ballard, E.D., Vande Voort, J.L., Luckenbaugh, D.A., MachadoVieira, R., Tohen, M., Zarate, C.A.: Acute risk factors for suicide attempts and death: prospective findings from the STEPBD study. Bipolar Disord. **18**(4), 363–372 (2016)
20. Barzilay, S., et al.: The interpersonal theory of suicide and adolescent suicidal behavior. J. Affect. Disord. **183**, 68–74 (2015)
21. Boonstra, T.W., Larsen, M.E., Christensen, H.: Mapping dynamic social networks in real life using participants' own smartphones. Heliyon **1**(3), e00037 (2015)

The Merits of Bitset Compression Techniques for Mining Association Rules from Big Data

Hamid Fadishei$^{(\boxtimes)}$, Sahar Doustian, and Parisa Saadati

Computer Engineering Department, University of Bojnord, Bojnord, Iran
fadishei@ub.ac.ir,
{sahar.doustian,parisa.saadati}@stu.ub.ac.ir

Abstract. The massive amounts of data being generated in human's world today may not be harnessed unless efficient and high-performance processing techniques are employed. As a result, continuous improvement in data mining algorithms and their efficient implementations is actively pursued by researchers. One of the widely applied big data mining tasks is the extraction of association rules from transactional datasets. ECLAT is an algorithm that can mine frequent itemsets as a basis for finding such rules. Since this algorithm operates on vertical representation of a dataset, its implementation may be significantly enhanced by employing sparse bitset compression. This paper studies the performance of four different bitset compression techniques proposed by researchers, using both real-world and synthetic big datasets. The effect of input data characteristics is analyzed for these compression methods in terms of energy consumption, performance, and memory usage behavior. Experimental results can guide the implementations to choose the proper compression method that best fits the problem requirements. The source code of this study is made available at https://github.com/fadishei/biteclat.

Keywords: Big data · Frequent patterns · Bitset compression

1 Introduction

The world today is experiencing an era of overwhelming abundance of data. Human is capable of generating and storing more data than ever imagined. Today, storage technologies are quite inexpensive and data comes from almost anywhere: IoT sensors, transactional records, search engines, social networks, etc. However, keeping the pace with this data generation rate in order to process them and extract beneficial knowledge from data has always been a challenge. The infamous triple V's of big data (Volume, Variety, and Velocity) [1] are the main reason behind this and behind the gap between available data and available data processing power today.

Extracting intrinsic patterns in large quantities of data is an essential part of many today's big data analysis tasks. One interesting analysis of this type is the extraction of association rules from transactional datasets. Association rules mining was first introduced by Agrawal et al. [2] for discovering relations between items in a dataset of transactions. For example, one may find that customers of a shop who buy a drink, are likely to include diapers in their basket too. In this paradigm, association rules can be

© Springer Nature Switzerland AG 2019
L. Grandinetti et al. (Eds.): TopHPC 2019, CCIS 891, pp. 119–131, 2019.
https://doi.org/10.1007/978-3-030-33495-6_10

used for market basket analysis, optimizing the shop layout and its catalog structure. The analysis of association rules and thus their application is not limited to commerce transactions and can include all types of transactional datasets.

An established method of extracting association rules from a dataset is to first discover the frequent itemsets. The term refers to the set of items whose rate of occurrence in a dataset is equal to or more than a specific threshold known as the minimum support. The rise of big data and the challenging requirements of big data analysis have stimulated seeking performance optimization of the data mining algorithms. As a result, sophisticated algorithms and/or implementations have been proposed for frequent pattern mining in recent years. Another approach for optimizing this task, which is the focus of this paper, is to keep the algorithm in its basic form and shift the sophisticated optimizations to the underlying building blocks. For example, ECLAT algorithm heavily relies on set operations such as cardinality and intersection calculations. Several optimized algorithms and compressed implementations exist for sparse set math which can be utilized for this purpose. In this paper, an analysis of the merits of some of the most popular sparse set compression algorithms for frequent pattern mining is performed on both real-world and synthetic big datasets in terms of performance, energy, and memory consumption. The rest of this paper is organized as follows. Section 2 overviews the related work and Sect. 3 introduces the concept of frequent pattern mining and describes ECLAT algorithm. Section 4 discusses several bitset compression techniques. Section 5 describes the implementation and Sect. 6 presents the experimental setup and discusses the results. Finally, Sect. 7 concludes the paper and presents the future work.

2   Related Work

The present study is an effort to characterize how different bitset compression methods perform when applied to a real-world big data analysis task like frequent itemset mining. Several algorithms for discovering frequent patterns from data have been proposed by the researchers. Apriori [3], FPGrowth [4], and ECLAT [5] are some examples of well-known frequent pattern mining algorithms. Apriori algorithm works by iteratively generating candidate itemsets of length $k + 1$ from frequent itemsets of length k and pruning the infrequent itemset from results. FPGrowth constructs a tree data structure by scanning the dataset and recursively extracts the frequent itemsets from that tree. ECLAT algorithm works by using a vertical representation of the dataset. Since this vertical representation of a dataset can be represented as a bitset, ECLAT performance can benefit significantly from bitset compression techniques. Several bitset compression techniques have been proposed by the researchers [6]. EWAH [7] and CONCISE [8] are two relatively simple methods that rely on Run-Length-Encoding (RLE) for compressing sparse parts of the bitsets. Roaring [9] and BitMagic [10] are more sophisticated methods which use a notion of container-based compression. These techniques are described in details in the Sect. 4.

Several researchers have tried to accelerate ECLAT frequent pattern mining algorithm by some form of bitset compression. Zaki et al. [11] proposed DiffSets for saving the storage space of ECLAT bitsets. Their method compresses the bitset data by keeping track of differences in item bitsets in order to compress them. In [12] the EWAH bitset compression algorithm is used for accelerating the performance of ECLAT algorithm. Another approach in [13] uses a hybrid data structure consisting of bitsets and arrays for compressing the ECLAT data table which is similar to the idea of hybrid containers in Roaring bitset compression method. The mentioned studies investigate the performance of their proposed method compared to the base uncompressed algorithm. Since various state of the art bitset compression techniques exist, it is necessary to investigate their performance and resource requirements in different conditions. This paper tries to perform an extensive comparative behavior analysis of a number of bitset compression methods in terms of execution time, memory usage, and energy consumption for both synthetic and real-world big datasets.

3 Mining Frequent Patterns

Frequent patterns can be used to mine association rules from transactional datasets. A transactional dataset is a set $T = \{T_1, T_2, ..., T_n\}$ of transactions where each transaction T_i is a subset of all possible items. The set of items is defined in the context of the problem. For example, it may be the set of items in a supermarket, or the set of words in a text document. Figure 1 shows an example of a transactional dataset.

The support count (or support percentage) of an itemset $I = \{I_1, ..., I_n\}$ in a transactional dataset T is the number (or the percentage) of transactions in T that I is a subset of. For example, the support of itemset ace in transactional dataset of Fig. 1 is 2 or 25%. An itemset is by definition a frequent one if its support is greater than or equal to a specific minimum support threshold.

Transaction	Items
1	ab
2	acd
3	c
4	aef
5	cde
6	abce
7	f
8	ace

Fig. 1. An example of a transactional dataset with 8 transactions

3.1 ECLAT Algorithm

ECLAT (Equivalent CLAss Transformation) algorithm [5] which is used in this paper, utilizes a vertical representation of the dataset as shown in Fig. 2(a). Instead of transactions, this representation uses items at each row along with the set of transactions that item belongs to. This set of transactions itself can be represented by a sequence of bits. Since the resulting bitset at each row is usually sparse, bitset compression techniques can help with the efficiency of computations and storage.

Figure 2 shows how ECLAT algorithm works by an example. It starts with single items that satisfy the required minimum support. At each recursion of the algorithm, this set of prefixes is augmented with the remaining items and the resulting itemsets are pruned according to the minimum support requirement. A pseudocode of ECLAT algorithm is presented in Algorithm 1. It can be noted that the algorithm depends heavily on set intersection and cardinality calculations over transaction sets T (the table rows in Fig. 2). The underlying bitset implementation should support these operations and it should be able to perform them efficiently.

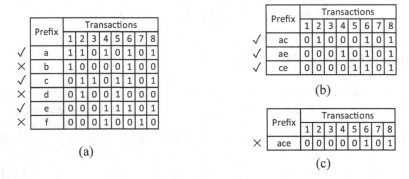

Fig. 2. An illustration of how ECLAT algorithm works using the dataset in Fig. 1

Algorithm 1. ECLAT algorithm for mining frequent patterns

```
1   ECLAT(P, minsup)
2   // Inputs:
3   //   minsup: minimum support threshold, I: set of items
4   //   T(i), i∈I: set of transactions each item belongs to
5   //   P: set of frequent prefixes, initially set to {i | i∈I, |T(i)| ≥ minsup}
6   // Output: F: global set of frequent itemsets, initially set to Ø
7   foreach prefix p∈P do
8       F ← F ∪ p
9       Ps ← Ø
10      foreach prefix q ∈ P where q > p do
11          s ← p ∪ q, T(s) ← T(p) ∩ T(q)
12          if |T(s)| ≥ minsup then
13              Ps ← Ps ∪ s
14      if Ps ≠ Ø then
15          ECLAT(Ps, minsup)
```

4   Bitset Compression

As stated earlier, ECLAT algorithm utilizes a vertical representation of the transactional dataset which can be effectively stored as a compressed bitset. A bitset is a data structure for presenting sets using a sequence of bits. Each bit of a bitset is 1 when the corresponding item belongs to the set and is 0 otherwise. The main reason behind the popularity of bitsets is that set operations often have similar counterparts in the instruction set of typical computing machines. For example, the intersection (or union) of two sets can be easily calculated as the AND (or OR) of the corresponding bitsets. Since such operations are performed by native machine instructions, bitset representations can significantly accelerate the computations. As a result, bitsets are widely used in several areas like search engines, database management, and data mining systems. When a bitset is sparse, which is the case in many of the mentioned applications of bitsets, it is better to compress the data structure in order to save space. Compression of bitsets is especially essential in big data systems where uncompressed representations are intractable due to the large amounts of data being processed by the system. Several techniques for bitset compression have been proposed by researchers [6]. The following paragraphs describe some of the well-known techniques in this area.

4.1   EWAH

Like many other bitset compression algorithms, EWAH (Enhanced Word-Aligned Hybrid) [7] uses a technique known as Run-Length Encoding (RLE) in which, instead of storing a bit sequence of identical values, the length of that sequence is stored. For this purpose, EWAH defines two type of words: dirty and marker words. The dirty word is for storing the uncompressible subsequences where a mix of zeros and ones are observed. The clean word is used for subsequences of identical bit values. The first bit of it stores the bit value of the sequence and the following 16 bits store the length of the sequence. The last 15 bits represent the number of dirty words following. Figure 3 illustrates how EWAH works in practice. The original bitset, as shown in Fig. 3(a), contains 5456 bits, all of which are zero except the first and the last 32 bits. The compressed form of the bitset starts with a marker word specifying the number of following dirty words (which is one), followed by the dirty word itself. The next marker word specifies the number of clean zero bits which is 5392.

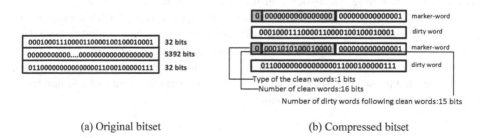

(a) Original bitset   (b) Compressed bitset

Fig. 3. An illustration of EWAH bitset compression

4.2 Concise

Compressed n-Composable Integer Set (ConCISe) [8] is another bitset compression method which is, like EWAH, based on word-aligned RLE technique. An improvement made to this algorithm is considering a "flipped" index of 5 bits for clean words which specifies a bit location that should be switched (from 1 to 0 or vice versa) in the clean sequence. It is shown this modification introduces an improvement in the space required for the compressed format without sacrificing the performance [8]. Figure 4 shows the encoding of bitset {2, 7, 31–93, 1026, 1029, 1040187421} in this method. The first word is a dirty one and represents the bit locations 0–30 for which bits 2 and 7 are 1. The second word is a pure one and represents one whole word of ones which corresponds to bits 31–92. The next dirty word represents 30 whole words of zeros in the range of 93–1022 with an exception (a flipped bit) at location 93 which is one. The next words constitute the rest of the example bitset in a similar manner.

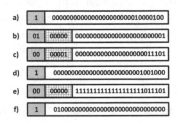

Fig. 4. Different type of words in a CONCISE bitset

4.3 Roaring

Roaring bitmaps [9] is a relatively new bitset compression technique which has gained significant popularity. Currently, numerous big data and database systems use this method at the core of their algorithms. The main idea of this technique is using hybrid containers. For sparse parts of the bitset, array containers are used, while the dense parts are stored in bitmap containers. The containers are organized in a two-level tree. The root index of this tree is typically so small that it can fit into the CPU cache. The containers are dynamic and their content share identical most significant 16 bits. Figure 5 shows an example of the roaring tree structure. The bitset in this example is sparse at the beginning and is dense at the end. Therefore, at the second level of the tree, it is composed of two array containers which are followed by a bitmap container. This sample bitset starts with the first 1000 multiples of 62 and continues with 100 consecutive ones. The last segment of the bitset contains all even numbers from 2×2^{16} to 3×2^{16} exclusive. Since the first two segments of the bitset are sparse, array containers are used for storing them. However, the last segment (sequence of odd numbers) is dense and goes into a bitmap container.

4.4   BitMagic

BitMagic [10] is another bitset compression algorithm which benefits from a two-level data structure similar to Roaring bitsets. However, the implementation details differ between these two algorithms. BitMagic does not use array containers. Unlike Bit-Magic, Roaring bitsets utilize heuristics to optimize container type selection for storing bitsets produced by operations on other bitsets. Roaring also uses binary search for calculating intersections for the parts of the bitsets where array containers are used. In overall, BitMagic has a more simplistic implementation than Roaring bitsets.

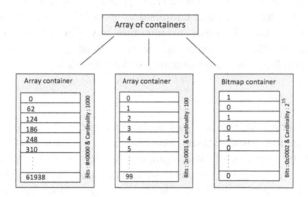

Fig. 5. An example of the tree data structure used in roaring bitmaps

5   Implementation

In order to evaluate the performance of the aforementioned bitset compression methods in different workload situations, a test framework was developed. At its core, the framework has an implementation of ECLAT algorithm in C language. This implementation uses one of the several bitset implementations discussed in previous section. For this purpose, wrapper libraries are developed that act as an interface to each specific bitset implementation. The desired interface is chosen at compile time. Input dataset file can be one of the real-world datasets or a synthetic dataset generated by the IBM quest generator [14]. This data generator accept a number of parameters and is able to generate workloads with different characteristics.

During the runtime of each test, the performance statistics are recorded for later analysis. Energy consumption measurement is performed by reading the Intel RAPL [15] interface counters which provide the energy consumption value of CPU and DRAM in units of microjoules (μJ). One pitfall of using this interface is the fact that the energy counters overflow more often than not and to avoid loss of data, they should be polled at a properly short interval (i.e. 60 s in our experiments). Open source implementations of the compressed bitsets EWAH, BitMagic, Roaring, and CONCISE are used from [10, 16–18] respectively.

6 Experiments

Several experiments were conducted on both real-world and synthetic datasets in order to evaluate the performance of the bitset compression methods in different workload situations.

6.1 Experimental Environment

All the experiments were performed on a system with an Intel Core i5-6600 CPU and 32 GB of RAM, running Ubuntu 16.04.2 LTS operating system with a Linux kernel of version 4.4.0. The source codes was compiled using GNU C/C++ compiler at the O2 optimization level setting. In order to minimize any side effects, all non-critical background tasks other than the test process were disabled. Furthermore, the CPU was run at its maximum frequency with dynamic scaling disabled.

6.2 Datasets

Since it was important for the analyses to be representative of real-world big data mining tasks, some real-world datasets were chosen for the experiments. On the other hand, analyzing the behavior of bitset compression algorithms in different workload situations and their sensitivity to various workload characteristics required the ability to vary workload parameters such as the size and sparsity level of input dataset. Therefore, besides the real-world datasets, we utilized a dataset generator known as IBM quest [14] to synthesize transactional datasets while varying some effective parameters. Table 1 shows a summary of the datasets used for the experiments.

Table 1. Datasets used in the experiments.

Dataset	Trans. count	Avg. trans. len.	Item count
Instacart (grocery shop) [19]	3.35M	10.11	49.7K
Kosarak (news portal) [20]	990K	8.10	41.3K
Quest (Synthetic) [14]	100K–1.2M	10–120	10K–120K

6.3 Results and Discussion

In order to study the behavior of bitset compression algorithms, four different experiments have been conducted each trying to analyze the effect of varying one of the following specific parameters:

- Total number of input transactions
- Minimum support count for mining frequent patterns
- Total number of distinct items that are present in transactional dataset
- Average length of transactions
- Number of generated frequent itemsets

While the first two experiments are possible to be performed on both real and synthetic datasets, the third and fourth are only applicable to synthetic data. Figure 6 shows the effect of dataset size (number of input transactions) on the performance of frequent itemset mining with different bitset compression methods. In order to properly emphasize the differences of the performance variables in the graphs, vertical axes are plotted in logarithmic scale. It can be observed that Time, energy consumption, and memory usage increases by dataset size. Energy and time exhibit a similar trend in the experiments and in general, container-based methods (Roaring and BitMagic) show a better performance than RLE-based methods. Although BitMagic shows the highest amount of memory usage, it consistently outperforms other methods in terms of energy and time. For the whole instacart dataset, BitMagic consumes half the energy and time of CONCISE in the cost of an order of magnitude higher memory usage.

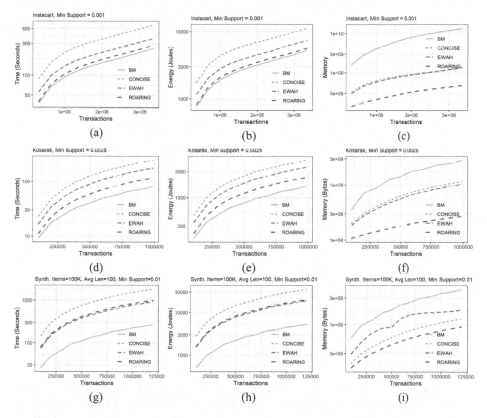

Fig. 6. Behavior of bitset compression methods vs dataset size for real and synthetic datasets

Figure 7 shows how the consumption of resources increases as the minimum support value decreases. BitMagic exhibits a better scalability in response to decreasing minimum support. Although it performs worse than other three methods at high minsup values (right side of the energy and time graphs), it soon outperforms others as the minsup value is reduced. It is worth noting that the memory consumption of EWAH and CONCISE methods increases exponentially and gets closer to BitMagic as the minimum support decreases.

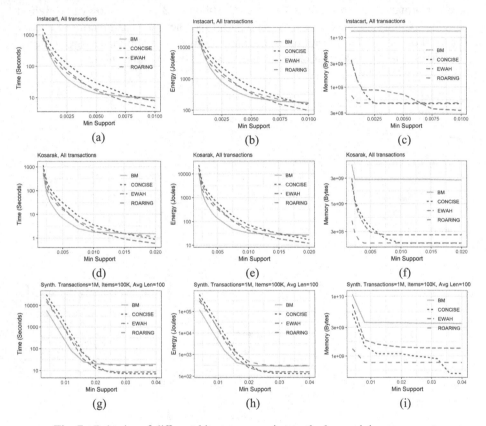

Fig. 7. Behavior of different bitset compression methods vs minimum support

The experiment on the effect of average transaction length on the resource consumption is presented by Fig. 8. This was only applicable to synthetic datasets and as shown in the plots, BitMagic consumes more memory but shows a greater scalability. Figure 9 depicts the effect of total item count in the dataset on the performance of bitset compression techniques. Increase in the count of items decreases execution time and energy usage as it decreases the chance of itemsets to be frequent.

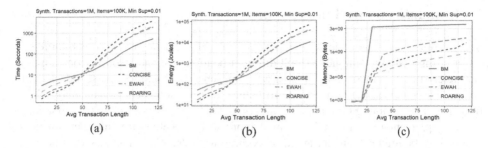

Fig. 8. Behavior of bitset compression methods vs average transaction length

All the experiments reported above study the direct effect of some parameter on the performance. One indirect relationship that may be of importance, is the relation between the number of frequent itemsets that are mined from a dataset and the amount of resource consumption for the mining task. Figure 10 shows a scatter plot of such relationship for all the experiments over both real-world and synthetic datasets. Figure 10(a) shows Energy-Delay Product (EDP) of the mining tasks as a single measure of both speed and energy-efficiency. It can be observed that memory usage pattern is more scalable than EDP. In addition, BitMagic occupies the area of the relatively highest memory usage and lowest EDP. Roaring shows a very good EDP while maintaining itself at the relatively lowest memory consumption area of the plot.

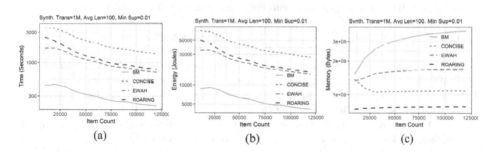

Fig. 9. Behavior of different bitset compression methods vs item count for synthetic datasets

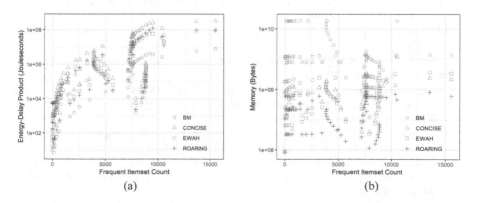

Fig. 10. Behavior of different bitset compression methods vs the count of frequent itemsets

7   Conclusions and Future Work

Sparse bitsets serve as underlying block of many big data processing tasks. Different bitset compression techniques exhibit different performance and resource consumption characteristics in practice. After a survey on some well-known sparse bitset compression techniques, their energy, performance, and memory behavior is analyzed for frequent pattern mining on real and synthetic datasets. The results show that container-based implementations generally show a more scalable behavior while their memory usage may be higher than run-length encoding methods. When memory cost is not a serious concern, BitMagic can be a good choice for performance and energy saving. Roaring bitmaps is another implementation with a very good trade-off between memory consumption and energy/performance. Notwithstanding, the proper selection of a bitset compression technique depends on workload-related parameters such as the size of the dataset, the average length of transactions and the minimum support threshold. Motivated by the dramatic differences in the behavior of sparse bitset implementations, the authors decide to develop a framework for energy/performance optimization of big data mining tasks by exploiting such variation and dynamically adopting the proper compression method at runtime.

References

1. Laney, D.: 3D data management: controlling data volume, velocity and variety, vol. 6, no. 70. META Group Research Note (2001)
2. Agrawal, R., Imielinski, T., Swami, A.: Mining association rules between sets of items in large databases. ACM SIGMOD Rec. **22**(2), 207–216 (1993)
3. Agrawal, R., Srikant, R.: Fast algorithms for mining association rules. In: 20th International Conference on Very Large Data Bases, VLDB, vol. 1215, pp. 487–499 (1994)
4. Han, J., Pei, J., Yin, Y.: Mining frequent patterns without candidate generation. ACM SIGMOD Rec. **29**(2), 1–12 (2000)
5. Zaki, M.J., Parthasarathy, S., Ogihara, M., Li, W.: New algorithms for fast discovery of association rules. In: KDD, vol. 97, pp. 283–286. ACM (1997)
6. Chen, Z., et al.: A survey of bitmap index compression algorithms for big data. Tsinghua Sci. Technol. **20**(1), 100–115 (2015)
7. Lemire, D., Kaser, O., Aouiche, K.: Sorting improves word-aligned bitmap indexes. Data Knowl. Eng. **69**(1), 3–28 (2010)
8. Colantonio, A., Di Pietro, R.: CONCISE: compressed n-composable integer set. Inf. Process. Lett. **110**(16), 644–650 (2010)
9. Lemire, D., et al.: Roaring bitmaps: implementation of an optimized software library. Softw. Pract. Exp. **48**(4), 867–895 (2018)
10. Kuznetsov, A.: BitMagic library. https://github.com/tlk00/BitMagic. Accessed 13 Jan 2019
11. Zaki, M.J., Gouda, K.: Fast vertical mining using diffsets. In: Proceedings of the Ninth ACM SIGKDD, pp. 326–335. ACM (2003)
12. Mimaroglu, S., et al.: Mining frequent item sets efficiently by using compression techniques. In: Proceedings of the International Conference on Data Mining (DMIN) (2011)
13. Dwivedi, N., Satti, S.R.: Set and array based hybrid data structure solution for frequent pattern mining. In: 10th International Conference on Digital Information Management, pp. 14–29. IEEE (2015)

14. Quest Synthetic Data Generator. http://almaden.ibm.com/cs/quest/syndata.html. Accessed 13 Jan 2019
15. Hahnel, M., Dobel, B., Volp, M., Hartig, H.: Measuring energy consumption for short code paths using RAPL. ACM SIGMETRICS Perform. Eval. Rev. **40**(3), 13–17 (2012)
16. Lemire, D.: EWAHBoolArray library. https://github.com/lemire/EWAHBoolArray. Accessed 13 Jan 2019
17. Roaring bitmaps. https://github.com/RoaringBitmap/CRoaring. Accessed 13 Jan 2019
18. Lemire, D.: CONCISE. https://github.com/lemire/Concise. Accessed 13 Jan 2019
19. Instacart Online Grocery Shopping. https://www.instacart.com/datasets/grocery-shopping-2017. Accessed 13 Jan 2019
20. Frequent Itemset Mining Dataset Repository. http://fimi.ua.ac.be/data/. Accessed 13 Jan 2019

Distributed-to-Centralized Data Management Through Data LifeCycle Models for Zero Emission Neighborhoods

Amir Sinaeepourfard[(✉)] and Sobah Abbas Petersen

Department of Computer Science,
Norwegian University of Science
and Technology (NTNU), Trondheim, Norway
{a.sinaee,sobah.a.petersen}@ntnu.no

Abstract. Data management and organization in context have been highlighted as a complex scenario during their entire life cycle (DLC) models by several research groups. Similarly, smart city has been faced several challenges and complexities to organize the obtained data from data sources across the city. Currently, there are two main references for the data management architecture in the smart city scenarios, Centralized Data Management (CDM) and Distributed-to-Centralized Data Management (D2C-DM). In this paper, we developed our proposed hierarchical D2C-DM architecture for Zero Emission Neighborhoods (ZEN) center in Norway. In the beginning, we extend the proposed Data LifeCycle model (DLC) for smart city scenario concerning the ZEN Key Performance Indicators (KPI) and their required business models. Afterward, we map the ZEN DLC model to our proposed D2C-DM for smart city, including the ZEN center. In addition, the fully hierarchical D2C architecture has the potential to organize all data life cycle stages from data production to data consumption across the city on the smart city scenarios. Finally, we discuss and conclude several capabilities of the proposed D2C-DM through the related DLC models in the ZEN center scenario, such as: (i) using all benefits of data management architectures from distributed to centralizes schema simultaneously and in one unified architecture; (ii) handling all different obtained data types (including real-time, last-recent, and historical data) in smart cities and the ZEN data types (consisting of the context, research, and KPI data) each cross-layer (from IoT devices to Cloud technologies) of the architecture.

Keywords: Smart cities · IoT · Data management · Fog-to-Cloud Data Management (F2C-DM) · Fog-to-cloudlet-to-Cloud Data Management (F2c2C-DM) · Distributed-to-Centralized Data Management (D2C-DM) · Centralized Data Management (CDM) · Data LifeCycle (DLC)

1 Introduction and Motivation

Data is one of the primary elements in the smart cities' perspective. Data makes a city smart, agile, and creative through several appropriate services across the city concerning user requirements and their related business models. Therefore, data prepare the

© Springer Nature Switzerland AG 2019
L. Grandinetti et al. (Eds.): TopHPC 2019, CCIS 891, pp. 132–142, 2019.
https://doi.org/10.1007/978-3-030-33495-6_11

expected information for usage of services concerning contextual parameters or some more important value knowledge obtained from complicated data analysis. Hence, the smart city provides an ideal scenario for generating multiple data from different sources, types, and formats in the city to merge with the available historical information.

Smart city solutions have grasped the attention of multiple researchers in recent decades. The primary contribution of smart city solutions is to update their citizens' quality of life by several different technology management solutions [1–6] (e.g., resource management and data management) between end-users, city planners, and technological devices (e.g., sensor) [1–3]. As a consequence, the data management strategies deal with an important and critical resource of the smart city environments, which is data. The obtained data in the smart city scenarios seem necessary to organize by a wise data management strategy and architecture from data production to data usage in the smart city.

In this research paper, we extended our proposal for the ZEN data management architecture first [7–9]. In addition, we are interested in using the potential of the D2C-DM architectures through Fog, cloudlet, and Cloud technologies [1–3, 10] (F2c2C-DM) in the smart cities, in particular, ZEN center [7]. Second, we extend the proposed the COSA-DLC (COmprehensive Agnostic Data LifeCycle) model [11] for the smart city (Smart City Comprehensive Data LifeCycle (SCC-DLC) model) [12, 13] concerning the ZEN center KPI and their related requirement [7, 14]. Finally, we envisage the advantages of the F2c2C-DM architecture for the smart city scenarios, e.g., ZEN center [7].

The rest of this paper is organized as follows. Section 2 introduces the main background about the proposed data management architectures in smart city scenarios, including CDM [2, 3, 15], and D2C DM [8, 9, 16, 17]. In Sect. 3, we explain some main contributions of the smart city environments, constituting different data types and main architectures for the data management. Section 4 describes the essential concepts in the ZEN center scenario in Norway, including distinct data types and all proposed data management architecture. Plus, our focus is to develop our proposed ZEN data management architecture [7–9] with available advanced technologies on the one hand. On the other hand, we extend the DLC model for our extended data management architecture for the ZEN center [7]. Section 5 concludes this paper and introduces the possible extension of the works.

2 Related Works

There are recently ongoing interests to propose different technology management proposal (for instance, management of resource, service, and data) for smart cities. Those proposed technology management strategies have been worked among users, city organizers, and technology devices to update their citizens' daily life and their requirements [1–6].

Concerning data management issues, the data management architectural models provide facilities by simple and assured access to the vast amount of the data sources and their related repositories to extract all possible novel values of the data sources for future usages and purposes. In addition, Big Data paradigms generate several new complexities for existing data management and organization systems [18, 19]. As an example, the current Relational Database Management Systems (RDBMS) and the Extract-Transform-Load (ETL) process for modeling the life cycles of data in data warehousing models [18, 20] are not sufficient for the Big Data paradigm. Therefore, data management and organization have been discussed as a complex scenario during their entire life cycle (DLC) models [21, 22].

Concerning data management issues in the smart city, the data management architecture has been illustrated through two main proposals in smart cities, CDM [2, 3, 15], and D2C-DM [8, 9, 16, 17]. On the one hand, the CDM-DM architectures have been offered by several studies in the smart city scenarios [2, 3]. The CDM-DM architectures draw architecture for data management as a centralized. It means that numerous distinct physical and non-physical sources in the citywide are generated data, but all obtained data can be saved and be accessible from a centralized technology. A centralized technology most of the time refers to the Cloud computing technologies [2, 3]. On the other hand, the D2C-DM architectures have been mentioned by a less group of research in the smart cities scenario [8, 9, 16, 17]. The D2C-DM architectures go beyond distributed strategies for data management through some advanced technology level of distributed solutions, for instance, Fog Computing [4, 23] and cloudlet [10]. Consequently, at the same time, D2C-DM architecture uses both facilities of centralized and distributed technologies. However, the majority of the D2C-DM architectures have been proposed by Fog-to-Cloud technologies (only using Fog and Cloud) [8, 9, 16, 17].

Concerning data management issues through DLC models, in order to analyze some important available DLC models behavior, in [12, 13] revisits most relevant contributions, analyzing their challenges and limitations through an evaluation process concerning the 6Vs challenges. Then, in [11–13] concludes that by tailoring DLC models to specific scenarios' needs. Later, in [11] proposed a global and comprehensive framework, from data creation to data consumption, to be widely utilized in different fields. Finally, the proposed COSA-DLC model [11] matched to the smart city scenarios, SCC-DLC [12, 13].

Within this paper, we extend a hierarchical D2C-DM architecture [7–9] by using Fog, cloudlet, and Cloud platform (F2c2C-DM) [27]. Then, we develop a new DLC model for the proposed F2c2C-DM architecture concerning the ZEN KPI and their required business models [7, 14]. As a result, this fully hierarchical architecture has the potential to match with the ZEN center scenario [7] and support all different ZEN data types and formats [7, 14].

3   Smart City Environments

Coined a few years ago, the significance of the "smart city" term has been highlighted by many scientific communities. The smart cities ideas are based on technological participation to people's daily life started from traditional life to modern life. Smart cities faced with numbers of the technological challenges and require a ubiquitous deployment of computing resources across the city (from IoT devices to centralized data centers). Several communication networks by different network topologies (including wireless sensor networks, Bluetooth, 4G, Wi-Fi, etc.) have been established to connect all data resources throughout the city. Therefore, advanced technology architectural approaches (e.g., Internet of Things, and others [24]) look important and mandatory to build the whole smart city idea. However, beyond all technologies, data can raise as a most precious resource for a city. The city can be smarter and more agile through smart services by using the appropriate data. Moreover, there are several challenges to discover the appropriate data for the services because there is a vast number of data sources (including sensors, smart devices, third-party applications and so on) through smart cities in the current smart city. As a result, there are several challenges to organize the big amount of obtained data in the smart cities in academia and industries nowadays [28].

A proposed model for conceptual levels of a neighborhood and smart city are shown in Fig. 1 [7, 14]. Micro, meso, and macro levels are designed in the conceptual model as described below.

- All city elements are at the micro-level. A building or house is an example of a city element.
- The neighborhoods are at the meso level. Buildings, houses, streets, and their neighbors are examples of elements in a neighborhood.
- All neighborhoods are at the macro level. Therefore, this level is a city scale.

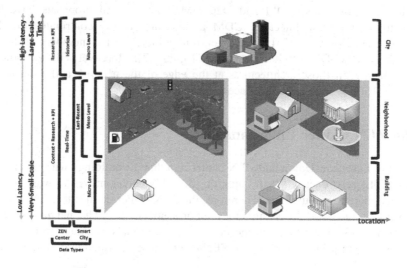

Fig. 1. ZEN center and smart cities data types [27]

Section 3 is divided into two subsections. First, multiple data types of smart city are defined. Second, distinct data management architectures for the smart cities are discussed, CDM and D2C-DM.

3.1 Data Types

Several multiple sources from physical to non-physical are in the citywide. These sources generate a huge number of data in a city, including structured, semi-structured, and non-structured. Figure 1 shows that the obtained data are coming by physical and non-physical sources concerning its data age in smart cities, as described in detail below.

- **Real-time data** is in the closest place to the users of a city and their sources. Then, the real-time data is utilized once generated. For instance, real-time data may be useful for critical shallow latency applications in smart cities.
- **Historical data** is a type of data once all data within beneath layer accumulated and stored on the data repositories in the top layer. In this scenario, historical data can be located on the ultimate position from the users and their sources in the city because an upper level of latency is required for finding data in the Cloud platform.
- **Last-recent data** is not as near as real-time is to the users and their sources in the city, but the last-recent data is not farther than historical data is to the users and their sources in the city.

3.2 Data Management

The data management architectures in smart city scenarios have been offered by two different main proposals, CDM and D2C-DM. First, several proposals have been referred to the CDM architecture. CDM architecture mainly contributes to Cloud computing technologies [2, 25]. Second, currently, there are few numbers of the proposal with the D2C-DM architecture [9, 12, 16, 26, 27]. Moreover, Fog technologies are used mainly on those architectures at the edge of the network and then moving forward to reach to the centralize place technology (Cloud).

4 The ZEN Center Environments[1]

In 2020, the European Union and ZEN center aim that all buildings will get closer to the idea of the zero energy [7]. In addition, the ZEN center moves forward the zero energy idea more than buildings. Therefore, the ZEN center defines the zero-emission neighborhoods idea for the smart cities environments [7]. The neighbourhood is located at the meso-level in the proposed conceptual models as shown in Fig. 1 [27]. Furthermore, eight pilot projects of the ZEN center are established in different cities of Norway [7].

Two subsections are defined in this section. First, distinct data types of the ZEN center are briefly described [7]. Second, our recent D2C-DM architecture proposal for the ZEN center is discussed [7–9], and eventually we extend the ZEN data management architecture with DLC models based on Fog, cloudlet, and Cloud.

---

[1] **Some part in [27].**

4.1   Data Types

Based on the report [7], three categories of the ZEN data types are defined, as shown below:

- **Context data:** Physical data sources are produced the context data. For instance, physical sensors produced data for energy consumption and noise conditions.
- **Research data:** Non-physical data sources (including third-party applications, simulation data) are generated the research data, including energy data from buildings, occupant behavior data, etc.
- **KPI data:** The ZEN center defines multiple Key Performance Indicators (KPIs) [7], such as emissions, economy, and spatial qualities. Three distinct ZEN data types have shown in the conceptual model in Fig. 1 [27]. Based on Fig. 1, we observe that the context data is available in every level of from micro to meso, and macro. Next, we see that the research data is accessible in two different levels, meso, and macro. Finally, we find that the KPI data exists in every level, which is similar to the context data.

4.2   Proposing a F2c2C-DM Through DLC Models

In the related works for the ZEN [8, 9], we designed the F2c2C-DM architecture first. Note that we only concentrated on organizing sensors and IoT data from distributed to centralized schema. Therefore, we realized that the ZEN faces with several data challenges based on our previous designed experience [8, 9]. On one side, the architecture must deal with eight different city pilots from distinct scale size to location in Norway. On the other side, this architecture must support several different ZEN data types, as we mentioned in Sect. 4.1 [7, 14].

In [12, 13] has been proposed an SCC-DLC model to demonstrate the idea of data management through their entire life cycles started with data acquisition to data preservation and processing blocks and also their related phases in a smart city. In addition, the primary structure of the SCC-DLC model is created by three blocks (including Data Acquisition, Data Processing, and Data Preservation). Every block defines a set of distinct phases as explained briefly below and in more details in [12, 13]:

- The Data Acquisition block provides the facility for data collection, data filtering (including data aggregation and some other related data optimization techniques), data quality, and data description phases.
- The Data Processing block can organize the tasks related to data process and data analysis phases.
- The Data Preservation block adds the extra facility for the data classification, data archive, and data dissemination techniques.

As we mentioned above in Sect. 4.1, this report [7, 14] defined three different ZEN data types. In addition, the numerous ZEN KPIs are considered for pilot projects. Therefore, we extend the SCC-DLC model with a new block (namely is "KPI Requirements") to support the ZEN KPI and all the required business policies of the ZEN center [7, 14]. It means the "KPI Requirements" block sets by a number of the

KPI (consisting of Energy, Power, Emissions, Economy, Mobility, and Spatial Qualities) [7, 14] to organize all tasks of the Data Acquisition, Data Preservation, and Data Processing blocks and their related phases through the ZEN KPI requirements as shown in Figs. 2 and 3.

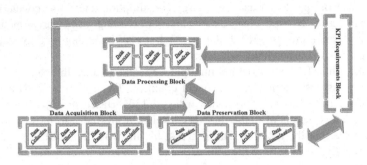

Fig. 2. The ZEN DLC model

KPI Requirements Block

Fig. 3. KPI requirements for ZEN center

Figure 4 proposes complete data management architecture from distributed to centralized schema for organizing the produced data in the smart cities. Two main axes are the main principle of our architecture. Time and Location are those axes which explain our data management opinion in smart cities through the four different concepts as describe briefly below. Note that more details about "technology solutions," "data types," and "data management" are available in [27]. Here we just explain "DLC models" for each technology layer. In addition, Fig. 4 depicts hierarchical F2c2C-DM architecture by three different technology layers as described each below.

- **Fog-Layer** is very close to users and all city data sources in a city and ZEN city pilots. The Fog-Layer can handle multiple tasks as shown one of those tasks below.
  - **DLC model:** First, this layer has been dedicated mainly for the duties of the Data Acquisition block (darker color level) because most of the physical data sources are located in this layer. Second, this layer makes the possible level of processing and storage concerning to the data sources capacities (lighter color level). Third and last, the KPI Requirements block can add some additional policies concerning the ZEN KPI and their related business model requirements.
- **Cloudlet-Layer** is positioned in a between of Fog and Cloud layers. Note that the cloudlet-Layer is available in the same pilot city. The cloudlet-Layer can organize some distinct tasks as shown one of those tasks below.

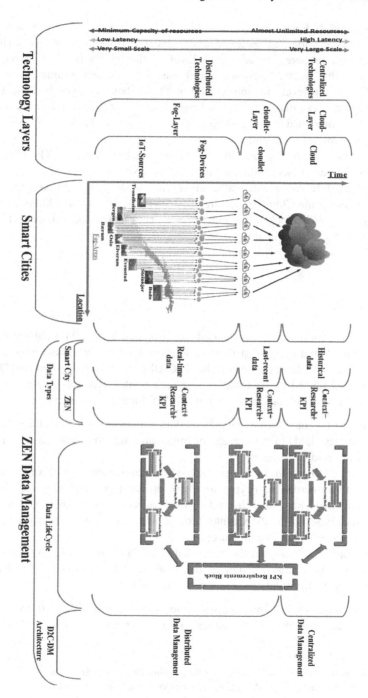

Fig. 4. The ZEN F2c2C-DM architecture (some part in [27])

- **DLC model:** First, cloudlet technologies are ready for high-level tasks of the Data Preservation and Data Processing blocks (medium color level). However, the Cloud prepares an advanced level of the processing and storage tasks. Second, the Data Acquisition block has fewer responsibilities in this layer (lighter color level) in comparison with the bottom layer because the data sources are less than the bottom layer. Last, the KPI Requirements block can provide some additional policies concerning the ZEN KPI and their related business model requirements.
- **Cloud layer** is at the top position of the proposed architecture. The Cloud-Layer provides different tasks as shown one of those tasks below.
  - **DLC model:** All sophisticated tasks can be done by the advance level of resources on the Cloud environment (darker color level). In addition, the "KPIs Requirements" block can apply some additional policies concerning the ZEN KPI and their related business model requirements.

5   Conclusion

In this research paper, we have extended our proposed D2C-DM architecture for the ZEN center based on different distributed and *centralized* technologies on the one hand. On the other hand, we have developed the SCC-DLC model for the proposed D2C-DM concerning the ZEN KPI requirements and their related business models. The main advantages of this development for our F2c2C-DM are the following:

- We depicted the upsides of using both data management architectures (including CDM and D2C-DM) in the context of the smart cities to handle data management complexities and challenges;
- We illustrated that our proposed D2C-DM architecture is a flexible facility to organize all numerous data types in smart city scenario (including real-time, last-recent, and historical data) from IoT devices to cloud technologies as well as distinct data types in the ZEN center (consist of context, research, and KPI data) in each cross-layer of our architecture;
- We developed and adopted the proposed SCC-DLC models for the ZEN center. Then we discussed thatD2C-DM (from distributed to centralized) can handle all data life stages (from data generation to data usage);

As a part of our future work, we will study more options related to extending our ICT architecture for the ZEN center, such as designing the service/software architecture.

Acknowledgment. This paper has been written within the Research Centre on Zero Emission Neighborhoods in smart cities (FME ZEN). The authors gratefully acknowledge the support from the ZEN partners and the Research Council of Norway.

References

1. Tei, K., Gurgen, L.: ClouT: Cloud of things for empowering the citizen clout in smart cities. In: IEEE World Forum on Internet of Things (WF-IoT), pp. 369–370. IEEE (2014)
2. Jin, J., Gubbi, J., Marusic, S., Palaniswami, M.: An information framework for creating a smart city through internet of things. IEEE Internet Things J. **1**, 112–121 (2014)
3. Gubbi, J., Buyya, R., Marusic, S., Palaniswami, M.: Internet of Things (IoT): a vision, architectural elements, and future directions. Futur. Gener. Comput. Syst. **29**, 1645–1660 (2013). Elsevier Journal
4. Bonomi, F., Milito, R., Zhu, J., Addepalli, S.: Fog computing and its role in the internet of things. In: Proceedings of the First Edition of the MCC Workshop on Mobile Cloud Computing, pp. 13–16. ACM (2012)
5. Hu, X., Ludwig, A., Richa, A., Schmid, S.: Competitive strategies for online cloud resource allocation with discounts: the 2-dimensional parking permit problem. In: 35th IEEE International Conference on Distributed Computing Systems (ICDCS), pp. 93–102. IEEE (2015)
6. Rao, T.V.N., Khan, A., Maschendra, M., Kumar, M.K.: A paradigm shift from cloud to fog computing. Int. J. Sci. Eng. Comput. Technol. **5**, 385 (2015)
7. https://fmezen.no/
8. Sinaeepourfard, A., Krogstie, J., Petersen, S.A.: A big data management architecture for smart cities based on fog-to-cloud data management architecture. In: Proceedings of the 4th Norwegian Big Data Symposium (NOBIDS) (2018)
9. Sinaeepourfard, A., Krogstie, J., Petersen, S.A., Gustavsen, A.: A zero emission neighbourhoods data management architecture for smart city scenarios: discussions toward 6Vs challenges. In: International Conference on Information and Communication Technology Convergence (ICTC). IEEE (2018)
10. Bilal, K., Khalid, O., Erbad, A., Khan, S.U.: Potentials, trends, and prospects in edge technologies: fog, cloudlet, mobile edge, and micro data centers. Comput. Netw. **130**, 94–120 (2018)
11. Sinaeepourfard, A., Garcia, J., Masip-Bruin, X., Marín-Torder, E.: Towards a comprehensive data lifecycle model for big data environments. In: Proceedings of the 3rd IEEE/ACM International Conference on Big Data Computing, Applications and Technologies, pp. 100–106. ACM (2016)
12. Sinaeepourfard, A., Garcia, J., Masip-Bruin, X.: Hierarchical distributed Fog-to-cloud data management in smart cities. Doctoral thesis. Departament d'Arquitectura de Computadors, Universitat Politècnica de Catalunya (UPC), Barcelona, Spain (2017)
13. Sinaeepourfard, A., Garcia, J., Masip-Bruin, X., Marin-Tordera, E., Yin, X., Wang, C.: A data lifeCycle model for smart cities. In: International Conference on Information and Communication Technology Convergence (ICTC), pp. 400–405. IEEE (2016)
14. Ahlers, D., Krogstie, J.: ZEN data management and monitoring: requirements and architecture (2017)
15. Rathore, M.M., Ahmad, A., Paul, A., Rho, S.: Urban planning and building smart cities based on the internet of things using big data analytics. Comput. Netw. **101**, 63–80 (2016)
16. Sinaeepourfard, A., Garcia, J., Masip-Bruin, X., Marin-Tordera, E.: Fog-to-cloud (F2C) data management for smart cities. In: Future Technologies Conference (FTC) (2017)
17. Sinaeepourfard, A., Garcia, J., Masip-Bruin, X., Marin-Tordera, E.: Data preservation through fog-to-cloud (F2C) data management in smart cities. In: 2018 IEEE 2nd International Conference on Fog and Edge Computing (ICFEC), pp. 1–9. IEEE (2018)

18. Hu, H., Wen, Y., Chua, T.-S., Li, X.: Toward scalable systems for big data analytics: a technology tutorial. J. Mag. IEEE Access **2**, 652–687 (2014)
19. Almeida, F.L.F., Calistru, C.: The main challenges and issues of big data management. Int. J. Res. Stud. Comput. **2**, 11–20 (2012)
20. Henry, S., Hoon, S., Hwang, M., Lee, D., DeVore, M.D.: Engineering trade study: extract, transform, load tools for data migration. In: IEEE Conference on Design Symposium, Systems and Information Engineering, pp. 1–8 (2005)
21. Demchenko, Y., Zhao, Z., Grosso, P., Wibisono, A., De Laat, C.: Addressing big data challenges for scientific data infrastructure. In: IEEE 4th International Conference on Cloud Computing Technology and Science (CloudCom), pp. 614–617. IEEE (2012)
22. Grunzke, R., et al.: Managing complexity in distributed Data Life Cycles enhancing scientific discovery. In: IEEE 11th International Conference on E-Science (e-Science), pp. 371–380. IEEE (2015)
23. Xiong, Z., Feng, S., Wang, W., Niyato, D., Wang, P., Han, Z.: Cloud/fog computing resource management and pricing for blockchain networks. IEEE Commun. Mag. (2018)
24. Kyriazopoulou, C.: Smart city technologies and architectures: a literature review. In: 2015 International Conference on Smart Cities and Green ICT Systems (SMARTGREENS), pp. 1–12. IEEE (2015)
25. Gubbi, J., Buyya, R., Marusic, S., Palaniswami, M.: Internet of Things (IoT): a vision, architectural elements, and future directions. J. Futur. Gener. Comput. Syst. **29**, 1645–1660 (2013)
26. Sinaeepourfard, A., Garcia, J., Masip-Bruin, X., Marin-Tordera, E.: A novel architecture for efficient fog to cloud data management in smart cities. In: IEEE 37th International Conference on Distributed Computing Systems (ICDCS), pp. 2622–2623. IEEE (2017)
27. Sinaeepourfard, A., Krogstie, J., Petersen, S.A.: F2c2C-DM: a fog-to-cloudlet-to-cloud data management architecture in smart city. In: IEEE 5th World Forum on Internet of Things (WF-IoT). IEEE (2019)
28. Sinaeepourfard, A., Garcia, J., Masip-Bruin, X., Marin-Tordera, E.: Estimating Smart City sensors data generation. In: Mediterranean Ad Hoc Networking Workshop (Med-Hoc-Net). IEEE (2016)

Analytical Comparison of Virtual Machine and Data Placement Algorithms for Big Data Applications Based on Cloud Computing

Seyyed Mohsen Seyyedsalehi and Mohammad Khansari[✉]

Faculty of New Sciences and Technologies, Tehran University, Tehran, Iran
{sseyyedsalehi, m.khansari}@ut.ac.ir

Abstract. Data generation proliferation and big data utilization increase in different areas, cause to highlight data and virtual machine placement problem in MapReduce framework. Since problem NP-hardness is proven, different algorithms using various techniques have been proposed in recent years and every solution targets some objectives in this regard. But, it has not proposed any troubleshooting solution to placement problem till now and it is still an issue for service providers. In this paper, we present a comprehensive evaluation of current researches and highlight the new paths for researchers by identifying the weaknesses of existing studies. To reach to this goal, we evaluate many of current researches about vm placement in cloud computing and big data applications on cloud computing and select some of them to be presented in this paper. Also, we propose our method to solve the problem and show how it could be more effective than available methods.

Keywords: MapReduce · VM placement · Data placement · Algorithm

1 Introduction

Massive amounts of data are generated everyday by various sources including sensors, Internet transactions, social networks, Internet of Things (IoT) devices, video surveillance systems, and scientific applications. Many organizations and researchers store such data to enable breakthrough discoveries in science, engineering, and commerce [1]. Such a way that big data alone will represent 30 percent of data stored in data centers by 2021, up from 18 percent in 2016. Big data is a significant driver of traffic within the data center. While much of big data traffic is rack-local, enough exits the rack that big data will be responsible for 20 percent of all traffic within the data center by 2021, up from 12 percent in 2016 [2].

Hadoop is an open source platform providing highly reliable, scalable, distributed processing of large data sets using simple programming models. It has been widely used by a large number of business companies for production purposes [4]. Hadoop ecosystem, has two vital tools including MapReduce and HDFS[1]. MapReduce is a distributed programming framework that allows service developers to write simple

---

[1] Hadoop Distributed File System.

© Springer Nature Switzerland AG 2019
L. Grandinetti et al. (Eds.): TopHPC 2019, CCIS 891, pp. 143–155, 2019.
https://doi.org/10.1007/978-3-030-33495-6_12

programming using a lot of computing resources to speed up the processing of massive amount of data [3]. HDFS splits a file into small data blocks and place them on different nodes in the cluster. Furthermore, HDFS provides an API that exposes the locations of file blocks and allows applications to schedule a task to the node which hosts the data block [5]. Figure 1 shows the hierarchical network topology of HDFS, where D represents the data centers, R indicates the racks and H shows the nodes.

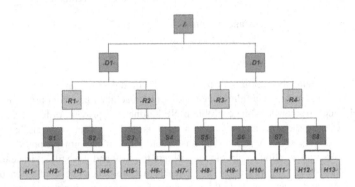

Fig. 1. Hierarchical Network Topology of HDFS [5]

The Hadoop framework can be deployed on both of physical and virtualized hardware. But, high economic costs and difficulties in maintaining of bare-metal hardware besides flexibility, pay per use and scalability of cloud infrastructure, increase the intention of Hadoop implementation on cloud, significantly. However, deploying MapReduce- like frameworks on cloud, has its own challenges. Virtual machine(VM) and data placement and data locality[2] are the big problems in optimizing the cost and time of performing big data tasks based on cloud computing.

In spite of wide spreading researches which have been done until now on placement problem, a perfect solution has not been achieved as a result of complexity and hardness of the problem. So, in our article, we present the investigation of different aspects of the subject and proposed solutions to help researchers to find the right target.

The rest of this article is organized as follows. In section two, the goals of investigated algorithms are mentioned. In section three, used techniques to solve the problem are presented. Section four is devoted to the results of algorithms deployment. In section five and six, we analyze reviewed algorithms and propose a new method to solve the problem. Our conclusion is presented in section seven.

2   Algorithms' Purposes

VM placement is one of the main challenges of cloud providers and its optimization can impress different components. Researchers in [6], categorize all the placement algorithms in two sections based on their objective:

---

[2] Scheduling a task close to the corresponding data is known as data locality [5].

- Power-based approach: a system that is energy-efficient with utmost resource utilization.
- QoS³-based approach: using a structure to ensure maximal fulfillment of quality of service requirements.

In another research, the goals are classified as below [7]:

1. Energy consumption minimization
2. Cost optimization
3. Network traffic minimization
4. Maximizing resource utilization
5. QoS maximization

Studying available researches in placement algorithms for big data applications shows the goal of the majority of them are the same as mentioned above. However, existence of massive amount of data, increases the complexity. We investigate the goals of placement algorithms in selected solutions in below.

2.1 Purlieus

This model was presented in 2011. It was one of the first researches considered the data placement and is the source of other studies. The goal of this investigation is reducing the network distance between storage and compute nodes for both map and reduce processing. To attain the goal, MapReduce tasks are divided into three classes [8]:

1. Map input heavy: These jobs read large amounts of input data for map, but generate only small map-outputs that is input to the reducers.
2. Map-and-Reduce-input heavy jobs: These jobs process large amounts of input data and also generate large intermediate data.
3. Reduce-input heavy applications: Jobs that are reduce-input heavy, read small sized map-inputs and generate large map-outputs that serve as the input to the reduce phase.

2.2 CAM

Researchers presented a platform, named CAM to avoid placement anomalies due to inefficient resource allocation [9]. These anomalies - loss of data and job locality- are supposed to be corrected by a three level approach [9] to maximize locality.

1. Data placement: Rather than accommodating an arbitrary data placement, strategically placing the data can significantly improve locality.
2. VM/job placement: For a given job, CAM selects the best possible physical nodes to place the set of VMs that represent the job.
3. Task placement: CAM exposes, otherwise hidden, compute, storage, and network topologies to MapReduce job scheduler such that it makes optimal task assignments.

---

³ Quality of Service.

The assumption in CAM are mostly based on Purlieus, for example, MapReduce jobs are divided into three categories as same as Purlieus.

2.3   VMPDN

The objective of this study is to determine the assignment of VMs for Data Node (DN)s such that the maximum access latency among all pairs of a DN and its assigned VM and all pairs of assigned VMs (the VMs that are assigned to DNs in the assignment) is minimized when the location of each DN is determined in advance and is fixed [10]. The result should not be bigger than twice the minimum time.

2.4   ACO Algorithm

The primary objective of this meta-heuristic algorithm is reducing cross network traffic and bandwidth usage, by placing required number of VMs and data in physical machine(PM)s which are physically closer [11].

2.5   Genetic Algorithm (GA)

Researchers in [12] presented an algorithm to determine the virtual machine allocation, virtual machine template selection, and file replica placement in order to minimize the power consumption, physical resource waste and file unavailability.

2.6   Algorithms' Purposes Conclusion

Table 1 shows a short view of mentioned researches objectives. As it is seen, in majority of studies, the VM placement optimization has been used in order to improve system performance and decrease jobs execution time. Decreasing physical resource waste and energy consumption are other considered subjects.

Table 1. Goals of placement algorithms in big data applications based on cloud computing

	Data placement	Reducing distance between nodes	Reducing access latency	Multiple distributed data centers	Reducing network traffic	Minimizing energy consumption	Minimizing data retrieval time	Minimizing power consumption
Purlieus	✓	✓						
CAM	✓		✓					
VMPDN			✓					
ACO algorithm	✓	✓		✓	✓			
Genetic algorithm	✓					✓	✓	✓

3   Problem Modeling

In this section, proposed algorithms and used techniques are investigated. Researchers in [14] show that different models are used to profile the resource provision problem due to the differences in objectives and resource granularity. Exact methods such as branch-and-bound are applicable only to small instances. Alternative methods include approximate algorithms or reductions of problems to easier versions, provided that loss of detail in the model is acceptable [13]. Bin packing (BP), graph theory (Graph), virtual network embedding (VNE), greedy and meta-heuristic (MH) are other samples of used methods. However, the most important defect in relation to heuristic algorithms is that they (in most cases) have been proposed, without any performance guarantee. This is problematic because even if they perform well in controlled experiments, they may yield poor results in real settings [15].

The algorithm used by the operator for re-optimizing the placement of VMs has large impact on multiple vital metrics like energy consumption, application performance and migration overhead [16]. Regarding to this importance, the proposed algorithm and problem-solving technique is examined in each selected article.

3.1   Purlieus

Researchers in [8] have been looking for a way to host VMs on the same physical server, or to store on the closest physical server with the necessary features, in order to be able to minimize costs in map and reduce phases. To do this, as described in the previous section, the types of tasks were divided into three categories, each with a different strategy. They used the graph to model the problem.

1. Map input heavy: The Purlieus platform in this case selects PMs for hosting that have appropriate storage capacity for data, while computing load on them is at least.
2. Map-and-Reduce-input heavy jobs: To achieve map-locality, data must be placed in PMs can host VMs locally. Also, this data placement should also support reduce-locality for which VMs need to be close together. Therefore, the researchers in this study tried to find a k-club of a graph G^4 which has an answer in the hierarchical data centers in polynomial time.
3. Reduce-input heavy applications: Since the reduction of distance between compute and data storage nodes in reduce phase is much more important than the map phase, map input data can be placed anywhere in the cluster to be transmitted to the corresponding VMs during the implementation of the phase. The data placement algorithm chooses the PM with the maximum free storage.

3.2   CAM

CAM is designed as an extension to IBM ISAAC product [9]. In this research, they used a min-cost flow graph that encodes the factors. To achieve this, the placement problem is divided into three sub-issues including [9]:

---

[4] A k-club of a graph G is defined as a maximal subgraph of G of diameter k.

- Guaranteeing VM closeness
- Hotspot factor
- Balancing physical storage utilization according to different job types

Min-cost flow graph is based on the characteristics of the three sets of assigned tasks.

3.3   VMPDN

For modeling the problem, the bipartite graph $G = (D \cup V, E)$ has been implemented and, based on the predefined assumption, the delay of access to data follows the triangular inequality.

To achieve the goal, two algorithms are presented. In the first one, the threshold and linear programming (LP) techniques are used. Initially, the threshold t is set based on the minimum weight of edges. In each round, this threshold increases. The algorithm employs a procedure to examine whether there is a feasible assignment of DNs to VMs under the threshold t. If an assignment of DNs to VMs can be made as the threshold t is not greater than the maximum weight of edges, algorithm 1 outputs the made assignment. The results have shown that there is no such a response.

In the second algorithm, the problem of placing DNs for VMs is based on an algorithm with a maximum approximation of 2 times the optimal value. The threshold t increases from the smallest to the largest value of the edges in the graph G. In a given value of threshold t, the issue is divided to $|V|$ smaller parts and the algorithm allocates resources by observing the conditions set for each small placement problem. Ultimately the best answer is chosen.

3.4   ACO Algorithm

The cloud environment is mapped to a weighted complete graph; while its vertices represent physical servers and weights on the vertices show the capacity of servers to place VMs. The weight of the edges also represents the distance between the corresponding servers. To find the optimal solution, a set of servers must be selected so that the total weight of the selected vertices is at least equal to the resources requested and the total weights of the edges between the vertices is minimal.

They used the Ant Colony Optimization (ACO) metaheuristic algorithm and clique pheromone strategy. Pheromone values are associated with every pair of vertices. The desirability of vertices are not independent. The selection of a vertex depends on the subset of already selected vertices. The quantity of pheromone on the edge (v_i, v_j) represents the learned desirability of selecting both the vertices V_i and V_j within the same solution [11].

3.5 Genetic Algorithm (GA)

In this study, researchers used the non-dominated sorting genetic algorithm-II, which is a multi-objective optimization algorithm. In this case, the individuals of the population represent VM allocation, VM type selection, and replica placement, considering a fixed number of replicas and a variable number of VMs [12].

Initial placement of the data replica is performed in a Round-robin distribution. The number of data chunks replica is constant and the number of VMs is variable. Each of these individuals is represented by two arrays: the VM-chromosome for the allocation and VM type relationships and the block-chromosome for the placement relationship. The length of the VM-chromosome can change between solutions since there will be solutions with different numbers of VMs [12].

The results are analyzed by comparing each solution of Pareto series in each of the scenarios implemented with all Pareto optimal sets.

3.6 Problem Modeling Conclusion

The importance of network related parameters like distance and traffic, causes to use graph as a suitable tool to model the problem more frequently. But, the range of selected techniques to solve the problem is very wide. Heuristics like ACO, genetic algorithms and PSO are widely used. Anyway, increasing the execution time like something happens in [12] eclipses their good points. Greedy is the other common method to work out the problem. However, its usage depends on the structure of optimization problem.

4 Simulation Results

In this section, the results of algorithms implementation are presented. This implementation is generally done in two levels: The first method is micro-modeling, in which the algorithm is implemented on a limited sample space and the effect of each task (depending on the tasks specified) is separately investigated. The second method is the macro modeling that evaluates the general capabilities of the algorithm and studies users' important criteria.

Table 2 shows the methods and results of implementation of the algorithms under consideration.

Table 2. Comparison of the results of the implemented algorithms

Model	Test bed	Compared algorithms/techniques	Experimental results
Purlieus [8]	100 & 1000 PM/10 to 200 VM	VM Placement Techniques: - Locality-unaware VM Placement (LUAVP) - Map-locality aware VM placement (MLVP) - Reduce-locality aware VM placement (RLVP) - Map and Reduce-locality aware VM placement (MRLVP) - Hybrid locality-aware VM placement (HLVP)	In this assessment, all kinds of the tasks are given to system and the algorithm must decide which strategy will have better results: - Overall, HLVP with LLADP shows 2x faster execution time when compared to RDP + MLVP schemes and a 9.1% improvement with most conservative policy of LLADP + MRVLP - The job execution time decreases with increasing number of VMs but that decrease almost stops beyond a certain number of VMs (100 VMsin this case)
CAM [9]	Run 100 jobs simultaneously on 192 VMs	Load and locality aware (LLA) data placement and Hybrid VM placement from Purlieus [8]	In this assessment, Map-intensive, MapReduce-intensive, and a workload with a mix of Map, MapReduce, and CPU-intensive jobs, is considered as input: - The combination of proposed data and VM placement with various tasks will speed up the execution of results up to 8.6 times
VMPDN [10]	1024 racks	- 3-Approximation Algorithm for VMPDN - 2-Approximation Algorithm for VMPDN	- The results show that the performance of two algorithms in relation to maximum access latency is similar in most cases - Broadly speaking, as the rack range increases, the maximum access latencies of algorithms tend to increase in all network architectures. The maximum access latencies are proportional to the maximum number of switches between two racks
ACO algorithm [11]	720, 2000 and 6000 servers	- First Fit Decreasing (FFD) - Distance-aware FFD	- The number of PMs selected by FFD is always less than or equal to the number of PMs selected by ACO and distance-aware FFD - Sum of distances between VMs of the selected PMs in is very much lesser in ACO than FFD
Genetic algorithm (GA) [12]	50 to 200 PMs/50 to 200 files	- NSGA-VM: the proposal based on the work of Adamuthe et al.	- In terms of file unavailability, proposed model obtained 407.41% more availability than NSGA-VM on average - Proposed model only saved 1.9% more power on average - The Solution wasted 170.39% less resources than NSGA-VM on average

5 Discussions and Open Issues

By analyzing different researches' dimensions, it is concluded that none of the algorithms and methods is chosen as a superior strategy; because the details of the objectives, the problem modeling, the scale of implementation, and the simulation criteria in each research is different to the others. But, some defects are evident in most of them:

- **Laboratory conditions:** One of the main objections of proposed solutions is algorithm implementation in a specific situation. There is some example in [10, 11] and [9].
- **Being operational:** Between examined articles, only algorithms presented in [8] and [9] were implemented on physical hardware. However, the deployment was on a very small scale and under specific conditions.
- **Scalability of algorithms:** In Table 2, the size of the test space of each algorithm is shown. In some instances, such as [10], the test space has expanded to a maximum of 1024 racks, while in some others like [12], the number of physical servers is at most 200. The important point in this regard is algorithm's scalability and compliance with real conditions [11].
- **One-dimensional solutions:** In the real world, a desirable system must meet a set of expectations simultaneously. Although studies such as [12] has considered more than one goal for optimal placement, which is better than others, selecting two or three chosen and aligned goals, does not meet the needs of providers and consumers.
- **Infrastructure capacity:** In most studies, it is assumed that there are sufficient resources for every request. However, this is not possible in reality.

A review of the research trends in this area has shown that the proposed methods have made significant improvements to the prototype, but Remained ambiguous and open issues make these suggestions inefficient. Hence, the following suggestions have been proposed for more practical solutions:

- **Collaboration with business units:** The majority of in-depth studies conducted by academic researchers, while the main destination of proposed solutions are business service providers. So, it is necessary to bring the views of these two groups, together. Conducting joint studies, prevent parallel work and will result in better and more practical results.
- **Implementation of algorithms in physical data centers:** It is better to test proposed algorithms by implementing on physical servers rather than testing them in laboratory conditions. In this way, we can find the impact of each method on different indicators.
- **Identifying the negative effects of implementing each one of the algorithms:** Since all the mentioned algorithms are tested in special environments with predetermined conditions, their implementation under normal circumstances may have negative effects on other important indicators. Specifying the negative effects, helps to provide solutions to eliminate negative aspects.

- **Considering modern topologies in implementation:** Network topology and architecture has considerable impact on performance factors. Nevertheless, most of the studies, used traditional topologies for problem solving, while service providers, need to decide about implementing novel ones.

6 Researchers Point of View

Based on out-and-out investigations on previous studies, two approaches can be considered for the VM and data placement problem in big data applications:

1. Given theoretical evidence and similar to the previous proposed solutions, we can accept this is a classical Knapsack problem and its NP-completeness has been proven, So we can solve it by using heuristic and approximate methods.
2. The second mode, is to go beyond theoretical assumptions and enter real-world cloud implementation facts in problem solving. For instance, empirical evidence suggests that the size of MapReduce VMs in every task is pretty flexible and their size can be considered dynamic. For example, a 50-VM MapReduce task is received and every VM needs 4 CPU cores. In our hypothesis, it will not disturb computing power, if we assign 3 or 5 CPU cores to some of the VMs. By using these suggestions, the problem NP-completeness state turns to a linear one. Even if using the mentioned method, creates a percentage of inefficiency and non-inference in physical resources managing, it is negligible according to cloud computing dynamism[5]

Definition: We use a weighted graph G = (V, E) to solve the problem where V and E represent the set of graph's vertices (PMs) and Edges, respectively. P_i denotes the processing capacity of each PM and x is the empty capacity percentage of a vertex. α_i demonstrates the cost of a processing unit at vertex i and D_{ij} exposes the weight of each edge which represents the distance between the two i and j PMs.

Our assumptions:

1. In (1), we show how the distance between two PMs is defined based on different scenarios of VMs positioning. In a datacenter with Leaf-Spine structure, the traffic between servers in different racks is always passed through two leaf switches and a spine switch [17].

$$D_{ij} = \begin{cases} i = j, & 0 \\ i \neq j, & DC(i) = DC(j), Rack(i) = Rack(j) \quad Delay_{ToR} \\ i \neq j, & DC(i) = DC(j), Rack(i) \neq Rack(j) \quad 2*Delay_{ToR} + Delay_{Spine} \\ i \neq j, & DC(i) \neq DC(j), \quad 2*Delay_{ToR} + 2*Delay_{Spine} + 2*Delay_{Router} \end{cases}$$

(1)

---

[5] It should be noted this evidence is based on the researchers' experiences in XaaS Cloud implementation. XaaS is the main cloud computing service provider in Iran. Ref: www.XaaS.ir.

2. S_i equals to total weight of outgoing edges from vertex i:

$$S_i = \sum_{\substack{i=i \\ j \in P}} D_{ij} \tag{2}$$

3. If x percent of empty capacity of a vertex is used, we will be charged $x_i S_i (0 \leq x_i \leq 1)$.
4. We define α_i as the cost of a processing unit at vertex i:

$$\alpha_i = \frac{S_i}{P_i} \tag{3}$$

Theorem: A Big Data Task (BDT) requirements is received. There is a subset of vertices as Cluster$_{alg}$, provided that:

$$BDT \leq \sum_{i \in Cluster} P_i \tag{4}$$

$$Minimizing \left(\sum_{i \in Cluster} S_i \right) \tag{5}$$

To find the Cluster$_{alg}$, We arrange the vertices according to α_i in ascending order $(\alpha_1 \leq \alpha_2 \leq \cdots \leq \alpha_n)$. Then, we add vertices to the Cluster$_{alg}$ -which is initially $\varnothing$ - until added vertices' P_i to the Cluster$_{alg}$ is as large as the BDT. For the last chosen vertex, we use of empty processing power just as much as needed by applying x_k coefficient.

$$P_1 x_1 + P_2 x_2 + \cdots + P_k x_k = BDT$$

In the proposed approach, CPU, RAM and Storage are mentioned.

Proof
Assume Cluster$_{opt}$ is the optimal solution of the problem and the Cluster$_{alg}$ is the result of the proposed algorithm. We consider the first contradiction point between Cluster$_{alg}$ and Cluster$_{opt}$. Since all the cheaper nodes have been already selected, algorithm chooses the cheapest node in remained choices. Therefore, Cluster$_{alg}$ is less expensive or equal to Cluster$_{opt}$. While the Cluster$_{opt}$ is assumed to be an optimal solution and according to the Proof by contradiction, no subset with BDT processing requirements can cost less than Cluster$_{alg}$. So, it is proven that Cluster$_{opt}$ and Cluster$_{alg}$ are the same, and Cluster$_{alg}$ is the optimal answer.

In this way, we proved that it is possible to turn the problem of VM and data placement from an NP-complete problem to a linear one. In upcoming article, we will examine the effect of the implementation of this algorithm and its results in comparison with the previous proposed solutions.

7 Conclusion

Optimizing the placement of data and virtual machines in big data applications is an issue that has become an important challenge regarding to increasing demand of big data services based on cloud platform. So far, different approaches using various techniques have been proposed, which have made significant advances. However, these solutions, have not yet been able to cover the various aspects of the problem well. Examining the examples of valid solutions show that solving the problem requires to take into account the real conditions of implementing cloud computing in physical data centers. Considering the prioritization of the initial assumptions and weighting the impact of each of them on the output, will greatly increase optimality and efficiency of the solutions. Also, we show that implementing our approach and omitting some cumbersome and useless terms, can decrease the problem complexity and raising solutions' efficiency.

References

1. Hall, L., Harris, B., Tomes, E., Altiparmak, N.: Big data aware virtual machine placement in cloud data centers. In: BDCAT 2017, Austin, Texas, USA, 5–8 December 2017 (2017)
2. Cisco Global Cloud Index: Forecast and methodology, 2016–2021, Cisco Systems
3. Yang, S.-J., Chen, Y.-R.: Design adaptive task allocation scheduler to improve MapReduce performance in heterogeneous clouds. J. Netw. Comput. Appl. **57**, 61–70 (2015)
4. Sakr, S., Liu, A., Fayoumi, A.G.: The family of MapReduce and large scale data processing systems. ACM Comput. Surv. (CSUR) **46**(1) (2013). Article no. 11
5. Xu, H., Liu, W., Shu, G., Li, J.: LDBAS: location-aware data block allocation strategy for HDFS-based applications in the cloud. KSII Trans. Internet Inf. Syst. **12**(1), 204–226 (2018)
6. Usmani, Z., Singh, S.: A survey of virtual machine placement techniques in a cloud data center. In: International Conference on Information Security & Privacy (ICISP 2015), Nagpur, India, 11–12 December 2015 (2015)
7. Attaoui, W., Sabir, E.: Multi-criteria virtual machine placement in cloud computing environments: a literature review. Cornell University, Computer Science, Networking and Internet Architecture (2018)
8. Palanisamy, B., Singh, A., Liu, L., Jain, B.: Purlieus: locality-aware resource allocation for MapReduce in a cloud. In: SC '11: Proceedings of 2011 International Conference for High Performance Computing, Networking, Storage and Analysis, WA, USA, 12–18 November 2011 (2011)
9. Li, M., Subhraveti, D., Butt, A.R., Khasymski, A., Sarkar, P.: CAM: a topology aware minimum cost flow based resource manager for MapReduce applications in the cloud. In: HPDC 2012, Delft, The Netherlands, 18–22 June 2012 (2012)
10. Kuo, J.-J., Yang, H.-H., Tsai, M.-J.: Optimal approximation algorithm of virtual machine placement for data latency minimization in cloud systems. In: IEEE INFOCOM 2014 - IEEE Conference on Computer Communications (2014)
11. Shabeera, T.P., Kumar, S.D.M., Salam, S.M., Krishnan, K.M.: Optimizing VM allocation and data placement for data-intensive applications in cloud using ACO metaheuristic algorithm. Eng. Sci. Technol. Int. J. **20**, 616–628 (2017)
12. Guerrero, C., Lera, I., Bermejo, B., Juiz, C.: Multi-objective optimization for virtual machine allocation and replica placement in virtualized hadoop. IEEE Trans. Parallel Distrib. Syst. **29** (11), 2568–2581 (2018)

13. Guzek, M., Bouvry, P., Talbi, E.-G.: A survey of evolutionary computation for resource management of processing in cloud computing. IEEE Comput. Intell. Mag. **10**(2), 53–67 (2015)
14. Zhang, J., Huang, H., Wang, X.: Resource provision algorithms in cloud computing: a survey. J. Netw. Comput. Appl. **64**, 23–42 (2016)
15. Mann, Z.A.: Allocation of virtual machines in cloud data centers – a survey of problem models and optimization algorithms. ACM Comput. Surv. **48**(1) (2015)
16. Mann, Z.A., Szabo, M.: Which is the best algorithm for virtual machine placement optimization? Concurr. Comput. Pract. Exp. **29**, e4083 (2017)
17. Li, X.: An energy aware green spine switch management system in spine-leaf datacenter networks. A thesis for the degree of Master of Applied Science in Electrical and Computer, Carleton University (2014)

A Survey on Load Balancing in Cloud Systems for Big Data Applications

Arman Aghdashi[✉] and Seyedeh Leili Mirtaheri

Department of Electrical and Computer Engineering,
Kharazmi University, Tehran, Iran
arman.aghdashi@gmail.com, Mirtaheri@khu.ac.ir

Abstract. Today's ever-growing information world, in which we witness the juggernaut of information explosion stemming from social networks, medical records, diverse medias, IoT, and so forth, has called for a solution—encompassing boundless resources for this voluminous information's storing as well as processing in a distributed manner. To do so, although cloud computing has come up with an applicable remedy, it has overwhelmingly required a well-defined load-balancing mechanism, lifeblood of any given distributed system; a load-balancing algorithm has consistently strove to pinpoint overloaded nodes so as to disseminate and shift the burden of extra workload towards the under-loaded ones—by which the overall system performance in terms of resource utilization, throughput, cost, and response time will be guaranteed after all. In the interests of placing a high premium on load-balancing issue in distributed systems, in this study, we have provided a review concerning load-balancing algorithms in cloud environment for Big Data environment.

Keywords: Cloud computing · Load balancing · Task scheduling · Distributed

1 Introduction

Cloud computing is a novel processing paradigm in which there is a pool of stupendous virtual resources, and it will be act as a unified service through a communication network, including LAN and the Internet [1]; the supreme goal of cloud computing is to cater to different clients by omission of any prerequisite to any particular hardware, let alone knowledge of resources' locations. As different cloud services like storage, servers, and applications for diverse organizations are attainable through the Internet, cloud computing is called a computing platform based on the Internet [2,3].

Cloud computing–a distributed storage and processing paradigm—enables clients to have on-demand access to manifold resources like memory, disk, network, application, and so on, and with regard to their consumption, they should pay accordingly. Today, cloud computing has entailed disparate services, including SaaS, PaaS, IaaS, Data as a Service, and Big Data as a Service [4]. Cloud

© Springer Nature Switzerland AG 2019
L. Grandinetti et al. (Eds.): TopHPC 2019, CCIS 891, pp. 156–173, 2019.
https://doi.org/10.1007/978-3-030-33495-6_13

computing encounters different issues like Big Data management, resource allocation, and load-balancing, to name but a few, the last of which, namely load-balancing acting as a deterrent to load-imbalance situation, is so challenging, since distributing workload among various systems nodes in an efficient manner to reach the fullest performance and minimum response time, namely Quality of Service (QoS) metrics are critical for any given system [5–7]; given this point, devising an efficacious load-balancing mechanism is of paramount importance to any given distributed system

With emergence of new trend—Big Data coupled with such a myriad of issues as load-balancing, it goes without saying that any given Big Data management and processing systems are in desperate need of cloud computing infrastructure, as cloud computing has offered scalability and provided swift access to data in order to fulfill optimum performance [8–11]; therefore, Big Data systems derive great benefits from cloud computing. [1,3] accompanying with an efficient load balancing unit Load-balancing is a key component in cloud computing domain. Workload should be dissipated among nodes so as to foster system performance [8,12], and if task allocation as well as migration are not conducted delicately, the systems will end up in a situation, in which some processes undergo elongated time as well as delay due mainly to task migration procedure in comparison to their run-time, causing performance degradation and prolonged response time respectively.

Cloud-based load-balancing algorithm has received client's request, and then it has made up their mind to map each task to the appropriate machine for processing. Data center controller and VM manager are in charge of task-management, and virtual machine management respectively. Furthermore, the main purpose of virtualization, predominant concept/technology of cloud systems, is to share hardware/software resources among virtual machines which are software implementation of the host computer in which miscellaneous operating systems as well as applications can run. More importantly, resource allocation is of indispensable importance in cloud environment, as it plays a critical role in system performance, whose clients, the request of which randomly runs by virtual machines, can reside anywhere in the world [13].

Architecture of a system dedicated for Big Data processing should entail diverse services for big-data management, decision-making, gathering, and analyzing the information [8,14]. In this regard, for transferring Big Data processing in cloud environment, utilizing the whole resources is so crucial in order to reach splendid performance, reduced response time, reduced run-time, and a plunge in energy consumption so as to realize green computing's aims [6,15]. To address these goals, new load-balancing algorithms or customized ones should be maintained [12]. All in all, due to immense value as well as importance of load-balancing in cloud environment, this paper has provided a review on some load-balancing mechanisms in cloud systems with concern of Big Data processing. The rest of the paper is organized as follows; the consecutive section provides a review encompassing load-balancing algorithms classification and its attendant policies as well as metrics—putting forth our own load-balancing classification into three imperative classes, namely General, Big Data-based, and Heuristic. Ultimately, in Sect. 2, we conclude the survey.

Effective Factors in Designing Load Balancing Algorithms. Load-balancer should perform resource management and task scheduling, both of which profoundly affect the consecutive factors: resource availability, resource performance, and cost-rate in resource allocation process. Therefore, by virtue of studying different research resources, we ascertain that drawing up a decent load-balancing urges for scrutinizing the following three groups—load-balancing algorithms' predominant purposes, metrics, and issues—in a detailed manner.

Load-Balancing Algorithm's Main Purposes [16]

- Fortifying Performance

- Job Equality

- System Stability

- Cost-effectiveness

- Incorporating task priority

- Fault-tolerance

- Considering Future system alternations

For assessing the algorithms based on the mentioned purpose, some Underlying Metrics [5,17] such as Performance, Throughput, Response Time, Scalability and Migration Time should be considered. In addition, In cloud-based load-balancing algorithms—upon receiving client's re-quest, load-balancer starts dispensing requests among virtual machines, various factors, including performance, response time, scalability, throughput, resource utilization, fault tolerance, migration time, coupled overhead, cloud elasticity have been considered. Furthermore, for devising efficient algorithms such parameters as energy consumption should also be included as well. In this regard, Green Computing takes into account some policies to reduce energy consumption [18], and preserves natural environment in order to provide a green computing environment [19]. Given this point, load-balancing can be considered as a viable solution for energy saving in cloud environment to reach the optimal usage of resources, and to do so, two below factors should be deemed:

- Reducing Energy Consumption

- Reducing Carbon Emission

Cloud-Based Load-Balancing Algorithms' Challenges

- Virtual machine migration

- Spatially distributed nodes in a cloud

- Single point of failure

- Algorithm complexity

- Emergence of small data centers in cloud computing

- Energy management

In light of the above-mentioned factors, when it comes to the load-balancing process, the following key concepts should be pondered and perused:

- In distributed environment, communication channels possess limited bandwidth, and processing units may be located far-away hence, load-balancer should decide whether or not task migration should be employed.

- One job may be considered indivisible so as to breaking down into a couple of small tasks.

- Each job can encircle some relevant tasks, each of which has serrated run-time, so considering all of them as a big-picture can enhance the overall systems performance.

- Workload of each processor can be fluctuating according to a client's job status.

- Each processor enjoys heterogeneous capacity in terms of disk, operating system, CPU speed, memory, and so on.

In light of the above-mentioned factors, dynamic load balancing mechanism in cloud computing comprises of four stages [20]:

- Monitoring load and processor status

- Exchanging load information and node status between processor

- Calculating new job distribution

- Deeming correlation among various tasks in any given job

1.1   Categorization of Load-Balancing Algorithms

To put it in a nutshell, load-balancer is a component receiving submitted tasks, and soon afterwards it embarks on assigning each task to a more appropriate VM. It goes without saying that if some VMs are swamped with loads of work, whereas other VMs are free or even under-loaded, experiencing performance degradation will be inevitable. Figure 1 has evidently demonstrated the classification of the existing load-balancing mechanisms. Some studies have classified load-balancing algorithms into two foremost groups, namely system-state, and process's initiator, the last of which is as either sender initiator (load-balancing decision making based on task initiation), receiver initiator (load-balancing decision making based on task completion), or symmetric one [17,21,22].

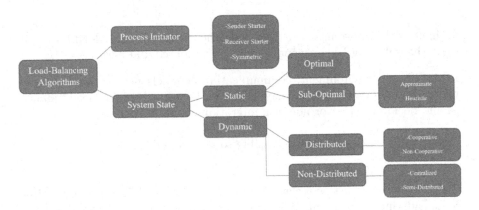

Fig. 1. Taxonomy of load-balancing algorithms.

Load-balancing algorithms predicated upon systems status have been classified as either static or dynamic. Static one has disseminated the workload equally among nodes [10], and they have asked for knowledge regarding system resources ahead of time [23].

Static algorithms such as Round Robin algorithm is based on master/slave concept in which each node's performance is determined before running real task, and this information is passed on to master node, and afterwards the master node has scheduled tasks to slave nodes in compile-time based on nodes' performance [24]; in fact, no alternation is applied during run-time. Static mechanisms confront lower overhead, in that task migration is not applicable during run-time—the pivotal features of static algorithms are strictly conformity to well-defined regulations, lack of pliability, and requisition to prior knowledge of memory, users' requirements, and node capacity With regard to load-balancing decision, static algorithms are either optimal if the load-balancer has been capable to make an optimal decision swiftly, or semi-optimal, if the other way around.

Dynamic load-balancing algorithms, tailored to satisfy cloud-based environments' requirements, have opted for the nodes whose load is the lowest one by considering the current systems status during run-time, thus ensuring better performance as compared to that of static load-balancing algorithms [21, 25]; however, monitoring workload of each node regularly imposes extra overhead for each CPU cycle. Dynamic algorithms are divided in two classes: distributed, and non-distributed, and in the former one, all system nodes run the load-balancing algorithm, and thus load-balancing duty will be shared among them, and with this in minds, prior to any task migration step, three steps have occurred, overseeing workload of resources, interchanging of current system status, and appraising re-balancing condition for any new task [21, 22, 26].

Grounded on nodes' relationships during their collaboration to accomplish load-balancing task, the existing dynamic algorithms are split up into two categories—cooperative or non-cooperative. In cooperative, nodes work together for fulfilling one shared goal such as response time reduction, whereas in non-cooperative one, each node has worked independently toward its local aim like response time reduction for a local task [17].

Non-distributed algorithms are classified as either centralized or semi-distributed. In centralized version, there is a central node, this exclusive node in the charge of load-balancing, and launches load-balancing algorithm. In semi-distributed one, all system nodes are distributed into some clusters, and each cluster has reached the load-balancing state.

1.2   Load-Balancing Algorithm's Policies

Dynamic load balancing algorithm has utilized the current system status by virtue of some policies [21, 27] as follows:

– Transfer Policy

T policy—incorporating migration as well as rescheduling of tasks is interlinked with the predefined regulations and the current amount of each node's workload in order to make a decision when a task migration should occur.

– Selection policy

By taking into account parameters, including plausible overhead, and task execution-time, this policy has opted for a task called for migration.

– Location Policy

The applicable node for shouldering the responsibility of task from overloaded nodes has been identified by this policy. This policy also verifies such services as task migration and task rescheduling in destined node as well.

– Information Policy

This policy has accumulated information regarding all system nodes, which in turn, utilizes by other policies as well. Relationship between diverse policies is as follows: Firstly, transfer policy has decided whether each entered task should be processed locally or migrated to a remote node for load-balancing purpose. If latter is true, the location policy has tried to find a remote node for task processing, and if the desired node has not been found, the task will be processed locally. Information policy furnishes the two previous policies with required information for decision making [28].

Both static and dynamic task allocation algorithms are highly based on the condition in which the workload has been scheduled during either runtime or compile-time. As study has revealed, there is not any perfect load-balancing mechanism, each of which should be applied based on the current needs and requirements, and all of which have their own merits as well as demerits. Status of any load balancing mechanism is interwoven with its run-time speed, and the imperative goal of any load-balancing mechanism is to lessen the interconnection between processes by virtue of optimizing the resource consumption.

In the following, we have surveyed some load-balancing algorithms by dividing them into three major groups, namely General Load-Balancing Algorithms—including both static as well as dynamic algorithms, Load-Balancing Algorithms for Big Data Environment, and finally Heuristic Load-Balancing Algorithms exploited in cloud computing systems:

– General Load-Balancing Algorithms

– Load-Balancing Algorithms for Big Data Environment

– Heuristic Load-Balancing Algorithms

1.3   General Load-Balancing Algorithms

Static Load-Balancing Algorithms

– Round Robin Algorithm

Round Robin realizes load-balancing in the static environment. It utilizes time-slice method by which every node in the cloud environment is fixed with time slice and every node has to wait for their turn to execute their job, and works in a round robin manner. To put it simply, it makes use of random sampling and in comparison to other existing mechanism, its complexity is low [29].

– Min-Min Algorithm

Min-Min mechanism, a static load-balancing mechanism, embarks with a set of all unassigned tasks, and afterwards it computes the least completion time for all tasks, and opt for the minimum by which the task is schedule on the related machine. Then the execution time for all other jobs is updated on that machine, and this is an iterative process; however, this approach has a major drawback leading to starvation [30,31].

Dynamic Load-Balancing Algorithms

– Active Clustering

It works based on the clustering concept in which the same type of nodes are grouped and work together. This method, a dynamic load-balancing mechanism, soars system throughput [32] uses the resources efficiently, by using re-sources effectively. There is a method called match-maker; when an execution launches in a network, the process starts finding for the next similar node, namely match-maker which should meet the criteria. This is an iterative process in the network to efficiently balance the load [33].

– Throttled Load balancing

In Throttled algorithm, a dynamic load-balancing mechanism, job manager will keep a list of node detail using index list by which it will discover the specific node for assigning the new job [34].

– Honey Bee Foraging

Dhinesh et al. have proposed a dynamic load-balancing algorithm grounded on the behavior of honey bees in finding their food. Once the food source has been found the hunter bees come back to the bee hive to announce the food source by a dance called "waggle dance". Quality, quantity, and distance of the food has been identified by the type of dance. With regard to load balancing, the servers are grouped into virtual servers, each server creates a re-quest, then it calculates its profit and assess it with the colony profit; if the profit was high, then the server live at the existing virtual server; otherwise, the server returns to the hunt or survey behavior, thus balancing the load with the server [35].

– Index Name Server Algorithm

This algorithm, utilizes Distribution Hash Table (DHT), is employed to identify data duplication as well as redundancy, and strives to ascertain whether the connection can manage additional clients [36].

– Randomized Algorithm

Randomized Algorithm (RA) takes advantage of random numbers, based on a statistic distribution, for electing computing nodes to process tasks, with the lack of owing any knowledge concerning both current and previous load status on any node [24, 37].

– Threshold Algorithm

In Threshold Algorithm (TA) tasks are assigned, when they are engendered for computation in which computing nodes are selected; each node maintains a private copy of the entire system's load information. Node load can be defined by virtue of two threshold parameters t_under and t_upper, and are classified as follows: under loaded, medium and overloaded [38]: • Under loaded: workload < t_under, • Medium: t_under workload t_upper, • Overloaded: workload > t_upper. Firstly, all computing nodes are deemed under-loaded; when a node becomes over-loaded due to exceeding the load threshold, it propagates messages concerning the new load state to all the other computing nodes, thereby regularly updating them and revealing the actual load state of the whole system to each system node. If the local node is not overloaded, then the process is allocated locally. Otherwise, a remote under-loaded processor is considered for processing.

– Equally Spread Current Execution Algorithm

Equally spread current execution algorithm, utilizing the spread spectrum technique in which it shares out the load over various nodes, allocates a job to each node with priority perspective. If the node is lightly loaded, the load balancer moves that job to that respective node and achieve high throughput. The load balancer maintains a job queue, which helps it to recognize which node is free as well as light-weighted, and need to be assigned with a new job [39].

Load-Balancing Algorithms in Big Data Environment. As researches have demonstrated, load-balancing mechanisms play a paramount role in modern distributed storage as well as processing systems, especially cloud computing systems, and by introduction of Big Data, their importance has been magnified as well. Subsequently, a thorough review of existing load-balancing mechanisms in Big Data platform, the important of which Hadoop has been given in this section.

Task scheduling should be applied in two stages: job-level, and task-level. In job-level scheduling, jobs have been opted for scheduling based on a specific strategy, while in task-level scheduling, all tasks belonging for a particular job have been scheduled. The default scheduling algorithm for Hadoop framework is FIFO strategy which can be replaced by new algorithms based on client's requirements [40]; existing scheduling algorithms in Hadoop is as follows:

Fair Scheduler: Fair Scheduler developed by Facebook has utilized a multiple-queue pool in order to schedule their tasks; as compared to FIFO algorithm preemptive one in which tasks face starvation and prolonged waiting time, Fair Scheduler is a non-preemptive algorithm [41].

Capacity Scheduler: Capacity Scheduler developed by Yahoo has ensured the fair allocation of resources among many clients, and it has used some customized queues, and in this regard, available resources have been devoted to each queue based on their priority. In fact, scheduler has selected tasks with regard to entrance time, waiting time, and Service Level Agreement (SLA)—thus, the tasks which possess the highest waiting time and lighter workload have been elected first by which the fair allocation of resources has been assured [41].

Delay Scheduler: Delay Scheduler an improved version of fair scheduler has sort-ed out the data locality problem of the fair scheduler, leading to starvation, by opt for another task to be scheduled when the node who owns the intended data is busy [42].

Longest Approximate Time to End: Longest Approximate Time to End (LATE) algorithm has scrutinized toward slow tasks—namely speculative tasks, the main reasons of which are due to high load of processor, race condition, or background process. LATE has found speculative tasks, and migrate them to other nodes by which both the whole systems performance has been enhanced and response time has been diminished—LATE has improved Hadoop's performance as well [43].

Deadline Constraint Scheduler: Deadline Constraint Scheduler is concocted with regard to limitations imposed from clients, and the pivotal goals of this scheduler are as follows:

- Providing the clients with an instant feedback regarding turnaround time for completing their task.

- Maximizing the number of tasks can be accomplished in order to meet their deadlines.

Some similar algorithms have also been proposed by companies, including Google, Microsoft, and Facebook in this area elaborated upon their functionality and architectures accordingly [44].

Cloud Scheduler: enjoying five pivotal scheduler architectures: monolithic, two-level, shared-state, distributed, and hybrid– has multiple goals, including using the cluster's resources effectively, working with user-supplied restraints, and meeting the acceptable level of fairness.

Monolithic schedulers: not supporting parallel tasks, like Google Borg are a static centralized schedulers in which scheduling decision-making process is done by merely one scheduler. Google Borg owns two priorities bands, namely high and low one, and hence all jobs are statistically scheduled [45].

Apache YARN and Mesos Schedulers: Apache YARN, two-level scheduler [46], and Mesos [47] are centralized as well as static schedulers splitting the resource allocation from the task management duties. In these systems, there is one resource manager that grants resources to multiple autonomous schedulers, in which each scheduler has specific policies for resources queuing and resource allocation pertaining to the user preferences. These type of schedulers are less complex as compared to the Monolithic ones, since they have delegated the pre-application scheduling work to the applications themselves, and in the meantime handling the dissemination of re-sources among applications parallel to striking fairness.

Shared-state schedulers: is a successor to the two-level schedulers, such as Google Omega [48] centralized and static schedulers owns a shared state of the cluster state; all available resources in the cluster are offered to the applications, and then conflicts are resolved at task execution-time. All applications schedulers enjoyed the similar view of the cluster resources surging the scheduling performance as well as resources utilization. Applications with high priority are allowed to preempt lower one.

Sparrow: distributed schedulers like Sparrow [49] is the same as multiple isolated schedulers to serve the incoming workload, each of which works with its local, and partial knowledge of the cluster, so it is hard to reach global fairness.

Microsoft Mercury: a hybrid schedulers whose goal is to address all flaws of the previous architectures, and involves two strategies to handle the workload [50]:

- Group of distributed schedulers for very short and/or low-priority tasks
- One centralized scheduler for the rest of workload (high priority tasks and/or long tasks like system services)

Table 1 has provided a brief comparison between the above-mentioned load-balancing mechanisms for Big Data category.

Table 1. Overview of load balancing scheduling mechanisms in big data systems

Algorithm	Advantages	Disadvantages
FIFO	-Simple implementation -Efficient	-Only one job at a time uses cluster resources -No considering priority or job size -Low data locality when running a small job on cluster
Fair	-Distributes resources among jobs fairly -Fast response time for both short and long jobs -Improves the QOS through job classification	-Does not consider the actual workload of nodes for task scheduling which may result imbalance
Capacity	-Improves resources utilization -Flexible and efficient -It can be used in large clusters	-Should acquire large amount of information about system to set a queue
Delay	-Job response time may decrease	-Killing speculative tasks is instantaneous and wastes the work performed by them -Does not ensure reliability
Longest Approximate Time to End	-Suitable for heterogeneous environment -Improves the performance by reducing response time	-It uses the past information -Is not suitable for environment with dynamic loading -It does not ensure reliability
Deadline Constraint	-Focus on increasing system -Works better for large clusters utilization	-Homogeneous nodes are assumed Dynamic -job deadline can affect the response time -the job may not execute in the system due to its deadline constraints

(continued)

Table 1. (*continued*)

Algorithm	Advantages	Disadvantages
Google Borg	-High utilization by combining admission control, efficient task-packing, over-commitment, and machine sharing with process-level performance isolation -It supports high-availability applications with run-time features that minimize fault-recovery time	-Jobs are restrictive as the only grouping mechanism for tasks
Yarn	-Better utilization of resources -Scheduling and resource management capabilities are separated from the data processing component -Permits simultaneous execution of a variety of programming models	-Resource allocation and management was poor that would make the whole system inconsistent
Mesos	-Simple: easier to scale and make resilient -Easy to port existing frameworks, support new ones	-Scheduling decision is not optimal
Google Omega	-Supporting diverse, independent scheduling frameworks on a single shared cluster -Uses an optimistic concurrency control model for updating shared cluster state about resource allocation	-Scalability has not yet been a problem
Sparrow	-Handles per-job constraints -Use multiple queues to handle different resource allocation policies	-It does not handle crashes of the schedulers or the worker (If a scheduler crashes all the tasks associated with it will be lost) -Does not handle precedence constraints
Microsoft Mercury	-Improved job latency and throughput -Works well even for heterogeneous workloads that share resources at finer granularity -supports the full spectrum of scheduling, from centralized to distributed	-There are fewer disadvantages than other cases -The problems of the previous cases have somewhat improved

1.4   Heuristic Load-Balancing Algorithms

In this section, we review some load-balancing mechanisms fallen within the heuristic category, these mechanisms have tried to find a near-optimum solution by virtue of Genetic Algorithms, PSO, Ant Colony, and Bee Colony strategies so as to culminate in a better result [51].

Dam et al. [52] has introduced an algorithm, based on Xen cloud technology (hypervisor), dynamically comparing and balancing resources as well as loads among servers. The algorithm regularly checks the CPU as well as RAM utilization, and when any resource is required, it will provide via scaling out; however, if the resources is not attainable then migration will occur.

Zhao et al. [53] has put forth a load-balancing mechanism grounded on GA as well as gel (gravitational emulation local search) to resolve load issues among

VMs—GA uses a global neural network as well as two fitness functions to apply mutation, crossover, and selection.

Zhang et al. [54] suggested an algorithm to assesses and balances load based on sampling to strike an equilibrium solution whereby it dwindles the migration time and fulfills the zero downtime.

Zhu et al. [55], proposed an algorithm based on PSO has allotted VMs grounded on their status to improve the response time; it uses CloudSim technology for the implementation.

Nishant et al. [56] has mentioned an algorithm in based on ACO a mixture of ant colony and complex network theory for OCCF (open cloud computing federation) to realize load balancing in a distributed environment, thus getting max performance.

Yao et al. [57], proposed a multi-agent Genetic Algorithm (maGA), is based on GA for reaching good performance, it is a hybrid Algorithm combining GA and multi-agent techniques. Algorithm ACO-based proposed in [57] distributes workload among nodes in which ants gradually establish the best result set by updating mechanism.

Aslanzadeh et al. [58] proposed an algorithm based on mimicking of behavior of honey bees so as to optimize the amount of nectar (throughput) to obtain the maximum throughput.

Sun et al. [59], has been proposed an algorithm grounded on PSO parallel to using endocrine method spurred from behavior of human's hormone system; load-balancing achieve by applying self-organizing method among overloaded VMs. This technique is structured and based on communications between VMs, helping the overloaded Vms to transfer extra tasks to another under loaded VMs by running the enhanced feed backing approach using PSO.

Wen et al. [60] has suggested a load-balancing mechanism grounded on ACO a self-adaptive ant colony optimization task-scheduling algorithm to improve calculation as well as updating of pheromone.

Pan et al. [61] has posited a ACO based VM migration algorithm in which local migration agent independently oversees resource utilization and run migration method by virtue of both old as well as new information regarding system condition. This algorithm, considered as a dynamic load-balancing strategy, operates two diverse traversing strategies for ants to pinpoint the near-optimal mapping connection between virtual machines (VM) and physical machines (PMs).

Babu et al. [62] has taken into consideration the features of complex net-works to define a resource-task allocation paradigm, this load-balancing mechanism is based on PSO, and takes advantages of definition of particle's position as well as velocity parallel to updating methods to amend its fitness value as well.

Wang et al. [63] With regard to foraging attitude of honey bees, has pro-posed a load-balancing algorithm according to BCO to balance load among miscellaneous VMs—under-loaded and over-loaded VMs are deemed as food resources, and honey bees respectively.

Rana et al. [64] has put forwarded a dynamic load-balancing algorithm on account of double-fitness adaptive method—job spanning time as well as load bal-ancing genetic algorithm (jlGA). Not only does this mechanism operate on task scheduling sequence, but it also strives to fulfill load-balancing issues between inter-nodes, and moreover, this method firstly applies a greedy algorithm to initialize the population, and utilize variance so as to define load intensive between diverse nodes to use multi-fitness function.

Gupta et al. [65] One mechanism has suggested to maximize resource utilization, and enhance the performance by identifying overloaded as well as underloaded servers, and then run load-balancing mechanisms with regard to destined servers.

Li et al. [66] The algorithm proposed is utilized to balance the whole system performance, and it is grounded on ACO—cloud-task scheduling policy—based on ant colony optimization load balancing technique (lbACO), a dynamic load-balancing mechanism.

Kaur et al. [67] has put forth an algorithm, mainly focusing on tasks and resources, pertaining to ACO in order to wane Makespan balancing. In comparison to previous algorithms, this mechanism has divulged better processing time, and it is more cost-effective.

2   Conclusion

Stupendous machine age in which we live has faced a drastic upsurge in information generation, so storing as well as processing this massive information has insisted a rightful load-balancing algorithm in order to rest assured the whole system performance along the way. Given this fact, and due to substantial importance of load-balancing algorithms, this paper has surveyed existing load-balancing mechanisms as well as its co-founding parameters in cloud environment.

References

1. Graham-Rowe, D., et al.: Big data: science in the petabyte era. Nature **455**(7209), 8–9 (2008)
2. Neves, P.C., Schmerl, B.R., Cámara, J., Bernardino, J.: Big data in cloud computing: Features and issues. In: IoTBD, pp. 307–314 (2016)
3. Mell, P., Grance, T., et al.: The NIST definition of cloud computing (2011)
4. Job, M.A.: Big data-as-a-service (BDaaS) in cloud computing environments
5. Patel, N., Chauhan, S.: A survey on load balancing and scheduling in cloud computing. Int. J. Sci. Res. Dev. **1**, 185–189 (2015)
6. Singh, A., Juneja, D., Malhotra, M.: Autonomous agent based load balancing algorithm in cloud computing. Procedia Comput. Sci. **45**, 832–841 (2015)
7. Yadav, V.K., Yadav, M.P., Yadav, D.K.: Reliable task allocation in heterogeneous distributed system with random node failure: load sharing approach. In: 2012 International Conference on Computing Sciences, pp. 187–192. IEEE (2012)
8. Fox, G., Qiu, J., Jha, S., Ekanayake, S., Kamburugamuve, S.: Big data, simulations and HPC convergence. In: Rabl, T., Nambiar, R., Baru, C., Bhandarkar, M., Poess, M., Pyne, S. (eds.) WBDB -2015. LNCS, vol. 10044, pp. 3–17. Springer, Cham (2016). https://doi.org/10.1007/978-3-319-49748-8_1

9. Garg, S.K., Yeo, C.S., Anandasivam, A., Buyya, R.: Environment-conscious scheduling of HPC applications on distributed cloud-oriented data centers. J. Parallel Distrib. Comput. **71**(6), 732–749 (2011)
10. Katyal, M., Mishra, A.: A comparative study of load balancing algorithms in cloud computing environment. arXiv preprint arXiv:1403.6918 (2014)
11. Mata-Toledo, R., Gupta, P.: Green data center: how green can we perform. J. Technol. Res. Acad. Bus. Res. Inst. **2**(1), 1–8 (2010)
12. Khiyaita, A., El Bakkali, H., Zbakh, M., El Kettani, D.: Load balancing cloud computing: state of art. In: 2012 National Days of Network Security and Systems, pp. 106–109. IEEE (2012)
13. Hwang, K., Dongarra, J., Fox, G.C.: Distributed and Cloud Computing: From Parallel Processing to the Internet of Things. Morgan Kaufmann, Burlington (2013)
14. Lohr, S.: The age of big data. New York Times **11**, 2012 (2012)
15. Kansal, N.J., Chana, I.: Cloud load balancing techniques: a step towards green computing. IJCSI Int. J. Comput. Sci. Issues **9**(1), 238–246 (2012)
16. Escalante, D., Korty, A.J.: Cloud services: policy and assessment. Educause Rev. **46**(4) (2011)
17. Rastogi, G., Sushil, R.: Analytical literature survey on existing load balancing schemes in cloud computing. In: 2015 International Conference on Green Computing and Internet of Things (ICGCIoT), pp. 1506–1510. IEEE (2015)
18. Kabiraj, S., Topkar, V., Walke, R.C.: Going green: a holistic approach to transform business. arXiv preprint arXiv:1009.0844 (2010)
19. Baliga, J., Ayre, R.W.A., Hinton, K., Tucker, R.S.: Green cloud computing: balancing energy in processing, storage, and transport. Proc. IEEE **99**(1), 149–167 (2010)
20. Kushwaha, M., Gupta, S.: Various schemes of load balancing in distributed systems–a review. Int. J. Sci. Res. Eng. Technol. (IJSRET) **4**(7), 741–748 (2015)
21. Jafarnejad Ghomi, E., Masoud Rahmani, A., Nasih Qader, N.: Load-balancing algorithms in cloud computing: a survey. J. Netw. Comput. Appl. **88**, 50–71 (2017)
22. Rathore, N., Chana, I.: Load balancing and job migration techniques in grid: a survey of recent trends. Wirel. Pers. Commun. **79**(3), 2089–2125 (2014)
23. Shah, N., Farik, M.: Static load balancing algorithms in cloud computing: challenges & solutions. Int. J. Sci. Technol. Res. **4**(10), 365–367 (2015)
24. El-Zoghdy, S.F., Ghoniemy, S.: A survey of load balancing in high-performance distributed computing systems. Int. J. Adv. Comput. Res. **1** 2014
25. Mirtaheri, S.L., Grandinetti, L.: Dynamic load balancing in distributed exascale computing systems. Cluster Comput. **20**(4), 3677–3689 (2017)
26. Khaneghah, E.M., Nezhad, N.O., Mirtaheri, S.L., Sharifi, M., Shirpour, A.: An efficient live process migration approach for high performance cluster computing systems. In: Pichappan, P., Ahmadi, H., Ariwa, E. (eds.) INCT 2011. CCIS, vol. 241, pp. 362–373. Springer, Heidelberg (2011). https://doi.org/10.1007/978-3-642-27337-7_34
27. Sharma, G.: A review on different approaches for load balancing in computational grid. J. Glob. Res. Comput. Sci. **4**(4), 82–85 (2013)
28. Arab, M.N., Mirtaheri, S.L., Khaneghah, E.M., Sharifi, M., Mohammadkhani, M.: Improving learning-based request forwarding in resource discovery through load-awareness. In: Hameurlain, A., Tjoa, A.M. (eds.) Globe 2011. LNCS, vol. 6864, pp. 73–82. Springer, Heidelberg (2011). https://doi.org/10.1007/978-3-642-22947-3_7
29. Samal, P., Mishra, P.: Analysis of variants in round robin algorithms for load balancing in cloud computing. Int. J. Comput. Sci. Inf. Technol. **4**(3), 416–419 (2013)

30. Al Nuaimi, K., Mohamed, N., Al Nuaimi, M., Al-Jaroodi, J.: A survey of load balancing in cloud computing: challenges and algorithms. In: 2012 Second Symposium on Network Cloud Computing and Applications, pp. 137–142. IEEE (2012)
31. Wang, S.-C., Yan, K.-Q., Liao, W.-P., Wang, S.-S.: Towards a load balancing in a three-level cloud computing network. In: 2010 3rd International Conference on Computer Science and Information Technology, vol. 1, pp. 108–113. IEEE (2010)
32. Sahu, Y., Pateriya, R.K.: Cloud computing overview with load balancing techniques. Int. J. Comput. Appl. **65**(24) (2013)
33. Kokilavani, T., Amalarethinam, D.I.G., et al.: Load balanced min-min algorithm for static meta-task scheduling in grid computing. Int. J. Comput. Appl. **20**(2), 43–49 (2011)
34. Jamuna, P., Kumar, R.A.: Optimized cloud partitioning technique to simplify load balancing. Int. J. Adv. Res. Comput. Sci. Softw. Eng. **3**(11), 820–822 (2013)
35. LD, D.B., Krishna, P.V.: Honey bee behavior inspired load balancing of tasks in cloud computing environments. Appl. Soft Comput. **13**(5), 2292–2303 (2013)
36. Wu, T.-Y., Lee, W.-T., Lin, Y.-S., Lin, Y.-S., Chan, H.-L., Huang, J.-S.: Dynamic load balancing mechanism based on cloud storage. In: 2012 Computing, Communications and Applications Conference, pp. 102–106. IEEE (2012)
37. Mühlenbein, H., Schlierkamp-Voosen, D.: Predictive models for the breeder genetic algorithm i. Continuous parameter optimization. Evol. Comput. **1**(1), 25–49 (1993)
38. Grosu, D., Chronopoulos, A.T.: Noncooperative load balancing in distributed systems. J. Parallel Distrib. Comput. **65**(9), 1022–1034 (2005)
39. Lingawar, R.P., Srode, M.V., Ghonge, M.M.: Survey on load-balancing techniques in cloud computing. Int. J. Advent Res. Comput. Electron. **1**(3), 18–21 (2014)
40. Bareen, S., Shinde, K., Borde, S.: Challenges of big data processing and scheduling of processes using various hadoop schedulers: a survey. Int. J. Multifaceted Multilingual Stud. **3**(12) (2017)
41. Zaharia, M.: Job scheduling with the fair and capacity schedulers. Hadoop Summit **9** (2009)
42. Zaharia, M., Borthakur, D., Sen Sarma, J., Elmeleegy, K., Shenker, S., Stoica, I.: Delay scheduling: a simple technique for achieving locality and fairness in cluster scheduling. In: Proceedings of the 5th European Conference on Computer Systems, pp. 265–278. ACM (2010)
43. Lee, K.-H., Lee, Y.-J., Choi, H., Chung, Y.D., Moon, B.: Parallel data processing with mapreduce: a survey. AcM sIGMoD Rec. **40**(4), 11–20 (2012)
44. Kc, K., Anyanwu, K.: Scheduling hadoop jobs to meet deadlines. In: 2010 IEEE Second International Conference on Cloud Computing Technology and Science, pp. 388–392. IEEE (2010)
45. Verma, A., Pedrosa, L., Korupolu, M., Oppenheimer, D., Tune, E., Wilkes, J.: Large-scale cluster management at Google with Borg. In: Proceedings of the Tenth European Conference on Computer Systems, p. 18. ACM (2015)
46. Schwarzkopf, M., Konwinski, A., Abd-El-Malek, M., Wilkes, J.: Omega: flexible, scalable schedulers for large compute clusters (2013)
47. Ousterhout, K., Wendell, P., Zaharia, M., Stoica, I.: Sparrow: distributed, low latency scheduling. In: Proceedings of the Twenty-Fourth ACM Symposium on Operating Systems Principles, pp. 69–84. ACM (2013)
48. Karanasos, K.: Mercury: hybrid centralized and distributed scheduling in large shared clusters. In: 2015 {USENIX} Annual Technical Conference ({USENIX}{ATC} 15), pp. 485–497 (2015)
49. Vavilapalli, V.K., et al.: Apache hadoop yarn: yet another resource negotiator. In: Proceedings of the 4th Annual Symposium on Cloud Computing, p. 5. ACM (2013)

50. Achar, R., Thilagam, P.S., Soans, N., Vikyah, P.V., Rao, S., Vijeth, A.M.: Load balancing in cloud based on live migration of virtual machines. In: 2013 Annual IEEE India Conference (INDICON), pp. 1–5. IEEE (2013)

51. Daraghmi, E.Y., Yuan, S.-M.: A small world based overlay network for improving dynamic load-balancing. J. Syst. Softw. **107**, 187–203 (2015)

52. Dam, S., Mandal, G., Dasgupta, K., Dutta, P.: Genetic algorithm and gravitational emulation based hybrid load balancing strategy in cloud computing. In: Proceedings of the 2015 Third International Conference on Computer, Communication, Control and Information Technology (C3IT), pp. 1–7. IEEE (2015)

53. Zhao, Y., Huang, W.: Adaptive distributed load balancing algorithm based on live migration of virtual machines in cloud. In: 2009 Fifth International Joint Conference on INC, IMS and IDC, pp. 170–175. IEEE (2009)

54. Zhang, Z., Zhang, X.: A load balancing mechanism based on ant colony and complex network theory in open cloud computing federation. In: 2010 The 2nd International Conference on Industrial Mechatronics and Automation, vol. 2, pp. 240–243. IEEE (2010)

55. Zhu, K., Song, H., Liu, L., Gao, J., Cheng, G.: Hybrid genetic algorithm for cloud computing applications. In: 2011 IEEE Asia-Pacific Services Computing Conference, pp. 182–187. IEEE (2011)

56. Nishant, K., et al.: Load balancing of nodes in cloud using ant colony optimization. In: 2012 UKSim 14th International Conference on Computer Modelling and Simulation, pp. 3–8. IEEE (2012)

57. Yao, J., He, J.: Load balancing strategy of cloud computing based on artificial bee algorithm. In: 2012 8th International Conference on Computing Technology and Information Management (NCM and ICNIT), vol. 1, pp. 185–189. IEEE (2012)

58. Aslanzadeh, S., Chaczko, Z.: Load balancing optimization in cloud computing: applying endocrine-particale swarm optimization. In: 2015 IEEE International Conference On Electro/Information Technology (Eit), pp. 165–169. IEEE (2015)

59. Sun, W., Ji, Z., Sun, J., Zhang, N., Hu, Y.: Saaco: a self adaptive ant colony optimization in cloud computing. In: 2015 IEEE Fifth International Conference on Big Data and Cloud Computing, pp. 148–153. IEEE (2015)

60. Wen, W.-T., Wang, C.-D., Wu, D.-S., Xie, Y.-Y.: An ACO-based scheduling strategy on load balancing in cloud computing environment. In: 2015 Ninth International Conference on Frontier of Computer Science and Technology, pp. 364–369. IEEE (2015)

61. Pan, K., Chen, J.: Load balancing in cloud computing environment based on an improved particle swarm optimization. In: 2015 6th IEEE International Conference on Software Engineering and Service Science (ICSESS), pp. 595–598. IEEE (2015)

62. Babu, K.R.R., Joy, A.A., Samuel, P.: Load balancing of tasks in cloud computing environment based on bee colony algorithm. In: 2015 Fifth International Conference on Advances in Computing and Communications (ICACC), pp. 89–93. IEEE (2015)

63. Wang, T., Liu, Z., Chen, Y., Xu, Y., Dai, X.: Load balancing task scheduling based on genetic algorithm in cloud computing. In: 2014 IEEE 12th International Conference on Dependable, Autonomic and Secure Computing, pp. 146–152. IEEE (2014)

64. Rana, M., Bilgaiyan, S., Kar, U.: A study on load balancing in cloud computing environment using evolutionary and swarm based algorithms. In: 2014 International Conference on Control, Instrumentation, Communication and Computational Technologies (ICCICCT), pp. 245–250. IEEE (2014)

65. Gupta, E., Deshpande, V.: A technique based on ant colony optimization for load balancing in cloud data center. In: 2014 International Conference on Information Technology, pp. 12–17. IEEE (2014)
66. Li, K., Xu, G., Zhao, G., Dong, Y., Wang, D.: Cloud task scheduling based on load balancing ant colony optimization. In 2011 Sixth Annual ChinaGrid Conference, pp. 3–9. IEEE (2011)
67. Kaur, R., Ghumman, N.: Hybrid improved max min ant algorithm for load balancing in cloud. In: Proceedings of the International Conference on Communication, Computing and Systems (CCS 2014), pp. 188–191 (2014)

Internet of Things

A Comprehensive Fog-Enabled Architecture for IoT Platforms

Malihe Asemani[1(✉)], Fatemeh Jabbari[1], Fatemeh Abdollahei[1], and Paolo Bellavista[2]

[1] Iran Telecommunication Research Center, Tehran, Iran
{asemani, f.jabbari, f.abdollahei}@itrc.ac.ir
[2] Alma Mater Studiorum - Università di Bologna, Bologna, Italy
paolo.bellavista@unibo.it

Abstract. IoT platform has a key infrastructure role in IoT. Recently, many IoT reference models and architectures have been proposed for the IoT ecosystem, typically organized in three layers, i.e., sensor, network, and platform layers. Moreover, there are various IoT platform products in the market offering different components and features, based on the solution they are providing for industry. However, to the best of our knowledge, there is not any reference architecture for IoT platform, which gathers all of the different aspects of the platform (covered by products, or proposed in academic works, and standardization efforts) in a comprehensive architecture, that also determines all of the IoT platform functional blocks, and their detailed features executed on the cloud/fog, despite of the vertical solutions which are hosted on them. In this paper, a novel comprehensive architecture is proposed, which describes components of an IoT platform executed on both cloud and fog computing resources. The proposed architecture reflects the architectural and functional attributes of existing products, as well as the components of reference models. Moreover, the functionality associated with the fog side for IoT scenarios is discussed, with specific attention to highlight the specific related attributes of primary commercial products.

Keywords: Internet of Things (IoT) · IoT platform · Architecture · Data analytics · Cloud computing · Fog/edge computing

1 Introduction

Internet of Things (IoT) is a technological concept, effective on all aspects of life. In 2016, IoT was at the top of hype cycle of Gartner. In addition, Gartner forecasts that in 2020 more than 20.8 billion connected devices will exist all around the world. ITU-T Y.2060 has defined IoT as a global infrastructure for the information society, enabling advanced services by interconnecting (physical and virtual) things based on existing and evolving interoperable information and communication technologies. Through the exploitation of identification, data capture, processing and communication capabilities, the IoT makes full use of things to offer services to all kinds of applications, whilst ensuring that security and privacy requirements are fulfilled [1].

© Springer Nature Switzerland AG 2019
L. Grandinetti et al. (Eds.): TopHPC 2019, CCIS 891, pp. 177–190, 2019.
https://doi.org/10.1007/978-3-030-33495-6_14

An IoT platform is one of the main elements of any IoT ecosystem. There are many different products in the market that implement this core element. Given the central role of IoT platforms in this context, it is very relevant to know IoT platform's strengths or weaknesses, define a comprehensive standard architecture for it, and describe its components. Although, despite of existence of some reference architecture for IoT, there is not any standard architecture for IoT platform, which discuses on the component and detail attributes of platform on three-layer architectures, that consider fog/cloud hosting of the execution, result from some standardization efforts, that also reflects IoT reference architectures building blocks, and IoT platform products' attributes, in an integrated, comprehensive, reference architecture for IoT platform. Proposing a comprehensive architecture, not only describes IoT platform and its components in detail, but also it can be used as a strategic map in implementation and evaluation of IoT platforms evolution.

In this paper, the proposed definition of IoT platform in our last article [2] is improved by providing a multi-layer architecture for IoT platform, including detailed description of the components features. First, commercial and open source IoT platform projects/products, and IoT reference architectures are discussed. Afterwards, a novel comprehensive architecture is proposed based on analyzing the discussed architectures, which describes the main components and attributes of an IoT platform, executed on both cloud and fog computing resources. Furthermore, the proposed functionalities for fog layer is discussed to highlight the specific related attributes of primary commercial products. Section 2 includes an overview of IoT reference architectures, commercial IoT platforms architectures, and open source IoT platforms architectures. In Sect. 3, the comprehensive architecture for IoT platforms is proposed, and one of the main units of the proposed architecture is analyzed in detail. Finally, Sect. 4, summarizes the main consequences of this work and describes related future work.

2 Literature Review

Many various definitions and architectures are presented in academic papers for IoT platforms. Gubbi et al. point out IoT platform should be able to read data streams from sensors or DB, process and analyses data streams, visualize events on visualization screens, as well as create and deploy IoT applications on cloud, rapidly. Cloud computing provides the virtual infrastructure for this scenario, and integrates monitoring, storage, analytics, and visualization platforms, as well as end- to-end service provisioning. However, authors do not describe all the components of IoT platform, as well as the detailed description of the pointed components or features of a cloudy IoT platform [3]. Köhler et al. determine data management, application integration, device management, identity management, development, lifecycle management, and core application, as the building blocks of an IoT platform. Although a brief list of features for some of the building blocks is provided, a detail and complete set of features is not described for all of the building blocks. In addition, some other building blocks, and other functionalities are considerable, which are not discussed in the paper. [4]. Ganchev et al. introduce key design elements of a generic IoT platform. Although an overall view of IoT platform in different fields is provided, the descriptions do not provide a detailed, technical view of IoT platform components, and features [5]. Mohan et al. present a comparison of ten cloud-based IoT platforms, based on supporting protocols, type of analytics, security protocols, integration to cloud ability, as well as the applications are used mostly on them. While lots of other criteria can be discussed in comparison of IoT platforms, and authors have not determined why the mentioned criteria are selected [6]. Guth et al. provide an analysis of IoT solutions to propose simple and general reference architecture (RA) for IoT eco-system. The core functionality of the RA is IoT Integration Middleware, which serves as an integration layer for devices, and applications, and is responsible for managing devices and users, and for receiving, processing, aggregating, and utilizing data. The proposed architecture in this paper is a simple five layers RA, which does not include IoT platform components or features [7, 8].

In addition to academic definition of IoT platform, there are some other important IoT platform architectures, includes Reference Architectures(RA), commercial products' architectures, and open source projects' architecture, which IoT platform definition and components can be extracted from them, as well.

Reference architectures of IoT ecosystems are categorized into standards and research projects. The most relevant standard is ITU-T Y.2060 [1]. This standard proposes some components such as data processing and storage, special data analytics methods, device management, network topology management, traffic congestion, and security features as the main components of IoT ecosystem. Beside standards, research projects around the world represent some other reference architectures such as Korean RA, WSO2 RA, Mulesoft RA, and Intel RA. Main aspect of Korean RA is considering development, deployment and test environment alongside other functionalities, such as service and resource management, security, connection and device management, and knowledge extraction from raw data [9]. WSO2 RA emphasis on event processing and analysis, resource management and service discovery, message aggregation service and enterprise Service Bus (ESB), API management and service sharing, and finally security and privacy. The most notable features of WSO2 RA is introducing a specific API layer and three kinds of PaaS for IoT platform (aPaaS, mPaaS, and iPaaS). Moreover, this RA considered some layers for device management, data and knowledge management, rules execution on data flows, and intelligent data processing [10–12]. Intel RA divides the architecture into two separate sections, cloud and edge. In this RA, some agents are located on environment's gateway, which process data, and manage devices on edge. In addition, there are some components on cloud, which are responsible for service orchestration, data analytics, configuration, security and device management [13].

From a wide range of commercial IoT platforms in the market, six dominant products have been chosen, which their fundamental functionalities are described briefly, in the following. **Watson IoT Platform** on IBM Bluemix (IBM's cloud solution) emphasis on PaaS, building application, run, deployment, device management, connection management, data management, analysis, and visualization. The major strength of Watson is data analytics component(s). Watson provides device registration, scalable connection, secure connections, data storage and analysis [14]. **Intel** is one of the first companies, which discusses on service management in IoT platforms. The company comes up with a platform fulfills data analysis, access control, API control, device management, IoT application build and development, orchestration, integration into cloud services, integration into big data analysis products, and finally data processing on edge nodes [13]. **Google** cloud platform has a significant role in Google's IoT solutions. It includes infrastructure, data storage and analytic services, network, and development APIs and services. Describing the process, which is applied on IoT data, is one of the interesting concepts that Google's IoT platform addressed. The process includes data/event ingestion, device's stream data processing, time series data storage, query process on data (related or not-related to IoT), and development kit for connecting devices to the platform (in order to send data) [15, 16]. **Microsoft IoT platform** is based on Microsoft cloud (Azure), and focuses on central control, and real-time big data analysis and management. Four integration levels (on data, platform, service, and cloud), and three security layers (devices, connections, cloud) are defined on this platform. Connections and devices management, analysis and visualization, remote monitoring, predictive maintenance, providing development kit, and integration capability with Azure PaaS are the other functionalities of this platform [15, 17]. **Amazon IoT Platform** is a cloud-based platform, which can be integrated into all of the Amazon services and products. The platform supplies big data analysis, devices management, connection management, security on multiple layers including device, services, connections, at-rest and in-motion data, and finally hardware and software development kit [15, 18]. **Cisco** provides cloud-based and edge-based products, but its main aspect is data analysis on edge/fog nodes. Significant functionalities of cisco platform, in cloud layer are device and connection management, real-time big data process and analysis, and business based intelligent analysis extendable to the edge. Cisco's edge analysis products host fog applications, manage application life cycle (includes development and distribution, deployment, hosting, monitoring, and application management), provide edge analysis, fog data security and control, and finally debugging mechanisms on edge. In addition, cisco delivers some open APIs, and cloud application development environment on cloud and fog, at any scale [19, 20].

There are many open source projects in IoT platform area, which some of them are used in industrial products, or sometimes are used as a basic component of a commercial product. **OpenIoT** is an infrastructure for IoT services' integration. It collects, filters, combines, and semantic markups data, and transfers them from environment to the platform. OpenIoT supplies data storage, data visualization, sensors management and configuration, and services management. Service management comprises service consumption tracking, data flow composition based on workflow, and determination of platform's service specifications via applications [21]. **OpenRemote** is a middleware which its aim is overcoming integration challenges among various types of protocols, and available solutions, in order to use for home automation. Home automation includes remote user, resources, policies, and notifications management, protocols integration, providing configurable policy engine, and providing APIs for management systems or mobile application client's management [22]. **SiteWhere** is a platform for integration with external systems, which supports development on most of the cloud-based platforms, and provides device management, data collection, and big data processing [23]. **WSO2** focuses on integration, data processing, and visualization. Integration component integrates cloud services with legacy software, and data resources, and provides a solution for API creation, distribution, and life cycle management. Data processing and analysis component provides a solution for data integration, and analysis, and some tools for visualization of business operations, and complex event processing [10–12, 24]. **Mulesoft** main functionality is providing an iPaaS, which integrates applications and APIs, and provides a solution for API management, design, and distribution, and some tools for projects design, build, management, and test. This integration platform creates a strong network of data, devices, and applications via API connections [11, 25]. **Kaa** is a multi-tenancy cloud-based platform, which supports data, tenants, applications, users, and devices management and monitoring, data collection and processing, and integration, and configuration management. In addition, it delivers some development SDK-s, which comprises device and user APIs for connection to platform and using it. SDKs enable real-time two-sided data transfer between device and server [26].

By categorizing all of the features, functionalities, and attributes of the mentioned platform's architecture, a set of high level features is obtained for IoT platforms. Table 1 represents the high level features, and existence of them in commercial products[1], open source projects, or reference architectures (which we call them all solutions, here after)[2]. If a high level feature is emphasized in architecture of a solution, the solution is underlined in the following table. It should be mentioned that almost, all of the solutions have emphasized on security and scalability. These two non-functional requirements are not mentioned in the following table because common to all solutions.

From Table 1, it can be concluded that "Device & Connection Management", "Data Analysis" and "Data Management & Storage" are the most notable components, among all of the discussed solutions. On the other hand, "Development Tools & Environment" and "Integration, API Management, Interoperability" have the next level of importance. Finally, "Service Management", "Distributed Processing (Edge/Fog Computing)", "Data Visualization & Business Intelligence", and "Accounting, Billing" are ranked as third level of privilege.

---

[1] The discussion on the features of company's products is based on the company's open documents for the products.

[2] Table 1 represents a revised version of the Tables 3 & 4 in the reference [2] (in Sect. 3). The current tables is different from the mentioned tables in some of the discussed IoT solutions, and the derived conclusion.

Table 1. Mapping IoT platform high level features to the solutions

High level feature			Solutions
Device and Connection Management			**Commercial:** Cisco, IBM, Intel, Google, Microsoft, Amazon, **Open-Source:** SiteWhere, WSO2, MuleSoft, Kaa, OpenRemote, **Standards/RA/RM:** ITU-T, Korean RA, WSO2, MuleSoft, Intel
Data	Data Management and Storage		**Commercial:** Cisco, IBM, Intel, Google, Microsoft, Amazon, **Open-Source:** OpenIoT, IoT Toolkit, SiteWhere, WSO2, MuleSoft, Kaa, OpenRemote, **Standards/RA/RM:** ITU-T, Korean RA, WSO2, MuleSoft, Intel
	Data Analysis		**Commercial:** Cisco, IBM, Intel, Google, Microsoft, Amazon, **Open-Source:** SiteWhere, WSO2, MuleSoft, Kaa, SiteWhere, OpenIoT, **Standards/RA/RM:** ITU-T, Korean RA, WSO2, MuleSoft, Intel
	Data Visualization and Business Intelligence		**Commercial:** Intel, IBM, Microsoft, **Open-Source:** OpenIoT, WSO2
Development Tools and Environment			**Commercial:** Cisco, IBM, Intel, Google, Microsoft, Amazon, **Open-Source:** OpenIoT, SiteWhere, OpenRemote, Kaa, WSO2, **Standards/RA/RM:** Korean RA, MuleSoft
Distributed Processing (Edge/Fog Computing)			**Commercial:** Cisco, IBM, Intel, Microsoft, Amazon, **Standards/RA/RM:** MuleSoft, Intel
Integration, API Management, Interoperability			**Commercial:** Cisco, IBM, Intel, Google, Microsoft, Amazon, **Open-Source:** SiteWhere, WSO2, MuleSoft, Kaa, **Standards/RA/RM:** WSO2, MuleSoft,
Service Management			**Commercial:** Cisco, Intel, Google, IBM, **Standards/RA/RM:** Korean RA, Intel
Accounting, Billing			**Commercial:** Intel, IBM, Microsoft, Cisco

3   Proposed Architecture

Our analysis of IoT standards, reference architectures, open source and enterprise-oriented IoT platform projects/products has allowed us to extract and propose a comprehensive architecture, as well as to identify the main components and features envisioned for IoT platforms. Here our comprehensive architecture is introduced; then, one of the main units of the proposed architecture is analyzed in detail by presenting how it is included in primary commercial IoT platforms.

In our architecture, IoT platform components are distributed on two physically separated sections/parts: Cloud (datacenter) and Fog (edge of the network). Most of the open source and all of the enterprise IoT platforms, which are considered in this paper, are based on Cloud Computing. Of course, some companies support both cloud-based and on-premises installation, simultaneously. The main purpose of cloud in IoT platforms is to achieve scalability. Number of supported devices by platform, data processing scalability, data analytics scalability, and computing scalability, are some of the key criteria, which are satisfied by cloud-based IoT platforms. Therefore, providing a cloud-based solution is the first attribute of IoT platform solutions. On the other hand, fog computing or processing data at the edge of network is a new concept, which some companies are starting to consider as crucial for their IoT platforms.

Based on Table 1, the main components of IoT platforms on the cloud can be classified in six unites, "Connection and Device management", "Data Management and Processing", "Service Management", "Development Tools", "Integration", and finally "Billing System". The units are categorized in three horizontal and two vertical layers. Figure 1 represents these layers of the proposed IoT platform in an IoT ecosystem. IoT platform components, which their background is yellow in the following figure, are distributed on Fog and Cloud. The lowest horizontal layer, in the Cloud part, is the platform interface to the environment's sensors and gateways. It contains "Connection and Device Management" unit, which manages connection to the fog layer or the edge devices, also controls the devices in the environment. The middle layers is the main layer of IoT platform which manage all of the services, processes data, delivers some development tools to IoT developers, and provides a "Billing System", as well. The highest layer is Vertical Service, which runs on top of IoT Platform. IoT Platform delivers the processed data and all of its services and capabilities to this layer. Vertical Service layer provides a specific service (such as temperature monitoring) to end-user, using Mobile Applications, Dashboards, Desktop Applications, and so on. The vertical "Security" layer is applied on all of the architecture's components. Finally, "Integration" unit, the other vertical layer, provides integration capabilities on different assets of the various layers of IoT Platform.

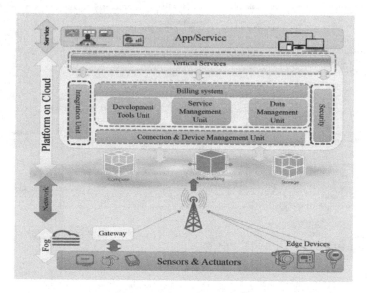

Fig. 1. Vertical & horizontal Layers of IoT Platform distributed on Fog and Cloud

The proposed units are divided into some components, sub-components, and features. Figure 2 represents the main components of the architecture. In this view, each layer's units are decomposed to some main components. As an Example, "Data Management & Process" unit in the Middle layer is decomposed to "Data Management & Storage", "Data & Event Analysis", and "Data Visualization". On the other hand, the gray arrows between the components in the following figure show data flow from environment to end-user. Sensor and actuators sense data, and send it to the fog layer. Fog layer components process data and transfer it, using network, to the Cloud section/part of IoT platform. In the Cloud part, "Data management and Processing" unit processes and analysis data deeply, visualizes it, and sends the results to the "Vertical Service", which delivers the result to the end-user device.

To explain the components of the proposed architecture, the following methodology is applied on the components, in 3 steps: (1) Description: a detail description is provided, (2) Features: some main features and components are explained (based on studying the main products in the concept of the component), (3) Features/Components Priority: some commercial products are studied according to the features, and the features priority is determined based on the results. Whilst, the methodology is applied only on the fog layer in this paper, we are working on applying it on all of the architecture's units.

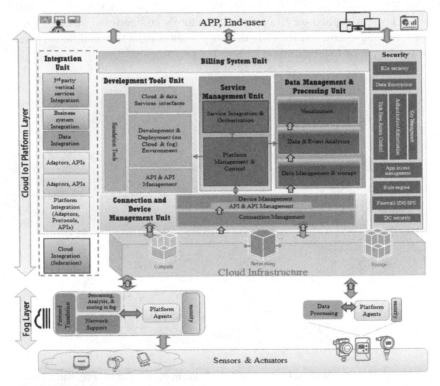

Fig. 2. The layers' components of the proposed architecture for IoT platform

3.1   Step1: Description of Edge Layer

Fog is a deployment layer of emerging relevance for IoT platforms, on the edge of the network, close to the IoT sensors and actuators, implemented on the Gateway (GW).

GW is a point on a network that transfers information on the network, and maintains sessions and subscriptions for all connected devices in the IoT solution (The router used at the home, on the internet is one type of GW, which mostly use MQTT, WebSockets, & HTTP). There are some wired and wireless communication technologies employed by sensors and GW in IoT solutions, such as BLE, ZigBee, LORA, 6LoWPAN, MQTT, COAP, and SigFox. These protocols are different in some aspects like power usage, coverage range, typical topology, and data transfer rate. The communication technology should be chosen based on the IoT scenario [27].

Often, fog nodes in IoT scenarios are gateways or edge devices connected to a power resource, which have some computation, networking, and storage capabilities. Fog nodes connect environment to Cloud, also store & processes data partially, and control environment using some software agents. Agents are located on the Fog layer, which connects to a specific component on IoT platform, and make some decision (such as data process, or security policy application). As an example, a data management agent on the Fog is a tiny data process unit, which decides about data process on fog or transferring it to the cloud, regarding its connection to the main "Data Management & processing" unit on the Cloud. Initial process on fog nodes reduces data transfer rate, and consequently reduces traffic, and occupied bandwidth. On the other hand, agents on the fog nodes execute platform commands related to device management, security, application management, and data process on edge of the network. As an example, cloud sends config-update command to a fog node, and the device management agent executes config-update on the specified devices. Agents have limited decision making and execution power. Policies and rules deal with device status (such as active, sleep, shutdown, etc.) are received from the Cloud part, stored on a local databases on the fog nodes, and decisions are made based on them.

3.2 Step2: Features of Edge Layer

To determine the main components and features of the edge layer, the features of IoT-Edge products of dominant companies in the market have been analyzed. A concise summary of this study is provided in this section.

AWS IoT Greengrass is edge product of Amazon, that can act locally on the data, execute predictions based on machine learning on connected devices, keep device data in sync, communicate with other devices securely, use languages and programming models to create and test device software in the cloud, and then deploy it to the devices. It filters device data and only transmits necessary information back to the cloud, responds to local events in near real-time, operates offline, contains IoT devices testing tools, and includes some pre-built connectors to extend edge devices functionality by connecting to 3rd party applications, and AWS services [28]. *Watson IoT platform Edge* of IBM provides a gateway for connection to the cloud and devices, processes data locally before sending it to the cloud, provides the Core IoT default services, provides edge analytics, stores data locally, handles alerts and device actions, filters data, automates deployment of workloads and services to the edge, manage infrastructures of edge nodes, maintain and update software on edge nodes, and deploy custom IBM and third party services to the Edge [29]. *Azure IoT Edge Runtime* and *Azure IoT Hub* are the products of Microsoft for edge computing in IoT platform. *Azure IoT Edge* is open source and provides security management, device provisioning, device management, local storage, data process and filter, and process real-time streaming data. It delivers cloud intelligence locally by deploying CEP (Complex Event Processing), machine learning, image recognition, and other AI functionalities. *Azure IoT Hub* connects, monitors, provisions, and manages edge applications and devices [30]. Intel provides some edge gateway devices, also offers several IoT solutions and reference architectures with its ecosystem of partners. *Intel® IoT Gateway* is a device which offers computing for in-device, near real time analytics, pre-process filtering of selected data and local decision making, connectivity of legacy industrial devices and other systems, enterprise-grade security (a hardware root of trust, data encryption, and software lockdown), and easy manageability. It is built on open architecture to ensure interoperability, and enable application development, and services deployment. McAfee integrated with the Intel IoT Gateway platform maintains system integrity, and Wind River provides runtime environments and tools to build intelligent systems for applications. Reference architecture from Intel and SAP includes data acquisition and device control, security, edge-data collection, storage, and analysis, and real-time business intelligence for decision-making [31]. Cisco provides several Edge products. *Edge Fog Fabric (EFF)* includes real-time edge/fog processing, local storage, streaming analytics, graphical development (defining streaming data transformations and analytic logic) and management tools (provisioning, deployment, lifecycle management, configuration, and administration), advanced monitoring and diagnostics in real time, and connectivity. *Cisco Fog Data Services* are software services that deliver edge analytics, control, and security for data in the fog. They provide streaming analytics at the network edge, security and privacy, application control, develop and deploy. *Cisco IOx* provides applications lifecycle management (development, distribution, deployment, hosting, monitoring, and management) for direct deployment to IoT gateways. *Fog Director* delivers IOx application lifecycle control, REST APIs, consistent management through fog application lifecycle (such as start/stop, version control, configuration, resource consumption, backup/restore, and so on), continuous application monitoring, and remote application debugging and troubleshooting. *Cisco Kinetic* is built on IOx. It adds automatic IoT gateway management as a cloud service to IOx, and provides tools for Cisco IoT gateway auto-provisioning, edge and fog processing, and data control and transfer [32, 33]. *Google Cloud IoT Edge* is composed of two runtime components, Edge Connect and Edge ML, and also takes advantage of Google's Edge TPU ASIC chip. Edge Connect connects edge devices to the cloud securely, enables software and firmware updates, and manages the exchange of data with Cloud IoT Core. Edge TPU Google's purpose-built AI chip designed to run TensorFlow Lite ML models at the edge. Edge ML is built on a TensorFlow Lite runtime, uses CPUs and hardware accelerators (Edge TPUs and GPUs) to run offline on-device machine learning models that provide real-time predictions and decision-making [34].

Table 2 Components and features of Fog layer shows the components and features of the proposed Fog layer, regarding the discussed products and architectures.

Table 2. Components and features of Fog layer

Component	Features
Data processing agent	Data synchronization with the Cloud, and act on data in offline mode, as well Local data store, filter, process and analysis. Sending some filtered, processed and necessary data back to the cloud, instead of the whole raw data Advanced analysis based on Artificial Intelligence such as predictions on devices, and CEP. Real time response to local events
Application life cycle Management agent	Development, distribution, deployment, install and hosting, update, monitoring and management of device applications. Development and test of applications can be done on the cloud, and the results deploy on the devices, or can be done directly on the edge. Maintain and update device applications
Device Management agent	Local device configuration, health monitoring, event and logging alerting, provisioning, sending commands (based on local analysis or received commands from cloud) and manage device actions, real-time devices abnormalities detection, self-healing, and device testing tools. Provide APIs for connecting to 3rd party services for extending edge device functionality
Security agent	Security and certificate management, authentication, access control to data and devices, access control of devices to the network, secure device/cloud communication, and security policy management and application
Gateway's basic features	Protocol translation: manage IP device and legacy devices, with different connection protocols, Connection to the cloud, Data Ingestion to the Cloud

3.3   Step3: Features/Components Priority

The mentioned components in Table 2 are studied in the products of 10 top IoT commercial platforms; the results are shown in Table 3. Orange and red cells in the table indicate the platforms including the component, in 2016 and 2018, respectively. Also the total number of companies, with a product satisfying a mentioned component/feature, is shown in the last column of the table. As it is depicted in the table, all of the companies started to deliver Fog Computing as a part of their IoT platform products, in recent years. Moreover, Cisco, and Intel (and its partner ecosystem) have implemented more fog-oriented features if compared with the others in 2016. Furthermore, "Data Process and Storage", and "Device Management" are the main priority of companies in implementation of Fog features on gateway. Data advanced analytics and security management, are the second priority, and application life cycle management is the last priority in delivering Fog attributes for IoT platforms.

To sum up, Fog Computing is a novel crucial paradigm for IoT platforms, which Cisco (as fog leader, who introduced fog computing in 2014), Intel and its partner ecosystem, Amazon, Google, and GE are **its current pioneers** in the market. In addition, the main implemented components of the Edge Layer in the current edge products are about data processing, storage, Device Management, and then advanced analytics and security.

Table 3. Implemented Fog Layer's components in the products of 10 companies

Feature/Company	Amazon	GE	Intel	Google	IRM	Microsoft	Arm	HP	Cisco	Oracle	Total
1. Data process and storage											10
2. Data Advanced Analysis											7
3. Device Management											9
4. App Life Cycle Management											6
5. security Management											7
6. Gateway Features											10

4 Conclusion

This work hopes to be a first step toward presenting a mature architecture for IoT platform. In this paper, components of IoT platform have been identified and a comprehensive architecture for IoT platforms is proposed, based on reviewing and discussing on architecture of commercial and open source IoT Platform products. Functionalities of Fog layer of the proposed architecture is explained regarding the fog components of some commercial products. The work will be continued by applying the explained methodology on all the architecture's units.

References

1. Recommendation ITU-T Y.2060, June 2012
2. Asemani, M., Abdollahei, F., Jabbari, F.: Understanding IoT platforms: towards a comprehensive definition and main characteristic description. In: 5th International Conference on Web research, ICWR, Tehran, Iran (2019)
3. Gubbi, J., Buyya, R., Marusic, S., Palaniswami, M.: Internet of Things (IoT): a vision, architectural elements, and future directions. Future Gen. Comput. Syst. **29**(7), 1645–1660 (2013)
4. Köhler, M., Wörner, M., Wortmann, D.: Platforms for the internet of things: an analysis of existing solutions. In: 5th Systems and Software Engineering, Bosch (2014)
5. Ganchev, I., Ji, Z., O'Droma, M.: A generic IoT architecture for smart cities. In: 25th IET Irish Signals and Systems Conference, pp. 196–199 (2014)
6. Mohan, S., Kavitha, K.: A survey on IoT platforms. Int. J. Sci. Res. Modern Educ. (IJSRME) **1**(1), 2455–5630 (2016)
7. Guth, J., Breitenbücher, U., Falkenthal, M., Leymann, F., Reinfurt, L.: Comparison of IoT platform architectures: a field study based on a reference architecture. In: Cloudification of the Internet of Things (CIoT), pp. 1–6 (2016)
8. Guth, J., et al.: A detailed analysis of IoT platform architectures: concepts, similarities, and differences. In: Di Martino, B., Li, K.-C., Yang, L.T., Esposito, A. (eds.) Internet of

Everything. IT, pp. 81–101. Springer, Singapore (2018). https://doi.org/10.1007/978-981-10-5861-5_4

9. Korean RA: ISO/IEC JTC 1/SWG 5 (IoT): Study Report on IoT Reference Architectures and Frameworks. ITU (2015)

10. Buyya, R., Dastjerdi, A.: Internet of Things: An Overview. Internet of Things: Principles and Paradigms, 1st edn. Elsevier, Amsterdam (2016)

11. Sharma, S.: Planning an architecture for the Internet of Things. In: IoT Expo (2014). https://www.slideshare.net/sumitcan/iot-architecture. Accessed: 08 Feb 2019

12. A Reference Architecture for the Internet of Things (2016). http://wso2.com/wso2_resources/wso2_whitepaper_a-reference-architecture-for-the-internet-of-things.pdf. Accessed 08 Feb 2019

13. The Intel IoT Platform: Architecture Specification. Intel (2016). https://www.intel.com/content/dam/www/public/us/en/documents/white-papers/iot-platform-reference-architecture-paper.pdf. Accessed 08 Feb 2019

14. IBM IoT Foundation Documentation (2016). http://docplayer.net/17428394-Ibm-iot-foundation-documentation.html. Accessed 08 Feb 2019

15. Nakhuva, B., Champaneria, T.: Study of various internet of things platforms. Int. J. Comput. Sci. Eng. Surv. (IJCSES) 6(6), 61–74 (2015)

16. Overview of Internet of Things (2019). https://cloud.google.com/solutions/iot-overview. Accessed 08 Feb 2019

17. What is Azure Internet of Things. https://docs.microsoft.com/en-us/azure/iot-suite/iot-suite-what-is-azure-iot. Accessed 08 Feb 2019

18. AWS IoT Developer Guide (2017). https://docs.aws.amazon.com/iot/latest/developerguide/what-is-aws-iot.html. Accessed 08 Feb 2019

19. Getting started with Cisco IoT. http://www.cisco.com/c/en/us/solutions/internet-of-things/overview.html. Accessed 08 Feb 2019

20. Fog computing and the internet of things: Extend the Cloud to Where the Things Are (2015). https://www.cisco.com/c/dam/en_us/solutions/trends/iot/docs/computing-overview.pdf. Accessed 08 Feb 2019

21. OpenIoT Architecture. https://github.com/OpenIotOrg/openiot/wiki/OpenIoT-Architecture. Accessed 08 Feb 2019

22. OpenRemote Wiki Home Page, https://github.com/openremote/Documentation/wiki. Accessed 08 Feb 2019

23. SiteWhere architecture. https://sitewhere1.sitewhere.io/architecture.html. Accessed 08 Feb 2019

24. An Introduction to WSO2 IoT Architecture. (2017). https://wso2.com/library/articles/2017/07/an-introduction-to-wso2-iot-architecture. Accessed 08 Feb 2019

25. Anypoint Platform Architecture and Components. https://www.slideshare.net/RajeshKumar404/anypoint-platform-architecture-and-components. Accessed 08 Feb 2019

26. Kaa Architecture. https://kaaproject.github.io/kaa/docs/v0.10.0/Architecture-overview/. Accessed 08 Feb 2019

27. Devare, M.: Preparation of raspberry Pi for IoT-enabled applications. In: González García, C., García-Díaz, V., García-Bustelo, B., Lovelle, J. (eds.) Protocols and Applications for the Industrial Internet of Things, pp. 264–308. IGI Global, Hershey (2018)

28. AWS IoT Greengrass. https://aws.amazon.com/greengrass/. Accessed 08 Feb 2019

29. Watsonn IoT Platform, Edge Technical Preview Now Available. https://developer.ibm.com/iotplatform/2018/03/28/edge-technical-preview-now-available. Accessed 08 Feb 2019

30. Azure IoT Edge. https://azure.microsoft.com/en-us/services/iot-edge/. Accessed 08 Feb 2019

31. Intel IoT Gateway. https://www.intel.com/content/dam/www/public/us/en/documents/product-briefs/gateway-solutions-iot-brief.pdf. Accessed 08 Feb 2019

32. Cisco Fog Data Services, Cisco Fog Director, Cisco Edge Fog Fabric. https://www.cisco.com/c/en/us/products/cloud-systems-management/(fog-data-services,fog-director,edge-fog-fabric). Accessed 08 Feb 2019
33. https://www.cisco.com/c/en/us/solutions/internet-of-things/iot-kinetic.html. Accessed 08 Feb 2019
34. Google Cloud IoT Edge. https://cloud.google.com/iot-edge/. Accessed 08 Feb 2019

The Effect of Internet of Things on E-Business and Comparing Friedman's Results with Factor Analysis (Case Study: Business in Online Stores Member of Iranian E-Mail Symbol)

Leila Abdollahzadeh Ramhormozi[1]($\boxtimes$), Amir Houshang Azh[2]($\boxtimes$), and Mehdi Khaki[3]($\boxtimes$)

[1] Islamic Azad University, North Tehran Branch, Tehran, Iran
leila.abdollahzadeh92@gmail.com
[2] Islamic Azad University, Yadeqar Imam Khomeini, Tehran, Iran
ah.azh@cbi.ir
[3] University of Tehran Alborz Campus, Tehran, Iran
mehdikhl332@gmail.com

Abstract. This research studies the factors affecting the Internet of Things. The research community is a survey of 277 Internet shoppers (Digi Commodity, Kimia Online, Digi Style, Alyar). The research tool was a questionnaire with a Likert scale of 7 which has been verified by experts and reliability has been confirmed by Cronbach's alpha higher than 0.7. For our research questions, descriptive and deductive analyzes (Friedman, T single-sample and analytic factor) have been used, and at the end, the outputs of the two models of the same are compared. The results showed that according to Friedman rankings, the price management system with the highest 6.38 and advertising and marketing management with the lowest of 4.28 in priority of one and nine in the number of factors influencing the Internet of Things in the electronic business have been affected. But in the ECRM factor analysis, top priority and product management are top priority. Also, the two components of post-sales/post-sales/post-delivery management are equally important in the third rank and pay management/ advertising and marketing management are equally ranked fourth in importance.

Keywords: Internet of Things · Electronic business · Online stores

1 Introduction

The Internet of Things is a system related to computing equipment, digital and mechanical machines, things, animals or humans that have unique identifiers and the ability to transfer data over a network without the need for human-to-human or human-to- With-computer.

The Internet of Things is an evolved convergence of wireless technologies,

Micro electromechanical (MEMS) systems, micro-services and the Internet. This convergence between Operational Technology (OT) and Information Technology (IT) has helped to analyze the data produced by the machine and improve it. Today,

© Springer Nature Switzerland AG 2019
L. Grandinetti et al. (Eds.): TopHPC 2019, CCIS 891, pp. 191–219, 2019.
https://doi.org/10.1007/978-3-030-33495-6_15

computers, and ultimately the Internet, are almost entirely dependent on humans for the production of content. Approximately 50 petabytes (each petabyte is 1024 terabytes). Data that is available on the Internet has been generated through humans by typing, recording, capturing and scanning barcodes.

On the other hand, today the importance of e-commerce is not overlooked, and in Iran, too, it has been emphasized, but in terms of the extent of enjoyment of this type of trade in the global arena is not as satisfactory and the development of e-commerce in Iran There are many obstacles and challenges; therefore, efforts should; be made to advance with other developed countries. (Ghorbani et al. 2009).

Therefore, this research seeks to examine the impact of the Internet of Things on the electronic business in Iran.

2   Theoretical Background

2.1   Statement of the Research Question

The advent of the Internet and the applications of e-commerce in the business have dramatic effects on how the work and the structure of various are industries.

Because of the many benefits these innovations bring to organizations, many companies tend to use it. On the other hand, according to Michael Porter (2001), some e-commerce has been doing all the work using computer communication networks, especially the Internet. From the perspective of others, e-commerce is a kind of paperless business.

But the definition that the Commerce Net offers from e-commerce is more appropriate:

E-commerce uses computers from one or more networks to create and transfer business information that is more relevant to the purchase and sale of information, goods and services via the Internet. The Internet of Things is a phenomenon that seeks to change our lives, similar to the change that the Internet itself created in 1990. (Keskin and Kennedy 2015).

The Internet of Things integrates a large number of heterogeneous things that continuously generate information about the physical world (Atzori et al. 2010)

Most of this information is accessible through standard browsers, and several platforms offer functional programming interfaces for access to sensors and activators. Therefore, the Internet technology of things enables the provision of new services for end-users in various fields (Nitti et al. 2014).

The search for each specific service provided by the Internet devices of the things represents an important point; the number of connected things in the network is increasing, which leads to a very large search space. Interactive models are based on the interaction of Man-objects, but in the near future, interaction will be objects -objects, so that things seek to provide complex services for the benefit of human beings.

Analysts estimate that by 2020, the number of connected devices will reach 24 billion Chan (2015).

Marketers are always looking for data and information. Further connectivity will lead to more interactions and more accurate data generation. Thus, marketers will face new streams of data; consumer habits that have not yet been discovered will discover the Internet of things and improve marketing. In recent years, some researchers have looked at the various services of the Internet of Things (Nitti 2014; Atzori et al. 2011).

The term Internet of Things was first introduced by Quinn Ashton in 1999. (Yu and Jiangtao 2015).

The Internet of Things is a global, interconnected network of things and humans that can interact with each other and collaborate with their neighbors through a unique address plan. The primary purpose of the Internet of Things is to share information things. Which represents production, transportation, consumption and other details of people's lives. (Hirsch 2019)

The components of the Internet of Things are:

- The things within which sensors are embedded.
- A network that connects these things.
- Input and Output Data Processing Systems (Ali 2018).

This feedback loop enables the monitoring and control of things remotely and over the Internet. As a result, decisions are made on the basis of correct, up-to-date, and real-time data; and the decision-maker's person or entity will have the intelligence necessary to make the best decision and respond promptly to the issues. It is worth looking at the issues and methods. By dealing with them, it is possible to achieve predetermined goals with greater confidence and confidence (Pinochet 2017).

2.2 Theoretical Framework of the Research

Electronic business, integration and mobility are more than plans, processes, equipment and business systems that meet the changing needs of customers.

Multi-stage Model of Electronic Business Development

A multi-stage electronic business development model was developed by Rao et al. 2003. In this model, factors of facilitator and deterrent factors have been investigated. This model includes four stages of presence, portals, integration of exchanges and organization. Of course, in the next steps, the costs and complexity of technology issues will increase. Although the stage model is in succession, it is not necessary for companies to start from the first stage (stage one) and then go to the next steps. Instead, the company can enter at any stage. As much technology and e-commerce awareness increases, as much as a company can go straight to the next steps to speed up its goals. In this model, facilitators and obstacles or inhibitors are identified at each stage.

The Basic Model for Barriers to Electronic Business Adoption

This model was presented by Mac Gregor and Vrazalic in 2005. In this study, 477 Australian and Swedish companies have been studied. E-commerce can be considered as a competitive advantage, as it reduces costs and reaches potential customers around the world. Governments around the world have identified this issue and have set up policies to facilitate e-commerce. In this research, barriers and inhibitors of e-commerce were grouped in a basic model. In this model, barriers and intruders were classified into two distinct, very difficult and disproportionate groups. In the very tough group, there are barriers that make e-commerce more complicated. In the dispropor-tionate group, barriers such as e-commerce disproportion with products and services of the organization, the way businesses do business with customers and suppliers, and lack of awareness of the benefits of e-commerce.

Comprehensive Model of Electronic Business Implementation and Development
Chong (2005), provided a conceptual model that examined the factors affecting the success of electronic business implementation. This research has been implemented in several Australian and Singaporean companies and has provided a comprehensive model for the implementation and development of e-business. In this study, the degree of satisfaction of e-commerce as a determinant of success in the implementation of electronic business is considered. Has been. Of course, to measure success, it is important to determine the type of satisfaction of individuals. Therefore, in this research, management satisfaction is considered as the measure of electronic business success.

Amit and Zott (2001), have four main drivers in electronic business; this leads to the creation of new value or increase in the value of the company's performance in the electronic business:

(1) Innovative (2) Customer retention (3) Completion (4) Performance

An e-business seeks to add revenue streaming through the Internet and its capa-bilities to create and enhance relationships with customers and business partners and to improve productivity by using customer behavior analysis strategies. Electronic busi-ness solution solutions allow the process of internal and business integration (Arora et al. 2008).

2.2.1 Application of the Internet of Things

The Internet concept of things refers to unique identifiable things and virtual repre-sentations in a structure similar to the Internet.

retail;

Supply Chain Control: Monitor the warehouse status along the supply chain and trace products.

– NFC payment: payment based on the processing (production and processing) in place or duration of activity for public transport, sports halls, parks and ….
– Intelligent shopping software: Get advice at the point of sale based on customer habits, priorities, presence of allergic components for them, or expiration dates.
– Intelligent Product Management: Product rotation control on sales shelves and warehouses and automated procurement processes.

Examples of needs:

– Increasing productivity; which exists within many companies and affects the suc-cess and profitability of companies.

- Distinction of the market; In a market saturated with similar products and solutions, "distinction" is very important and the Internet of Things is one of the most important and possible differences.
- Cost Efficiency: Reducing running business costs is a "slogan" for many Executive directors. The better use of resources, the best information used in the decision making process, or the shortening of the time the plant does not work, are possible ways to achieve this (Fig. 1).

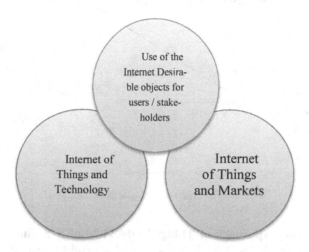

Fig. 1. The Internet Cycle of Things. Daniele Miorandi, Sabrina Sicari, Francesco De Pellegrini, Imrich chlamtac (2014)

2.2.2   How the Internet Changes Things and Interacts with the Physical World

The Internet of Things is called the next Industrial Revolution. The Internet of Things will change the way in which all businesses, governments, and consumers will interact with the physical world. For more than two years, business intelligence has followed closely the growth of the Internet of Things. Specifically, the agency has analyzed how the Internet of Things enables entities and institutions such as governments, businesses, and customers to connect and connect to their Internet devices. To control. This analysis was conducted in environments such as manufacturing, smart home, transportation and agriculture (Fig. 2).

- By 2020, a total of 34 billion devices will be connected to the Internet.
- Nearly $ 6 trillion for the Internet of Things will be spent over the next 5 years.
- Business will be the main component of the Internet of Things; businesses can improve their business by: (1) reducing operating costs. (2) Increased productivity. (3) Expansion of the market and new products.
- Governments will focus on increasing productivity to reduce costs and improve the quality of life of their citizens.
- Consumers will act as IOT users alongside businesses and governments. (I.T.U 2005).

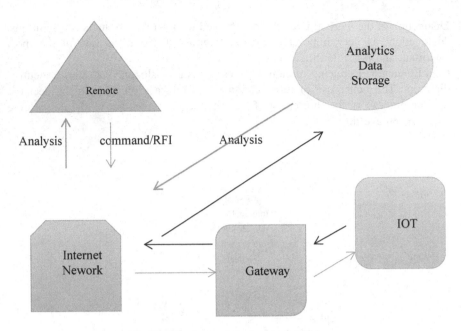

Fig. 2. The Internet Ecosystem of Things

2.2.3   The Internet Approach of Things in the Field of E-Business

The Internet of Things helps in marketing on social networks to get information about customers, their habits, buying process, frequency, major purchases, and so on.

The Internet of Things helps in marketing on social networks to get information about customers, their habits, the buying process, major purchases, and so on. This information is in the form of large data. Many data are collected by advertisers. Through these data, customer history, type of purchase and time of purchase are obtained. Often Internet purchases are a matter of urgency and needs. Connecting things over the Internet and collecting data in an efficient way brings an advantage in e-commerce (Maze berry Blog 2015).

Challenges in the field of human resources (low productivity, high costs, increased errors), competitive market (inadequate communication between suppliers, manufacturers, customers and vendors, incomplete market information), products (product specifications, product quality), Product specifications, product delivery paths, controls, etc.), product cuts, customer information (customer specifications, customer requirements, privacy, social relationships, etc.), further revise the management of interactions between globalization and e-commerce and the Internet things have been. Therefore, the study of these communications is very important and undoubtedly the use of information technology is one of the necessary loops to increase business efficiency in the national economy (Pinochet 2017).

2.2.4   Configuring the Internet Business Model

Van Catherine (2008) provides a framework for business re-configuration, which shows that IT grants offer the best potential for creating new and innovative mechanisms for creating competitive advantage. Five levels of IT-based re-configuration based on the two dimensions of business transformation and the potential benefits of information technology are shown in Fig. 3.

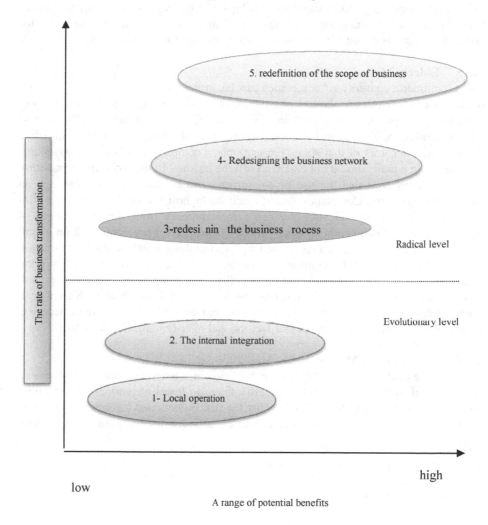

Fig. 3. Five levels of re-engineering of business-based information technology (Zheng et al. 2004)

"Local exploitation" improves the efficiency of tasks, and IT has little impact. "Integrity of the internal", efficiency and effectiveness through the integration of technical and organizational systems of business, improvement.

"Local exploitation" improves the efficiency of tasks, and IT has little impact. "Integrity of the internal", efficiency and effectiveness through the integration of technical and organizational systems of business, improvement gives. Van Kateram (2002) calls these "evolutionary" processes, which include gradual changes in business activities. The three "radical" levels require a major process change to achieve a competitive is advantage. "Business Process Design" requires reorganization of processes to leverage IT effectively. The "redesign of the business network" addresses the uses of information technology to manage relationships with colleagues at the organizational-level boundaries. "Redefining the business domain" is the ultimate challenge for companies that use information technology to change their business (Zheng et al. 2004).

2.2.5   Elderly Maturity

The e-commerce maturity of companies can be attributed to the degree of electronic readiness of companies in the use of e-commerce. Applying the maturity model to e-commerce also provides an opportunity for evaluating and evaluating the effectiveness of e-commerce, so that it can be tailored to the stage of electronic evolution to define the specific indicators of that stage. For example, a company that is at an early stage, such as using e-mail, should not expect to sell the Internet. In the e-commerce maturity model, it is believed that the use of e-commerce can be done in stages through an evolutionary process. Companies cannot reach the highest levels of e-commerce from the start, but they should start the deployment process from the early stages of developing this phenomenon. So far, various studies have been conducted on e-mail maturity in the field of e-commerce. The English consulting firm has provided e-mature e-commerce modeling for e-commerce in companies based on two components of the degree of change and complexity and the benefits of the company in five stages: e-mail, website, e-commerce, Electronic business and the developed organization (Karimi et al. 2009). Effective factors on electronic maturity can be divided into three categories: technical requirements, organizational factors, and inter-organizational systems.

2.2.6   E-Commerce Models

E-commerce models describe the roles and relationships among customers, consumers, partners, and suppliers, who seek to identify the main flows of products, information, and money, and identify the major benefits for stakeholders and business participants. And works with the Internet to interact and create value for the customer and other stakeholders.

1. Business with the consumer. (B2C)
2. Business with business. (B2B)
3. Consumer with the consumer. (C2C)

2.3   Research Objectives

The main objective

Examining the creation and role of Internet things in electronic business

Sub-goals

(1) The effect of information satisfaction on the behavior of purchases in the electronic business.
(2) Impact of profit on the behavior of purchases in the electronic business.
(3) The impact of site commitment on the behavior of purchases in the electronic business.

2.4   Research Questions

According to the concepts presented in the introduction sections, the theoretical framework of the research and the objectives, the research questions are formulated as follows:

The main question:

(1) What role does the Internet have on electronic business?

Sub Question:

(1) How does advertising and marketing management system affect e-commerce shopping behavior?
(2) How does trust and security management system affect the behavior of purchases in e-business?
(3) How does the online management system and customer relationship affect the buying behavior of e-business?
(4) How does the post-sales service management system affect e-commerce shopping behavior?
(5) How does the delivery management system affect e-commerce behavior?
(6) How does the payment management system affect the buying behavior of c-business?
(7) How does the price management system affect e-commerce shopping behavior?
(8) How does brand management affect e-commerce shopping behavior?
(9) How does the product management system affect the buying behavior of e-business?

2.5   Conceptual Model and Research Variables

The main research model is derived from the variables of the following resources (Fig. 4 and Table 1).

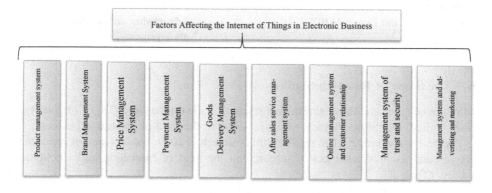

Fig. 4. Conceptual model of research: creation of a researcher

Table 1. Selection of components of the main research model

Scientific source	Marketing Strategy Factors
Chen and Hsu (2010)	Product management system
Ou and Ling (2010), Jun and Lun (2011)	
Cheung and Liao (2003)	
Pavlou et al. (2007)	
Zhang et al. (2011), Bente et al. (2012)	Brand or brand management system
Chen and Hsu (2010), Jun and Lun (2011)	Price Management System
Hinz et al. (2011)	
Ranganathan and Ganapathy (2002), Chen and Hsu (2010)	Payment Management System
Zhao (2011)	Shop and Delivery Management System
Chen and Hsu (2010)	
Chen and Hsu (2010), Svantesson and Clarke (2010)	After sales service management system
Cheung and Liao (2003), Jones and Spence (2008) Jun and Lun (2011)	
Zhao (2011)	
Bahrami (2010) and Ghaffari Estiani (2009)	ECRM
Jun and Lun (2011), Pavlou et al. (2007), Meuter et al. (2005), Jones and Spence (2008), Cheung and Liao (2003)	Online Customer Relationship Management System
Ranganathan and Ganapathy (2002), Ik et al. (2013), Casalo et al. (2008)	
Jing and Ying (2010)	
Ik et al. (2013), Rose et al. (2011) Jing and Ying (2010)	
Casalo et al. (2008), Jun and Lun (2011)	Trust and Security Management System
Chen and Hsu (2010)	
Ranganathan and Ganapathy (2002)	
Shaharudin and Mansor (2011)	
Jun and Lun (2011) Chang and Tao (2010)	
Jamshidi (2002) and Doaee (2010)	Advertising and marketing management

2.6   The Scope of Research Studies

The realm of research: The realm of this research is the online store (Digi Commodity, Kimia Online, DigiStyle, Alyar), which is active in the field of electronic business.

The realm of research: The realm of this research is 2019. Subject area of research: This research is in the subject of research in the field of marketing and electronic business (Table 2).

Table 2. The history of research on the Internet of Things

Results	Title of the article	Author Name	Line
Expanding the IOT Business Model with the help of the Canvas Business Model by identifying Building Block types	The business model for the Internet of Things	Eindhoven Deloitte July 2015	1
Introducing View and Control, Large Data and Business Analysis and Sharing and Collaboration as Classifications for IOT Business Applications	Internet of Things: An Investment Plan and Its Challenges for Organizations	In Lee Kyoochun Lee 2015	2
How can a business model be used for an IOT template?	Business models and the Internet of Things (application in the environment)	Jozef Glova Tomas Sabol Viliam Vajda 2014	3
By reviewing the IOT scope, identifying the current trend, describing the challenges	Check out the threads and trends of the Internet of Things	Andrew whitemore Anurag Agarwal Li Da Xu, 2014	4
Express issues related to the development of IOT services and technologies	Internet of Things: Perspectives, Plans and Challenges of Research	Daniele Miorandi Sabrina Sicari 2012	5
The basic features of the Internet of Things, the advising of the technologies required and the prediction of applications	The Perspective of the Internet of Things: The Key of Features, Programs and Open Issues	Eleonora 2014	6
Active dialogue between key community components, including government, industry and universities, will ensure that these timely challenges and the Internet will have the most impact from things.	Challenges of Internet of Things (computing systems)	Rajeev Alur Emery Berger Daniel Lopresti John A, 2015	7

2.7 An Overview of the Research Background

– Hirsch (2019), in an article titled "The Ghazi Who Made Golden Eggs: Personal Information and the Internet of Things", addresses the issue of monetization of personal information and the dangers of companies that do not understand this process. This article reviews recent literature on the use and misuse of personal information to identify trends and issues. The research findings suggest that, probably through Balochin technology or some other way, individual consumers will be able to earn money through their information, as they are rapidly changing privacy and data security. It is expected that the findings may be due to sudden events such as violations of the privacy of new global information. While there are numerous studies on how to index the efforts to create revenue-generating systems for consumer data, there is not much research about the risks of corporate marketing in the management of data from the Internet of Things.

– Ali (2018) in his studies the dimensions of individual privacy concerns relating to the Internet of Things used. The integration of Internet tools (IOT) into everyday life presents challenges to the privacy of users and those affected by these devices. This article will affect the factors affecting the use of IOT and examine it on the dynamics of privacy management with the presence of IOT. The four focus groups of IOT individuals and experts were reviewed to understand concerns about the privacy of the groups. The authors adopted qualitative research methodology based on component-based theory to find relevant dimensions of the privacy concerns of the situation under the IOT usage conditions. The findings showed that fourteen are concerned with the privacy concerns of individuals about IOT and can be categorized under four key factors: collecting, IOT devices, data storage and use of data Collected. The authors also analyzed the focus groups using the theory of disclosure theory and examined how privacy concerns affect the privacy practices of the individual. This study can help service providers and IOT manufacturers provide design principles and reduce concerns based on the information they need to provide their users. This study represents the first attempt to review the process of people's experience in managing their privacy.

– Pinochet (2017) in his article "The impact of the "Internet of Things" features on the practical and emotional experiences of buying intent. "Internet of Things" is a broad term used to describe physical network connectivity. Connected things or smart devices are referred to as electronic circuits and software that are capable of identifying, collecting and transmitting data and information. This study uses a consistent model of (2014) Yaping et al., By setting to build "emotional experience," "performance experience," and "intention to buy." This review includes an example of 747 valid questionnaires in relation to product users The Internet, which was structured using a structured questionnaire, was answered based on the Likert scale, a quantitative research approach was followed by an exploratory descriptive step and then the application of structural equation modeling. The results have proven the highest model relationships with a high level of importance. More emphasis was placed on the emotional impact more than the purchasing intention of the selected sample, which was mostly composed of young people. In short, this study confirmed the statistical significance of the structural paths and showed that the proposed model is consistent and With proper adjustment it can be used in future research.

– Leminen (2016) explores the future of the Internet of Things: Towards heterarkhic ecosystems and business model services. The purpose of this study was to understand the emergence and variety of business models in Internet Ecosystem things (IOT) is. This paper is based on the systematic literature review of IOT ecosystems and business models for constructing a conceptual framework in IOT business models and the use of qualitative research methods for analyzing seven industries. This study identifies four types of IOT business models: Value chain efficiency, industry collaboration, horizontal market and platform. In addition, three evolutionary paths for the advent of the new business model are discussed: Opening ecosystems for industry collaboration, repeating the solution in multiple services, and returning to the ecosystem as technology maturity.

– Murugesan (2015) warned in an article entitled "Generation 2 Ecommerce and Emerging in Emerging Markets" about changing e-commerce patterns and states that businesses need to have a fresh attitude in order to benefit from e-commerce. And it's more usable, friendlier, and economically feasible to make e-commerce available to most, as part of the initiative of organizations, developers, and applications that e-commerce is the most beneficial to businesses and Community.

– Industries (2014) in the article "E-commerce in Developing Countries; Opportunities and Challenges" argued that e-commerce is a phenomenon driven by the advancement of technology in the twentieth century, and in addition to being an additional marketing channel for many companies to promote sales It has been described as a unique tool for many companies like eBay, Amazon, etc. It addresses the challenges of developing countries in order to benefit from this new technology and develop it: (1) equipping these countries The Internet has been considered as a development tool, which is a development challenge and involves Hong Kong investments (2). Realizing and managing the growth and development of the Internet as a public facility for development and enhancement, which is, in their view, an international challenge that encompasses the growth and adjustment of the Internet and its facilities at a globally acceptable level. The logic in this article is that online upgrade and development is facilitating the delivery of information, goods and services, and this process creates more traffic in the data network that is needed to invest in development incentives Most infrastructure will work.

3   Research Method

This research is based on the nature and purpose of the application and seeks to explore the creation of the Internet of Things in the electronic business. The method of data collection is descriptive-survey.

3.1   The Society

Considering the above mentioned points, in order to determine the appropriate sample size for the research, according to the population of the statistical society, the sample size at the significance level of 95% and sampling error of 5% will be used from the Cochran formula (Rafipour 1995: 383).

$$n = \frac{Z_{\frac{\alpha}{2}}^2 \times p(1-p)}{\varepsilon^2} \tag{1}$$

n	= sample number
Z	= Confidence
P	= ratio of success in the statistical society
1–P	= Failure ratio in the statistical society
ε	= Estimation accuracy

In this research, the success rate in the statistical society is considered to be 0.5. Estimation accuracy is also considered to be 0.08. The sample size is 293, which is considered as the sample size (Table 3).

Table 3. Classification of the target statistical population

Number of statistical population	The name of the statistical society group	Line
101	Digi-Kala	1
58	Kimia Online	2
41	Digi Style	3
100	Alyar	4
300	total	5

3.2   Data Collection Tool

The collection of information in this research is based on two librarian methods (the use of books, scientific journals and published articles in this research, previous theses in this field in the universities and scientific centers and scientific resources of the Internet) and field.

3.2.1   Research Tool

The questionnaire consists of two parts. The first part of the questionnaire includes personal information that has demographic questions such as gender, age, and educational level of respondents. The second part contains questions related to the research variables (Table 4).

Table 4. Combining questionnaire questions

Scientific source	Indicator	Marketing strategy factors	Questions
Chen and Hsu (2010)	Products categorized or categorized	Product management system	4
Ou and Ling (2010) Jun and Lun (2011)	Quality of products		
Cheung and Liao (2003)	Confrontation or non-touch		
Pavlou et al. (2007)	Variety of products to match the tastes of different people		
Zhang et al. (2011) Bente et al. (2012)	Good repute and reputation of company, website or brand	Brand or brand management system	1
Chen and Hsu (2010)	Seasonal auctions	Price Management System	3
Chen and Hsu (2010) Jun and Lun (2011)	The right pricess		
Hinz et al. (2011)	Dynamic pricing policy based on customer purchases of purchased purchases or high purchases		

(continued)

Table 4. (*continued*)

Scientific source	Indicator	Marketing strategy factors	Questions
Ranganathan and Ganapathy (2002) Chen and Hsu (2010)	Different payment methods, including internet, payment when receiving a product, and so on	Payment Management System	2
Chen and Hsu (2010)	Provide receipt for follow up payment		
Zhao (2011)	Quality of delivery of goods (safe delivery)	Shop and Delivery Management System	5
Zhao (2011)	Sufficient information on how to submit and the ability to choose how to deliver the goods		
Chen and Hsu (2010)	Send timely goods		
Chen and Hsu (2010)	Delay notification system for the provision and delivery of goods		
Chen and Hsu (2010)	Delay notification system in the transportation of goods		
Chen and Hsu (2010) Svantesson and Clarke (2010)	Frequently Asked Questions	After sales service management system	3
Cheung and Liao (2003) Jones and Spence (2008) Jun and Lun (2011)	Warranty, repayment and replacement of defective goods		
Zhao (2011)	Timely response to calls, emails and phone buyers		
Bahrami (2010) and Ghaffari Estiani (2009)	Collecting customer information with respect to privacy	ECRM Online Customer Relationship Management System	7
Bahrami (2010) and Ghaffari Estiani (2009)	Proper classification and clustering system of customers		
Bahrami (2010) and Ghaffari Estiani (2009)	Answering customer problems		
Jun and Lun (2011), Pavlou et al. (2007), Meuter (2005), Jones and Spence (2008), Cheung and Liao (2003)	Interacting with clients in different ways for positive psychological effects on customers		
Ranganathan and Ganapathy (2002) Ik et al. (2013) Casalo et al. (2008)	Collecting, processing and analyzing customer problems and criticisms		
Jing and Ying (2010)	Customer satisfaction		
Ik (2013) Rose (2011) Jing and Ying (2010)	Customer Loyalty		

(*continued*)

Table 4. (*continued*)

Scientific source	Indicator	Marketing strategy factors	Questions
Casalo (2008) Jun and Lun (2011)	Trust in the company	Trust and Security Management System	5
Chen and Hsu (2010)	Provide details of the rights and obligations of the parties to the transaction		
Ranganathan and Ganapathy (2002)	Provide enough information on how to use personal information		
.Shaharudin and Mansor (2011)	Product trust (efficiency and benefits of using it)		
Jun and Lun (2011) Chang and Tao (2010)	Trust in infrastructure such as electronic payments, security systems and so on		
Jamshidi (2002) and Doaee (2010)	Online marketing and advertising	Advertising and marketing management	2
Jamshidi (2002) and Doaee (2010)	Marketing and Outdoor Advertising		
–	–	Total	32

3.2.2   Validity and Reliability of Research

Responses were calculated based on the CVR formula (Content Validity) and adapted to the Lawshe table. Numbers above 0. 59 were accepted. Thus, the number obtained for the CVR of this research was higher than 0.62, all confirmed (Table 5).

Table 5. Validity and reliability of researcher-made questionnaire

The minimum acceptable CVR is based on the number of expert scorers					
Quantity CVR	Number of professionals	Quantity CVR	Number of professionals	Quantity CVR	Number of professionals
0.37	25	0.59	11	0.99	5
0.33	30	0.56	12	0.99	6
0.31	35	0.54	13	0.99	7
0.29	40	0.51	14	0.75	8
		0.49	15	0.78	9
		0.42	20	0.62	10

3.2.2.1. Content Validity Index (CVI)

The content validity index is used by Waltz CF, Bausell (1981) to determine the relevance, clearness, and simplicity of each item based on a 4-part Likert scale. Specialists are not related to each item in their own terms, 2 are "relatively related", 3 are "relevant", and 4 are "totally relevant." The simplicity of the class is also 1, not "simple", 2 is "relatively simple", 3 is "simple", 4 is "simple", and the clarity of the word is also 1 "not clear", 2 "relatively Obvious", 3 is "obvious", up to 4 is "obvious".

$$CVI = \frac{\text{The number of specialists who scored grade 3 and 4}}{\text{Total number of specialists}}$$

The minimum acceptable value for the CVI index is 0.92, and if the CVI index is less than 0.92, then the item should be deleted. The CVI results indicated that all CVI questions were above 0.92 and therefore were appropriately recognized. Therefore, in order to measure reliability, Cronbach's alpha method was performed using SPSS $_{23}$.

3.2.2.2. Reliability of Research

Using the formula below, we calculate the Cronbach's alpha coefficient.

$$r_{\alpha} = \frac{J}{J-1}\left(1 - \frac{\sum_{j=1}^{n} s_j^2}{S^2}\right) \tag{2}$$

where in:

The number of sub scores of the questionnaire or test questionnaire = J

The variance of the following J-test = S_j^2

The variance of the whole questionnaire or test = S^2

For this purpose a preliminary sample of 32 pre-test questionnaires was used. Then, using the data obtained from these questionnaires, using SPSS software, the confidence coefficient was calculated using Cronbach's alpha. Cronbach's alpha coefficient for the research variables and its dimensions are described in the Table 6. Since the obtained value of all research variables is higher than 0.7, it can be said that the questionnaire has an acceptable reliability.

Table 6. Results of the reliability test of the researcher-made questionnaire

Cronbach's alpha coefficient estimated in this research	Research variables
0.82	Product management system
0.74	Brand or brand management system
0.66	Price Management System
0.79	Payment Management System
0.73	Shop and Delivery Management System
0.87	After sales service management system
0.84	ECRM Online Customer Relationship Management System
0.91	Trust and Security Management System
0.81	Advertising and marketing management
0.79	The entire questionnaire

As Table 6 shows, the reliability coefficient for the dimensions of the questionnaire is higher than 0.70 (0.79). Therefore, reliability is confirmed.

4   Analyze the Research Data

Data analysis in this study is done using SPSS software and has two descriptive and inferential levels. Descriptive data were collected using descriptive statistics (frequency tables and percentages). The Kolmogorov-Smirnov test was used to examine whether the variables and inferential statistics such as T-single and two-sample T were normal or not, as well as Friedman's in order to measure the assumptions and according to The measurement level of the variables was used (Table 7).

Table 7. The result of the test of the normal variables of the research

Conclusion	Confirm hypothesis	Significant level	Number of samples	Research variables
Normal	H_0	0.508	277	Product management system
Normal	H_0	0.703	277	Brand or brand management system
Normal	H_0	0.189	277	Price Management System
Normal	H_0	0.082	277	Payment Management System
Normal	H_0	0.913	277	Shop and Delivery Management System
Normal	H_0	0.303	277	After sales service management system
Normal	H_0	0.242	277	Online Customer Relationship Management System
Normal	H_0	0.300	277	Trust and Security Management System
Normal	H_0	0.348	277	Advertising and marketing management

According to the results of the above table, since the level of significance for all components is larger than the error value of 0.05, these variables have a normal distribution (Tables 8 and 9).

Table 8. Testing the average effective factors on the Internet of Things in an electronic business

T-statistic of a sample				
Concept	Number of samples	Mean	Standard deviation	Mean standard error
Factors affecting the Internet of Things	277	4.9779	0.80938	0.04863

Table 9. T Test is an example of factors affecting the Internet of Things in an electronic business

A test sample

Concept	Test value					
	T	Degrees of freedom	Standard deviation	Average difference	95% confidence interval Difference	
					High	Low
Product management system	20.109	276	0.000	0.097793	0.08822	1.0737
Brand or brand management system	26.403	276	0.000	1.12555	1.0416	1.2095
Price Management System	24.501	276	0.000	0.09976	0.09921	1.0031
Payment Management System	19.436	276	0.000	1.00935	0.08833	1.1354
Shop and Delivery Management System	22.119	276	0.000	1.4009	0.07326	1.0346
After sales service management system	23.453	276	0.000	1.90225	0.09265	1.4390
Online Customer Relationship Management System	21.789	276	0.000	1.94	1.2318	1.3241
Trust and Security Management System	27.222	276	0.000	1.60385	0.6349	1.2864
Advertising and marketing management	28.505	276	0.000	1.3154	0.5832	1.0238

4.1 Friedman Ranking Test

According to Friedman rankings, the price management system with the highest of 6.38 and advertising and marketing management with the lowest of 4.28 at the top of one and nine were influenced by factors affecting the Internet of Things (Tables 10 and 11).

Table 10. Ranking based on Friedman test

Rank	Average rating	Research variables
6	4.83	Product management system
2	6.12	Brand or brand management system
1	6.38	Price Management System
3	5.92	Payment Management System
8	4.67	Shop and Delivery Management System
5	5.33	After sales service management system
7	4.7	Online Customer Relationship Management System
4	5.87	Trust and Security Management System
9	4.28	Advertising and marketing management

Table 11. Friedman test

Friedman test	
277	Number of samples
153.076	The statistics amount x^2
8	Degrees of freedom
0	Significance level

4.2 Evaluation of Structural Modeling Research Using Confirmatory Factor Analysis

LISREL method, while estimating the unknown coefficients of the linear structural equation, estimates the fitting of models that include the variables, the measurement errors in each of the dependent and independent variables, two-sided causality, time and interdependence. In the present study, after drawing the model based on the data, the size of the model parameters was obtained using LISREL software. Therefore, using the (γ) coefficients, the hypothesis (t) test has been tested. The standard coefficient is the binary correlation values (between the two variables) and to compare the effects of the components of the model, and the higher the coefficient, the more effective the independent variable affects the dependent variable, and the meaning of the coefficient is that The meaningful number should be greater than 1.96 or less than -1.96 and generally used to confirm or reject the research assumptions. Confirmatory factor analysis is used to determine the suitability of selected questions for the investigated factors. At this stage, we first examine the confirmatory factor analysis indicators. In the following, we examine the significance of the relationship between each question and the factor under study using the LISREL charts in two significant and standardized states (Tables 12 and 13).

Table 12. Symbols characteristics affecting the Internet of Things

Symbol	Research variables
PRO	Product management system
BA	Brand or brand management system
PRI	Price Management System
PAY	Payment Management System
DEL	Shop and Delivery Management System
AFT	After sales service management system
ECRM	Online Customer Relationship Management System
TRU	Trust and Security Management System
ADV	Advertising and marketing management

Table 13. Confirmatory factor analysis indicators the Components Under Review

Amount (Factor of the Internet of Things in e-business)	Amount (infrastructure capacity components)	Statistic
1056.51	1396.23	**CHI SQUARE**
–	67.3	**NC**
–	0.95	**NFI**
–	0.088	**RMSEA**
–	0.095	**NNFI**
–	0.94	**GFI**
–	0.5	**RMR**

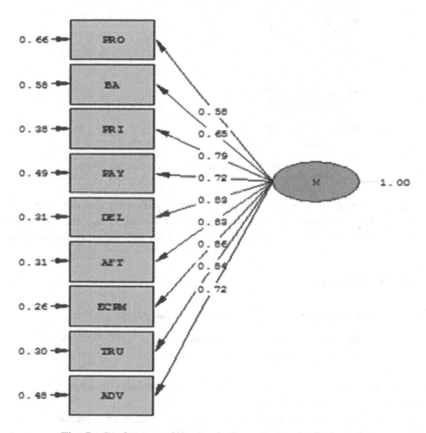

Fig. 5. Confirmatory factor analysis results in standard mode

Figure 5 Because the units of measurement of variables are the same, it is possible to compare the observer variables with questions related to a hidden variable. Through this output and according to the factor loads, it can be concluded that:

- Among the variables related questions, the characteristics of the Internet on things in the electronic business, the ECRM index has the highest correlation and the index of the product (factor one) has the lowest correlation.

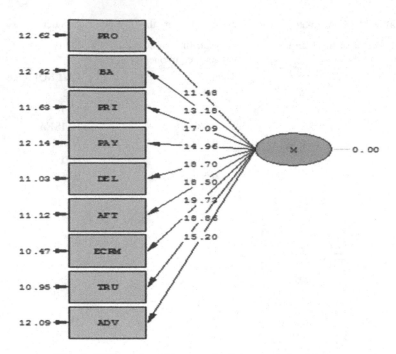

Fig. 6. The results of factor analysis in a meaningful state

Figure 6 examines the significance of each single parameter and the coefficients of the model error. For a parameter to be meaningful, a meaningful number with the value of T must be smaller than 1.96, and is larger than 1.96 and therefore meaningful.

5 Results Statistics

Examining the demographic data of customers indicates that:

- For age, the highest age group was in the age group of 36–40 years with 78 persons (28.2%) and the lowest frequency was over the age of 45 years with 17 persons (6.1%).
- In terms of gender, 233 people (84.1%) of respondents in men and 34 in the equivalent (12.3%) are women.
- The highest level of education was related to the level of education of persons related to the level of baccalaureate with 159 persons (57.4%), and the lowest frequency was related to the level of education for a master's degree with 12 persons (4.3%). Because the distribution of educational levels is disproportionate and because of the small size of the sample at the masters and diploma levels, it is not possible to very much be the results of analyzing the hypotheses related to the level of education.
- The highest frequency was related to the use of the Internet with 146 people (52.7%) and the lowest frequency of non-use of the Internet 13 persons (4.7%).

– The highest frequency of Internet shopping was found with 246 people (88.9%) and the least frequent was the lack of Internet shopping with 31 persons (11.1%).

5.1 Conclusion of the Research

Among the questions related to the variable, the features affecting the Internet of Things in the electronic business, the ECRM index has the highest correlation and the index of the product (the first factor) has the lowest correlation. The significance of each single parameter and the error coefficients of the model is examined. For a parameter to be meaningful, a meaningful number with the value of T must be smaller than −1.96, and is larger than 1.96 and therefore meaningful. The results between Friedman and confirmatory factor analysis are because they pursue separate goals, so when we use the factor load for ranking, the variables are based on the correlation with the relation that is associated with the factors affecting The Internet of things is prioritized in the electronic business, and the variable that has the highest correlation (more factor load factor) has a higher priority. Accordingly, the results of factor analysis show that online management of customer relationship management in the first place has the highest correlation with the factors affecting the electronic business (and the product management system at the final rank), the least correlation with the factors affecting the electronic business Is. On the other hand, using Friedman's approach to prioritizing variables, after reviewing the literature, and determining the factors affecting the Internet of Things in an electronic business, is carried out to rank variables. In this way, variables are prioritized by considering their rank and without examining their relationship. That is, each variable is assigned a rating based on the score obtained with the average value of the rating, which indicates the rank of that variable. The results of Friedman's analysis and confirmatory factor analysis are shown in Table 14.

Table 14. Comparison of Friedman's results and confirmation factor analysis

Ranking based on factor analysis	Ranking based on Friedman test	Priority
ECRM Online Customer Relationship Management System	Price Management System	1
Trust and Security Management System	Brand or brand management system	2
Shop and Delivery Management System	Payment Management System	3
Advertising and marketing management	Trust and Security Management System	4
Price Management System	After sales service management system	5
Brand or brand management system	Product management system	6
Product management system	Online Customer Relationship Management System	7
	Shop and Delivery Management System	8
	Advertising and marketing management	9

To use statistical techniques, it must first be determined that the collected data is normal or abnormal. Because if the distribution of the collected data is normal, the parametric tests can be used to test the hypotheses, and if non-parametric tests are not normal. For this purpose, at this stage, the results of Kolmogorov-Smirnov test for each of the variables were investigated. According to the results of the above table, since the level of significance for all components is larger than the error value of 0.05, these variables have a normal distribution.

According to Friedman rankings, the price management system with the highest of 6.38 and advertising and marketing management with the lowest of 4.28 priority number one and nine have influenced the factors affecting the Internet of Things in the electronic business. But in the ECRM factor analysis, top priority and product management are top priority. Also, the two components of post-sales/post-sales management/delivery and delivery management are the same, with equal importance in the third rank and pay management/advertising and marketing management alike, with equal importance in the fourth place.

5.2   Concluding Remarks

According to the findings of this research, it can be admitted that in order to achieve true success in the business world of electronics, knowing about the target customers is a competitive advantage based on knowledge extracted from their needs and beliefs, business owners Online can organize investment in different sectors according to the priorities they have, organize and improve the business management of the electronic business.

As a result of online management, customer relationship with the highest ranking correlation coefficient in this research should be prioritized by electronic business owners. In the virtual world when communicating with customers Though we have many tools for day-to-day technology achievements, but still challenging the performance of executives is that conversations and correspondence, although most ideal in the face-to-face manner, take place online, It would be very difficult to establish mental and psychological influence on the Internet, but in the business world, it needs to spend heavily on tracking each customer's clicks on each link and web page as well as creating large databases and conversions. This raw data is a useful knowledge for business owners who have data mining and advisory systems Greater will be more intense.

Empirical studies in recent years indicate that information technology is an important factor affecting different disciplines and has improved the quality of employee training. With increasing training, empowerment and value are created in a person, and positive changes such as progress in work, increasing efforts to achieve success appear in person. The results of this study were consistent with the results of the research Ali (2018). As information technology can eliminate computational time-consuming activities, and speed up summarization in performance.

Advertising, branding and marketing are among the factors affecting Internet businesses. Each of these three items has different mechanisms and operating methods in Internet businesses. The main factor that is very effective in sales and success is the attention to branding, the selection of a proper name that is relevant to the work or can remind the audience of the elements of the activity, making the name and signature worthy of being, High-quality graphics, and the presence of a specific color palette of the right branding measures.

We suggest that the following factors be put on the agenda of electronic business companies:

- Create value for the customer
- Unique service offering
- Encouragement and encouragement to buy and return to the site
- Prepare an attractive site
- Create virtual communities
- Create security and trust
- Full attention to customer relationship
- Simple and effective business flow
- Allow customer assistance to each other.
- Help customers get their work done
- Design a basic and defective model for electronic business
- The company's online preparedness for fast-response photography against sudden and moderate changes

5.3   Suggestions for Future Researchers

  - Although in this research we have pointed to several indicators that mention these indicators in literature as "factors affecting the Internet of Things in the electronic business" (both positive factors and factors that act as a barrier to action But they did not address how these barriers were removed, and this could be a good thing for future research.
- This research was conducted in private e-business companies. The research can be done in state-owned e-business companies and compared the results.
- The importance of these indicators separately in different stages of implementation and implementation of business-electric systems can be examined.
- In this research, we sought to determine the relationship between the factors affecting the Internet of Things in the business of electronics and the success of the electronic business in the companies with the electronic symbol, but it should be noted that intermediary agents may have an effect on this relationship (for example, Various management styles). So, in the future, you can do this.
- Ranking factors affecting the Internet of Things in the electronic business by methods other than Friedman tests and factor analysis in private companies.
- Comparing the priority of factors affecting the Internet of Things in the electronic business in Iranian companies with non-Iranian companies such as Amazon.
- This research can be carried out in a broader and wider statistical community at the provincial or national level or between countries.

References

Agarwal, A., Shankar, R., Tiwari, M.K.: Modeling the metrics of lean, agile and leagile supply chain: an ANP-based approach. Eur. J. Oper. Res. **173**, 211–225 (2005)

Alavi, M., Leidner, D.E.: Knowledge management and knowledge management systems: conceptual foundations and research issues. MIS Q. **25**(1), 107–136 (2001)

Padyab, A.: Exploring the dimensions of individual privacy concerns in relation to the Internet of Things use situations. Digit. Policy Regul. Govern. **20**(6), 528–544 (2018). https://doi.org/10.1108/DPRG-05-2018-0023

Amit, R., Zott, C.: Value creation in e-business. Strategic Manag. J. **22**, 493–520 (2001)

Anderson, B.B., Hansen, J.V., Lowry, P.B., Summers, S.L.: Model checking for design and assurance of e-Business processes. Decis. Support Syst. **39**, 333–344 (2005)

Belanger, F., Hiller, J.S., Smith, W.J.: Trustworthiness in electronic commerce: the role of privacy, security, and site attributes. J. Strategic Inf. Syst. **11**(3–4), 245–270 (2002)

Bente, G., Baptist, O., Leuschner, H.: To buy or not to buy: influence of seller photos and reputation on buyer trust and purchas e-behavior. Int. J. Hum. Comput. Stud. **70**(1), 1–13 (2012)

Bokma, A.: E-business and virtual enterprises: Integrated information and knowledge management and its use in virtual organizations, 2nd edn. IFIP (2000)

Bremser, W.G., Chung, Q.B.: A framework for performance measurement in the e-business environment. Electron. Commerce Res. Appl. Electron. Commerce Res. Appl. **4**(4), 395–412 (2005)

Casalo, L., Flavián, C., Guinalíu, M.: The role of perceived usability reputation, satisfaction and consumer familiarity on the website loyalty formation process. J. Comput. Hum. Behav. **24** (6), 2927–2944 (2008)

Cassidy, A.: A Practical Guide to Planning for E-business Success: How to Enable Your Enterprise. St. Lucie Press, Boca Raton (2002)

Kim, S.C., Tao, W., Shin, N., Kim, K.S.: An empirical study of customers' perceptions of security and trust in e-payment systems. J. Electron. Commerce Res. Appl. **9**, 84–95 (2010)

Chen, S.: Strategic Management of E-business, 2nd edn. Wiley, Hoboken (2005)

Cote, L., Sabourin, V., Vezina, M.: New electronic business models of small and medium sized enterprise development, 24 February 2005. www.cefrio.gc.ca

Deise, M.V., Nowikow, K., King, P., Wright, A.: Executive's Guide to e-business. Wiley, Hoboken (2000)

Doaee, H., Jafariyan, H.: A 3D Analysis of the Role of Culture, Policy and Technology in E-Commerce, Computer Science Research Center. Cultural Engineering Magazine No. 47 – 48 (2010)

European Commission Enterprise Directorate General, The European e-business report, 15 January 2005. www.europa.eu.int/comm/enterprise/ict/policy/watch/index.htm

Fahey, L., Srivastava, R., Sharon, J.S., Smith, D.E.: Linking e-business and operating processes: the role of knowledge management. IBM Syst. J. **40**(4), 889–907 (2001)

Filos, E., Banahan, Eoin P.: Will the organisation disappear? The challenges of the new economy and future perspectives. In: Camarinha-Matos, L.M., Afsarmanesh, H., Rabelo, R.J. (eds.) PRO-VE 2000. ITIFIP, vol. 56, pp. 3–20. Springer, Boston, MA (2001). https://doi.org/10.1007/978-0-387-35399-9_1

Firestone, J.M.: Enterprise knowledge portals and e-business solutions. White Paper No. Sixteen, Executive Information Systems, Inc., 1 October (2000)

Ghaffari Estiani, P., Mousavi Basri, S.M., Ghahari, B., Mahmudvandi, Z.: The Role and Effectiveness of Oral-To-Mouth Advertising and Viral Marketing on Purchasing Consumers of the Computer Science Research Center, Magazine Mizagh Managers, No. 41 (2009)

Grigoria, D., Casatib, F., Castellanosb, M., Dayalb, U., Sayalb, M., Shan, M.C.: Business process intelligence. Comput. Ind. **53**, 321–343 (2004)

Hauge, E., Eikebrokk, T.R., Olsen, D.H., Moe, C.E., Braadland, F.: SMEs competence for e-business success: exploring the gap between the needs for e-business competence and training offered. Edgar Research Serviceboks 415, N-4604 Kristiansand (2002)

Hayes, J., Finnegan, P.: Assessing the potential of e-business models: towards a framework for assisting decision–makers. Eur. J. Oper. Res. **160**, 365–379 (2005)

Hinz, O., Hann, I., Spann, M.: Price discrimination in e-commerce? An examination of dynamic pricing in name- your-own price markets. MIS Q. **35**(1), 81–98 (2011)

Chen, Y.H., Hsu, I.C: Website attributes that increase consumer purchase intention: a conjoint analysis. J. Bus. Res. **63**, 1007–1014 (2010)

Shin, J., Chung, K.H., Oh, J.S., Lee, C.W.: The effect of site quality on repurchase intention in Internet shopping through mediating variables: the case of university students in South Korea. Int. J. Inf. Manag. **33**(3), 453–463 (2013)

Jamshidi, M.: Viral Marketing on the Internet, Computer Science Research Center, Tadbir Magazine, No. 121 (2002)

Jing, P., Ying, C.: Theoretical and empirical study on success factors to enhance customer trust in e-commerce. In: IEEE 3rd International Conference on Information Management, Innovation Management and Industrial Engineering, vol. 3, pp. 496–499 (2010)

Jones, M.Y., Spence, M.T., Vallaster, C.: Creating emotions via B2C websites. Bus. Horizons **51**, 419–428 (2008)

Glova, J.: Business Models for the Internet Things (2014)

Jun, S., Lun, J.: Credibility evaluation of B-to-C E-commerce enterprise. In: IEEE International Conference on Business Management and Electronic Information, vol. 2, pp. 497–499 (2011)

Kha, L.: Critical success factors for business-to-consumer e-business: lessons from Amazon and Dell. Master of Science in Management of Technology Thesis, Massachusetts Institute of Technology, June 2000

Kidwell, J.J., Linde, K.M.V., Johnson, S.L.: Applying corporate knowledge management practices in higher education. Educ. Q. **23**(4), 28–33 (2000)

Koh, J., Kim, Y.G.: Knowledge sharing in virtual communities: an e-business perspective. Expert Syst. Appl. **26**, 155–166 (2004)

Kong, W.C., Hung, Y.C.: Modeling initial and repeat online trust in B2C ecommerce. In: Proceedings of the 39th Hawaii International Conference on System Sciences, vol. 6. IEEE (2006)

Lal, K.: Determinants of the adoption of e-business technologies. Telematics Inform. **22**, 181–199 (2005)

Li, P.P., Chang, S.T.-L.: A holistic framework of e-business strategy: the case of Haier in China. J. Global Inf. Manag. **12**, 44–62 (2004)

Liao, S.: Knowledge management technologies and applications – literature review from 1995 to 2002. Expert Syst. Appl. **25**, 155–164 (2003)

Lin, C.T., Chiu, H., Tseng, Y.H.: Agility evaluation using fuzzy logic. Int. J. Prod. Econ. **101**, 353–368 (2005)

Pinochet, L.H.C., Lopes, E.L., Srulzon, C.H.F., Onusic, L.M.: The influence of the attributes of "Internet of Things" products on functional and emotional experiences of purchase intention. Innov. Manag. Rev. **15**(3), 303–320 (2017). https://doi.org/10.1108/INMR-05-2018-0028

Malhotra, Y.: Integrating knowledge management technologies in organizational business processes. J. Knowl. Manag. (Emerald) Spec. Issue Knowl. Manag. Technol. Q3 (2004)

Mallahotra, Y.: Knowledge management for e-business performance: advancing information strategy to internet time. Inf. Strategy Executive's J. **16**(4), 5–16 (2000)

Mason, J.: From e-learning to e-knowledge. In: Rao, M. (ed.) Knowledge Management Tools and Techniques, pp. 320–328. Elsevier, London (2005)

Mehta, M.R., Shah, J.R., Morgan, G.W.: Merging an e-business solution framework with CIS curriculum. J. Inf. Syst. Educ. **16**(1), 65 (2005)

Meuter, M., Bitner, M., Ostrom, A., Brown, S.: Choosing among alternative service delivery modes: an investigation of customer trial of self-service technologies. J. Market. **69**(2), 61–83 (2005)

Millar, T., Matthew, L.N., Shen, S.Y., Shaw, M.J.: E-Business management models (2003)

Ayub, M.: Review and analysis of the Internet effect of IOT objects on electronic business models. Publications. Tehran Third Printing (2017)

Moodley, S.: The challenge of e-business for the South African apparel sector. Technovation **23**, 557–570 (2003)

Murphy, G.B., Tocher, N.: Gender differences in the effectiveness of online trust building information cues: an empirical examination. J. High Technol. Manag. **22**, 26–35 (2011)

OECD: The economic and societal impacts of electronic commerce: Preliminary findings and research agenda. OECD (1999)

OECD: ICT, e-business and SMEs. OECD (2004)

Overtveldt, S.V.: Creating an Infrastructure for e-business: Computing in an e-business world, 19 August 2005. cremesti.com/articles/e-infrastructure-byIBM.pdf - Supplemental Result

Pavlou, P.A., Liang, H., Xue, Y.: Understanding and mitigating uncertainty in online exchange relationships: a principal– agent perspective. MIS Q. **31**(1), 105–136 (2007)

Hirsch, P.B.: The goose that laid the golden eggs: personal data and the Internet of Things. J. Bus. Strategy **40**(1), 48–52 (2019). https://doi.org/10.1108/JBS-10-2018-0176

Phan, D.D.: E-business development for competitive advantages. Inf. Manag. **40**, 581–590 (2003)

Piris, L., Fitzgerald, G., Serrano, A.: Strategic motivators and expected benefits from e-commerce in traditional organizations. Int. J. Inf. Manag. **24**, 489–506 (2004)

Plessis, M.: Drivers of knowledge management in the corporate environment'. Int. J. Inf. Manag. **25**, 193–202 (2005)

Plessis, M.D., Boon, J.A.: Knowledge management in e-business and customer relationship management: South African case study findings. Int. J. Inf. Manag. **24**, 73–86 (2004)

Quaddus, M., Xu, J.: Adoption and diffusion of knowledge management systems: field studies of factors and variables. Knowl.-Based Syst. **18**, 107–115 (2005)

Ranganathan, C., Ganapathy, S.: Key dimensions of business to consumer websites. J. Inf. Manag. **39**(6), 457–465 (2002)

Raub, S., Wittich, D.V.: Implementing knowledge management. Eur. Manag. J. **22**(6), 714–724 (2004)

Ren, J., Yusuf, Y.Y., Burns, N.D.: Organizational competitiveness: identifying the critical agile attributes using principal component analysis. In: 16th International Conference on Production Research, ID 0588, 29 July 3-August, Prague, Czech Republic (2001)

Rose, S., Hair, N., Clark, M.: Online customer experience: a review of the business-to-consumer online purchase context. Br. Acad. Manag. Int. J. Manag. Rev. **13**, 24–39 (2011)

Ross, J., Vitale, M., Weill, P.: From place to space: Migrating to profitable electronic commerce business models. MIT Sloan School of Management, Working Paper, November, no. 4358-01 (2001)

Sambamurthy, V., Bharadwaj, A., Grover, V.: Shaping agility through digital options: reconceptualizing the role of information technology in contemporary firms. MIS Q. **27**(2), 237–263 (2003)

Leminen, S., Rajahonka, M., Westerlund, M., Wendelin, R.: The future of the Internet of Things: toward heterarchical ecosystems and service business models. J. Bus. Ind. Market. **33**(6), 749–767 (2016). https://doi.org/10.1108/JBIM-10-2015-0206

Seshasai, S., Gupta, A., Kumar, A.: An integrated and collaborative framework for business design: a knowledge engineering approach. Data Knowl. Eng. **52**, 157–179 (2005)

Shaharudin, M.R., Mansor, S.W., Abu Hassan, A., Omar, M.W., Harun, E.H.: The relationship between product quality and purchase intention: the case of Malaysia's national motorcycle/scooter manufacturer. Afric. J. Bus. Manag. **5**(20), 8163–8176 (2011)

Shevchenko, A.A., Shevchenko, O.O.: B2B e-hubs in emerging landscape of knowledge based economy. Electron. Commerce Res. Appl. **4**, 113–123 (2005)

Stanford, S.B., Kidd, P.T.: How business models influence the development of e-business applications. In: Proceedings of e-Business and e-Work 2000, Madrid, Spain, 18–20 October 2000

Svantesson, D., Clarke, R.: A best practice model for E-consumer protection. J. Secur. Law Comput. Secur. **26**, 31–37 (2010)

Cheung, M.T., Liao, Z.: Supply-side hurdles in Internet B2C e-commerce: an empirical investigation. Trans. Eng. Manag. IEEE **50**(4), 458–469 (2003)

Vasarhelyi, M., Greenstein, M.: Underlying principles of the electronization of business. Int. J. Account. Inf. Syst. **4**, 1–25 (2003). **63**(9), 1007–1014 (2010)

White, A., Daniel, E.M., Mohdzain, M.: The role of emergent information technologies and systems in enabling supply chain agility. Int. J. Inf. Manag. **25**, 396–410 (2005)

White Paper, Business to Business on the internet: IBM's e-Business Philosophy, May 2002

Ou, C.X., Sia, C.L.: Consumer trust and distrust: an issue of website design. Int. J. Hum.-Comput. Stud. **68**(12), 913–934 (2010)

Yang, H., Xu, B., Zhou, Y., Zhang, J., Biao, D.: Return in B2C E- commerce enterprises. In: Third International Conference on Business Intelligence and Financial Engineering, pp. 95–98. IEEE (2010)

Zain, M., Rose, R.C., Abdullah, I., Masrom, M.: The relationship between information technology acceptance and organizational agility in Malaysia. Inf. Manag. **42**, 829–839 (2005)

Zhang, Y., Fang, Y., Wei, K., Ramsey, E., McCole, P., Chen, H.: Repurchase intention in B2C e-commerce- a relationship quality perspective. J. Comput. Hum. Behav. **48**, 199–200 (2011)

Zhao, Y.: Evaluation model of B2C ecommerce site based on consumer perspective, pp. 2230–2232. IEEE (2011)

Zheng, J., Caldwell, N., Harland, C., Powell, P., Woerndl, M., Xu, S.: Small firms and e-business: cautiousness, contingency and cost-benefit. J. Purchasing Supply Manag. **10**, 27–39 (2004)

Internet of Things and Iranian Companies; An Empirical Survey from Industrial Market Perspective

Hamze Sadeghizadeh[1], Amirhossein Davaei Markazi[2(✉)],
and Saeed Shavvalpour[1]

[1] School of Progress Engineering,
Iran University of Science and Technology, Tehran, Iran
h_sadeghizadeh@pgre.iust.ac.ir,
shavvalpour@iust.ac.ir
[2] School of Mechanical Engineering,
Iran University of Science and Technology, Tehran, Iran
markazi@iust.ac.ir

Abstract. As an emerging technology, the Internet of Things (IoT) is predicted to offer hopeful solutions to renovate the operation and role of many current industrial systems. In this paper we investigated the state of the art among Iranian Companies delivering IoT-based Services through an empirical Survey in years of 2017 and 2018. Based on survey results from more than 150 companies; Top 3 verticals receiving services from private sector are respectively (1) Environment, agriculture, and natural resources, (2) Health and well-being, (3) Transportation and urban development. Some verticals that have a little attention from public and private sector are Security and disaster management, and Education & Research. From business model view, B2G services are the main targeted segments of market among supply side of IoT services and applications. About challenges that are in front of private sectors, it seems the lake of external funding be the main constraint of Iranian service providers in the field of IoT.

Keywords: Internet of Things · Iranian companies · Macro challenges · IoT-based services · Empirical survey

1 Introduction

By the quick expansion of Internet and communications technology, our lives are increasingly led into a fictional space of virtual world. People can conversation, work, shop, and retain pets and flowers in the virtual world delivered by the network. For integrating this fictional space and real-world we need a new technology which is called as Internet of Things (IoT) [1]. As displayed in Fig. 1, IoT allows individuals and things to be linked anytime, anyplace, with anything and anyone, preferably using any path/network and any service [2].

© Springer Nature Switzerland AG 2019
L. Grandinetti et al. (Eds.): TopHPC 2019, CCIS 891, pp. 220–227, 2019.
https://doi.org/10.1007/978-3-030-33495-6_16

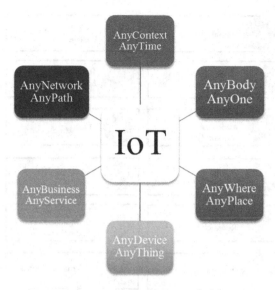

Fig. 1. IoT definition [2]

The elementary idea of this concept is the universal attendance around us of a diversity of things or objects such as Radio Frequency IDentification (RFID) labels, sensors, actuators, mobile phones, etc. which, through distinctive addressing schemes, are able to interrelate with each other and collaborate with their neighbors to reach shared objectives [3].

Rest of the paper is structured as follows: Sect. 2 gives an overview of Iran's industries and survey's domain. After presenting research process in 3rd section, the state of IoT in Industries is discussed in Sect. 4. Section 4 concludes survey study with references at the end.

2   An Overview of Iranian Verticals Operating in IoT

Iran country has a central government and 18 ministries [4]. In 2017 and in order to initiate National Enterprise Architecture (NEA) the Iranian ministry of Information and communication Technology (ICT) asked a number of Iranian public agencies to provide a catalog of services and finally The Iranian government sectors fit into 14 verticals that are shown in Fig. 2 [5]. It was interesting that there were some IoT-based applications and more interesting for being a representative of private sector for delivering aimed service.

Potentialities presented by the IoT make likely the growth of a huge number of applications, of which only a very small portion is currently accessible to our society. Many are the areas and the situations in which new applications would likely improve the quality of our lives: at home, while travelling, when sick, at work, when jogging and at the gym [6]. Among the current applications, there are some famous applications that are operating now and presented in Table 1.

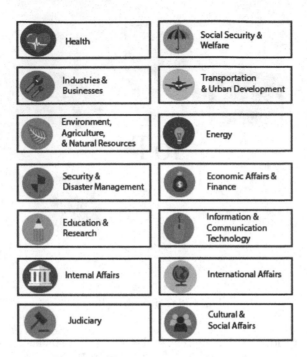

Fig. 2. The Iranian government sectors [5]

Table 1. Sectors and main IoT-based applications

Sector	Main application
Environment, agriculture, and natural resources	Industrial Plants
Cultural and social affairs	Social networking, Historical queries, Losses, Thefts
Health and well-being	Tracking, Identification, Authentication, Sensing
Transportation and urban development	Logistics, Assisted driving, Mobile ticketing, Environment monitoring, Augmented Maps
Industries and businesses	Comfortable offices
Energy	Smart Homes
Social security and welfare	Data Collection
Judiciary	Smart Bracelet
Security and disaster management	City Information Model

3 Research Process

Through investigating indications in literature, we applied a 4-steps research process shown in Fig. 3.

Fig. 3. The research methodology adopted in this work.

By analyzing related surveys mainly conducted and published between 2010 and 2017 in scientific databases, we derive a characterization of the literature (See Table 2), aiming at showing in a qualitative way the behavior of the research, and shape the foundation for the next steps.

Table 2. Chronological Summary of Previous Surveys in the IoT field.

References	Description	Year
Atzori et al. [3]	Vision, Apps	2010
Miorandi et al. [6]	Vision, Challenges	2012
Vermesan et al. [7]	Vision, Apps, Governance	2013
Zanella et al. [8]	Smart Cities	2014
Xu et al. [9]	Industries	
Stankovic [10]	Directions	
Al-Fuqaha et al. [11]	Protocols, Challenges, Apps	2015
Granjal et al. [12]	Security Protocols, Challenges	
Ray [13]	IoT Architectures	2016
Sethi et al. [14]	IoT Architectures, Apps	2017
Qiu et al. [15]	Challenges, Apps	2018
Van Der Zeeuw et al. [16]	Social aspects	2019

Empirical survey in this work started in the middle of 2017 and finished in the middle of 2018. Research tools were questionnaire and interview. At the end of deadline more than 150 companies participated in our research work and their results analysis presented in the following.

4 Results

4.1 Participants Background

First of all, it is necessary to understand the background of participated and interviewed companies. This issue presented in Table 3.

Table 3. Background of companies in survey (N = 153)

Category	Main filed	Companies (%)
Information Technology (IT)	Software	61
	Hardware	49
	Internet	12
	Network	9
Communication Technology (CT)	Infrastructure	38
	Mobile Phone	34
	Radio Communication	13
Information Community	E-Government	29
	E-Learning	18
	E-Commerce	14
Manufactures & Contractors	Telecommunication	47
	EPC	39
	ICT Equipment	26
Consultants and R&D	IT	52
	CT	38
	Security	29
	Standards	16
Commercial Services	Imports	31
	Exports	19

According to Table 3, the survey's community comprised of companies from private sector and have a good level of methodological standards for the next analysis.

4.2 Verticals Ranking Form IoT View

One of the main questions that survey asked participants, was about the main verticals they delivered IoT-based services. The result of Integrating answers to this question presented in Fig. 4.

As it can be visible, top 3 Iranian verticals on the demand-side of IoT-based services are respectively (1) Environment, agriculture, and natural resources, (2) Health and well-being, (3) Transportation and urban development. Among 14 verticals, there are 2 critical verticals; Security and disaster management, and Education & Research, that they haven't yet received sufficient attention from public and consequently private sector.

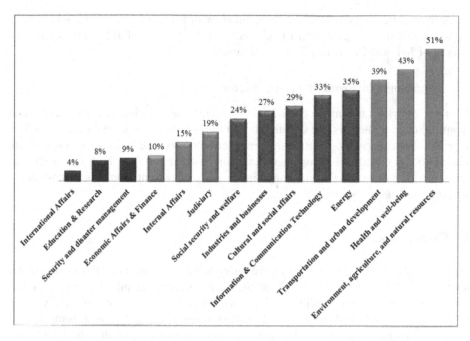

Fig. 4. State of demand for IoT-based Services among Iranian Verticals

4.3 Business Model: Market Segments

After collecting and integrating survey data, the main part of company's business model was delivering services to government bodies in a Business to Government manner (B2G = 69%) (See Fig. 5).

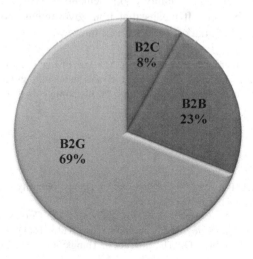

Fig. 5. Targeted Market by Iranian IoT-based Service Providers

Although B2G services have a dominant share of IoT Market (69%) in Iran, but it is very possible that by evolution of IoT value chain, growth of B2B services in Iran market of IoT will be occurred in next years.

4.4 Challenges in Front of Private Sector

After all interviews and at the end of questionnaires, the participants answer a question about the main challenges and constraints that they passed or they are passing to deliver their IoT-based services. The results are interesting because more of challenges are related to the outside of companies; Lake of External funding, high cost IoTbased activities, lake of external consultants, inflexibility of current laws, uncertainty about demand market.

5   Conclusion

There is no doubt that IoT technology/industry had a big impact on Iranian companies specially ICT companies/industry that they are operating in different layers of this technology: Sensors and equipment, Network layer, platform layer, and finally application layer. Although there are good understanding about IoT among some Iranian verticals such Environment, agriculture, and natural resources, Health and well-being, Transportation and urban development. Among 14 verticals, 2 critical verticals; Security and disaster management, and Education & Research, haven't yet received sufficient attention from public and consequently private sector. Although B2G services have a dominant share of IoT Market in Iran, but it is very possible that by evolution of IoT value chain, growth of B2B services in Iran market of IoT will be occurred in near future. Challenges in front of private sector, in delivering IoT-based, such as Lake of External funding, High cost IoT-based activities, Lake of external consultants, Inflexibility of current laws, Uncertainty about demand market, are entirely related to outside of companies borders. It means that Iran government should consider some facilities for catalyzing IoT market formation in future.

Acknowledgement.   We thank Dr. Majid Rasouli, Manager of Iran IoT Roadmap project in Iran Telecommunication Research Center (ITRC) and their assistance with giving us parallel data, Dr. Paolo Bellavista from University of Bologna, Italy, for his comments that greatly improved this paper.

References

1. Kumar, J.S., Patel, D.R.: A survey on Internet of Things: Security and privacy issues. Int. J. Comput. Appl. **90**(11), 20–26 (2014)
2. Perera, C., Zaslavsky, A., Christen, P., Georgakopoulos, D.: Context aware computing for the Internet of Things: a survey. IEEE commun. surv. Tuto. **16**(1), 414–454 (2014)
3. Atzori, L., Iera, A., Morabito, G.: The Internet of Things: a survey. Comput. Netw. **54**(15), 2787–2805 (2010)
4. Ministries of Iran. https://en.wikipedia.org/wiki/Ministries_of_Iran. Accessed 12 Dec 2018

5. Aliee, F.S., Bagheriasl, R., Mahjoorian, A., Mobasheri, M., Hosieni, F., Golpayegani, D.: A classification taxonomy for public services in Iran. ICEIS **2**, 712–718 (2018)
6. Miorandi, D., Sicari, S., De Pellegrini, F., Chlamtac, I.: Internet of Things: vision, applications and research challenges. Ad Hoc Netw. **10**(7), 1497–1516 (2012)
7. Vermesan, O., Friess, P. (eds.): Internet of Things: Converging Technologies for Smart Environments and Integrated Ecosystems. River Publishers, Gistrup (2013)
8. Zanella, A., Bui, N., Castellani, A., Vangelista, L., Zorzi, M.: Internet of things for smart cities. IEEE Internet Things J. **1**(1), 22–32 (2014)
9. Da Xu, L., He, W., Li, S.: Internet of Things in industries: a survey. IEEE Trans. Ind. Inf. **10**(4), 2233–2243 (2014)
10. Stankovic, J.A.: Research directions for the Internet of Things. IEEE Internet Things J. **1**(1), 3–9 (2014)
11. Al-Fuqaha, A., Guizani, M., Mohammadi, M., Aledhari, M., Ayyash, M.: Internet of Things: a survey on enabling technologies, protocols, and applications. IEEE Commun. Surv. Tutor. **17**(4), 2347–2376 (2015)
12. Granjal, J., Monteiro, E., Silva, J.S.: Security for the Internet of Things: a survey of existing protocols and open research issues. IEEE Commun. Surv. Tutor. **17**(3), 1294–1312 (2015)
13. Ray, P.P.: A survey on Internet of Things architectures. J. King Saud Univ.-Comput. Inf. Sci. **30**(3), 291–319 (2016)
14. Sethi, P., Sarangi, S.R.: Internet of Things: architectures, protocols, and applications. J. Electr. Comput. Eng. **2017**, 1–25 (2017)
15. Qiu, T., Chen, N., Li, K., Atiquzzaman, M., Zhao, W.: How can heterogeneous Internet of Things build our future: a survey. IEEE Commun. Surv. Tutor. **20**(3), 2011–2027 (2018)
16. Van Der Zeeuw, A., Van Deursen, A.J., Jansen, G.: Inequalities in the social use of the Internet of things: a capital and skills perspective. new media & society (2019)

Modeling and Performance Evaluation of Map Layer Loading in Mobile Edge Computing Paradigm

Reza Shojaee$^{(\boxtimes)}$ and Nasser Yazdani

School of Electrical and Computer Engineering, University of Tehran,
Tehran, Iran
{r.shojaee,yazdani}@ut.ac.ir

Abstract. Mobile Edge Computing (MEC) is a promising technology to prepare cloud services in the vicinity of mobile radio network. Mobile users could
offload their tasks to the MEC servers to acquire results in minimum latency. In
this paper, we analyze the performance of map layer loading in the MEC
paradigm. We show that how workload, connection failure and service rate
could influence on the mean response time and job rejection probability. We
extract the service architecture of map layer loading in the MEC platform. Each
phase is mapped into an M/M/1/C or M/M/K/C queue. The cyclic inter-
dependencies among sub-models are resolved by fixed-point iteration technique.
Discrete Event Simulation (DES) is conducted to find numerical results for each
sub-model. Finally, the behavior of mean response time and job rejection
probability as two performance metrics is studied.

Keywords: Mobile Edge Computing · Performance evaluation · Radio access
network · Discrete Event Simulation

1 Introduction

Cloud computing is recognized as a technology which facilitates resource provisioning
for their subscribers [1]. Transforming computing to the 5th utility is the principle target
of its designers and theoreticians [2]. Pay-per-use or pay-as-you-go is the main slogan
of cloud computing [3]. Nowadays, popular websites prefer to contract by the cloud
providers instead of deploying their own data centers. Therefore, cloud providers
guarantee the Quality of Service (QoS) based on the negotiated Service Level
Agreement (SLA) [4, 5].

Mobile Cloud Computing (MCC) is the deploying cloud technology for mobile
devices to overcome their constraints in computing power, memory and energy [6]. MCC
helps mobile devices to offload their heavy tasks to the cloud data centers. However, due
to the long latency between mobile devices and cloud, the MCC for real-time software is
not applicable. Therefore, Mobile Edge Computing (MEC) has been introduced in 2014.
MEC is known as one of the fundamental technologies in 5G. MEC will be formed by the
collaboration between mobile operators and cloud providers. In MEC, Base Stations
(BSs) are equipped by small data centers which are called MEC servers.

© Springer Nature Switzerland AG 2019
L. Grandinetti et al. (Eds.): TopHPC 2019, CCIS 891, pp. 228–239, 2019.
https://doi.org/10.1007/978-3-030-33495-6_17

MEC helps to reduce the congestion through the network backbone because it solves the local problems locally [7]. Furthermore, MEC provides new opportunities for the developers by the extra information such as location-awareness and signal strength [8].

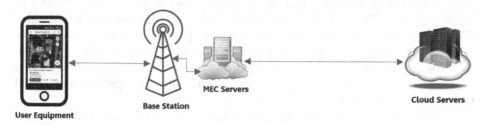

Fig. 1. Map layer loading service architecture in MEC paradigm[1] (The screenshot of map application is taken from *Google Maps* [16])

In this paper, we concentrate to the map layer loading process on the MEC platform. We show that how MEC technology facilitates map layer loading in terms of performance metrics. Based on Fig. 1 the User Equipment (UE) sends layer loading request to the nearest BS. The result should be provided for users by assistance of MEC servers. We show that how workload variation, connection failure and service rate could impact on the job rejection probability and mean response time. In contrast to MCC, the result (for processing or caching requests) will be provided by MEC servers in more cases.

The main contributions of this paper can be categorized as follows:

- The servicing steps for the map layer loading in the MEC paradigm are proposed
- Various parameters such as workload, connection failure and service rate are considered
- Discrete Event Simulation (DES) is conducted by MATLAB [9] to obtain numerical results
- The cyclic inter-dependencies among sub-models resolved by fixed-point iteration [10] technique

The rest of the paper is organized as follows. Related works are brought in Sect. 2. System description and assumptions are presented in Sect. 3. In Sect. 4, each phase is mapped by an M/M/1/C or M/M/K/C queue. Numerical results are provided in Sect. 5. Conclusion and future works are presented in Sect. 6.

2 Related Works

Khazaei et al. [11] analyzed the performance for the cloud data centers by an approximated analytical model. They found an accurate estimation for the probability of response time by their proposed model. In [12], performance evaluation is accomplished when high degree of virtualization is existed. They supposed that arrival requests formed batch tasks with Poisson distribution. They illustrate that the performance of cloud data centers may be improved if incoming tasks are partitioned on the basis of the coefficient of variation of service time and batch size. Ghosh et al. [13] proposed a framework to analyze the performance of Infrastructure as a Service (IaaS) in the cloud data centers. They categorized servers to the hot (ready), warm (standby) and cold (off servers) pools. They model each pool behavior by Markov Reward Model (MRM). Raei et al. [14] proposed an analytical model to evaluate performance of cloudlet in MCC. They supposed that mobile users want to create a Virtual Machine (VM) on the cloudlet or public cloud servers. They presented service architecture for the mentioned process. They also modeled their service architecture by MRM. Maheshwari et al. [15] presented a framework to performance analysis of augmented reality applications in city scale. They studied augmented reality application on hybrid cloud and edge networks.

Despite the mentioned works, our paper devotes to the performance evaluation of a tangible service in the map application where benefits from MEC technology. This work could help the designers to adjust the number of servers, buffer lengths and processing power to achieve the desirable performance metrics.

3   System Description and Assumptions

As mentioned before, MEC could help mobile applications to provide better services in terms of performance metrics. In this paper, we focus on smart map application. We show that how the MEC technology could impact on the layer loading process in map application. Layer loading is the basic service of every map application. Layers are consisted of satellite view, street view, traffic information and so on. As we illustrated in Fig. 2, the map application in the User Equipment (UE) transfers the layer loading request to the nearest Base Station (BS).

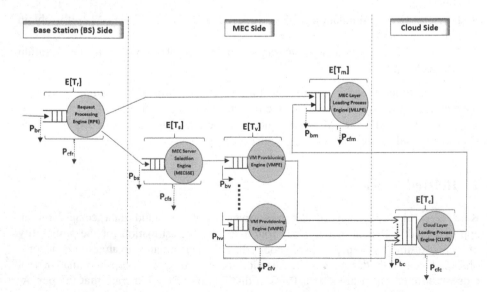

Fig. 2. Servicing steps for the map layer loading in MEC platform

Request Processing Engine (RPE) in the BS detects the kind of each request. It conducts various application requests to their servicing steps. BSs are equipped by MEC servers. Usually, BSs are connected to the MEC servers via fiber optics. Therefore, the latency between BSs and MEC servers is infinitesimal in comparison to broadband networks. As we shown in Fig. 2, if the map VM already located on the MEC servers and the requested layer is already existed on the VM, then the corresponding layer will be transferred to the UE by MEC Layer Loading Process Engine (MLLPE).

MLLPE analyzes the request id and returns the corresponding layer. Another scenario is the unavailability of Map VM on the MEC servers. In this situation, proper server should be chosen to host the map VM. MEC Server Selection Engine (MECSSE) is responsible for this important task. The resource allocation method for map VM is supposed as sequential.

In each phase of servicing steps, connection failure is likely. We capture this probability in our model and we study the effect of connection failure in the mean response time and job rejection probability. Moreover, request blocking may be occurred when the queue of each step is full. Virtual Machine Provisioning Engine (VMPE) allocates the resources for map VM in each server. The created VM could be deployed in the future requests. If no server has sufficient capacity for hosting map VM, then the request will be forwarded to the cloud server. Note that, after VM creation on the MEC server, because the new map VM is empty, therefore the corresponding layer will be transferred from cloud server to the new VM before servicing the UE request. The last phase is the Cloud Layer Loading Process Engine (CLLPE) which is responsible for the layer caching on the cloud side.

In this paper, we assume that map layer loading requests are identical and they need the same requirements. Furthermore, map VM properties on each server in terms of processing and storage capabilities are equal. We also supposed that MEC servers are connected to the only one cloud provider for the mentioned application. Moreover, connection failure is probable for a request in the queue or in the engine. We assume that request arrivals are determined by Poisson process and request service times have an exponential distribution. We also supposed that capacity of the servers is the same. All queues in the Fig. 2 follow the First-Come-First-Serve (FCFS) manner. Table 1 describes the symbols which we are used throughout the paper.

Table 1. Symbols and their descriptions

Symbols	Descriptions
λ_r, λ_m, λ_s, λ_c and λ_v	Effective request arrival rates to the RPE, MLLPE, MECSSE, CLLPE and VMPE sub-models respectively
$1/\mu_r$, $1/\mu_m$, $1/\mu_s$, $1/\mu_c$ and $1/\mu_v$	Mean service time in the RPE, MLLPE, MECSSE, CLLPE and VMPE sub-models respectively
δ_h	Connection failure rate
P_s	Probability that at least one of the servers in MEC accept the request
P_{OA}	Probability that a request is pertaining to the other applications
P_{vm}	The probability in which the map VM already existed on the MEC server

(*continued*)

Table 1. (*continued*)

Symbols	Descriptions
P_{br}, P_{bm}, P_{bs}, P_{bc} and P_{bv}	The probability of blocking the requests in the RPE, MLLPE, MECSSE, CLLPE and VMPE queues respectively
P_{cfr}, P_{cfm}, P_{cfs}, P_{cfc} and P_{cfv}	The dropping probability of the requests due to connection failure in the RPE, MLLPE, MECSSE, CLLPE and VMPE sub-models respectively
N	Maximum number of virtual machines on each MEC server
M_v	Maximum length of VMPE queue
$E[T_r]$, $E[T_m]$, $E[T_s]$, $E[T_c]$ and $E[T_v]$	Mean delay time for a request in the RPE, MLLPE, MECSSE, CLLPE and VMPE sub-models respectively
$E[T_{Response}]$	Mean response delay
$P_{Rejection}$	Job rejection probability

4   Stochastic Modeling

In this section, we consider the end-to-end performance evaluation for layer loading of map application in the MEC architecture. In previous section, we illustrated the servicing steps of layer loading by five sub-models. Each sub-model contains a queue and its corresponding engine. RPE, MECSSE, MLLPE, VMPE and CLLPE are those sub-models. We supposed that requests arrivals are determined by Poisson process and job service times have exponential distribution. Therefore, we can describe these sub-models by Continuous Time Markov Chain (CTMC).

4.1   Request Processing Engine (RPE) Sub-model

This sub-model is responsible for the processing of the arrival requests. This queue is the M/M/1/C because the arrival rates follow the Poisson process, service time is having the exponential distribution and the buffer of queueing is restricted. We have two ways for computing the outputs of RPE sub-model. We can draw its corresponding CTMC which is like the Birth-death Markov chain. By writing the balance equations, we can obtain the outputs in steady state probability. However, we conducted a Discrete-Event Simulation in MATLAB [9] for each sub-model. P_{br}, P_{cfr} and $E[T_r]$ are the outputs of RPE sub-model.

4.2   MEC Server Selection Engine (MECSSE) Sub-model

As mentioned before, this sub-model chooses the appropriate server of MEC to host map VM. Effective request arrival rate of MECSSE can be obtained by the following equation:

$$\lambda_s = \lambda_r * \left(1 - P_{br} - P_{cfr} - P_{OA} - P_{VM}\right) \tag{1}$$

Among the above variables P_{OA} and P_{VM} assumed be given while P_{br} and P_{cfr} can be obtained from RPE sub-model. Same as RPE, the MECSSE queue can be modeled by an M/M/1/C.

4.3 Virtual Machine Provisioning Engine (WMPE)

This sub-model imitates the VM provisioning steps on the top of only one server. Figure 3 illustrates the corresponding CTMC for the VMPE sub-model. Every state in VMPE is labeled by *(i,j)*. *i* is the representative of queued requests and *j* is the number of running VM on the server.

VMPE goes from (0,0) to (1,0) by arrival of new request for the VM provisioning in the *k*th server. However, after completion of VM provisioning service the VMPE moves from (1,0) to (0,1). Effect of connection failure in the model is shown by δ_h.

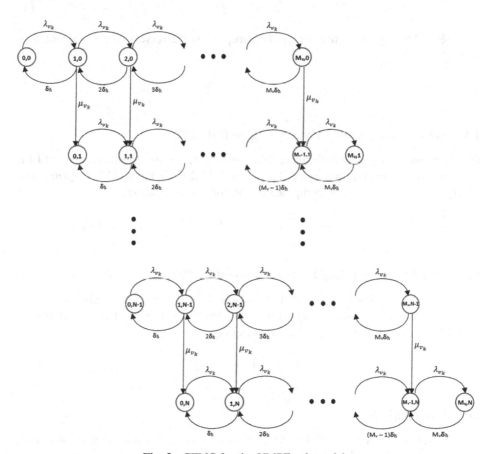

Fig. 3. CTMC for the VMPE sub-model

The effective arrival rate to VMPE can be obtained by the following equation:

$$\lambda_v = \lambda_s * \left(1 - P_{bs} - P_{cfs}\right) \tag{2}$$

We define the Q_i as the steady state probability where the requests will be blocked by all servers before the ith server.

$$Q_i = \begin{cases} 1 & if\ i = 1 \\ \prod_{j=1}^{i-1} P_{bv_j} & \forall\ 2 \leq i \leq k \end{cases} \tag{3}$$

Therefore, the effective request arrival rate for the kth server can be obtained by the following equation:

$$\lambda_{v_k} = \lambda_v * Q_k \tag{4}$$

The probability that a request can be accepted by the servers can be computed as follow:

$$P_s = 1 - Q_k * P_{bv_k} \tag{5}$$

4.4   MEC Layer Loading Process Engine (MLLPE)

This sub-model is responsible for caching map layers in the MEC servers. MLLPE works in the Software-as-a-Service (SaaS) level. MLLPE is the M/M/K/C queue. The effective arrival rates can be computed as the following equation:

$$\lambda_m = \lambda_r * \left(1 - P_{br} - P_{cfr} - P_{OA} + P_{vm}\right) + \left(1 - P_{cfc}\right) * \lambda_c * P_s \tag{6}$$

4.5   Cloud Layer Loading Process Engine (CLLPE)

CLLPE is working same as the MLLPE sub-model. CLLPE queue is also M/M/K/C. However, CLLPE caches the map layer in the cloud side. The effective arrival rate for the CLLPE can be obtained by the following equation:

$$\lambda_c = \lambda_v * \left(1 - P_{cfv}\right) \tag{7}$$

4.6   Overall Model

The overall model consists of the five mentioned sub-models and the relation among them. Figure 4 shows the relation among sub-models.

To resolve the cyclic inter-dependencies among sub-models, we apply the fixed-point iteration [10] method. The following algorithm is applied to resolve the mentioned issue:

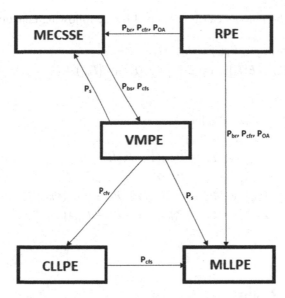

Fig. 4. The interaction graph among sub-models

Algorithm 1. Fixed-point iteration technique to resolve the cyclic inter-dependencies among sub-models

---

Inputs: RPE, MECSSE, VMPE, MLLPE and CLLPE sub-models

Outputs: The steady state probabilities for P_{br}, P_{cfr}, P_{bs}, P_{cfs}, P_{bm}, P_{cfm}, P_s, P_{cfv}, P_{bc} and P_{cfc}

declare P_{br}, P_{cfr}, P_{bs}, P_{cfs}, P_{bm}, P_{cfm}, P_s, P_{cfv}, P_{bc} and P_{cfc}

declare P'_{bs} and P'_{cfs}

declare E_f // It defines the upper bound for error

P_{cfr}, P_{br} ← RPE()

P_{bs} ← 0

P_{cfs} ← 0

repeat:

    P'_{bs} ← P_{bs}

    P'_{cfs} ← P_{cfs}

    P_{bs}, P_{cfs} ← MECSSE(P_{br}, P_{cfr}, P_{OA}, P_s)

    P_s, P_{cfv} ← VMPE(P_{bs}, P_{cfs})

    P_{bm}, P_{cfm} ← MLLPE(P_{br}, P_{cfr}, P_{OA}, P_s, P_{cfc})

    P_{bc}, P_{cfc} ← CLLPE(P_{cfv})

    P_{bs} ← (Min(P_{bs}, P'_{bs}) + (|P_{bs} − P'_{bs}|/2))

    P_{cfs} ← (Min(P_{cfs}, P'_{cfs}) + (|P_{cfs} − P'_{cfs}|/2))

until (|P_{bs} − P'_{bs}| < E_f and |P_{cfs} − P'_{cfs}| < E_f)

Print the output values for P_{br}, P_{cfr}, P_{bs}, P_{cfs}, P_{bm}, P_{cfm}, P_s, P_{cfv}, P_{bc} and P_{cfc}

Furthermore, the outputs can be obtained by the following equations:

$$E[T_{Response}] = E[T_r] + P_{vm}E[T_m] +$$
$$(1 - P_{vm})\{E[T_s] + P_s(E[T_v] + E[T_c] + E[T_m]) + (1 - P_s)E[T_c]\}$$
(8)

$$P_{Rejection} = (P_{br} + P_{cfr}) +$$
$$(1 - P_{br} - P_{cfr})(P_{vm})(P_{bm} + P_{cfm}) +$$
$$(1 - P_{br} - P_{cfr})(1 - P_{vm})(P_{bs} + P_{cfs}) +$$
$$(1 - P_{br} - P_{cfr})(1 - P_{vm})(1 - P_{bs} - P_{cfs})(1 - P_s)(P_{bc} + P_{cfc}) +$$
$$(1 - P_{br} - P_{cfr})(1 - P_{vm})(1 - P_{bs} - P_{cfs})P_s(P_{bv} + P_{cfv}) +$$
$$(1 - P_{br} - P_{cfr})(1 - P_{vm})(1 - P_{bs} - P_{cfs})P_s(1 - P_{bv} - P_{cfv})(P_{bc} + P_{cfc}) +$$
$$(1 - P_{br} - P_{cfr})(1 - P_{vm})(1 - P_{bs} - P_{cfs})P_s(1 - P_{bv} - P_{cfv})(1 - P_{bc} - P_{cfc})(P_{bm} + P_{cfm})$$
(9)

5 Numerical Results

We evaluated the performance of map layer loading phases on the MEC architecture by conducting DES for each sub-model. To find the final results, we implemented the fixed-point iteration technique which is brought in Algorithm 1 on the Python. Two important metrics (Mean response time and job rejection probability) were presented in the various situation of workload, service rate and connection failure. All results of DES are based on 96% confidence level. Moreover, the upper bound for error (E_f) in Algorithm 1 is set as 10^{-7}. Note that, the number in the parentheses shows the number of servers in the MEC small data center.

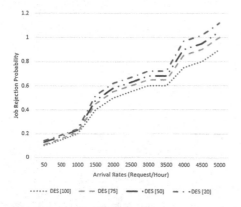

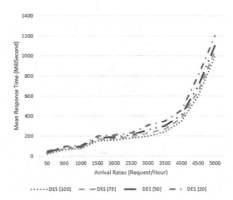

Fig. 5. Job rejection probability vs arrival rates   **Fig. 6.** Mean response time vs arrival rates

Figures 5 and 6 illustrate the influence of arrival rates on the job rejection probability and mean response time respectively. The increasing of arrival rates augments the blocking probability in the queues. Therefore, as shown in Eq. 9 the increasing blocking probability rises the job rejection probability.

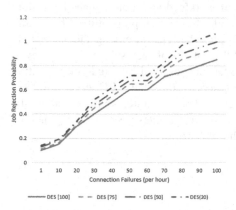

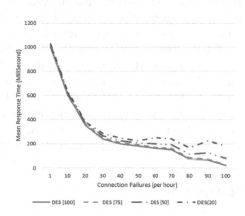

Fig. 7. Job rejection probability vs connection failure

Fig. 8. Mean response time vs connection failure

Figures 7 and 8 display the impact of connection failure in the job rejection probability and mean response time respectively. Note that, as shown in Fig. 8, increasing connection failure drops more requests and diminishes the mean response time for remaining requests.

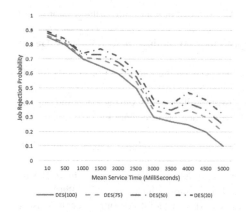

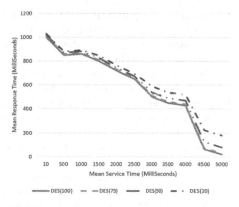

Fig. 9. Job rejection probability vs mean service time

Fig. 10. Mean response time vs mean service time

Figures 9 and 10 illustrate the effect of mean service time on the job rejection probability and mean service time respectively. Note that, $1/\mu_m$ is considered for mean service time. Because μ_m is the most important factor in the servicing rate of whole model.

6   Conclusion and Future Works

In this paper, we modeled and analyzed the performance of map layer loading service in the MEC paradigm. We study the impact of workload variation, connection failure and service rate on the mean response time and job rejection probability. The overall model consists of five stochastic sub-models plus the relation among them. Discrete Event Simulation (DES) is conducted to find the outputs of each sub-model. However, the cyclic inter-dependencies among sub-models are resolved by fixed-point iteration method. The results could help the system designers to adjust the MEC server resources to trade-off between cost and performance metrics. For the future, we plan to apply the general distribution for the arrival requests and service rates instead of exponential distribution to make the problem closer in the real world.

References

1. Mell, P., Grance, T.: The NIST definition of cloud computing. Commun. ACM **53**(6), 50 (2010)
2. Buyya, R., Yeo, C.S., Venugopal, S., Broberg, J., Brandic, I.: Cloud computing and emerging IT platforms: vision, hype, and reality for delivering computing as the 5th utility. Futur. Gener. Comput. Syst. **25**(6), 599–616 (2009)
3. Wang, Q., Ren, K., Meng, X.: When cloud meets ebay: towards effective pricing for cloud computing. In: INFOCOM, 2012 Proceedings IEEE, pp. 936–944 (2012)
4. Roy, N., Dubey, A., Gokhale, A.: Efficient autoscaling in the cloud using predictive models for workload forecasting. In: IEEE International Conference on Cloud Computing (CLOUD) 2011, pp. 500–507 (2011)
5. Tak, B.-C., Urgaonkar, B., Sivasubramaniam, A.: To move or not to move: the economics of cloud computing. In: HotCloud (2011)
6. Fernando, N., Loke, S.W., Rahayu, W.: Mobile cloud computing : a survey. Futur. Gener. Comput. Syst. **29**(1), 84–106 (2013)
7. Zhang, J., Xie, W., Yang, F., Bi, Q.: Mobile edge computing and field trial results for 5G low latency scenario. China Commun. **13**(2), 174–182 (2016)
8. Nunna, S., et al.: Enabling real-time context-aware collaboration through 5 g and mobile edge computing. In: 2015 12th International Conference on Information Technology-New Generations (ITNG), pp. 601–605 (2015)
9. Hanselman, D.C., Littlefield, B.: MATLAB; Version 4: User's Guide. Prentice Hall PTR, Upper Saddle River (1995)
10. Mainkar, V., Trivedi, K.S.: Sufficient conditions for existence of a fixed point in stochastic reward net-based iterative models. Softw. Eng. IEEE Trans. **22**(9), 640–653 (1996)
11. Khazaei, H., Misic, J., Misic, V.: Performance analysis of cloud computing centers using m/g/m/m + r queuing systems. IEEE Trans. Parallel Distrib. Syst. **23**(5), 936–943 (2012)
12. Khazaei, H., Misic, J., Misic, V.B.: Performance of cloud centers with high degree of virtualization under batch task arrivals. IEEE Trans. Parallel Distrib. Syst. **24**(12), 2429–2438 (2013)
13. Ghosh, R., Longo, F., Naik, V., Trivedi, K.: Modeling and performance analysis of large scale IaaS clouds. Futur. Gener. Comput. Syst. **29**(5), 1216–1234 (2013)

14. Raei, H., Yazdani, N., Shojaee, R.: Modeling and performance analysis of cloudlet in mobile cloud computing. Perform. Eval. **107**, 34–53 (2017)
15. Maheshwari, S., Raychaudhuri, D., Seskar, I., Bronzino, F.: Scalability and performance evaluation of edge cloud systems for latency constrained applications. In: 2018 IEEE/ACM Symposium on Edge Computing (SEC), pp. 286–299 (2018)
16. Google Maps: maps.google.com. Accessed 21 Oct 2017

Extracting Effective Features for Descriptive Analysis of Household Energy Consumption Using Smart Home Data

Hadiseh Moradi Sani, Soroush Omidvar Tehrani, Behshid Behkamal, and Haleh Amintoosi[✉]

Computer Engineering Department, Faculty of Engineering,
Ferdowsi University of Mashhad, Mashhad, Iran
{hadise.moradisani, omidvar}@mail.um.ac.ir,
{behkamal, amintoosi}@um.ac.ir

Abstract. The household energy consumption has a large share of global energy consumption. To have better understanding of energy generation, management and surplus storage, we need to discover implicit patterns of consumers' behavior and identify the factors affecting their performance. The main goal of this paper is to descriptively analyze the pattern of household energy consumption using RECS2015 dataset. To this end, we focus on selecting the most effective subset of features from high dimensional dataset that leads to a better understanding of data, reducing computation time and improving prediction performance. The result of this study can help decision makers to investigate the living conditions of families in different levels of society to ensure that their life style is well enough or should be improved.

Keywords: Feature selection · Smart home · Household energy consumption · Correlation analysis

1 Introduction

With Rapid growth in world population, the development of urbanization and increased living needs, we are facing a sharp increase in energy consumption, specially the urban energy consumption which accounts for a large share of produced energy. Based on the reports [1], 31 Percent of U.S. households have problems paying their energy bills. Recent US Energy Information Administration (EIA) reports[1] also show that about one in three households reduces its basic needs, such as food and medicine, to pay its own energy costs, and 14 percent of them have received disconnection alerts due to not paying their bills [1]. Such issues and reports reflect the importance of monitoring energy consumption in the household sector, because applications such as demand-side management, energy management and demand-response management require an energy-monitoring. Another survey on energy consumption patterns of users in South Africa shows that citizen's income level has been identified to be an influential factor in

---

[1] https://www.eia.gov/consumption/residential/.

© Springer Nature Switzerland AG 2019
L. Grandinetti et al. (Eds.): TopHPC 2019, CCIS 891, pp. 240–252, 2019.
https://doi.org/10.1007/978-3-030-33495-6_18

energy consumption, and it has been determined that more than 70 percent of low-income families rely on energy sources rather than electricity. It also mentions that there exists a negative relationship between the electricity cost and its consumption [2].

Therefore, energy generation, management and surplus storage requires understanding consumers' behavior and the factors affecting it. Analysis of such factors and their impact on power consumption are among the important tasks in understanding the consumption patterns. Authors in [3] have investigated these factors and the role of each in urban residential energy consumption in China. Figure 1 categorizes these factors.

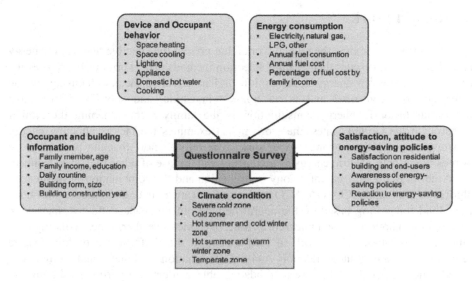

Fig. 1. Factors affecting the urban residential energy consumption [3]

As shown in Fig. 1, there is a wide range of factors affecting energy consumption. Thus, a main challenge for researchers in this field is selecting the most important factors in order to reduce the computational overhead. Lots of work has been done to increase the energy savings. Since the extracted data has high volume and diversity, big data techniques can be performed. Moreover, different data clustering algorithms, data analysis techniques and correlation between them have also been used in these works [4–7]. Other benefits of dimensionality reduction include avoiding over-fitting, resisting noise, and strengthening prediction performance in learning algorithms [8].

In this article, we consider RECS2015[2] dataset containing the data of residential energy consumption of US households in 2015, identify the most important features and remove non-relevant ones using feature selection techniques. Specifically, we first perform preprocessing with the aim of categorizing the features in a way to remove the unnecessary ones and select those features that are more relevant to the amount of

---

[2] https://www.eia.gov/consumption/residential/index.php.

energy consumed per year and the cost paid to the energy supplier. Once effective features are selected, clustering is done in two different ways on the data pertaining to these features in order to partition it to meaningful and accurate clusters. These clusters are then analyzed to infer useful information about the households' patterns of use, and its relation with their income and living spaces.

The structure of the paper is as follows. Section 2 discusses the related work. Section 3 presents proposed approach in three sub sections of introducing the dataset, preprocessing, and descriptive analysis. Section 4 presents a detailed analysis of the obtained results. Finally, Sect. 5 concludes the paper.

2   Related Work

In this section, we review the related works that present the effective features in energy consumption and discuss their feature selection methods. The work in [9] explores the relationship between energy consumption and its effective factors (focusing on the difference between rural and urban homes) in China, and shows that the most influential factor in energy consumption is the family income. Doing the similar research in the United States, the work in [1] examines the factors affecting energy consumption and their changes over time. The results show that the percentage of homes using only electric energy is increasing and the use of other fuels has declined.

Authors in [8] utilized ant colony optimization and the combination of the genetic algorithm (GA) and ACO (GA-ACO) to select the features in order to predict the short-term load forecasting based on the neural network. In this paper, effective factors on energy consumption are summarized in four categories: weather, time, economy, and random disturbances. The work in [10] evaluated the performance of four feature selection methods (autocorrelation, mutual information, RReliefF and correlation-based) and showed that all these methods are able to identify a portion of the highly-correlated features. Authors in [11] have used a two-step approach for choosing effective features in order to predict the cost of short-term electricity consumption, from the electricity market managers point of view. In the first step, using the Relief algorithm, the irrelevant candidate inputs are removed from the dataset, and then in the second step, using cross-correlation, redundant candidates are filtered. The selected features are then used in the forecast engine, implemented with the neural network. The work in [12] has used a filter-wrapper combination for feature selection with the aim of forecasting the short-term load. Taking the advantages of both filter and wrapper methods, this method first removes irrelevant and redundant properties based on the filter technique and then, by leveraging the wrapper technique and based on the Firefly algorithm, removes the extra features in a way not to decrease the accuracy. Also, in this paper, the need to use a different model to predict the energy consumption in special days has been emphasized.

3 Proposed Approach

The aim of our research is to select the most effective subset of features from high dimensional RECS2015 dataset of the household energy consumption. We believe that this leads to a better understanding of data, reducing computation time and improving prediction performance. There are three general categories for feature selection, namely, wrapper, embedded and filter methods [13]. Wrapper methods measure the "usefulness" of features based on the classifier performance. Naïve Bayes, SVM and K-means are some of the most famous examples of this group. Embedded methods are quite similar to wrapper methods since they are also used to optimize the objective function or performance of a learning algorithm. Decision tree algorithms such as CART and C4.5 are among the known examples of this category. In contrast, filter methods pick up the intrinsic properties of the features (i.e., the "relevance" of the features) measured via univariate/multivariate statistics instead of cross-validation performance. information gain, chi-square test, fisher score, correlation coefficient and variance threshold are well-known examples of filter methods.

In our research two attributes: 'the annual amount of energy consumption in thousand BTU' (known as TOTALBTU) and 'the total annual cost paid in dollars' (TOTALDOL) were considered as target attributes. The goal is to select the most effective features based on these target attributes. So, filter methods are known to be the best choice. To this end, we are taking the "relevance" of the features into account in order to analyze the relationship between features and target attributes. In the following, the dataset is described first. Then, two feature selection methods including Correlation Coefficient and Information Gain are explained and the results are compared. Finally, the results of descriptive analysis using clustering algorithms are presented.

3.1 Introduction of Dataset

The dataset investigated in this paper (RECS2015 (see footnote 2)) has been collected in 2015 by EIA [14] from American households and includes 740 attributes and 5,686 data records. Each data record represents a US household in 2015 based on data properties such as number and type of household appliances, patterns of energy consumption, structural characteristics of residential buildings, characteristics of household members and information of energy supplier. More detailed information is available on the EIA website [15].

3.2 Feature Selection

Generally raw data is normally vulnerable to noise, inconsistency and missing values, which may affect the result of data mining process. Hence, preprocessing is required to increase the data quality and the outcome of mining process. Data preprocessing includes four steps: data cleansing, data aggregation, data reduction and discretization. The first step is data clearing which is the process of detecting and removing inaccurate records within the dataset. The dataset used in this paper was first investigated for such records and it was observed that all records have valid values within their specified range, no needing for cleaning. The second step is data aggregation, which is the process of combining data from different sources into a single coherent data store. This process had also been carried out by the EIA since the data collected from households and energy suppliers were all integrated in one single dataset. The major step of the pre-processing is the data reduction step, which is providing a reduced representation of dataset without compromising its integrity. Data reduction in normally considered as either reducing dimensions or reducing the number of records.

Since the dataset is highly dimensional containing 740 attributes, we need to apply an appropriate feature selection method to select the most effective features and reduce irrelevant dimensions. First, we analyze the correlation between all features and target attributes using Pearson's correlation method. This method measures the statistical relationship, or association between two quantitative variables. It is known as the best method of measuring the association between variables of interest because it is based on the method of covariance. Then, we categorize features and analyze the most relevant feature in each category. We also use Information Gain as a metric to analyze the relevancy of categorized data. In the following, feature selection methods are expressed with more detail.

Correlation Analysis on All Dimensions: In this method, the correlation between every feature and the two target attributes (TOTALBTU and TOTALDOL) is calculated. This process is done for all 740 features and the features with highest correlation are going to be selected as the most important attributes. In should be noted that features expressing a single concept with several measurement units were removed.

Table 1 demonstrates these features and their correlation values with target attributes, ordered by the third column (labeled as 'Correlation with Class Label #1 TOTALDOL').

As shown in Table 2, the 12 features obtained by Correlation on Categorical Data method are among the features obtained when the correlation is computed for all features. This means that by categorizing the features first and selecting the most important attributes next, it is possible to identify and select the most important features out of 740 features.

Correlation Analysis on Categorized Data: In this step, we categorize dimensions of the dataset in order to select the most important features in each group. This process is performed with the help of experts. The result is partitioning 740 features into 8 categories, briefly described below.

1. *Home Characteristics*: In this category, reside the characteristics of the residential house such as: the geographical area, type of house (flat, apartment, etc.), number of floors, number of bedrooms, materials used in the construction of windows and ceilings, climate of that region, etc., constituting 47 features.
2. *Human Characteristics:* In this category, reside the characteristics of family members living in the house, such as: number of family members, number of members in/below the legal age, the level of education, annual income, number of days in a week they are at home, etc., There exist 20 features in this category.

Table 1. Selected features from all dimentions

No.	Feature	↓Corr. with TOTALDOL	Corr. with TOTALBTU
1	TOTALDOLSPH	0.66	0.65
2	TOTROOMS	0.52	0.56
3	WINDOWS	0.52	0.55
4	TOTALDOLNEC	0.51	0.42
5	TOTALBTUSPH	0.51	0.85
6	DOLELLGT	0.47	0.38
7	LGTOUTNUM	0.47	0.48
8	SWIMPOOL	0.46	0.45
9	DOLELCOL	0.45	0.24
10	TOTALBTUNEC	0.44	0.42
11	LGTINNUM	0.44	0.44
12	DOLELAHUCOL	0.43	0.25
13	TOTALDOLWTH	0.43	0.14
14	CELLAR	0.42	0.54
15	TOTALBTUWTH	0.41	0.52
16	SOLAR	0.4	0.42
17	STORIES	0.4	0.48
18	DRYUSE	0.39	0.37
19	WASHLOAD	0.39	0.36
20	DOLELAHUHEAT	0.38	0.55
21	MONEYPY	0.37	0.34

Table 2. Selected features from categorized data

No.	Category	Feature	Corr. with TOTALDOL	Corr. with TOTALBTU
1	Appliances	LGTINNUM	0.44	0.44
2	Characteristics	LGTOUTNUM	0.47	0.48
3	Energy	TOTALBTUSPH	0.51	0.85
4	Characteristics	TOTALBTUWTH	0.41	0.52
5	Behavioral	WASHLOAD	0.39	0.36
6	Characteristics	DRYUSE	0.39	0.37
7	Financial Billing	DOLELAHUHEAT	0.38	0.55
8		TOTALDOLSPH	0.66	0.65
9	Fuel Consumption Profile	SOLAR	0.4	0.42
10	Home	TOTROOMS	0.52	0.56
11	Characteristics	WINDOWS	0.52	0.55
12	Human Characteristics	MONEYPY	0.37	0.34

3. *Financial Billing:* In this category, reside the characteristics related to the bills paid for electricity consumption such as: the total amount paid, the amount paid for each electric appliance, etc. which constitute 57 features.

4. *Behavioral Characteristics:* This category includes those behavioral characteristics of family members which are related to energy consumption. These characteristics include: number of times household appliances such as stove, oven, microwave oven, washing machine, dishwasher, etc. are being used in a week, home temperatures in summer and winter, home temperatures when no one is at home, number of days cooling or heating appliances are being used, etc. 37 features are selected related to this category.

5. *Energy Consumption Characteristics:* In this category, features related to the way energy is consumed are located. These features define whether electricity has been used for cooling, heating or air conditioning, whether or not gas is being used for cooling, heating, cooking, etc., total amount of electricity consumption in 2015, the amount of electricity consumed by each appliance, and so on. The total number of feature residing in this category are 126 features.

6. *Fuel Consumption profile:* In this category, features related to the type and amount of fuel consumed within the house are located. They include the type of fuel being used to heat the bathroom, jacuzzi, and other household devices, the amount of solar energy used, and so on. 14 features related to this category are selected.

7. *Appliances Characteristics*: In this category, the characteristics of home appliances such as: number of fridges and their specifications (e.g., the year of production) and so on are described. The total number of features residing in this category is 114.

8. *Flag:* There are features with the role of flag for other features, that are considered to be in this category. Total number of flag features is 325.

Since the flag category did not provide useful information for data analysis, features belonging to this category were removed from the dataset. The remaining process was thus performed on features in categories 1 to 7.

Then, by performing the correlation function on each category, 12 features are identified with the highest correlation to two target attributes, as expressed in Table 2.

Information Gain (IG): One of the schemes for creating a decision tree is to use the notion of entropy. In this method, in each step, the attribute that mostly reduces the entropy is selected as the node of the tree (the root or the middle node). The reason behind this intuition is that attributes which reduce the entropy more are able to give more information about the data. In this section, the information gain method was performed on each of the 7 categories described above and in each category, features providing more information about the two target attributes are selected as the most important and effective ones. Table 3 shows these features and their information gain (IG) values for two target attributes.

Using information gain, 13 important features, most of which are similar to the features selected by two previous methods are identified that used the correlation criterion.

Table 3. Selected features by information gain

No.	Category	Feature	IG with TOTALDOL	IG with TOTALBTU
1	behavioral	WASHLOAD	0.12	0.11
2	characteristics	DRYUSE	0.12	0.11
3	appliances	LGTOUTNUM	0.16	0.17
4	characteristics	H2OHEATAPT	0.12	0.14
5	energy	TOTALBTUSPH	0.17	0.6
6	characteristics	BTUELAHUHEAT	0.13	0.34
7	Financial billing	TOTALDOLSPH	0.3	0.32
8		DOLELAHUHEAT	0.14	0.31
9	Fuel consumption profile	SOLAR	0.12	0.14
10	Home	WINDOWS	0.18	0.21
11	characteristics	TOTROOMS	0.18	0.2
12	Human	MONEYPY	0.06	0.05
13	characteristics	NHSLDMEM	0.06	0.04

After applying three different methods for feature section, we understood that most of the features obtained from these methods are common. Thus, the intersection between the attributes selected by the above-mentioned three methods are 10 features as shown in Table 4.

Table 4. The feature suite selected by all methods

No.	Feature	Mean	SD
1	WASHLOAD	3.6	4.0
2	DRYRUSE	3.5	4.1
3	LGTOUTNUM	0.7	1.5
4	TOTALBTUSPH	33039.2	33492.6
5	DOLELAHUHEAT	24.2	37.5
6	TOTALDOLSPH	511.1	482.3
7	SOLAR	−0.4	0.8
8	TOTROOMS	6.2	2.4
9	WINDOWS	35.8	11.3
10	MONEYPY	3.7	2.2

3.3 Descriptive Analysis

The result of preprocessing was the selection of 10 features which have the highest correlation with the target attributes. Since data records are un-labelled, unsupervised clustering schemes have been used to analyze the data.

Before going through the clustering process, we first investigated the information related to 10 selected features (expressed in Table 4). According to the data presented in Table 4, on average, the annual income of the households is $40,000 to $60,000. Houses are normally large with an average of 6 bedrooms. Solar energy is rarely used to generate electricity. The energy required to warm the living space is high, therefore, there is the possibility of large amounts of energy being wasted in heating/cooling the residential space. In the following section, the behavior of households in energy consumption is investigated using two clustering algorithms.

Distance-Based Clustering: The first clustering algorithm is *k-means* which is a distance-based clustering. Via this technique, data is divided into k clusters in a way that the sum of the square of distances between the clusters is minimal [16]. *k-means* was first executed with different k values to find the optimal value for k, i.e., the value that results in the optimal number of clusters. The result was $k = 4$. Then, *k-means* with $k = 4$ was performed to partition the data to four clusters. The size of clusters equal to 1541, 286, 899, and 2960 respectively. Figure 2.a represents a view of the data in each cluster.

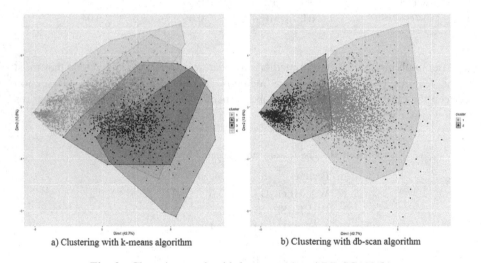

a) Clustering with k-means algorithm   b) Clustering with db-scan algorithm

Fig. 2. Clustering result with k-means (a) and DB-SCAN (b)

Based on the results obtained, the data residing in the first cluster belongs to the families that frequently use the washing machine and the dryer during in a week and use a moderate number of bulbs. The households of this cluster use an average and even lower amount of energy to warm up their home space, so as to pay less for electricity. Solar panels are not used to generate energy. Also, these families have a relatively low annual income (generally between $20,000 and $60,000). This cluster includes 27.10% of the households.

Compared to the first cluster, families residing in the second cluster use higher amount of energy to warm the living spaces and water. So, they pay higher for electricity. Solar panels are not used in these households either. Households in this cluster have a high annual income (generally more than $140,000). Their houses are large and have more rooms and windows compared to other three clusters. This cluster includes 5.02% of all households.

The characteristics of the data in the third cluster is a combination of the first two clusters. About 15.81% of households reside in this cluster.

Families residing in the fourth cluster generally do not have a washing machine and dryer, and the number of lamps used in their houses is very low. These households consume a very low amount of energy to heat their home space. The annual income of these households is very low (less than $4,000). This cluster contains 52.05% of households.

Density-Based Clustering: In density-based algorithms, the clustering is done based on the data density. In other words, the goal is to divide the dataset into subgroups of highly dense regions. Dense regions are then called clusters. We selected *DB-SCAN*[3], a density-based clustering algorithm which is also able to detect outliers -the data points that are in low density regions [16].

Running *DB-SCAN* on the dataset resulted in partitioning the data into two clusters, one with 4464 and the other with 1164 members. 58 records were also identified as outliers. Figure 2.b represents a view of the clusters made by *DB-SCAN*.

Analyzing the data of clusters show that families residing in the first cluster use washing machine and dryer. They also use a quite enough number of lamps in their houses. They also consume medium-to-high level of energy to warm their living spaces, resulting in a fairly high cost to be paid for electricity. The number of bedrooms and windows within the houses are quite large. Households in this cluster have an average-to-high annual income.

As for the second cluster, washing machines and dryers are not used. Moreover, no light bulbs are used outside the houses. The energy consumption is low, resulting in low cost to be paid. Houses in this cluster have a small number of bedrooms and windows. The annual income of families is very low.

Households in outlier data include those with very high or very low income. The amount of using washing machines and dryers in these two groups is very different, as they are used many times during the week. The amount of energy consumed in these households is also different. In general, households in this group show an unfamiliar behavior in energy consumption or their annual income differs from the rest of the community.

4 Discussion

In the previous section, two clustering algorithms were performed on the dataset. Here, the results obtained from these two methods are compared.

According to the results from *k-means* algorithm, low-income households (poor people), most of which are in cluster 4, generally do not use high electricity-consuming devices such as washing machines, and minimize the use of other electrical devices such as bulbs. This can be because they are either unable to pay the bills or they cannot

---

[3] Density-Based Spatial Clustering and Application with Noise.

use their own devices due to the debt to energy supplier. On the other hand, wealthy households which are generally in cluster 2, are not concerned about their energy consumption and their bills, because of their high income, Therefore, they consume energy without worrying about the cost. High energy loss (via windows for example) can also be observed in these households, which could be due to not caring about the cost to be paid.

Looking deeper at the results of *k-means* clustering, it is observed that *k-means* has not clustered the data in an effective way. In fact, only households in two clusters of 2 and 4 are different in their behaviors; the two other clusters, i.e., clusters 1 and 3, are a combination of both poor and wealthy households.

Next, we investigated the clusters provided by *DB-SCAN*. The obtained results are as follows:

- It is observed that low-income families are in cluster 2, accounting for about 20.4% of the total households. These households have small houses since the number of their windows and rooms are few. Therefore, in these houses, less energy is consumed for heating, and less energy is lost.
- Families with higher income reside in cluster 1, accounting for about 78.5% of the total households. They have big houses, causing energy to be wasted more.
- Outlier data is also about 1.02% of the data. Households residing in this cluster behave quite differently from the rest. So, putting them in a separate cluster as outlier prevents them from negatively affecting the behavioral analysis of ordinary households.

To conclude, results show that *DB-SCAN* clustering algorithm outperforms *k-means* in partitioning the data into more accurate clusters, as well as resulting valid pattern of the energy consumption.

5   Conclusion

The main purpose of this study is to analyze the pattern of household energy consumption using RECS2015 dataset on energy consumption which is collected from 5686 American households in 2015. Initially, a feature selection process was performed on all data to identify features that are more relevant to the amount of energy consumed per year and the cost paid to the energy supplier. By applying different filter methods 10 out of 740 features are selected. After dimensionality reduction, the dataset is clustered using two clustering algorithms of k-means and DB-SCAN in order to achieve accurate clusters and analyzing the life routine of households in terms of energy consumption. The results showed that density-based clustering provides more accurate and meaningful clusters compared to distance-based clustering.

It is worth mentioning that clustering helps to investigate the low-income community to know whether their living conditions (related to the energy sector) are well enough or should be improved. For example, a household may be forced not to consume the required energy due to severe financial problems or it may not be able to repair or purchase the necessary electric appliances. Similar results can also be concluded on high-income households to know whether or not they save energy even though they are not concerned about paying electricity bills.

Appendix A. Description of the Features Mentioned in the Paper

No.	Features	Description
1	BTUELAHUHEAT	Electricity usage for air handlers and boiler pumps used for heating, in thousand Btu at 2015
2	CELLAR	Housing unit over a basement
3	DOLELAHUCOL	Electricity cost for air handlers used for cooling, in dollars, 2015
4	DOLELAHUHEAT	Electricity cost for air handlers and boiler pumps used for heating, in dollars, 2015
5	DOLELCOL	Electricity cost for air conditioning (central systems and individual units), in dollars, 2015
6	DOLELLGT	Electricity cost for indoor and outdoor lighting, in dollars, 2015
7	DRYUSE	Frequency of clothes dryer use
8	H2OHEATAPT	Water heater in apartment or other part of building
9	LGTINNUM	Number of light bulbs installed inside the home
10	LGTOUTNUM	Number of light bulbs installed outside the home
11	MONEYPY	Annual gross household income for the last year
12	NHSLDMEM	Number of household members
13	SOLAR	On-site electricity generation from solar
14	STORIES	Number of stories in a single-family home
15	SWIMPOOL	Swimming pool
16	TOTALBTU	Total usage, in thousand Btu, 2015
17	TOTALBTUNEC	Total usage for other devices and purposes not elsewhere classified, in thousand Btu, 2015
18	TOTALBTUSPH	Total usage for space heating, main and secondary, in thousand Btu, 2015
19	TOTALBTUWTH	Total usage for water heating, main and secondary, in thousand Btu, 2015
20	TOTALDOL	Total cost, in dollars, 2015
21	TOTALDOLNEC	Total cost for other devices and purposes not elsewhere classified, in dollars, 2015
22	TOTALDOLSPH	Total cost for space heating, main and secondary, in dollars, 2015
23	TOTALDOLWTH	Total cost for water heating, main and secondary, in dollars, 2015
24	TOTROOMS	Total number of rooms in the housing unit, excluding bathrooms
25	WASHLOAD	Frequency of clothes washer use
26	WINDOWS	Number of windows

References

1. U.S. Energy Information Administration, 2015 Residential Energy Consumption Survey. https://www.eia.gov/consumption/residential/reports.php
2. Bohlmann, J.A., Inglesi-Lotz, R.J.R.: Analysing the South African residential sector's energy profile. Renew. Sustain. Energy Rev. **96**, 240–252 (2018)
3. Hu, S., Yan, D., Guo, S., Cui, Y., Dong, B.: A survey on energy consumption and energy usage behavior of households and residential building in urban China. Energy Build. **148**, 366–378 (2017)
4. Dinesh, C., Makonin, S., Bajic, I.V.: Incorporating time-of-day usage patterns into non-intrusive load monitoring. In: Signal and Information Processing (GlobalSIP), 2017 IEEE Global Conference, pp. 1110–1114. IEEE (2017)
5. Gajowniczek, K., Ząbkowski, T.: Data mining techniques for detecting household characteristics based on smart meter data. Energies **8**(7), 7407–7427 (2015)
6. Perez-Chacon, R., Talavera-Llames, R.L., Martinez-Alvarez, F., Troncoso, A.: Finding electric energy consumption patterns in big time series data. Distributed Computing and Artificial Intelligence, 13th International Conference. AISC, vol. 474, pp. 231–238. Springer, Cham (2016). https://doi.org/10.1007/978-3-319-40162-1_25
7. Zhang, P., Wu, X., Wang, X., Bi, S.: Short-term load forecasting based on big data technologies. CSEE J. Power Energy Syst. **1**(3), 59–67 (2015)
8. Sheikhan, M., Mohammadi, N.: Neural-based electricity load forecasting using hybrid of GA and ACO for feature selection. Neural Comput. Appl. **21**(8), 1961–1970 (2012). https://doi.org/10.1007/s00521-011-0599-1
9. Zhang, M., Bai, C.: Exploring the influencing factors and decoupling state of residential energy consumption in Shandong. J. cleaner prod. **194**, 253–262 (2018)
10. Koprinska, I., Rana, M., Agelidis, V.G.: Correlation and instance based feature selection for electricity load forecasting. Knowl. -Based Syst. **82**, 29–40 (2015)
11. N. Amjady, F. J. E. C. Keynia, and Management, "Day-ahead price forecasting of electricity markets by a new feature selection algorithm and cascaded neural network technique," vol. 50, no. 12, pp. 2976–2982, 2009
12. Hu, Z., Bao, Y., Xiong, T., Chiong, R.: Hybrid filter–wrapper feature selection for short-term load forecasting. Eng. Appl. Artif. Intell. **40**, 17–27 (2015)
13. Jović, A., Brkić, K., Bogunović, N.: A review of feature selection methods with applications. In: 38th International Convention on Information and Communication Technology, Electronics and Microelectronics (MIPRO), 2015, pp. 1200–1205. IEEE (2015)
14. What's New in How We Use Energy at Home: Results from EIA's 2015 Residential Energy Consumption Survey (RECS). https://www.eia.gov/consumption/residential/reports.php
15. Residential Energy Consumption Survey (RECS) 2015 Household Characteristics Technical Documentation Summary. https://www.eia.gov/consumption/residential/reports.php
16. Han, J., Pei, J., Kamber, M.: Data Mining: Concepts and Techniques. Elsevier, Amsterdam (2011)

A Partitioning Scheme to Route X-Folded TM Topology Deadlock-Free

Mehrnaz Moudi[1($\boxtimes$)] and Mohamed Othman[2]

[1] Department of Computer Engineering, University of Torbat Heydarieh,
Razavi Khorasan Province, Iran
mmoudi@torbath.ac.ir
[2] Department of Communication Technology and Network,
Universiti Putra Malaysia, 43400 UPM Serdang, Selangor D.E., Malaysia
mothman@upm.edu.my

Abstract. In this paper, a partitioning scheme is proposed to improve the performance of x-Folded TM topology. We put forward x-Folded TM as an efficient topology in interconnection networks. Then, the proposed partitioning scheme applied by dividing x-Folded TM topology into three critical partitions. To evaluate its performance, adaptive and deterministic routing algorithms used under different traffic patterns. The obtained results reveal that the average delay at x-Folded TM network is obviously lower than other topologies. In contrast, network throughput for x-Folded TM is higher than Torus and TM. Consequently, the relatively large improved performance for x-Folded TM topology compared other topologies.

Keywords: Partitioning scheme · Routing algorithm · Deadlock · x-Folded TM topology

1 Introduction

One significant feature of communication performance in interconnection networks is the routing algorithm, which should be simple and logical. Routing algorithms provides the shortest path for each route in the given topology by avoiding deadlock. Efficient routing is critical for low latency and high network throughput, and to improve the performance of the interconnection network. In particular, when designing a fast router with low delay, many network properties such as connectivity, fault tolerance and deadlock freedom are of direct importance. Over the years, various algorithms in different studies have been designed and applied to interconnect topologies to remove deadlocks and support communication. Basically, these algorithms can be classified as deterministic and adaptive features of interconnection topologies [1–4].

Topology refers to how the various nodes are mapped to each other in an interconnection network. In the related works including [10–14], a number of

© Springer Nature Switzerland AG 2019
L. Grandinetti et al. (Eds.): TopHPC 2019, CCIS 891, pp. 253–264, 2019.
https://doi.org/10.1007/978-3-030-33495-6_19

interconnect topologies have been proposed to provide bandwidth within a certain amount of time and ensure improved performance. Torus, TM and x-Folded TM (Fig. 1) have been applied in this paper. Torus is a simple and popular topology consisting of a set different components connected together. Next simple topology, TM, is the proposed solution by [15,16] in interconnection networks. Its key feature is the removal of the cycles of each dimension of the torus, producing the TM topology. Third one is x-Folded TM which is a topology by folding the TM topology based on an imaginary x-axis and removing extra links [17].

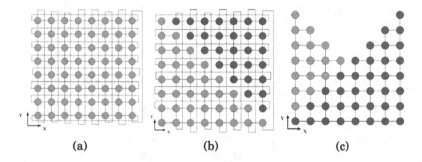

Fig. 1. Different interconnection topologies where $k = 8$ and $n = 2$:(a) Torus, (b) TM and (c) x-Folded TM

Based on the removal of several links and the utilization of shared nodes in x-Folded TM topology, the limited connectivity of the higher level links makes an increase in injection rate, which leads to congestion. The outcome of this relocation and use of shared nodes is network congestion, resulting in an earlier saturation time for this topology in larger network's size. Consequently, these limitations induced us to propose a new scheme in this paper to route the topology deadlock-free. The rest of this paper is organized as follows: deadlock problem is described in Sect. 2. Section 3 presents the partitioning scheme with its verification and applied routing algorithm. Section 4 shows the simulation environment with details. The evaluation results of the routing algorithms for different topologies are compared in Sect. 5. Finally, Sect. 6 gives the conclusions and direction for future work.

2   Deadlock Problem

Deadlock is a common problem and critical issue in interconnection networks. Deadlock between two messages happens when the required resources are in use by another message and the deadlocked packets stop completely. In particular, one necessary condition for possible deadlock is the cyclic dependency which happens due to the possible turns made by the packets [6,7]. Limitations in the interconnection network resources make deadlock of data packets unavoidable, resulting in network congestion [5]. Figure 2 depicts an example for deadlock situation in an interconnect network. Four packets $P_0, ..., P_3$, destined for nodes and each of them holds three buffers, requesting the fourth, but this is always held by another also waiting packet.

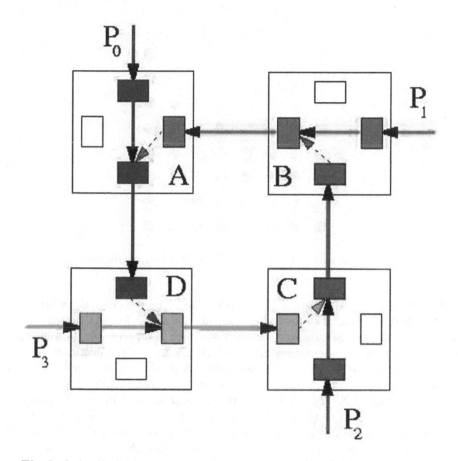

Fig. 2. A deadlock situation in an interconnect network (Color figure online)

The ability to guarantee the deadlock-free routing packets across a network topology is essential because congestion can block communication between cores and reduce the packet delivery rate, which in turn, drastically reduces communication performance [8,9]. With this in mind, effective design and an efficient routing algorithm are critical to maximizing the performance of any interconnection network.

3   Partitioning Scheme

To design a new routing in the x-Folded TM topology, partitioning scheme is introduced as the first step. All nodes in x-Folded TM topology were found in three different colour triangles; purple, blue and red. To present the partitioning scheme, we assign the x-axis and y-axis of the x-Folded TM topology and split it into four partitions based on the axes directions. The four partitions in the x-Folded TM topology are illustrated in Fig. 3.

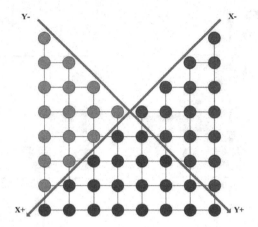

Fig. 3. An x-Folded TM topology with four partitions

The partitions are called X+Y+, X-Y-, X+Y- and X-Y+. By folding the TM topology to design an x-Folded TM, all the nodes in partition X-Y- are shared with the nodes in partition X+Y+, which represent purple, blue, and red triangles respectively. According to the coordinates of the destination node, both partitions X+Y+ and X-Y- are assigned to the purple triangle. If the destination node coordinate along the y-axis (y_D) belongs to $[0, \frac{k}{2} - 1]$, the selected partition is X+Y+. Otherwise the destination node belongs to partition X-Y- and is relocated to the partition X+Y+ while its new coordinate is $(x_D, |\ y_D - (k-1)\ |)$. If the destination node is located at the blue or red triangle, the assigned partition is X+Y- or X-Y+ respectively. Algorithm 1 represents how to assign each partition in x-Folded TM topology.

3.1   Scheme Verification

The topology is partitioned into four specific directions, ensuring that each partition is deadlock-free. For this reason, we must prove that no cyclic dependency exists. The deadlock situation is prevented by removing cyclic dependency in each of the proposed partitions. The direction for each partition is presented from left to right (X+), right to left (X-), down to up (Y+) and up to down (Y-). The directions in the four partitions are separately defined:

Definition 1. In partition X+Y+, the direction of packet transmission is from left to right along the x-axis and down to up along the y-axis.

Algorithm 1. Partitioning Scheme

1: **Input:** (x_D, y_D) // It is the destination node coordinate.

2: **if** Destination node $\in$ Purple Triangle **then**
3:      **if** $0 \leq y_D \leq (\frac{k}{2} - 1)$ **then**   // k is number of nodes in each topology dimension.
4:          Partition $= X + Y+$;   // Purple nodes are located in this partition.
5:      **else**
6:          $y_D = \mid y_D - (k - 1) \mid$;   // New coordinate along y-axis for the relocated destination.
7:          Partition $= X - Y-$;   // Purple nodes are located in this partition.
8:      **end if**
9: **end if**
10: **if** Destination node $\in$ Blue Triangle **then**
11:      Partition $= X + Y-$;   // Blue nodes are located in this partition.
12: **else**
13:      Destination node $\in$ Red Triangle;
14:      Partition $= X - Y+$;   // Red nodes are located in this partition.
15: **end if**

16: **Output:** Assigned partitions; X+Y+, X-Y-, X+Y- and X-Y+.

Definition 2. In partition X+Y-, the direction of packet transmission is from left to right along the x-axis and up to down along the y-axis.

Definition 3. In partition X-Y+, the direction of packet transmission is from right to left along the x-axis and down to up along the y-axis.

Definition 4. In partition X-Y-, the direction of packet transmission is from right to left along the x-axis and up to down along the y-axis.

Following this, a theorem is proved to avoid cyclic dependency in the x-Folded TM topology based on the definitions and presented lemmas.

Lemma 1. *Partitions X+Y- and X-Y+ are deadlock-free in x-Folded TM topology.*

Proof. We draw Fig. 4(a) to establish simply there is not cyclic channel dependency in partition X+Y- of the x-Folded TM topology. In partition X-Y+, it has been shown in Fig. 4(b) and is proven in this way. Therefore, there is no deadlock in the partitions X+Y- and X-Y+ and they are deadlock-free.

Lemma 2. *Partitions X+Y+ and X-Y- are deadlock-free in x-Folded TM topology.*

Proof. In Figs. 5(a) and (b), cyclic channel dependency has been removed in the x-Folded TM topology using the partitions X+Y+ and X-Y-. In these partitions, the topology is proven without deadlock.

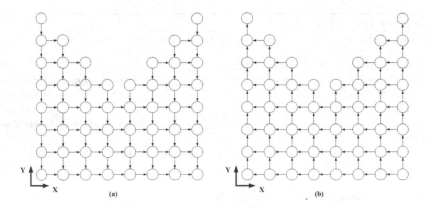

Fig. 4. In x-Folded TM topology: (a) Partition X+Y- (b) Partition X-Y+

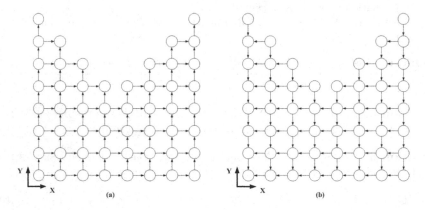

Fig. 5. In x-Folded TM topology: (a) Partition X+Y+ (b) Partition X-Y-

Theorem 1. *x-Folded TM topology is deadlock-free with partitioning scheme.*

Proof. Using the partitioning scheme, the cyclic channel dependency is broken. We proved no deadlock existence in the partitions X+Y-, X-Y+, X+Y+ and X-Y- separately. Consequently, no deadlock exists in the x-Folded TM topology.

3.2 Applied Routing Algorithm

Many studies about efficient routing algorithms without deadlock in interconnection networks can be found in literature [3,9]. After presenting the partitioning scheme in Algorithm 1 as the first step, we now apply the routing algorithm to find the shortest path and transfer the packets deadlock-free between each source and destination in the x-Folded TM topology. Through this algorithm, the routing starts from the source node. Accordingly, this algorithm can provide deadlock-free paths in the x-Folded TM topology. The routing algorithm applied in this study is of the adaptive type, which is a popular form of algorithm in interconnection networks, used to transfer packets to their final destination. In Adaptive algorithm, the routing decision is based on the addresses and other information about network traffic and channel status.

Correct partitioning of the x-Folded TM topology is a vital step in finding deadlock-free routes. The key difference between the adaptive algorithm and the applied deterministic algorithm in [17] is in the method of selection of the proper assigned partition. Simulation results will confirm the improvement in terms of average delay and network throughput.

4   Simulation Environment

The evaluation of performance will be performed by using Booksim 2.0 [18] that simulates a cycle-accurate interconnection network and was originally introduced in "Principles and Practices of Interconnection Networks". The simulations have been carried out for three traffic patterns including:

- **Uniform:** A node sends a packet with the same probability to each other node in the network. Source and destination are randomly selected in a packet-by-packet basis.
- **Bit-Complement:** With a binary representation of the node, a node $(a_{\beta-1}, a_{\beta-2}, ..., a_1, a_0)$ communicates with node $(\bar{a}_{\beta-1}, \bar{a}_{\beta-2}, ..., \bar{a}_1, \bar{a}_0)$.
- **Bit-Reversal:** With a binary representation of the node, a node $(a_{\beta-1}, a_{\beta-2}, ..., a_1, a_0)$ sends a packet to communicates with node $(a_0, a_1, ..., a_{\beta-2}, a_{\beta-1})$.

The configurations for the different topologies based on the Booksim2.0 simulator are presented in Table 1. The packet size is 8 flit and they are transmitted for 10,000 cycles from the source to the destination. There are two virtual channels per physical channel and two flit buffer sizes have been simulated while the transfer method of the packet is the wormhole strategy.

Table 1. Simulation configuration

Parameters	Value
Message length	8 flits, each message is split into fixed-length packets, every packet consists of flits
Switching strategy	Wormhole strategy [19]
Buffer Size	2 flits
Virtual channel	2 virtual channels
Simulation time	Simulation running for 10000 cycles

In all presented results, the performance of three different topologies has been shown in terms of average delay and network throughput. In addition, it is evaluated based on deterministic and adaptive routing algorithms.

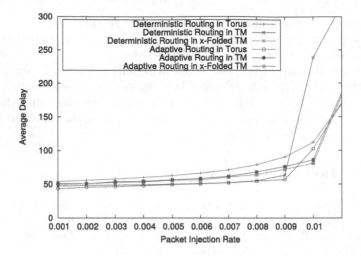

Fig. 6. Average delay performance under uniform traffic pattern

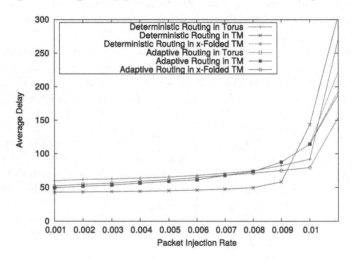

Fig. 7. Average delay performance under bit-complement traffic pattern

5   Simulation Results

A comparative study of routing algorithms in this section presents that the adaptive routing algorithm with the proposed scheme gives higher performance on x-Folded TM topology compared with the deterministic algorithm. It is based on average delay and network throughput for different topologies under various traffic patterns. Figure 6 depicts the x-Folded TM topology with low delay because of applying new routing algorithm. It is saturated in 0.009 (flit/node/cycle) which is faster than the other topologies.

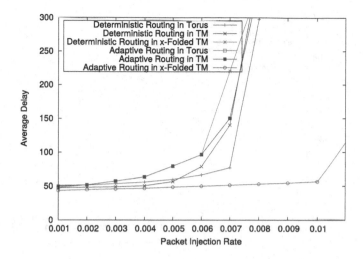

Fig. 8. Average delay performance under bit-reversal traffic pattern

Figure 7 shows the average delay for different topologies under a bit-complement traffic pattern using both algorithms, presenting the low delay x-Folded TM topology. It shows the average delay for all topologies are similar while the TM topology is saturated sooner using a deterministic algorithm with an increasing injection rate.

In Fig. 8, the applied traffic pattern is bit-reversal. As the figure shows, there is considerable difference in the average delay for the x-Folded TM topology. The presence of the new adaptive routing in the x-Folded TM topology suspends the saturation point compared to the Torus and TM topologies, while the improvement of the adaptive routing algorithm is obvious in comparison with the deterministic algorithm.

Next, we compare the network throughput of these topologies using the adaptive routing algorithm. The same set of conclusions hold with respect to the throughput performance of all topologies. The results are shown in the following figures.

Figures 9 and 10 show the network throughput for the different topologies under uniform and bit-complement traffic patterns. This data shows that all topologies - Torus, TM and x-Folded TM - exhibit a similar network throughput and become saturated at the same injection rate. Furthermore, the maximum network throughput of the Torus and TM is greater with the deterministic algorithm compared to the adaptive algorithm, although the difference is not considerable.

Figure 11 shows the worse network throughput rate under a bit-reversal traffic pattern for the Torus, TM and x-Folded TM topologies. It can be noted that the saturation point is significantly different for the x-Folded TM at 0.009 (flit/node/cycle). The Torus and TM topologies achieve a lower network throughput and are saturated sooner. It is proven that both the deterministic

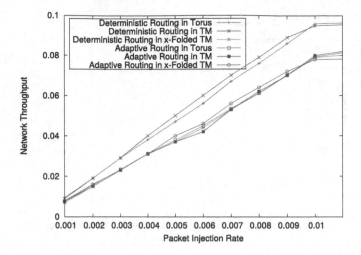

Fig. 9. Network throughput performance under uniform traffic pattern

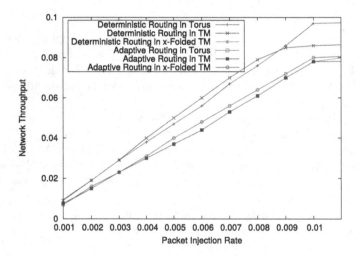

Fig. 10. Network throughput performance under bit-complement traffic pattern

and adaptive routing algorithms improve the performance of the x-Folded TM topology compared to the other two options.

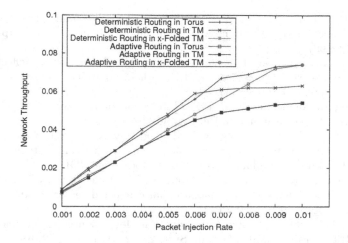

Fig. 11. Network throughput performance under bit-reversal traffic pattern

6   Conclusion

The contribution of this paper is studying the performance of x-Folded TM topology via design of a partitioning scheme based on the routing algorithm. The proposed scheme used to route the x-Folded TM topology without deadlock in terms of the two performance metrics; average delay and network throughput. The performance has been evaluated in comparison with the Torus and TM topologies, using a variety of different traffic patterns. The achievements reveal a performance enhancement over the average delay and network throughput in the x-Folded TM topology and prove the improved performance of x-Folded TM topology using the partitioning scheme.

Acknowledgment. This research was supported by the Universiti Putra Malaysia under High Impact Putra Grant: UPM/700-2/1/GPB/2017/9557900.

References

1. Borhani, A.H., Movaghar, A., Cole, R.G.: A new deterministic fault tolerant wormhole routing strategy for k-ary 2-cubes. In: IEEE International Conference on Computational Intelligence and Computing Research (ICCIC), pp. 1–7 (2010)
2. Ma, S., Enright Jerger, N., Wang, Z.: DBAR: An efficient routing algorithm to support multiple concurrent applications in networks-on-chip. In: 38th Annual International Symposium on Computer Architecture (ISCA 11), pp. 413–424. ACM, New York (2011)
3. Seiculescu, C., Murali, S., Benini, L., Micheli, G.D.: A method to remove deadlocks in networks-on-chips with wormhole flow control. In: Design, Automation Test in Europe Conference Exhibition (DATE), pp. 1625–1628 (2010)
4. Su, K.M., Yum, K.H.: Simple and effective adaptive routing algorithms in multi-layer wormhole networks. In: IEEE International Performance, Computing and Communications Conference (IPCCC), pp. 176–184 (2008)

5. Xiang, D., Zhang, Y., Pan, Y.: Practical deadlock-free fault-tolerant routing in meshes based on the planar network fault model. IEEE Trans. Comput. **58**(5), 620–633 (2009)
6. Feng, C., Li, J., Lu, Z., Jantsch, A., Zhang, M.: Evaluation of deflection routing on various NoC topologies. In: IEEE 9th International Conference on ASIC (ASICON), pp. 163–166 (2011)
7. Duato, J., Yalamanchili, S., Ni, L.M.: Interconnection Networks: An Engineering Approach. Morgan Kaufmann, Burlington (2003)
8. Chao, H.L., Chen, Y.R., Tung, S.Y., Hsiung, P.A., Chen, S.J.: Congestionaware scheduling for NoC-based reconfigurable systems. In: Design, Automation Test in Europe Conference Exhibition (DATE), pp. 1561–1566 (2012)
9. Somasundaram, K., Plosila, J., Viswanathan, N.: Deadlock free routing algorithm for minimizing congestion in a hamiltonian connected recursive 3dNoCs. Microelectron. J. **45**(8), 989–1000 (2014)
10. Camara, J.M., et al.: Twisted torus topologies for enhanced interconnection networks. IEEE Trans. Parallel Distrib. Syst. **21**(12), 1765–1778 (2010)
11. Liu, Y., Li, C., Han, J.: RTTM: a new hierarchical interconnection network for massively parallel computing. In: Zhang, W., Chen, Z., Douglas, C.C., Tong, W. (eds.) HPCA 2009. LNCS, vol. 5938, pp. 264–271. Springer, Heidelberg (2010). https://doi.org/10.1007/978-3-642-11842-5_36
12. Yu-hang, L., Ming-fa, Z., Jue, W., Li-min, X., Tao, G.: Xtorus: An extended torus topology for on-chip massive data communication. In: IEEE 26th International Parallel and Distributed Processing Symposium Workshops PhD Forum (IPDPSW), pp. 2061–2068 (2012)
13. Tatas, K., Siozios, K., Soudris, D., Jantsch, A.: Designing 2D and 3D Network-on-Chip Architectures. Springer, New York (2014). https://doi.org/10.1007/978-1-4614-4274-5
14. Wang, K., Zhao, L., Gu, H., Yu, X., Wu, G., Cai, J.: ADON: a scalable AWG-based topology for datacenter optical network. Opt. Quant. Electron. **47**(8), 2541–2554 (2015)
15. Wang, X., Xiang, D., Yu, Z.: TM: a new and simple topology for interconnection networks. J. Supercomput. **66**(1), 514–538 (2013)
16. Moudi, M., Othman, M.: Mathematical modelling for TM topology under uniform and hotspot traffic patterns. Automatika **58**(1), 88–96 (2017)
17. Moudi, M., Othman, M., Lun, K.Y., Abdul Rahiman, A.R.: X-Folded TM: an efficient topology for interconnection networks. J. Netw. Comput. Appl. **73**, 27–34 (2016)
18. Jiang, N., et al.: A detailed and flexible cycle-accurate Network-on-Chip simulator. In: IEEE International Symposium on Performance Analysis of Systems and Software (ISPASS), pp. 86–96 (2013)
19. Ni, L.M., McKinley, P.K.: A survey of wormhole routing techniques in direct networks. IEEE Comput. **26**(2), 62–76 (1993)

An Overview on Technical Characteristics of Blockchain Platforms

Zahra Moezkarimi, Reza Nourmohammadi, Sima Zamani,
Fatemeh Abdollahei, Zahra Golmirzaei, and Abuzar Arabsorkhi[✉]

Iran Telecommunication Research Center (ITRC), Tehran, Iran
{moezkarimi, s.zamani, f.abdollahei, z.golmirzaei,
Abuzar_Arabs}@itrc.ac.ir,
Rezanourmohammadi583@gmail.com

Abstract. The blockchain is an emerging technology which has a web of applications and potentials. A wide range of blockchain platforms has been developed to meet different technical and non-technical requirements and issues. Due to the diverse type of these platforms and the rapid evolution of blockchain technology, it would be necessary to have a big picture of existing blockchain platforms. Determining the characteristics of technology is an important factor to have such a total view and to standardize and expand technology. Despite the research has already been done on blockchain, due to the agile growth, rapid expansion and immaturity of this technology, determining its features is still challenging. In this paper, the results of studying and assessing the current state of blockchain technology from a technical point of view are presented. Furthermore, technical aspects of the blockchain are categorized as a two level taxonomy of characteristics and features.

Keywords: Blockchain technology · Cryptocurrency · Distributed systems · Taxonomy · Technical characteristic

1 Introduction

Determining the characteristic of a technology is an important factor to standardize and expand the technology in order to become a part of daily lives of people. Despite many research has already been done on the blockchain, due to the rapid expansion and immaturity of this technology, determining its features is still challenging. The blockchain is an emerge technology which has many undiscovered applications and potentials. On the other hand, as its agile growth and updates, its features and capabilities are still growing. Thus, characteristics and features gained through the research are results of studying and assessing the current state of blockchain technology and we put all our effort to comprehensively cover all blockchain features.

Characteristics and features of a technology can be classified using two methods: top-down and bottom-up classification. In the top-down method, the capabilities and general features of the technology can be considered. In contrast, in the bottom-up method, features of the technology are evaluated in the lowest level and it's going to be categorized based on mutual features. In the first part of the research related to this

© Springer Nature Switzerland AG 2019
L. Grandinetti et al. (Eds.): TopHPC 2019, CCIS 891, pp. 265–278, 2019.
https://doi.org/10.1007/978-3-030-33495-6_20

work, general features of blockchain technology examined and categorized from the top-down point of view along with considering its different usages. In the next step, characteristics categorized from a bottom-up point of view. Owing to the consideration of technical aspects of the blockchain, the proposed taxonomy is concentrated on the technical aspect of the blockchain platform as the core component of this technology.

The next section of this paper investigates the research background and related works to the blockchain characteristics and features. Then, third section overviews the high level characteristics of blockchain from different aspects. After that, in the fourth section a classification of the blockchain's technical features is presented. Finally, conclusion and future works of this research are discussed in the last section.

2   Research Background

Since Bitcoin publicly introduced in 2009, a wide range of blockchain platforms has been developed to meet different technical and business requirements and also legal issues. Due to the diverse type of the blockchain platforms and its rapid evolution, it would be necessary to have a big picture of existing blockchain platforms. In addition, the diversity of blockchain platforms can also result in duplication of efforts to find and establish a standard architecture to map these platform parameters which has major impact on this area. Accordingly, some research has been done to determine blockchain architectures. One of these researches conducted by Xu et al. around performance and non-functional attributes of blockchain [1]. In terms of software architecture, blockchain platforms have a diverse structure and parameters which plays a significant role in its configuration. A blockchain as a software, considered to express the main impact of architectural considerations on the resulting performance [2]. These researchers also propose a decision tree on whether to employ a blockchain or other software solutions. It is needed to consider the features of blockchain platforms when developing applications based on blockchain technology. Also, it is necessary to assess the impact of these features on functional and non-functional parameters to propose a taxonomy which captures the main architectural parameters of blockchain platforms [1].

There are some academic works include literature [3] and technical reports [4] that defined features and classified blockchain platforms which help to compare blockchain platforms and enable research into blockchain and their applications. Also, some research classified blockchain platforms based on their applications [5]. There was a shift in 2013 from focusing on currency and its payment aspects to other applications of blockchain platforms such as crowd funding and games which called "crypto 2.0". All blockchain platforms and related technologies classified into the "crypto 2.0" category. Then, "crypto 3.0" is appeared based on approaches to move around the existing limitation to mainstream adoption such as lightning network and sharing [5].

Moreover, a conceptual overview of blockchain platforms is proposed by Sultan et al. [6] due to the functions and potential business applications of blockchain. This overview provides a formal definition and proposes a classification of existing and emerging blockchain applications based on key functional features. It classified blockchain business applications into four categories: Blockchain as a Development Platform, Blockchain as a Smart Contract Utility, Blockchain as a Marketplace and a

Trusted-Service Application. A comprehensive survey on the development of block-chain technologies regarding the designing methodologies and related studies of consensus protocol as the main part of blockchain platforms provided in [7, 8]. Wang et al. have provided an overview of the organization of blockchain platforms by focusing on parameters of incentivized consensus. They have proposed an overview of the blockchain consensus mechanisms including BFT-based protocols, hybrid protocols and many more [7]. Glaser et al. [8] have proposed a taxonomy for the development purpose of blockchain applications. Analyzing new solutions would be more feasible if there are classification and an overview of how blockchain technology applies to the business requirements. The proposed taxonomy considered to be helpful tool for researchers and developers to quickly make a suitable decision in businesses which are built on blockchain platforms [8].

3 Characteristics Classification

Characteristics of blockchain can be assessed from the different aspects such as its usage type, services or features. As we mentioned, in this section blockchain features will be evaluated from a top-down point of view. This category focuses on application and services of blockchains [11]. In terms of application, four groups are defined for existing blockchain platforms: digital currencies, application stacks, asset-centric technologies, and asset-registry technologies. The mentioned items show the general applications of blockchain technology. The digital currencies consider blockchain applications which relate to cryptocurrencies. Examples of this category are Bitcoin, Dash, Monero and many other cryptocurrencies. Application stacks refer to non-currency applications of the blockchain platforms used to develop and run applications such as smart contracts or DAOs in distributed systems. Ethereum, NXT, and COMIT are examples of this category. The concept of asset registry technology refers to the applications of the blockchain which create links between cryptocurrencies and other digital currencies. ColorCoin and NameCoin are examples of this category. The concept of asset-centric technologies also refers to the applications of blockchain technology in which a group of contributors seeks to transfer data or value in a decentralized manner. Some famous samples in this category include Ripple, Hyper-ledger, and Stellar [12, 13]. The classification based on the level of service, along with the classification of applications, is also highlighted which classifies characteristics and features of blockchain platforms focusing on the service level provided on different platforms, their implementation methods, and services provided by different companies in this area. The possible values in this categorization are specific as blockchain as a service, blockchain first, development platforms, horizontal solutions, and APIs and overlays [7]. In this case, blockchain as a service or BaaS refers to the services provided to end users in a blockchain platform. As an example, Microsoft and ConsenSys can be pointed out that developed the Ethereum blockchain as a service (EBaaS), on Microsoft Azure with the goal of providing a Cloud-Based blockchain development environment. The second one involves the deployment of a blockchain platform or development of some applications based on the existing blockchain platforms such as Bitcoin and Ethereum. The development platform category refers to a series of blockchain platforms which provide platforms and tools needed to develop solutions based on blockchain which can be used by developers. Vertical solutions are types of blockchain platforms have been developed for specific applications in specific industries. R3 and Chain are some examples of this category [14].

Another important aspect in classifying blockchain platforms is the classification based on the main characteristics in the design and development of blockchain platforms, which in general are divided into two main parts: technical and non-technical features. In turn, the technical characteristics section includes two functional and non-functional feature sets [9]. Functional features include characteristics where explicitly affecting the design of a platform or software. In contrast, non-functional characteristics include parameters that are not quantitatively evaluated and their values influenced by other features. These characteristics are more qualitative rather than quantitative and aren't directly influenced by a specific parameter defined at design level [1]. These types of characteristics can also be the basis for the blockchain platforms classification. Performance, scalability, security, usability, robustness, and reliability are samples of non-functional characteristics [1, 9, 10]. As mentioned, the other classification of technical features includes functional ones. Some of the functional features of platforms are: permission level, consensus protocol, governance, anonymity level and trust level [1, 9, 25].

Based on research conducted on various resources in this area, these characteristics are commonly used as the basis for many classifications on blockchain platforms. Based on the range of possible values for each parameter, various platforms have been formed. The variety of existing platforms shows a wide range of possible values for characteristics. This research is intended to propose a comprehensive classification of blockchain platforms based on functional features as an extension of the research conducted by Tasca and his colleagues [31]. It drills down to a deep layer of features in addition of the other research, so, the classification can be used when comparing and evaluation different blockchains and it would be helpful in the design procedures.

4   Comprehensive Functional Features Classification

The categories proposed in the previous section offer the highest possible level of classification characteristics of blockchain platforms. Further, focusing on technical and functional features, in this section, their details will be classified at lower levels. In this part of the paper, the bottom-up point of view has been used to study technical and functional features and aims to design a hierarchical structure to categorize a variety of existing platforms. In the bottom-up classification, low-level features are first studied and then, based on the results obtained from this study, high-level features and ultimately applications are investigated. All four types of blockchain applications mentioned in the previous section will be covered in the comprehensive classification of the blockchain technology in this section, and the focus will not be directed only on cryptocurrencies [24]. Accordingly, in follows, the functional features of blockchains are introduced in eight categories including: consensus, structure of blockchain, block and transaction, owning assets and tokenization, extensibility, security and privacy, base code, identity management, and charging and reward system.

4.1 Consensus

In blockchain network which relies on participants to validate transactions, add them into blocks and append blocks to the blockchain, the whole networks try to obtain an agreement about the latest block to be included into the chain. The consensus mechanism ensures the validity of transactions and the duplication of them in a decentralized way without any centralized authority [1]. The consensus is the first and most important characteristic of the blockchain, which refers to the set of rules and procedures used to update the blockchain and distributed ledger [33]. This characteristic has four features as follows, each of which has a different range of possible values [28]: consensus network topology, types of consensus algorithms, gossiping and consensus agreement.

Consensus Mechanism. This feature describes how network's nodes communicate with each other and determines the type of information flow between them to transfer transactions or main network processes such as consensuses. Systems, in the past, were usually designed in a centralized manner [7]. This type of design would dramatically reduce the maintenance, updating, and modification cost, and all of these activities were carried out in just one central point. Despite its high performance, this design always faced with problems such as limited scalability. Consortium architectures, in addition to this centralized structure, were also used which had redundancy with high efficiency, and also greatly resolved the problem of scalability [7]. In this structure, the role of nodes is not the same and normally a sub set of nodes contributes in the vital activities of network. Consortium is a simplified structure of the hierarchical structure [7]. Banking systems are examples that are designed with this structure. The decentralized structure is along with the two mentioned structures. In the decentralized structure, all existing nodes have the same power. Therefore, it can generally be said that this parameter is correlated with the degree of decentralization in the consensus mechanism, although this parameter is not the only parameter to measure the degree of decentralization in the consensus mechanism [21].

The process of reaching to an agreement or consensus in a blockchain among unreliable nodes is a redefinition of Byzantine Generals' Problem [7]. The consensus network topology determines the way these nodes communicate to reach the final consensus in the blockchain platforms, so the network topology determines the type of connection and communication between the nodes and the type of current information between them to handle transactions in the validation process [33].

Consensus Algorithm. Consensus is one of the fundamental processes in blockchain that ensures a shared ledger with many consistent copies is stored in the network. During the consensus process, the blockchain network nodes work together to obtain an agreement on the validity and order of transactions which are candidates to be added to the ledger [36]. In general, consensus algorithm can be divided in two classes including proof-based and voting-based algorithms [30]. In the proof-based consensus model, the nodes which are contributing in the consensus process need to prove that they are more qualified than other nodes to validate and record a new transaction or block. Proof of Work (PoW) and Proof of Stake (PoS) are two samples of commonly used proof-based consensus algorithms. In the voting-based consensus model, all or sub set of nodes in the network needs to exchange their results (decision) on validating or recording a new transaction or block in order to make a final decision. Practical byzantine fault tolerant and RPCA (Ripple Consensus algorithm) are examples of voting-based consensus [27, 29, 31].

Gossiping. In addition to decentralization, blockchain is also a redundant data storage system making it more difficult to tamper information stored in the blockchain. In the blockchain structure, the transmission of information and data between nodes in the network varies from platform to platform. In contrast to the traditional banking network, in a blockchain without a centralized routing system, nodes send information to their neighboring nodes (the nodes directly linked to them). Hence, the nodes will have a list of neighbor nodes. When a new block is added to the local ledger stored in a node, a new block will be broadcast to all its neighbors. Then, the neighbor nodes send the new block to their neighbors using the gossiping protocols. Therefore, it can be said that Gossiping specifies how to send data in a blockchain network [24].

Consensus Agreement. This feature refers to the set of rules and mechanisms used in the consensus process in distributed systems in order to obtain agreement on the content of the ledger and make the system resistant to the byzantine faults.

4.2 Structure of Blockchain, Block and Transaction

The features corresponding to this parameter are used to indicate the capabilities of the blockchain platform, blocks, and transactions in the blockchain-based applications and subsequently the overall system scalability. This includes the following five parameters: data structure in the block header, transaction model, ledger storage model, block storage model and design constraints in terms of scalability.

Block Structure in the Block Header. Different blockchain platforms have different types of transaction data storage in their block header. Most of the platforms used Merkle tree mechanism which uses concatenated hashes of transactions in a tree. The original application of Merkle tree schema was implemented in Bitcoin [21]. Another type of data storage in block header is Patricia tree [22].

Transaction Model. This feature is used to prove ownership of tokens. Transaction Model demonstrates the way of storing and updating the transaction information in blockchain nodes. Also, it considers how the transactions are processed in the blockchain platform. Account based and UTXO (Unspent Transaction Output) models are two transaction models commonly used to store transactions in blockchain.

Ledger Storage. The main feature of any blockchain-based system is its decentralized nature so that the linked nodes across the P2P network cannot be distinguished from one another. Given the existing restriction in terms of storage space, processing power or bandwidth, this feature may not be realized perfectly. Therefore, in different models of blockchain platforms, different nodes may have access to different layers of information, out ruling the need for storing the entire deal of information. In the context of the blockchain technology, this model of data storage is referred to as light node capability-based ledger data storage [14]. As such, this feature is used to determine processing levels and storing ledger information at blockchain network nodes.

Block Storage. This parameter determines the information that is stored for each block along the blockchain. Depending on the model used to design a particular blockchain, scalability of the blockchain evaluated in different dimensions and identified the accessible user's information.

Design Limits in Terms of Scalability. Distributed nature of blockchain technology along with data redundancy in this structure, imposes limitations on system's size that can prevent the scalability. Here, system size means the number of nodes connected to the network, the number of service's users, active connections, the amount of network's traffic and the number of network's transactions. It should be noticed that these parameters are very close in the real world. Moreover, according to increasing demand of using blockchain and sustainable infrastructure development, existing limitations of blockchain are improving. In emerging technologies such as blockchain, these changes are mostly carried out by developer teams. Improvements like Segwit or Lightning Network in Bitcoin platform are two samples that was presented to overcome the scalability problem [15]. Limitations of Bitcoin scalability can be considered as a criterion for blockchain platforms design.

4.3   Owning Assets and Tokenization

Bitcoin and similar platforms represent a new set of assets. At the same time, some other applications of blockchain can be mentioned in which assets have meaning apart from currency, for example, convertible values can be mentioned that are created by tokens [20].

Native Assets. Assets and tokens are significant added values of blockchain that are exchanged daily on the blockchain platform [24].

Tokenization. A token acts as a digital bond which is transferable, and its ownership, is defined by the embedded data in the blockchain. In the context of blockchain, the ownership transfer process does not require any verification of the middle authenticity [36].

Asset Management. This parameter focuses on the token supply sources. The procedure of digital token creation varies from one blockchain platform to another one. This parameter also has an impact on user's motivation to participate in the validation process.

4.4   Extensibility

The purpose of this characteristic is to evaluate the extensibility of the blockchain and its ability to communicate with the other related technologies. Therefore, this parameter investigates the possibility of combination and communication of blockchain technology with the others. This characteristic consists of four subsection includes interoperability, intraoperability, governance and script language.

Interoperability. This feature represents the ability to transmit blockchain information to/from the other systems outside the blockchain world. This information includes the data which is not necessary to be recorded in blockchain and does not affect the consensus mechanism. For example, a smart contract may need to receive information (e.g. value of a specific coin or fiat money) from outside of the blockchain world [15].

Intraoperability. This feature, unlike the previous one, refers to the ability to transmit the information between the blockchain's components as well as communicating between various blockchains [16].

Governance. Blockchain platforms that have the ability to change, adapt and interact have more chance to be successful in the real world. As the blockchain developments are different, their control and management methods are also different. The governance is a set of rules related to the control and management mechanisms of the blockchain platforms which are crucial for the successful adoption of them.

In general, there are two categories of governance rules, including technical and regulatory rules. Technical rules include software policies, algorithms and protocols, and regulatory rules indicate rules that are enforced by external entities [17]. According to the focus of this paper on the technical aspect of blockchain technology, the technical rules will be considered only by focusing on investigating rules related to the design and development of blockchain platforms.

Script Language. General purposes languages have the possibility to perform any arbitrary computations. Turing completeness is one of the main requirements for the languages to be a general purposes language. Many programming languages are Turing Complete. The languages that are not Turing-complete are restricted to do some specific actions in particular situations at run-time. In some blockchain-based applications such as smart contracts, sometimes it is necessary to change the transaction information according to the specific conditions. Conducting these changes should take place under a routine or algorithm which is embedded in the structure of the languages used to design the blockchain. This feature allows users to develop and run their own algorithms. The freedom degree of action and the limitations in this area are important features that should be considered at design level [18].

In addition, the execution of some smart contracts requires the availability of information from outside of the blockchain. The external environment is not deterministic. Thus, the need for using external data should be addressed in the early steps of blockchain design, or the possibility of using Oracles should be considered at design level. Oracles are responsible for collecting the external/outside information needed for smart contracts [35]. In general, the investigation of this feature requires examining the capabilities of Turing machine used in the blockchain.

4.5   Security and Privacy

Some blockchain platforms have technical and operational risks in terms of security and Privacy. Security of the blockchain based systems is one of the most important concerns about this technology. Cryptocurrencies, as one of the most important and widely used applications of blockchain technology, has suffered from over the past years, which is mostly due to the lack of design and/or the lack of proper management of the user's data and sensitive information [19]. Potential security issues in blockchain systems include the lack of proper management of data and information (alternation, destruction, disclosure and deletion), vulnerabilities in implementation (vulnerability of implementation of cryptographic mechanisms and information leakage at runtime), mismanagement of cryptographic mechanisms (use of weak algorithms and disclosure of cryptographic key) or lack of proper management of user permissions [23]. Due to the importance of this characteristic, in this section, the concepts of security and privacy will be examined in two sections of data encryption and data privacy [32].

Data Encryption. Cryptographic algorithms are used to ensure the authenticity and integrity of the data. For example, in some blockchain platforms ECDSA digital signature scheme used for authenticity and integrity, and also hash functions such as SHA-2 used for integrity and order of incoming transactions [21].

Data Privacy. In general, cryptographic mechanisms, like pairs of keys as well as hash functions, should provide assurances such as the only receiver of a transaction on the network have access to the message content. However, research has shown that on some platforms, such as Bitcoin, it is possible to get more information about users and their real identity by combining related transactions' information. To address this issue, some solutions have been proposed to encrypt data in public networks, so the important information of transactions can be transmitted in the network, vaguely [23]. As can be seen, this feature and the previous one are closely related.

4.6   Base Code

This section focuses exclusively on the applied programming language and examines it from the perspective of programmers. The features which are investigated under this characteristic include: Programming language, source code licenses and software architecture.

Programming Language. Development of blockchain platforms are extensive and has many dimensions. Hence, in some designs for ease of use, as well as increasing the speed of development, several programming languages are used to develop these platforms. In addition to the core of the blockchain, additional side services are also developed based on a blockchain, in which various programming languages may also be used.

Source Code License. The source code license provides the possibility of the modification and changes and also deletion of the source code of the mentioned technology. Possible values for this feature include open source and closed source which is not necessarily opened.

Software Architecture. Software architecture refers to high-level structures of the blockchain. Each software includes the software elements and the relationship between them, as well as the features that each of these elements provided. This feature is very important in the design of blockchain platforms due to its impact on future modification and extension management. Integrated and modular design is two main variations of this feature.

4.7   Identity Management

This parameter provides the essential security for access to sensitive data in order to establish a proper data governance model. Despite its great importance, it has a lot of complexity which is due to the direct relationship of this characteristic with the digital identities of different network participants contributing in various blockchain processes such as validation, mining and auditing. In general, this parameter includes a set of defined rules and mechanisms within the blockchain that are inherently enforced to it. The most important elements for identifying management that provide secure access to sensitive information for users is discussed in this section [18]. Tow features are identified under this characteristic that involve: control and access layer, and identity layer.

Control and Access Layer. When it comes to controlling rights and permissions on the network, blockchains' structures and models have important roles. For example, with regard to expected performance, the blockchain may be controlled by a single entity, or managed by a set of rules that are distributed in the blockchain and participants are enforced to the rules. These rules provide required access levels for users to interact with the blockchain.

This feature also controls read and write access in the blockchain network and on the other hand, accesses and participates in the core network processes. The network membership permissions and permission to participate in the consensus mechanism can be mentioned as some of the blockchain access control policies [10].

Identity Layer. This feature is talking about user's authentication. Generally, users are either known in the network or have an anonymous identity. In some applications such as banking, it is necessary to know the real-world identity of transacting parties to satisfy Know-Your-Customer (KYC) regulation [1]. KYC is an essential procedure to verify user identities to prevent users to transfer money without governmental control [26].

4.8   Charging and Reward System

This characteristic focused on motivational mechanisms of users including rewarding mechanisms and blockchain fees. [32].

Reward Mechanism. Due to their decentralized nature, the blockchains suffer from operating and maintenance costs, which should be largely compensated by participants in the network. One of the most important part of the costs is the transaction validation overhead. Hence, it is necessary to use suitable mechanisms in the blockchain network to motivate the main network participants to provide network costs and continue the maintenance of the blockchain. These mechanisms, known as the reward mechanism, guarantee the sustainability and security of the network. The reward mechanism is created and controlled by the blockchain to compensate the costs imposed on network participants that perform tasks such as storing distributed ledger, transaction validation and verification. Additionally, the reward mechanism is used to motivate participants who have placed part of their assets in the blockchain to ensure network stability. An example of this kind of reward is established in some platforms such as Waves, which, in proportion to the amount of capital provided to the network, will receive a fixed amount of profit in certain time periods [34].

Fee System. This feature describes the mechanism of fee reward systems and for users to contribute when using blockchain platforms. Fee system describes the fees which are necessary for users when using a blockchain. It used to incentivize miners in a transaction [21]. Fee is another kind of reward which provided by the users to other participants of the blockchain when send any request in the network [28].

5   Conclusion and Future Works

In this paper, the features of blockchain that have effective roles in the design and development of the platforms have been investigated. First of all, this paper introduced the number of existing categories of the characteristics and features expected from the blockchains. The investigation of different categories indicated that the blockchain features can be generally classified as the technical and non-technical features. Due to the extensiveness of the number of features expected from the blockchain platforms, this paper focused on reviewing the technical features which are divided into functional and non-functional features. The functional features have a significant impact on the design step of blockchain platforms. Accordingly, this paper specifically investigates the functional features and provides a comprehensive ontology on this set of blockchain features.

The functional features can be categorized in different ways depending on their purposes and effects they may have on blockchain design. Regards to the studies conducted in this research, the functional features can be generally categorized under a set of characteristics including consensus, blockchain structure, native currency and tokenization, extensibility, security and privacy, basic code, and charging and rewarding system. The features under each characteristic are summarized in Table 1.

Table 1. An overview of blockchain functional characteristics

Characteristic	Feature
Consensus	Consensus mechanism
	Consensus algorithm
	Gossiping
	Consensus agreement
Structure of blockchain, block and transaction	Block structure in the block header
	Transaction model
	Ledger storage
	Block storage
	Design limits in terms of scalability

(*continued*)

Table 1. (*continued*)

Characteristic	Feature
Owning assets and tokenization	Native asset
	Tokenization
	Asset management
Extensibility	Interoperability
	Intraoperability
	Governance
	Script language
Security & privacy	Data encryption
	Data Privacy
Base code	Programming language
	Source code license
	Software architecture
Identity management	Control and access layer
	Identity layer
Charging and reward system	Reward mechanism
	Fee system

Different goals and functions could be expected from the platforms due to the different values for each of these features. In the next step of this research, besides reinvestigating the proposed category for the characteristics and features, the granularity of the obtained features will be further expanded. Moreover, the possible values of the obtained features can be examined and evaluated on a set of famous blockchain platforms.

References

1. Xu, X., et al.: A taxonomy of blockchain-based systems for architecture design. In: Proceedings of the 2017 IEEE International Conference on Software Architecture (ICSA), pp. 243–252, Gothenburg, Sweden, 3–7 April 2017
2. Xu, X., Pautasso, C., Zhu, L., Gramoli, V., Ponomarev, A., Tran, A.B., Chen, S.: The blockchain as a software connector. In: WICSA (2016)
3. Decker, C., Wattenhofer, R.: Information propagation in the Bitcoin network. In: P2P, Trento, Italy (2013)
4. Distributed ledger technology: beyond blockchain. Technical report, UK Government Chief Scientific Adviser (2016)
5. Mougayar, W.: The business blockchain: promise, practice, and application of the next Internet technology. Wiley, 2016
6. Sultan, K., Ruhi, U., Lakhani, R.: Conceptualizing blockchain: characteristics & applications. In: 11th IADIS International Conference Information Systems, pp. 49–57 (2018)
7. Wang, W., et al.: A survey on consensus mechanisms and mining management in blockchain networks. arXiv preprint arXiv:1805.02707 (2018)

8. Glaser, F., Bezzenberger, L.: Beyond cryptocurrencies-a taxonomy of decentralized consensus systems. In: European Conferences on Information Systems, pp. 1–18 (2015)
9. http://www.ict.griffith.edu.au/network/Mark%20Staples_SDLT2017.pdf
10. https://blockchainhub.net/blockchains-and-distributed-ledger-technologies-in-general/
11. McWaters, R., Galaski, R., Chaterjee, S.: The future of financial infrastructure: an ambitious look at how blockchain can reshape financial services. In: World Economic Forum (2016)
12. Schwartz, D., Youngs, N., Britto, A.: The ripple protocol consensus algorithm, vol. 5. Ripple Labs Inc White Paper (2014)
13. Cachin, C.: Architecture of the hyperledger blockchain fabric. In: Workshop on Distributed Cryptocurrencies and Consensus Ledgers (DCCL 2016), Chicago, IL, July (2016)
14. Brown, R.G.: R3, Introducing r3 cordaTM: a distributed ledger designed for financial services (2016)
15. Xu, Q., Aung, K.M.M., Zhu, Y., Yong, K.L.: A blockchain-based storage system for data analytics in the internet of things. In: Yager, R.R., Pascual Espada, J. (eds.) New Advances in the Internet of Things. SCI, vol. 715, pp. 119–138. Springer, Cham (2018). https://doi.org/10.1007/978-3-319-58190-3_8
16. Tessone, C.J., Tasca, P.: A parsimonious model for blockchain consensus: Scalability and consensus collapse (2017)
17. Dilley, J., Poelstra, A., Wilkins, J., Piekarska, M., Gorlick, B., Friedenbach, M.: Strong federations: an interoperable blockchain solution to centralized third-party risks. CoRR, abs/1612.05491 (2016)
18. Chen, Z.D., Zhuo, Y., Duan, Z.B., Kai, H.: Inter-blockchain communication. In: DEStech Transactions on Computer Science and Engineering (2017)
19. Kim, H., Laskowski, M.: A perspective on blockchain smart contracts: reducing uncertainty and complexity in value exchange. In: 26th International Conference on Computer Communication and Networks (ICCCN) (2017)
20. Wright, C.: Turing Complete Bitcoin Script. White Paper, SSRN Electronic Journal (2016)
21. Bonneau, J., Miller, A., Clark, J., Narayanan, A., Kroll, J.A., Felten, E.W.: Research perspectives on Bitcoin and second-generation cryptocurrencies. In: Symposium on Security and Privacy. IEEE, San Jose (2015)
22. Wood, G.: Ethereum: a secure decentralized generalized transaction ledger. Ethereum Project Yellow Paper 151, 1–32 (2014)
23. Lin, I., Liao, T.: A survey of blockchain security issues and challenges. Int. J. Netw. Secur. 19(5), 653–659 (2017)
24. Luther, W.J.: Is Bitcoin intrinsically worthless?', American Institute for Economic Research Sound Money Project Working Paper no. 7, Great Barrington, Massachusetts (2018)
25. Hopwood, D., Bowe, S., Hornby, T., Wilcox, N.: Zcash protocol specification. Zerocoin Electric Coin Company, Technical Report, December 2017
26. Tsukerman, M.: The block is hot: a survey of the state of Bitcoin regulation and suggestions for the future. Berkeley Tech. LJ 30, 1127 (2015)
27. Nakamoto, S.: Bitcoin: a peer-to-peer electronic cash system (2008)
28. Carlsten, M., The impact of transaction fees on bitcoin mining strategies, Ph.D. thesis, Princeton University (2016)
29. Tasca, P., Liu, S., Hayes, A.: The evolution of the Bitcoin economy: extracting and analyzing the network of payment relationships (2016)
30. Nguyen, G.T., Kim, K.: A survey about consensus algorithms used in blockchain. J. Inf. Process. Syst. 14(1), 101–128 (2018). https://doi.org/10.3745/JIPS.01.0024
31. Tasca, P., Tessone, C.: Taxonomy of Blockchain Technologies. Principles of Identification and Classification, 31 March 2018. Available at SSRN: https://ssrn.com/abstract=2977811. https://doi.org/10.2139/ssrn.2977811

32. Zheng, Z., Xie, S., Dai, H., Chen, X., Wang, H.: An overview of blockchain technology: architecture consensus and future trends. In: Proceedings of IEEE Int. Congr. Big Data (BigData Congr.), pp. 557–564, June 2017. http://ieeexplore.ieee.org/document/8029379/
33. Wright, A., De Filippi, P.: Decentralized blockchain technology and therise of lex cryptographia (2015)
34. Eyal, I.: The miner's dilemma. In: Security and Privacy (SP), IEEE Symposium, pp. 89–103. IEEE (2015)
35. Henry, K., Marek, L.: A perspective on blockchain smart contracts: reducing uncertainty and complexity in value exchange. In: 26th International Conference on Computer Communication and Networks, ICCCN (2017)
36. NIST standard draft, Retrieved from https://csrc.nist.gov/CSRC/media/Publications/nistir/8202/draft/documents/nistir8202-draft.pdf July 2018

Optimal Weight Design for Decentralized Consensus-Based Large-Scale Data Analytics over Internet of Things

Reza Shahbazian[1(⊠)] and Francesca Guerriero[2]

[1] Department of Mathematics and Computer Science,
University of Calabria (UniCal), 87036 Rende, CS, Italy
[2] Department of Mechanical, Energy and Management Engineering,
University of Calabria (UniCal), 87036 Rende, CS, Italy
Reza.Shahbazian@unical.it

Abstract. We are witnessing an increasing demand for smarter systems requiring more sensory data for decision makings. The main data sources are connected objects to the network forming the Internet of things (IoT). Processing such huge amount of data, coming back from IoT devices forces big data challenges in both storage and processing. Different solutions exist for data storage and analysis over IoT including cloud, Edge and Fog platforms that mainly focus on decentralization of the data. To process the decentralized data with no need to a center, optimized distributed decision making algorithms are required. In this paper, we study one of the promising distributed algorithms named consensus to make a decision over decentralized data generated by IoT devices. It is assumed that the data sources are capable to analyze limited portion of data while share their decisions and partially of their data with neighbor sources. Sharing the information, the network objects combine their decisions with neighbor sources and reach to a consensus using iterative optimization algorithms. We propose an optimal weight design for IoT devices over the practical networks. We assume that the data transmission is noisy and the weights are chosen so that the network cost function is minimized. Using the proposed weighting the network reaches a faster convergence while the resource consumption decreases.

Keywords: Big data · Distributed · Internet of Things · Consensus · Optimal weight design

1 Introduction

Big data existed long before arrival of Internet of things (IoT). Big data is a term applied to data where its characteristics is beyond the ability of traditional relational databases to capture, manage, and process while keeping the latency in the acceptable level. Big data is known with several characteristics including veracity, velocity, variety and volume as depicted in Fig. 1 that further can be categorized into unstructured and structured [1].

ⓒ Springer Nature Switzerland AG 2019
L. Grandinetti et al. (Eds.): TopHPC 2019, CCIS 891, pp. 279–288, 2019.
https://doi.org/10.1007/978-3-030-33495-6_21

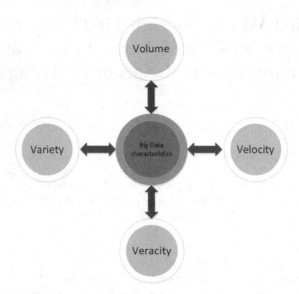

Fig. 1. Characteristics of big data summarized as 4Vs.

IoT takes a broad range of things and convert them into smart objects, from watches and fridges to cars and railway tracks. The data from IoT devices lies in big data and will significantly touch our lives in smart-homes [2], manufacturing [3], transportation [4], wearables [5], smartphones [6] and many other aspects. Everyday, more devices and objects are linked to the network, transmitting the data they gather back for the analysis and further decision makings. The goal is to utilize this data to learn more and make effective decisions that can be utilized to make a positive impact on lifestyle, energy consumption, health care, transportation and more [7]. This data could be generated from many sensors in smartphones or the ones attached to cyber-physical systems [8].

In a closer look, IoT may be considered as one of the different techniques that are used in order to store and collect the data. The connected objects work well in terms of different fields that makes it convenient to extract data. Many solutions have been proposed for data storage and analysis for IoT devices including cloud-based, Edge-based and Fog platforms [9]. All mentioned structures propose advantages while the user should deal with some drawbacks of the system. In this paper, we are not trying to compare the performance of edge, fog or cloud platforms but focus on how we can benefit form their decentralized large-scale data.

To illustrate more, lets consider the following futuristic example. Consider a connected autonomous vehicle network as depicted in Fig. 2 in which many sensors form the vehicles or surroundings (for instance traffic lights) report to a center in each vehicle. The process-limited center makes a decision on the speed, the driving path, driving pattern and more so that the vehicular network performs in its best performance. The best performance is a complex function including maximizing the throughput, minimizing the fuel/energy consumption and the time spent in the traffic. This example shows a big data problem in which it is almost impossible to transfer the whole data from all connected vehicles and surrounding sensors to a center that analyzes such huge amount of data, makes a decision and broadcasts the final decision to the network objects. As can be understood, the processing time is needed to be kept almost near real-time while the final network decision is optimal. Addressing the solution to the above mentioned problem as a use case of big data over IoT, many parameters should be considered, simultaneously. The vehicle itself should be able to make a robust and reliable decision, assuming that the network can not contribute. Many of researchers are trying to find the optimized solutions for the such smart autonomous vehicle using artificial intelligence algorithms [10].

Fig. 2. Diagram of a sample connected smart vehicular network.

In mentioned example, using a centric decision making is not a solution because of following reasons:

- The centric solution in very expensive, because the big data infrastructures both in storage and processing are needed. In the other words, the whole data from different sources should be processed, simultaneously in a real-time manner that makes it very expensive or impossible to be processed in a center.
- The velocity, variety and volume of the data, makes it hard to transfer while considering the bandwidth, latency, energy consumption, security and privacy challenges.

– The network reliability is dependent on the center and therefore, any failure in the center leads to a chaos.

Considering these challenges, researchers are trying to find optimized distributed decision making algorithms. In such algorithms, it is assumed that the decentralized data from variety of sources could be utilized to make an optimized decision for the network while all the contributers benefit. The main idea of this solution lies behind several assumptions, listed as follows:

– All data sources can make a primary decision by themselves (can process a limited amount of data).
– All data sources are capable to share their own decisions and partial of their data (in vehicular network, it is sensory data) with some neighbors in the network.
– The decision making objective is accepted among all the data sources. Each source makes its decision only gathered data while contributes in the final decision that satisfies.

In this paper we first study the related works followed by presenting consensus based distributed decision making algorithm that could be used to make a converged network decision. We further propose an optimal weighting algorithm to maximize the performance of distributed decision making algorithms over IoT.

The rest of this paper is organized as follows: In Sect. 2 we present the related works. Section 3 presents the system model of consensus-based distributed decision making algorithm. In Sect. 4, we propose an optimal weighing algorithm and the evaluation results. Finally, Sect. 5 concludes the paper.

2   Related Works

The consensus based distributed decision making optimization algorithms are mainly based on iterative solutions including gradient descent and newton methods [11]. The main concept of consensus algorithms is not new and goes back to 1974 where degroot for the first time commented on reaching a consensus [12]. However, the first efforts on using distributed consensus for big data analytics is rather new, starting from 2013 where authors in [13] used a consensus optimization technique for hypergraph-partitioned vertex programming. They showed primary 50% of improvement in solutions to the big data partitioning problem. In 2014, authors in [14] reviewed the consensus problems in big data clustering. Also in [15] the authors proposed a random double clustering based cluster ensemble framework (RDCCE) to perform tumor clustering using fuzzy consensus algorithms. In [16] authors use sampling combined with consensus strategy to dissemble the whole Big Data into small subsets. They proposed a partial data clustering (PDC) for sampling part according to different nodes. Consensus clustering for heterogeneous big data is investigated in [17]. The authors in this research propose an algorithmic framework for FCC with flexible choice of utility functions, and speeds fuzzy consensus clustering (FCC) significantly with a FCM-like iterative process of piFCM. In [18], authors study distributed big-data non-convex optimization in multi-agent networks by using a method named alternating direction method of multipliers (ADMM). They consider the minimization of the agents' sum-utility and propose a distributed solution whereby at each iteration agents optimize and then communicate only a subset of their decision variables. In [19], authors study Hyperparameter optimization in Gaussian processes (GPs). They propose an alternative distributed GP hyperparameter optimization scheme using the efficient proximal alternating direction method of multipliers, and derive the closed-form solution for the local sub-problems. Their proposed scheme can work in either a synchronous or an asynchronous manner, thus flexible to be adopted in different computing facilities.

All above mentioned algorithms are based on consensus in which the network's cost function is distributed among objects. Each object makes its own decision while in general contributes to the network decision. In an iterative manner, each object shares its decision and partial of its data (for instance, sensory data) with some of neighbor objects. After a few iterations, the network reaches a consensus which is a sub-optimal solution to the problem.

Finding the optimal solution that grantees the best convergence rate, convergence area and minimum cost is still an open issue. In this paper, we try to improve the performance of the existing algorithms by proposing optimal weight design over Internet of things applications that normally face with some practical challenges including imperfect and noisy communication channel.

3   System Model

In consensus based method it is assumed that the network cost function is dividable among data sources in such a way that the combination of all decisions is equivalent to the network decision [20]. In mathematical presentation we have:

$$J^{Network}(\omega) = \sum_{k=1}^{N} J_k(\omega) \tag{1}$$

where N is number of different data sources and J represents the cost function. Data source l is said to be a neighbor of data source k if they can communicate and cooperate with each other. We denote the set of all neighbors of data source k by $\mathcal{N}_k$. The objective of the data sources in the network is to make a decision in a fully distributed manner. The solution is an estimate of an unknown vector ω^o. We assume that the cost function J is twice differentiable and satisfies the conditions in (2) for some positive parameters $\alpha \leq \beta$.

$$0 < \alpha I_M \leq \nabla_\omega^2 J(\omega) \leq \beta I_M \tag{2}$$

The conditions presented in (2) makes J to be α strongly convex. The gradient vector is also β-Lipschitz. The gradient based solution to minimize $J(\omega)$ by considering the conditions in (2) could be presented as (3).

$$\omega_i = \omega_{i-1} - \mu \nabla_{\omega^T} J(\omega_{i-1}) \tag{3}$$

where $i \geq 0$ is the iteration index and $\mu > 0$ is the step size that could be constant or time varying. The absolute gradient vector (∇) is not available in practical cases and therefore, instantaneous approximation is used to replace it ($\hat{\nabla}$).

$$\omega_i = \omega_{i-1} - \mu \hat{\nabla}_{\omega^T} J(\omega_{i-1}) \tag{4}$$

It is also possible to employ iteration dependent step-size sequences $\mu(i)$ instead of the constant step-size μ, that satisfies $\sum_{i=0}^{\infty} \mu^2(i) < \infty$, $\sum_{i=0}^{\infty} \mu(i) = \infty$.

At every time instant, i, each source measures a scalar random process d and a vector random process u_i. We may define the cost function $J(\omega) = \mathbf{E}|\mathbf{d}(i) + u_i\omega|^2$ [20]. The parameter d could be defined as $d(i) = u_i\omega^o + n(i)$ where n is the measurement noise. Therefore, the consensus strategy to estimate the network decision could be written as follows [21]:

$$\omega_{k,i} = \omega_{k,i-1} - \mu_k \sum_{l \in N_k \backslash \{k\}} b_{l,k}(\omega_{k,i-1} - \omega_{l,i-1}) + \mu_k u^*_{k,i}(d_k(i) - u_{k,i}\omega_{k,i-1}) \tag{5}$$

In (5), the k shows the index of data source and by some simplifications we have:

$$\omega_{k,i} = \sum_{l \in N_k} a_{l,k}\omega_{l,i-1} + \mu_k u^*_{k,i}(d_k(i) - u_{k,i}\omega_{k,i-1}) \tag{6}$$

where the coefficient $a_{l,k}$ denotes the weight that the data source k assigns to the estimate $\omega_{l,i-1}$ that is received from its neighbor, l. The weights $a_{l,k}$ are nonnegative for sufficiently small step sizes. The collected weights form an $N \times N$ matrix with $a_{l,k}$ as its elements showed by A_i.

In next section we would propose the optimal weight selection scheme in which IoT devices communicate over noisy channels and the network tries to reach a consensus from different data sources.

4   Optimal Weighting

Considering $a_{l,k}(i) = \gamma_{l,k}\mathcal{I}_{l,k}(i)$ where $\gamma_{l,k}$ are positive fixed combination weights that data source k assigns to its neighbors l and $\mathcal{I}_{l,k}(i)$ is equal to 1 when the two data sources are neighbors (can communicate) and 0 otherwise. Therefore, the weights could be presented as (7) where $\sum_{l \in \mathcal{N}_{k,i}} \gamma_{l,k} < 1$.

$$a_{l,k}(i) = \begin{cases} \gamma_{l,k}\mathcal{I}_{l,k}(i), if\, l \in \mathcal{N}_{k,i} \backslash \{k\} \\ 1 - \sum_{l \in \mathcal{N}_{k,i} \backslash \{k\}} a_{l,k}(i), if\, l = k \\ 0, otherwise \end{cases} \tag{7}$$

As long as the A_i is left stochastic (LF) meaning that $\sum_{l \in \mathcal{N}_{k,i} \backslash \{k\}} a_{l,k}(i) = 1$ we have:

$$\mathbf{E}[A_i] = \mathbf{E}\left[\sum_{l \in \mathcal{N}_{k,i}} a_{l,k}(i) \mathcal{I}_{l,k}(i)\right]$$

$$\mathbf{E}[A_i] = \lim_{i \to \infty} \frac{1}{i} \sum_{j=1}^{i} \sum_{l \in \mathcal{N}_{k,i}} a_{l,k}(i) \mathcal{I}_{l,k}(i) \tag{8}$$

$$\mathbf{E}[A_i] = \lim_{i \to \infty} \frac{1}{i}\left[\sum_{l \in \mathcal{N}_{k,i}} a_{l,k}(1) \mathcal{I}_{l,k}(1) + \cdots + \sum_{l \in \mathcal{N}_{k,i}} a_{l,k}(i) \mathcal{I}_{l,k}(i)\right] = 1$$

The optimal weight should be chosen in such a way that the network error is minimized. The upper bound for network error is achieved in [22]. Therefore, to find the optimal weight we need to solve (9).

$$Min \sum_{k=1}^{N} \sum_{l \in N_k} a_{lk}^2 \left[\mu_l^2 \sigma_{n,l}^2 Tr(R_{u,l}) + Tr(R_{n,lk})\right] \tag{9}$$

where $Tr(.)$ shows the trace operator and $\sigma_{n,l}^2$ is the variance of zero mean measurement noise for data source l The solution to (9) is presented in (10).

$$a_{l,k}(i) = \begin{cases} \frac{\varsigma_{l,k}^{-2}(i)}{\sum_{o \subset \mathcal{N}_{k,i}} \varsigma_{c,k}^{-2}(i)}, if \, l \in \mathcal{N}_{k,i} \\ 0, otherwise \end{cases} \tag{10}$$

where the parameter ς is defined as in (11).

$$\varsigma_{l,k}^2(i) = \begin{cases} \left(\mu_l^2 \sigma_{n,l}^2 Tr(R_{u,l}) + N\sigma_{n,lk}^2\right), if \, l \in \mathcal{N}_{k,i} - \{k\} \\ \mu_l^2 \sigma_{n,l}^2 Tr(R_{u,l}), if \, l = k \end{cases} \tag{11}$$

4.1   Evaluation Results

To evaluate the performance of proposed weighting, we consider a network with 50 data sources as depicted in Fig. 3. The noise power, network error and the noise power regression (trace function) of this diagram is presented in Fig. 4.

5   Conclusion

The number of devices linked to the network that transmit their data for the further decisions is increasing dramatically, causing the network to face with big data challenges. Different solutions exist to storage data and analyze it over IoT including cloud-based, Edge-based and Fog platforms. Such platforms decentralize the data. Therefore, to make a decision over all data sources, distributed decision making algorithms are needed. One the promising solutions

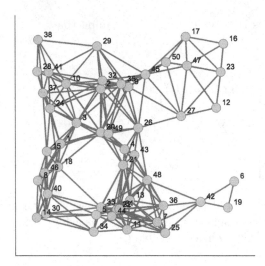

Fig. 3. Diagram of a network with 50 data sources

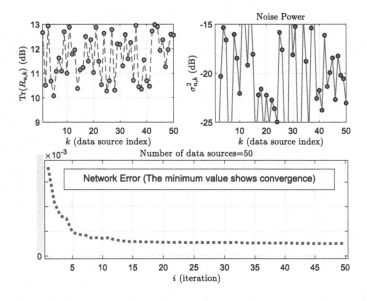

Fig. 4. The network error and power noise for the simulated diagram with optimal weight

is consensus-based algorithms in which the data sources are capable to analyze limited amount of data while share their decision and partially of their data with neighbor sources. Using such algorithms, the network may reach a consensus over an optimized decision while the data sources decide separately, with no need to a central control. In this paper, we assumed that the data transmission is noisy and propose the optimal weight design for consensus-based algorithms. Using the proposed weighting algorithm, the network can reach a faster convergence while the resource consumption decreases.

References

1. Gandomi, A., Haider, M.: Beyond the hype: big data concepts, methods, and analytics. Int. J. Inf. Manage. **35**(2), 137–144 (2015)
2. Ghayvat, H., Mukhopadhyay, S., Gui, X., Suryadevara, N.: WSN-and IOT-based smart homes and their extension to smart buildings. Sensors **15**(5), 10350–10379 (2015)
3. Tao, F., Zuo, Y., Da Li, X., Zhang, L.: Iot-based intelligent perception and access of manufacturing resource toward cloud manufacturing. IEEE Trans. Ind. Inf. **10**(2), 1547–1557 (2014)
4. Mishra, S., Patel, S., Panda, A.R.R., Mishra, B.K.: Exploring IoT-enabled smart transportation system. In: The IoT and the Next Revolutions Automating the World, pp. 186–202. IGI Global (2019)
5. Gia, T.N., et al.: Energy efficient wearable sensor node for iot-based fall detection systems. Microprocess. Microsyst. **56**, 34–46 (2018)
6. Froytlog, A., et al.: Ultra-low power wake-up radio for 5G IoT. IEEE Commun. Mag. **57**(3), 111–117 (2019)
7. Provost, F., Fawcett, T.: Data science and its relationship to big data and data-driven decision making. Big data **1**(1), 51–59 (2013)
8. Da Xu, L., Duan, L.: Big data for cyber physical systems in industry 4.0: a survey. Enterpr. Inf. Syst. **13**(2), 148–169 (2019)
9. Mohan, N., Kangasharju, J.: Edge-fog cloud: a distributed cloud for internet of things computations. In: 2016 Cloudification of the Internet of Things (CIoT), pp. 1–6. IEEE (2016)
10. Bouton, M., Nakhaei, A., Fujimura, K., Kochenderfer, M.J.: Scalable decision making with sensor occlusions for autonomous driving. In: 2018 IEEE International Conference on Robotics and Automation (ICRA), pp. 2076–2081. IEEE (2018)
11. Sra, S., Nowozin, S., Wright, S.J.: Optimization for Machine Learning. MIT Press, Cambridge (2012)
12. DeGroot, M.H.: Reaching a consensus. J. Am. Stat. Assoc. **69**(345), 118–121 (1974)
13. Miao, H., Liu, X., Huang, B., Getoor, L.: A hypergraph-partitioned vertex programming approach for large-scale consensus optimization. In: 2013 IEEE International Conference on Big Data, pp. 563–568. IEEE (2013)
14. Fahad, A., et al.: A survey of clustering algorithms for big data: taxonomy and empirical analysis. IEEE Trans. Emerg. Topics Comput. **2**(3), 267–279 (2014)
15. Zhiwen, Y., et al.: Adaptive fuzzy consensus clustering framework for clustering analysis of cancer data. IEEE/ACM Trans. Comput. Biol. Bioinf. **12**(4), 887–901 (2014)
16. Zoghlami, M.A., Hidri, M.S., Ayed, R.B.: Sampling-based consensus fuzzy clustering on big data. In: 2016 IEEE International Conference on Fuzzy Systems (FUZZ-IEEE), pp. 1501–1508. IEEE (2016)
17. Junjie, W., Zhiang, W., Cao, J., Liu, H., Chen, G., Zhang, Y.: Fuzzy consensus clustering with applications on big data. IEEE Trans. Fuzzy Syst. **25**(6), 1430–1445 (2017)
18. Notarnicola, I., Sun, Y., Scutari, G., Notarstefano, G.: Distributed big-data optimization via block-iterative convexification and averaging. In: 2017 IEEE 56th Annual Conference on Decision and Control (CDC), pp. 2281–2288. IEEE (2017)

19. Xie, A., Yin, F., Xu, Y., Ai, B., Chen, T., Cui, S.: Distributed gaussian processes hyperparameter optimization for big data using proximal ADMM. IEEE Sig. Process. Letters **26**(8), 1197–1201 (2019)
20. Sayed, A.H.: Adaptive networks. Proc. IEEE **102**(4), 460–497 (2014)
21. Shahbazian, R., Grandinetti, L., Guerriero, F.: A new distributed and decentralized stochastic optimization algorithm with applications in big data analytics. In: Nicosia, G., Pardalos, P., Giuffrida, G., Umeton, R., Sciacca, V. (eds.) Machine Learning, Optimization, and Data Science. LOD 2018, vol. 11331. Springer, Cham (2019). https://doi.org/10.1007/978-3-030-13709-0_7
22. Zhao, X., Tu, S.Y., Sayed, A.H.: Diffusion adaptation over networks under imperfect information exchange and non-stationary data. IEEE Trans. Sig. Process. **60**(7), 3460–3475 (2012)

Data Mining, Neural Network and Genetic Algorithms

Automatic Twitter Rumor Detection Based on LSTM Classifier

Amin Saradar Torshizi$^{(\boxtimes)}$ and Adel Ghazikhani$^{(\boxtimes)}$

Imam Reza International University, Mashhad, Iran
{a.saradar,aghazi}@imamreza.ac.ir

Abstract. With the rapid growth of Online Social Networks (OSNs), information can spread rapidly and widely more than ever. Besides, rumors could also be easily posted and propagated in OSNs which can lead to serious social issues. The problem of identifying rumors on social networks has received considerable attention in recent years. In this research, we propose a novel automatic rumor detection method based on a Long Short-Term Memory (LSTM) classified. Our proposed method not only achieves higher accuracy, F1 score and precision but also has lower false positive rate value. Extensive experiments conducted on Twitter show that the accuracy of the proposed method is 92.45% and F1 score is 89.95%. Meanwhile, false positive rate is less than 5.01%.

Keywords: Rumor detection · Classification · Bag of hashtag · LSTM · Neural network · Anomaly detection · Social networks · Twitter · Clustering

1 Introduction

A statement whose true value is unverified or deliberately false is commonly defined as a rumor. The rumor may be misinformation (false information) or disinformation (deliberately false information). A false rumor may spread fear, hate, or even euphoria. It also may lead to defamation, protests or other undesirable responses.

Social media is currently a place where large amounts of unverified information are generated continuously. Social media users not only read the news but also may propagate rumors that seem to be news.

With the rapid growth of microblogging websites, the amount of rumors has increased remarkably. One of the most popular microblogging platforms is Twitter. In fact, breaking news appears first on Twitter, before making it through to traditional media. Therefore, the design of systems that automatically detect misinformation and disinformation seems to be crucial.

In this paper, we address the problem of detecting rumors on Twitter. We treat the problem of rumor detection as an anomaly detection task by considering rumor as an anomaly. In fact, the user behaviors of posting rumors will diverge from those of posting genuine facts [1]. We employed a new clustering method in order to cluster twitter data and extract recent trends. Then each cluster is analyzed by a novel LSTM based anomaly detection method in order to check if the cluster topic is rumor or not. To the best of our knowledge, this is a first study using LSTM based approach in order to detect rumors on twitter based on anomaly detection.

© Springer Nature Switzerland AG 2019
L. Grandinetti et al. (Eds.): TopHPC 2019, CCIS 891, pp. 291–300, 2019.
https://doi.org/10.1007/978-3-030-33495-6_22

The rest of the paper is organized as follows: in Sect. 2 we give an overview of related work and review prior works. In Sect. 3 we describe our proposed model. In Sect. 4 we present the experimental results and Sect. 5 concludes this paper and propose future works.

2 Related Work

Researchers have analyzed rumors from different aspects, including psychological studies and computational analyses. In this section, we briefly discuss these models.

Liang et al. [2] proposed a rumor detection method based on behavioral features of users. They observed behavioral divergence of rumor spreaders with normal users and tried to distinguish the rumor-mongers from normal users.

Takayasu et al. [3] study the diffusion and convergence of a rumor in the 2011 Japan Earthquake on twitter. They investigated rumors that appeared in the immediate aftermath of the Great East Japan Earthquake on March 11, 2011.

Kumar et al. [4] Examined the use of psychological theories to estimate the spread of misinformation social networks.

They also investigated the rumor diffusion mechanism in scale free networks and a new rumor model proposed with a new compartment of stifling nodes who reject the rumors.

2.1 Text Features

Text features are features derived from text used in natural language processing (NLP) tasks. Textual features can be extracted to show rumors in words, sentences, messages, topics, and events. Compared with non-rumors, rumors have certain patterns.

Zhao et al. [5] discovered two types of language patterns in rumors: the modifier type and query type. They determined the Trending rumors based on query patterns which given the content features in the early days of the rumor distribution process. They provided an approach for clustering single patterns, including tweets and controversial events that are likely to be a rumor.

Qazvinian et al. [6] used three sets of features that include content-based features, network-based features, and Twitter behavior patterns. They categorized tweets in two different patterns: lexical patterns and part-of-speech patterns.

2.2 Linguistic Features

Linguistic patterns express certain meanings or emotions that are important clues to the description of the text. The features extracted at the word level of rumor can be statistical patterns, lexical patterns or emotional patterns.

Catillo et al. [7] computed the statistics of rumor texts based on its words, including the total number of words and characters, the number of unique words and the average word length in a rumor text.

2.3   Subject Features

Subject features are extracted from sentence level which aims to understand the message and the underlying relationships between the set of sentences.

Wu et al. [8] defined a set of subject feature for a semantic summary to identify rumors in Sina Weibo. They trained a Latent Dirichlet Allocation by distributing 18 Subjects in all messages, and each message can belong to one or more topics. They converted an 18-dimensional vector into a binary vector by tuning the highest probability of topics to 1 and the rest of the issues turned 0.

2.4   Classification-Based Feature

In Classification-based feature selection, detecting of rumors has been considered as a binary classification issue. Researchers used a supervised learning approach to automatically determine a specific popular that speared topic is true or false.

Sun et al. [9] studied the characteristics of event rumors and extracted features which can distinguish rumors from ordinary posts and used multimedia and location-based features to detect rumors in Sina Weibo from ordinary posts.

Dayani et al. [10] applied machine Learning for detection spreading rumors on Twitter. Their proposed detection approach is based on Rumor-Knowledge-Base (RKB) which is a repository of tweets related to different rumor topics, manually pre-detected and pre-verified, along with sentiment polarities to suggest whether this tweet spreads this rumor or refutes it.

2.5   Pattern Matching-Based Feature

Ennals et al. [11] used pattern matching techniques to illustrate the debunking claims on the Web. Their method automatically searches for lexical patterns for claims on the web. Subsequently, claims are filtered by a classifier and show only one part of the opposite claim.

3   Proposed Method

The proposed method is shown in Fig. 1. Each step is described in the corresponding subsection. We assume that the data arrives from a social network service in a sequential manner through some API.

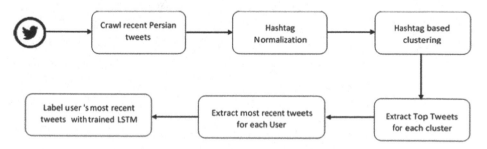

Fig. 1. Overall flow of the proposed method

3.1 Crawl Recent Persian Tweets

We used a data collection method relying on the API functions which are officially provided by twitter [12].

We assumed that rumors circulating on social networks can be considered as two types of the following:

1. long-standing rumors that circulate for long periods of time
2. short-standing rumors that spawned during fast-paced events such as breaking news, and often with an unverified status in their early stages.

We assume twitter rumors as short-standing rumors and thus we collect Persian tweets which posted in the last T hours. In fact, T is assumed as maximum rumor lifetime in twitter.

3.2 Hashtag Normalization

One of the main problems in Persian text processing is the existence of different character encodings in text documents. This problem is more obscurant when we deal with various character encodings in punctuations. We normalized the tweet hashtags using the following steps:

– Removed all symbols.
– Convert all digits to English characters.
– Removed all the stop words.
– Removed all extra whitespaces.
– Normalize Persian characters into a standard format

After using the above rules, tweet hashtags are cleaned and normalized and ready for the next phase.

3.3 Hashtag Based Clustering

Assume $T = \{t_1, t_2, t_3, \ldots, t_n\}$ a set of tweets and t_i is the i'th tweet of set T. Besides, we denote $H = \{h_1, h_2, h_3, \ldots, h_m\}$ as list of distinct hashtags of T. In order words, T contains n tweet and m distinct hashtags. Besides, for each tweet i and hashtag j, function $f_{i,j}$ defines as follow:

$$f_{i,j} = \begin{cases} 1 & \text{if tweet } t_i \text{ contains } h_j \\ 0 & else \end{cases} \tag{1}$$

Consider com_i, rep_i, ret_i denoting the number of comment, number of replies and number of retweet of tweet t_i respectively. Then weight of each tweet t_i defined as follow:

$$w_i = a * com_i + b * rep_i + c * ret_i \tag{2}$$

Where a, b and c are constants and denote the weights of comments, replies and retweets respectively.

Two hashtags are considered similar if they co-occurred in a tweet. Co-occurrence relation of two hashtag h_i and h_j can be formally defined as:

$$C_{i,j} = \sum_{k=0}^{|T|} w_i * f_{k,i} * f_{k,j} \tag{3}$$

In other words, $C_{i,j}$ is the sum of tweet weights which contains both h_i and h_j.

For each set of tweets T, we create a hashtag graph where nodes are hashtags. In other words, $V = \{h_1, h_2, h_3, \ldots, h_m\}$ There is an edge between two hashtag h_i and h_j if and only if

$$C_{i,j} > \epsilon \tag{4}$$

Weight of edge between two hashtag h_i and h_j shows the distance between two hashtags and is calculated as follow:

$$W_{i,j} = \frac{1}{C_{i,j}} \tag{5}$$

Using Floyd–Warshall algorithm [13], the distance between each two pair of hashtags will be calculated as follow:

$$w_{i,j}^0 = \begin{cases} 0 & \text{if } i = j \\ \frac{1}{C_{i,j}} & \text{else if } C_{i,j} > \epsilon \\ \infty & \text{else} \end{cases} \tag{6}$$

$$w_{i,j}^n = \min\left(w_{i,j}^{n-1}, w_{i,k}^{n-1} + w_{k,j}^{n-1}\right) \tag{7}$$

By knowing the distance between each two pair of hashtag, we employed k-means algorithm in order to cluster hashtags into K different clusters. Table 1 shows the pseudocode of the proposed clustering approach.

Table 1. Pseudocode of the proposed clustering method

$Function\ HashtagBasedClustering(W_{i,j})$
$Input{:}\ Weight\ of\ edge\ between\ each\ two\ hashtag$
$Output{:} Distance\ between\ each\ two\ pair\ of\ hashtags$

$let\ w_{i,j}^{0}\ =\ W_{i,j}$

$For\ k \leftarrow 1\ to\ n\ do$
    $For\ i \leftarrow 1\ to\ n\ do$
        $For\ j \leftarrow 1\ to\ n\ do$
            $w_{i,j}^{k} = \min\left(w_{i,j}^{k-1}, w_{i,k}^{k-1} + w_{k,j}^{k-1}\right)$
        $end\ For$
    $end\ For$
$end\ For$

$Initialize\ C_i$ to a random value between 1 and n (i $=$ 1, ..., k)
$While\ m_i\ has\ not\ converge$
    $For\ i \leftarrow 1\ to\ n\ do$
        $c_j = j$ $where\ w_{i,c_j}^{n} = \min_{\forall k} w_{i,c_k}^{n}$
    $end\ For$

    $For\ j \leftarrow 1\ to\ k\ do$
        $C_j = m$ $where$ $\sum_{\forall i \in C_j} w_{im}^{n}\ is\ minimum$
    $end\ For$
$end\ While$

3.4   Top Tweet Extraction

For each cluster K, we define cluster weight as the following:

$$CW_i = \sum_{\forall i \in K} \sum_{\forall j \in T} w_j * f_{j,k} \tag{8}$$

The weight of each cluster shows the influence and importance of each cluster. For each cluster i we select a set of tweets that contains at least one hashtag of that cluster and sort them by their weights in descending order. Then we take top P tweets which satisfy the following condition:

$$\sum\nolimits_{\forall j \in P} w_j \geq CW_i/\alpha \qquad (9)$$

Where α is the constant value and w_j is the weight of tweet t_j.

3.5 Extract Most Recent Tweets

For each tweet t_i, we crawl the most recent tweets posted by its author u_i and thus get its recent tweet set $R_{i,k} = \{t_{i1}, t_{i2}, \ldots, t_{ik}\}$. Assuming the original tweet t_i was posted at time, $R_{i,k}$ denotes the recent k tweets posted by user u_i before r.

For each tweet t_i, we assume the number of comments, replies and retweets as features.

3.6 LSTM Labeling

In this section we briefly describe the proposed LSTM model. Given an input sequence $R_{i,k} = \{t_{i1}, t_{i2}, \ldots, t_{ik}\}$, LSTM computes the hidden vector sequence $h_{i,k} = \{h_{i1}, h_{i2}, \ldots, h_{ik}\}$ and output vector sequence $Y_{i,k} = \{Y_{i1}, Y_{i2}\}$. In fact Y_{i1} is the probability of t_i to be a rumor and Y_{i2} shows the probability of t_i to not being a rumor.

At each time step, the output of the module is controlled by a set of gates as a function of the previous hidden state h_{ij} and the input at the current time step t_{ij}, the forget gate f_j, the input gate i_t, and the output gate o_t. These gates collectively decide the transitions of the current memory cell c_t and the current hidden state h_t. The LSTM transition functions are defined as follows:

$$i_t = \sigma(W_i.[h_{t-1}, x_t] + b_i) \qquad (10)$$

$$f_t = \sigma\left(W_f.[h_{t-1}, x_t] + b_f\right) \qquad (11)$$

$$l_t = tanh(W_l.[h_{t-1}, x_t] + b_l) \qquad (12)$$

$$o_t = \sigma(W_o.[h_{t-1}, x_t] + b_o) \qquad (13)$$

$$c_t = f_t \odot c_{t-1} + i_t \odot l_t \qquad (14)$$

$$h_t = o_t \odot \tanh(c_t) \qquad (15)$$

Here, σ is the sigmoid function that has an output in $[0, 1]$, tanh denotes the hyperbolic tangent function that has an output in $[-1, 1]$, and $\odot$ denotes the component-wise multiplication. The extent to which the information in the old memory cell is discarded is controlled by f_t, while i_t controls the extent to which new information is stored in the current memory cell, and o_t is the output based on the memory cell c_t.

LSTM is explicitly designed for learning long-term dependencies, and therefore we choose LSTM to learn dependencies in the sequence of extracted features.

In sequence-to-sequence generation tasks, an LSTM defines a distribution over outputs and sequentially predicts tokens using a softmax function.

$$P(Y|X) = \prod_{t \in [1,N]} \frac{\exp(g(h_{t-1}, y_t))}{\sum_{y'} \exp\left(g\left(h_{t1}, y_t'\right)\right)} \tag{16}$$

While g is the activation function.

4   Experimental Results

In this section the impacts of different experimental conditions are investigated we have crawled 600,000 Persian tweets including both rumors and non-rumors with their recent tweet sets. All experiments are run on a server with Xeon E3-1275 processor and 32 GB of RAM and 512 GB NVME hard disk.

4.1   Proposed Clustering Method Evaluation

In order to evaluate the performance of our proposed clustering method, we use a data set of 1.5 million Persian tweets. In fact, to the best on our knowledge, there are about 600,000 Persian tweets posted in tweeter in each day. We assume K = 10. In other words, we try to cluster hashtags into 10 different clusters.

Time taken for clustering with the different number of hashtags is shown the Fig. 2. Meanwhile, Fig. 3 shows the time taken for clustering with the different number of tweets.

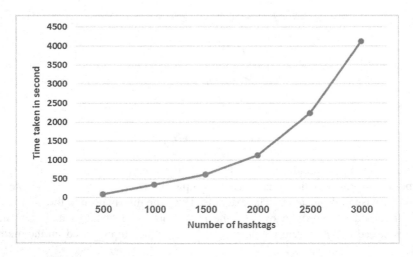

Fig. 2. Impact of number of hashtags on time taken for clustering

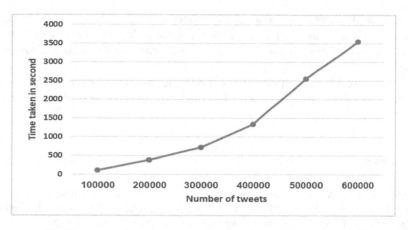

Fig. 3. Impact of number of tweets on time taken for clustering

4.2 Proposed LSTM Classifier Evaluation

1850 labeled rumors and 3200 non-rumors used in order to train the proposed LSTM classifier. We further compare the performances of the proposed LSTM classifier method proposed with methods proposed in [14] and [15]. Both [14] and [15] treat the rumor preemption task as an anomaly detection problem and try to detect rumors on Sina Weibo which is a Chinese microblogging website.

In fact, the proposed method in [15] is based on multi-layer structured autoencoder. The accuracy (Acc), F1 score (F1), precision (Prec), recall (Recall) and false positive rate (FPR) of the detection results are selected as an evaluation parameter. The summary of the comparison results is shown in Table 2.

Table 2. Comparison results

Method	Acc (%)	F1 (%)	Per (%)	Recall (%)	FPR (%)
Proposed LSTM classifier	92.45%	89.95%	88.41%	89.00%	5.01%
[15]	85.60%	77.9%	79.56%	81.22%	7.95%
[14]	82.50%	74.58%	78.41%	72.23%	12.2%

Form this table, we can see that the proposed LSTM not only achieve higher accuracy, F1 score and precision but also has lower false positive rate value.

5 Conclusions and Future Work

In this paper, In order to detect rumor to prevent public issues they may cause, we proposed an LSTM based rumor detection method. In fact, we assumed rumor detection problem on twitter as an anomaly detection task. Tweet features are analyzed in order to detect anomaly detection.

Extensive experiments conducted on Twitter show that the proposed model can significantly outperform other state-of-the-art models. In fact, the proposed method archives not only higher accuracy, F1 score and precision but also has lower false positive rate.

As future work, we propose to use LSTM as an autoencoder and convert our proposed supervised anomaly detection method into an unsupervised anomaly detection.

References

1. Kwon, S., Cha, M., Jung, K., Chen, W., Wang, Y.: Aspects of rumor spreading on a microblog network. In: Jatowt, A., et al. (eds.) SocInfo 2013. LNCS, vol. 8238, pp. 299–308. Springer, Cham (2013). https://doi.org/10.1007/978-3-319-03260-3_26
2. Liang, G., He, W., Xu, C., Chen, L., Zeng, J.: Rumor identification in microblogging systems based on users' behavior. IEEE Trans. Comput. Soc. Syst. **2**(3), 99–108 (2015)
3. Misako, T., Sato, K., Sano, Y., Yamada, K., Miura, W., Hideki, T.: Rumor diffusion and convergence during the 3.11 earthquake: a Twitter case study. PLoS one **10**(4), e0121443 (2015)
4. Singh, A., Kumar, R., Singh, Y.N.: Rumor dynamics with acceptability factor and inoculation of nodes in scale free networks. In: 2012 Eighth International Conference on Signal Image Technology and Internet Based Systems (SITIS), pp. 798–804 (2012)
5. Zhao, Z., Resnick, P., Mei, Q.: Enquiring minds: Early detection of rumors in social media from enquiry posts. In: Proceedings of the 24th International Conference on World Wide Web, International World Wide Web Conferences Steering Committee, pp. 1395–1405 (2015)
6. Vahed, Q., Rosengren, E., Radev, D.R., Mei, Q.: Rumor has it: identifying misinformation in microblogs. In: Proceedings of the Conference on Empirical Methods in Natural Language Processing, Association for Computational Linguistics pp. 1589–1599 (2011)
7. Castillo, C., Mendoza, M., Poblete, B.: Information credibility on twitter. In: Proceedings of the 20th International Conference on World Wide Web, pp. 675–684. ACM (2011)
8. Wu, K., Yang, S., Zhu, K.Q.: False rumors detection on sina weibo by propagation structures. In: 2015 IEEE 31st International Conference on Data Engineering (ICDE), pp. 651–662. IEEE (2015)
9. Sun, S., Liu, H., He, J., Du, X.: Detecting event rumors on sina weibo automatically. In: Ishikawa, Y., Li, J., Wang, W., Zhang, R., Zhang, W. (eds.) APWeb 2013. LNCS, vol. 7808, pp. 120–131. Springer, Heidelberg (2013). https://doi.org/10.1007/978-3-642-37401-2_14
10. Dayani, R., Chhabra, N., Kadian, T., Kaushal, R.: Rumor: Detecting Misinformation in Twitter. In: 3rd Security and Privacy Symposium (2015)
11. Ennals, R., Byler, D., Agosta, J.M., Rosario, B.: What is disputed on the web?. In: Proceedings of the 4th Workshop on Information Credibility, pp. 67–74. ACM (2010)
12. Twitter Inc.:Twitter (2009). http://www.twitter.com
13. Floyd, R.W.: Algorithm 97: shortest path. Commun. ACM **5**(6), 345 (1962)
14. Chen, W., Yeo, C.K., Lau, C.T., Lee, B.S.: Behavior deviation: an anomaly detection view of rumor preemption. In: 2016 IEEE 7th Annual Information Technology, Electronics and Mobile Communication Conference (IEMCON), pp. 1–7. IEEE (2016)
15. Zhang, Y., Chen, W., Yeo, C.K., Lau, C.T., Lee, B.S.: Detecting rumors on online social networks using multi-layer autoencoder. In: 2017 IEEE Technology & Engineering Management Conference (TEMSCON), pp. 437–441. IEEE (2017)

Density-Based Clustering by Double-Bit Quantization Hashing

Mahdieh Dehghani, Ali Kamandi$^{(\boxtimes)}$, and Ali Moeini

School of Engineering Sciences, College of Engineering,
University of Tehran, Tehran, Iran
{dehghani.mahdieh,kamandi,moeini}@ut.ac.ir

Abstract. Grouping data into the different parts, while the objects in the same part have the most similarity with each other and cannot belong to the other parts, called data clustering. Clustering used for data analysis in data mining, so far, many different algorithms for clustering have been offered. Density-based algorithms are one of the useful clustering approaches, which used for databases with different shapes. This algorithms have a short response time for small databases and also be able to extract clusters with arbitrary shapes. DBSCAN is a density-based clustering algorithm that can detect and extend clusters based on a restricted neighbor radius and the number of near objects in neighbor radius. The time complexity of this algorithm belongs to $O(n^2)$ for large datasets. We used data indexing technique Local Sensitive Hashing (LSH) to reduce the implementation time of the algorithm, this data structure can be used to found neighbor points in the DBSCAN algorithm, so, the response time of the algorithm, will be reduced, because LSH be able to approximate the K-nearest neighbors algorithm in linear time complexity. We used this data structure to detect neighbor points quickly by mapped data to a binary space. We used the influence space idea to detect clusters, to improve the response time of the algorithm, this concept can reduce the search space to expand the clusters. We evaluated our algorithm by two density-based clustering algorithms DBSCAN and BLSH-DBSCAN. We can improve both mentioned algorithms in terms of response time for large datasets.

Keywords: DBSCAN · BLSH-DBSCAN · Double-bit Quantization Hashing

1 Introduction

One of the most important branches of data mining is clustering, which is an unsupervised algorithm for dividing data into similar categories. The best clustering is that the members of the cluster has the lowest distance with each other and has large distance with members of other clusters so that current cluster points can not belong to others. Several algorithms have been presented for

© Springer Nature Switzerland AG 2019
L. Grandinetti et al. (Eds.): TopHPC 2019, CCIS 891, pp. 301–314, 2019.
https://doi.org/10.1007/978-3-030-33495-6_23

clustering, which are consist of four based categories partition-based, density-based, hierarchical-based and grid-based. Partition-based algorithms have the least time complexity between the others, but they can not detect clusters in different shapes. Hierarchical-based algorithms have large time complexity but can recognize clusters of any shape, and grid-based algorithms detected artificial clustering because extracted clusters depended on grids size. Density-based algorithms have the ability to extract clusters with different shapes, in these algorithms, at first densest data locations founded and then clusters will be expanded by some conditions. The most popular and the first density-based algorithm is DBSCAN algorithm [1], that founded denser regions based on two parameters $Minpts$ and ε, and expanded the clusters with these limitations.

One of the disadvantages of the DBSCAN algorithm is to determine the parameters of this algorithm, also the algorithm creates different clusters according to the initial start points. This algorithm only one time scan the database and builds clusters, but in high-dimensional databases, this algorithm does not behave linearly because it requires to found neighbor points among database points at each step. To solve this problem, four methods have been proposed. Partitioning, data sampling, parallelization, and data indexing. So far, many data structures have been designed to identify similar items, such as R-tree [2], R*-tree [3] and SS-tree [4], but they could not be improved the DBSCAN algorithm for high-dimensional databases, while Local Sensitive Hashing (LSH) has the ability to approximate the K-nearest neighbor algorithm for large databases in linear time complexity [5]. There are several types of LSH techniques that have been used to improved DBSCAN algorithm, for example, in [6] p-Stable Distribution LSH and [7] Binary LSH are used for data indexing. In this paper, we present a density-based clustering algorithm (BDBQLSH-DBSCAN) that improved the response time of DBSCAN for large databases by using Double-bit Quantization Hashing (DBQLSH) for data indexing. This method maps the data to the binary space, therefore, we can calculate similar items by Hamming distance measure.

The rest of the paper is organized as follows: Sect. 2 gives related work done in this field. Sect. 3 introduces the proposed algorithm. Experiments and results are presented in Sect. 4. Some conclusions are given in Sect. 5.

2 Background

DBSCAN is a density-based clustering algorithm, that can be detected clusters with different shapes, this algorithm found and expand clusters based on the number of neighbors of a point in the neighbor radius, for large or high-dimensional databases, this algorithm spends a lot of time to determine the clusters [1]. DBSCAN defines three relations between database points. Two parameters are defined to determines the relations, $Minpts$ and ε, which ε represents the maximum distance between a point and its neighbors, and $Minpts$ represents the minimum number of neighbor points that a point must have to be able to participate in cluster expansion. Suppose $\{x_1, x_2, ..., x_n\}$ represent d-dimensional database with n points, then for each pair of points x_i and x_j, three relations are defined.

Definition 1. *If $x_i \in N_\varepsilon(x_j)$ and $|N_\varepsilon(x_j)| \geq Minpts$, where $N_\varepsilon(x_j)$ shows x_j neighbor points with, radius less than ε, x_i and x_j are directly density reachable.*

Definition 2. *If there is a chain of database points such as $\{y_1, y_2, ..., y_l\}$, which for all $1 \leq i \leq l$, y_{i+1} and y_i are directly density reachable and $y_1 = x_i$, $y_l = x_j$ then x_i is density reachable to x_j (with the same ε and $Minpts$ values).*

Definition 3. *If there is a point such as y, and both x_i and x_j are density reachable from y then x_i is density connected to x_j (with the same ε and $Minpts$ values).*

According to Definitions 1, 2 and 3 database points are divided into three types:

1. *Core: If $|N_\varepsilon(x)| \geq Minpts$ then x is a core point.*
2. *Border: If $|N_\varepsilon(x)| \leq Minpts$ and x is directly density reachable from a core point, then x is a border point.*
3. *Noise: If a point x is neither a core point nor a border point, then it will be considered as a noise point.*

In the DBSCAN algorithm, first, all points are labeled to "unvisited". Then, at each step, an "unvisited" point is selected and labeled with cluster number, if it is a core point, then a new cluster will be created with it, and all density reachable points from that, which "unvisited" will be mapped to the current cluster and labeled to "visited". This will be repeated as long as no point exist that could be added to the current cluster. If it is a noise point, then it will be labeled to "noise" and if a border point nothing will be done. This work will be repeated until, all points labeled with cluster number or "noise".

ST-DBSCAN is a density-based algorithm that can be improved DBSCAN, recognized clusters with non-spatial, spatial and temporal values of the objects, is the advantage of this algorithm [8].

For high-denominational databases, DBSCAN's time complexity turned to $O(n^2)$. Divide the database into small parts and run the algorithm for each of them can improve the time complexity of the algorithm. [9] offers tow sampling-based DBSCAN (SDBSCAN) algorithms, one of them implements data sampling inside of DBSCAN algorithm and the other applies sampling procedure outside of this algorithm. Until the DBSCAN algorithm run, the database must be placed on the main memory. It is impossible to run this algorithm for large databases with small main memory, in [10], this problem has been solved with data sampling. Parallel techniques, divided input data into smaller parts, then, the algorithm runs on all parts concurrently, and the response of all parts will be mixed to achieve the final answer. These techniques can reduce the time complexity of the DBSCAN algorithm.

MR-DBSCAN defined a quick partitioning strategy by a 4-sections MapReduce model also, can be done for large databases [11]. [12] offered an algorithm based on DBSCAN, that performs parallel exploration of each cluster. [6] presented a parallel Dbscan algorithm (Pdsdbscan) based on graph algorithmic concepts. P-DBSCAN is a parallel version of the DBSCAN algorithm, which used PR-tree for data indexing in distributed environment [13].

G-DBSCAN is a parallel version of the DBSCAN algorithm based on graphics processing units (GPU), which used GPU instead of CPU [14]. The difference between partitioning and data sampling is that partitioning methods are implemented for total input data and the final answer is the mixture of all responses, but data sampling selected a part of the data as an agent and just run the algorithm on the selected part and the answer is considered as a final answer. Partitioning-based DBSCAN algorithm (PDBSCAN), performs better for databases with different density clusters, but this algorithm depended on the initial values [15].

Grid-based DBSCAN Algorithm (GRPDBSCAN), used grid-based techniques to partition the data into several parts then, based on the properties of each grid's data, execute multiple DBSCAN algorithms for clustering [16]. There are several data structures that, can be detected neighbor points quickly, this data structures, will be improved time complexity of DBSCAN algorithm if there used to find neighbor points by reduced the search space. R-tree is a dynamic index structure that retrieves neighbor points for high-denominational databases, this data structure has linear time complexity for rectangles databases [2]. R*-tree, can be used for different types of queries and operations, it can be supported both pomt and spattal concurrently, this data structure can improve the implementation cost of the R-tree [3] data structure.

Similarity search tree (SS-tree), generate data structure, that can be determined the same objects in the database. This data structure performs better than R*-tree and dynamically can updated [4]. Used tree data structures to find the neighbor points will be improved the execution time of the DBSCAN algorithm just for databases with less than ten dimensions. But Locality Sensitive Hashing (LSH) [17], iDistance [18], and iPoc [19] are used effectively for higher-dimensional databases for data indexing. iPoc can effectively prune query-unrelated data points by estimating the lower and upper bounds in both radial coordinate and angle coordinate, that can be improved KNN algorithm for high-dimensional databases.

Local Sensitive Hashing (LSH) is a hash-based data indexing technique, data indexing done based on several random hash functions. Then some hash functions are selected independently and uniformly at random. Let M be a measure distance for data set D, in any metric space for $p, q \in D$ and distance r, $Sim(p, r)$ is set of points whose distance from p is less than r, LSH is defined as below [5]:

Definition 4. *A family* $\mathcal{H} = \{h : \mathcal{S} \to U\}$ *is called* (r_1, r_2, p_1, p_2)-*sensitive for* M *if for any* $p; q \in D$

- *if* $q \in Sim(p, r_1)$ *then* $Pr_{\mathcal{H}}[h(p) = h(q)] \geq p_1$,
- *if* $q \notin Sim(p, r_2)$ *then* $Pr_{\mathcal{H}}[h(p) = h(q)] \leq p_2$.

Hash functions which defined for locality-sensitive hashing must satisfied the inequalities $p_1 > p_2$ and $r_1 < r_2$. The gap between probability p_1 and p_2 will be amplified by concatenating several random hash functions. Exactly, for k specified late, define a function family $\mathcal{G} = \{g : S \rightarrow \mathcal{U}^k\}$ such that $g(v) = (h_1(v); \ldots; h_k(v))$, where $h_i \in \mathcal{H}$. L functions $(g_1; \ldots; g_L)$ are selected from $\mathcal{G}$, independently and uniformly at random. Several number of buckets will be ditected, then, only the non-empty buckets are retained.

K-Nearest Neighbor search (KNNS) [20] algorithm is used to distinguish similar items, for large databases it will take a large time to detect K-Nearest Neighbor objects for all database objects, this method introduces K objects which are most similar with the target object.

LSH can be effectively approximate KNNS algorithm, based on data indexing. Double-Bit Quantization Hashing (DBQLSH) [21] creates binary code equals to the number of hash functions by assumption, that input data has a normal distribution. In this way, the database with d-dimensional points mapped to the binary space. The extracted binary numbers can be compared by the Hamming distance measure. In this algorithm, the initial data is divided into three parts, and then each of them is mapped to $\{01, 11, 10\}$.

Two threshold values are needed to divided the data into three sets $\{S_1, S_2, S_3\}$. First, placed the positive data in S_1 and the negative data in S_2, then this sets S_1 and S_3 will be sorted, and set S_2 will be determined by using two ordered sets in such a way that the mean value of S_2 be limited to zero, because Eq. 1 must be minimized (μ_i is the mean value of set S_i) to minimize the variance of each set. $S_1 = \{x | -\infty < x \leq a \quad x \in S\}$, $S_2 = \{x | a < x \leq b \quad x \in S\}$ and $S_3 = \{x | b < x \leq \infty \quad x \in S\}$ determines the two thresholds a and b for data segmentation.

$$E = \sum_{x \in S_1} (x - \mu_1)^2 + \sum_{x \in S_2} (x - \mu_2)^2 + \sum_{x \in S_3} (x - \mu_3)^2 \tag{1}$$

In Eq. 2, E simplified, according Eq. 1, it will be possible to maximize F in Eq. 3 instead of minimized E.

$$E = \sum_{x \in S}(x)^2 - 2\sum_{x \in S_1} x\mu_1 + \sum_{x \in S_1}(\mu_1)^2 - 2\sum_{x \in S_3} x\mu_3 + \sum_{x \in S_3}(\mu_3)^2$$
$$= \sum_{x \in S}(x)^2 - |S_1|\mu_1^2 - |S_3|\mu_3^2 \tag{2}$$
$$= \sum_{x \in S}(x)^2 - \frac{(\sum_{x \in S_1} x)^2}{|S_1|} + \frac{(\sum_{x \in S_3} x)^2}{|S_3|}$$

$$F = \frac{(\sum_{x \in S_1} x)^2}{|S_1|} + \frac{(\sum_{x \in S_3} x)^2}{|S_3|} \tag{3}$$

LSH, improved original DBSCAN algorithm, the KNNS algorithm based on hashing can reduce DBSCAN time complexity from $O(n \log n)$ to $O(n)$ [17]. In [22] locality-sensitive hashing used to rapid nearest neighbor searching, and reduced the size of the search space by mixing the nearest neighbor. Shared Nearest Neighbors clustering based on LSH (LSH-SNN), purpose to the field of metagenomics, to cluster high-dimensional sequence data and can be used for larger datasets [23]. In [6] p-Stable Distribution LSH used to detect neighbor points and influence space concept, to reduce the search space rather than the DBSCAN algorithm. [7] used Binary LSH, to applied Hamming distance measure instead of Euclidean, then, reached the clusters based on the influence space.

3   Proposed Algorithm

We proposed a density-based algorithm for clustering, which used the K-Nearest Neighbors algorithm to detect high-density regions. K-Nearest Neighbor algorithm approximated with DBQLSH technique which perused in the Sect. 2. Proposed algorithm works like the DBSCAN algorithm, but the ε and $Minpts$ parameters are transformed into a parameter K. To defined the core, border, and noise points, the number of K-Nearest Neighbors points is used, these points may be far from the target point so that, these points will be not clustered with the target point, for example, noise points like that. Therefore, we consider influence space, for each point, see Definition 5.

Definition 5. *Influence space for point x_i defined as $IS_k(x_i)$, that contains points which are member of $KNN(x_i)$ set and x_i is as a member of KNN set for all of them (see Eq. 4).*

$$x_j \in IS_k(x_i) \quad if \quad x_i \in KNN(x_j) \quad and \quad x_j \in KNN(x_i). \tag{4}$$

We used the KNN algorithm with data indexing techniques [20] to reduce the time for influence space computation. Then, influence space calculated based on Hamming distance instead of Euclidean distance measure, by matrix $\mathcal{D}$. Let $\{x_1, x_2, ..., x_n\}$ be the set of database points and for x_i, set $\{n_{i,1}, n_{i,2}, .., n_{i,k}\}$ represent the KNN points then $\mathcal{D}$ is $n \times n$ matrix, each member of $\mathcal{D}$ specified by Eq. 5. Each row of matrix $\mathcal{D}$ represents the corresponding KNN point of that index and each row of $\mathcal{D}$ transposed represents points that the point corresponds to the index is KNN of them.

$$d_{i,j} = \begin{cases} 1 & \text{if } x_i \in KNN(x_j) \quad and \quad x_j \in KNN(x_i) \\ 0 & \text{otherwise} \end{cases} \tag{5}$$

The influence space of each point $x_i \in \{x_1, x_2, ..., x_n\}$ can be calculated by Eq. 6.

$$x_j \in IS_k(x_i) \quad if \quad d_{i,j} \wedge d_{j,i} = 1 \quad \{j | x_j \in KNN(x_i)\} \tag{6}$$

According to $|IS_k(x_i)|$ value (x_i influence space points number), the database points are divided into three groups, core, border and noise points. If $|IS_k(x_i)| \geq \sigma K$ then x_i be a core point, if $0 < |IS_k(x_i)| < \sigma K$ then x_i be a border point, which $\sigma = 2/3$ [6] and if $|IS_k(x_i)| = 0$, x_i is considered as noise point. Our algorithm has two sections:

1. Create hash tables based on DBQLSH algorithm and determine binary KNN for each database points then we can estimate the influence space of each one.
2. specify the core, border and noise points, then cluster them with using the core and border points and remove outliers from the database by noise points, such as DBSCAN algorithm.

   For each core point $x_i \in \{x_1, x_2, ..., x_n\}$ that previously not been labeled with cluster number create a new cluster, then all the points in $KNN(x_i)$ will be label with x_i's cluster number, if they already not belong to the cluster, each core points in set $IS_k(x_i)$ consider as seed point, if it's KNN set has lowest intersect with other seed point sets. All points in seed point's KNN

Algorithm 1. Find Seed Points

Input : x_i
Output: Seed Point Set
for $x_j \in IS_k(x_i)$ **do**
  if $|IS_k(x_j)| \geq 2k/3$ **then**
    if $|KNN(x_j) \cap KNN(x_l)|$ $\{x_l \in SeedPointSet\}$ has the lowest value **then**
      Add x_j to $SeedPointSet$;
    end if
  end if
end for

Algorithm 2. BDBQLSH-DBSCAN

Input : Database=$\{x_1, x_2, ..., x_n\}$
Output: Set of Cluster Labels
$cluster_{number} = 0$;
for $x_i \in \{x_1, x_2, ..., x_n\}$ **do**
  if Cluster Label(x_i)=="Unclustered" **then**
    if $|IS_k(x_j)| == 0$ **then**
    Cluster Label(x_i)="Noise";
    else if $|IS_k(x_j)| \geq 2k/3$ **then**
    $cluster_{number} + +$;
    Cluster Label(x_i)=$cluster_{number}$;
    for $x_j \in KNN(x_i)$ **do**
      Cluster Lable(x_j)=Cluster Lable(x_i);
    end for
    for $x_j \in FindSeedPoints(x_i)$ **do**
      for $x_l \in KNN(x_j)$ **do**
        if Cluster Lable(x_l)=="Unclustered" **then**
        Cluster Lable(x_l)=Cluster Lable(x_j);
        end if
      end for
    end for
  end if
  end if
end for

set label as x_i cluster number, if they have not already been clustered, this will continue while all seed points of the x_i check. When all the seed points checked, it means the cluster cannot extend anymore, this cycle repeated for another "Unclustered" points of the database. Algorithm 1 determined seed points and Algorithm 2 represented the proposed algorithm.

4   Experimentation and Results

The proposed algorithm compared with DBSCAN and BLSH-DBSCAN [7] algorithms, which both of them are density-based algorithms for clustering and our approach, has a procedure like them. Our algorithm implemented in MATLAB platform, comparisons are done in terms of response time and accuracy. The computational times are measured on a PC with 2.4 GHz CPU and 8 GB RAM. We used synthetic datasets with a different number of points and fixed dimensions to compare the response time of three algorithms, for large databases. 2D- synthetic datasets are used to evaluate the algorithms for low-dimensional datasets Fig. 2 shows the clustering produced by three algorithms for 6 different datasets with 1500 points. The best parameters ($Minpt, \varepsilon$ and k) are reported for each dataset. Obviously, extracted clusters with three mentioned algorithms approximately are the same. BLSH-DBSCAN and BDBQLSH-DBSCAN are randomized algorithms and response time of them changed at each run. Also, the DBSCAN algorithm is depends on the initial start point and has a different response time at each run. So, the average response times of the algorithms for 10 times will be reported. Table 1 report the maximum, minimum, and average time values for 10 runs. Figure 1 show the average response times for each dataset, according to this plot, that DBSCAN performs better than BLSH-DBSCAN and BDBQLSH-DBSCAN for low-dimensional datasets. Figure 3 show the variations in the response time of the algorithms for six 2-dimensional datasets.

Table 1. Response time values for each datasets.

Name	DBSCAN			BLSH-DBSCAN			BDBQLSH-DBSCAN		
	Min	Max	Avg	Min	Max	Avg	Min	Max	Avg
Circles	0.078335	0.091949	0.081984	0.22981	0.35037	0.28366	0.18244	0.21055	0.20221
Moons	0.081552	0.086176	0.083327	0.25971	0.33445	0.29829	0.21928	0.27816	0.25243
Varied	0.082708	0.092983	0.086668	0.24867	0.34288	0.29872	0.19286	0.24835	0.22013
Aniso	0.073528	0.098663	0.078769	0.25815	0.34162	0.29275	0.18802	0.25559	0.2118
Blobs	0.089248	0.11576	0.095806	0.28038	0.36286	0.31928	0.20772	0.34343	0.24488
No-structure	0.18549	0.20882	0.19287	0.30429	0.39989	0.35222	0.21911	0.24453	0.22887

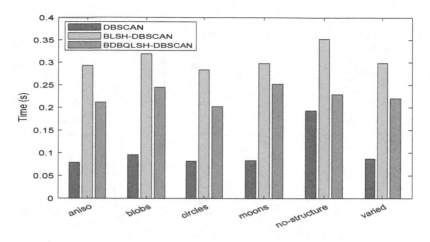

Fig. 1. Time comparison for 2-D datasets.

To compare the response time of three algorithms for large databases, we used synthetic datasets with 15 dimensions and a different number of points (5000–20000). Each synthetic dataset has three Gaussian distribution clusters with different variances. Table 2 display the maximum, minimum, and average time values for 10 different run times. Figure 4 show the time variation while the number of data points raised, can be concluded from this plot that BDBQLSH-DBSCAN performs better than BLSH-DBSCAN and DBSCAN for large datasets. Figure 5 show the variations of the response time for three algorithms on four large databases. According to box plots, always a distance exists between the response time of tow algorithms and BDBQLSH-DBSCAN, it means that this algorithm did not perform better just for a single run.

Table 2. Response time values for each Synthetic datasets ($K = 7, Minpts = 5$ and $\varepsilon = 10$).

Size	DBSCAN			BLSH-DBSCAN			BDBQLSH-DBSCAN		
	Min	Max	Avg	Min	Max	Avg	Min	Max	Avg
5000	4.2395	4.8414	4.3361	3.5457	3.9853	3.7602	1.9125	2.6912	2.3211
10000	34.0581	36.5024	35.0824	13.1951	16.5804	14.9997	7.4016	10.7697	8.7021
15000	82.7258	90.1161	86.0765	32.0736	37.7356	35.1679	17.9103	24.1803	20.5108
20000	143.5452	144.5372	144.1559	59.9442	73.1388	67.0695	33.9654	48.519	41.3201

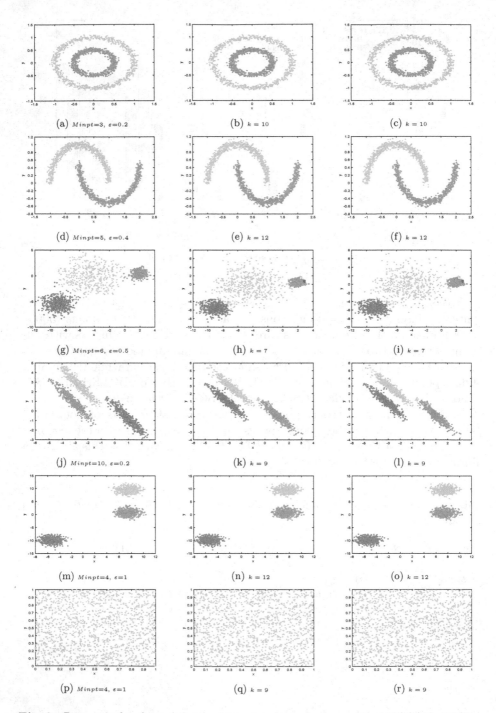

Fig. 2. Respectively, from left to right, detected clusters with DBSCAN, BLSH-DBSCAN, and BDBQLSH-DBSCAN algorithms displayed.

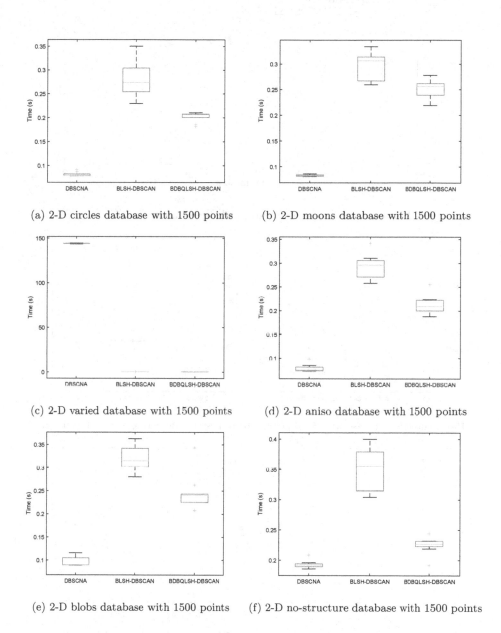

(a) 2-D circles database with 1500 points

(b) 2-D moons database with 1500 points

(c) 2-D varied database with 1500 points

(d) 2-D aniso database with 1500 points

(e) 2-D blobs database with 1500 points

(f) 2-D no-structure database with 1500 points

Fig. 3. Run time variation for algorithms

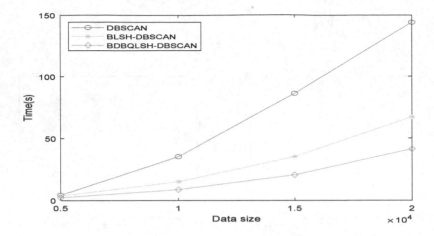

Fig. 4. Time comparisons for synthetic datasets (K = 7).

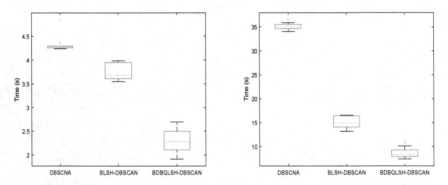

(a) 15-D synthetic database with 5000 points (b) 15-D synthetic database with 10000 points

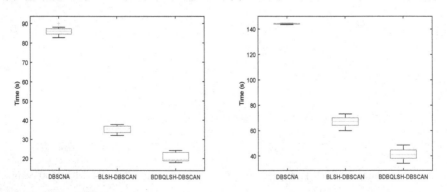

(c) 15-D synthetic database with 15000 points (d) 15-D synthetic database with 20000 points

Fig. 5. Run time variation for algorithms ($Minpt = 5, \varepsilon = 10$ and $k = 7$)

5 Conclusion

We proposed BDBQLSH-DBSCAN algorithm involved DBSCAN and Double-bit quantization hashing. This algorithm will be able to detect clusters with any shapes, and produced clusters which approximately has good accuracy. We used a data indexing method LSH to reduce the response time of the algorithm. This algorithm works better for high-dimensional databases rather than the DBSCAN algorithm and similar algorithms, that used data indexing structures such as BLSH-DBSCAN, which uses Binary LSH technique for data indexing. In BDBQLSH-DBSCAN, to reduce the algorithm run-time, hashing techniques are used to compared database members with Hamming distance instead of Euclidean. We can reduce the number of the DBSCAN algorithm parameters from ε and *Minpts* to K (number of neighbors). Also used LSH techniques led to mapped the dataset to a low-dimensional space, it means that memory consumption reduced rather than the DBSCAN algorithm. According to experiments on 2-D synthetic datasets, extracted clusters by BDBQLSH-DBCAN algorithm looking like detected clusters by tow density based algorithms DBSCAN and BLSH-DBSCAN. The response time of this algorithm for high-dimensional or large databases always will be smaller than the response time of the two mentioned algorithms.

References

1. Ester, M., Kriegel, H.P., Sander, J., Xu, X., et al.: A density-based algorithm for discovering clusters in large spatial databases with noise. In: KDD, vol. 96, pp. 226–231 (1996)
2. Guttman, A.: R-trees: a dynamic index structure for spatial searching, vol. 14. ACM (1984)
3. Beckmann, N., Kriegel, H.P., Schneider, R., Seeger, B.: The R*-tree: an efficient and robust access method for points and rectangles. In: ACM SIGMOD Record, vol. 19, pp. 322–331. ACM (1990)
4. White, D.A., Jain, R.: Similarity indexing with the SS-tree. In: ICDE, p. 516. IEEE (1996)
5. Indyk, P., Motwani, R.: Approximate nearest neighbors: towards removing the curse of dimensionality. In: Proceedings of the Thirtieth Annual ACM Symposium on Theory of Computing, pp. 604–613. ACM (1998)
6. Lv, Y., et al.: An efficient and scalable density-based clustering algorithm for datasets with complex structures. Neurocomputing **171**, 9–22 (2016)
7. He, Q., Gu, H.X., Wei, Q., Wang, X.: A novel DBSCAN based on binary local sensitive hashing and binary-KNN representation. Advances in Multimedia (2017)
8. Birant, D., Kut, A.: ST-DBSCAN: an algorithm for clustering spatial-temporal data. Data Knowl. Eng. **60**(1), 208–221 (2007)
9. Zhou, S., Zhou, A., Cao, J., Wen, J., Fan, Y., Hu, Y.: Combining sampling technique with DBSCAN algorithm for clustering large spatial databases. In: Terano, T., Liu, H., Chen, A.L.P. (eds.) PAKDD 2000. LNCS (LNAI), vol. 1805, pp. 169–172. Springer, Heidelberg (2000). https://doi.org/10.1007/3-540-45571-X_20

10. Borah, B., Bhattacharyya, D.: An improved sampling-based DBSCAN for large spatial databases. In: 2004 Proceedings of International Conference on Intelligent Sensing and Information Processing, pp. 92–96. IEEE (2004)

11. He, Y., et al.: MR-DBSCAN: an efficient parallel density-based clustering algorithm using mapreduce. In: 2011 IEEE 17th International Conference on Parallel and Distributed Systems (ICPADS), pp. 473–480. IEEE (2011)

12. Arlia, D., Coppola, M.: Experiments in parallel clustering with DBSCAN. In: Sakellariou, R., Gurd, J., Freeman, L., Keane, J. (eds.) Euro-Par 2001. LNCS, vol. 2150, pp. 326–331. Springer, Heidelberg (2001). https://doi.org/10.1007/3-540-44681-8_46

13. Kisilevich, S., Mansmann, F., Keim, D.: P-DBSCAN: a density based clustering algorithm for exploration and analysis of attractive areas using collections of geotagged photos. In: Proceedings of the 1st International Conference and Exhibition on Computing for Geospatial Research and Application, p. 38. ACM (2010)

14. Andrade, G., Ramos, G., Madeira, D., Sachetto, R., Ferreira, R., Rocha, L.: G-DBSCAN: a GPU accelerated algorithm for density-based clustering. Procedia Comput. Sci. **18**, 369–378 (2013)

15. Jiang, H., Li, J., Yi, S., Wang, X., Hu, X.: A new hybrid method based on partitioning-based DBSCAN and ant clustering. Expert Syst. Appl. **38**(8), 9373–9381 (2011)

16. Kellner, D., Klappstein, J., Dietmayer, K.: Grid-based DBSCAN for clustering extended objects in radar data. In: 2012 IEEE Intelligent Vehicles Symposium (IV), pp. 365–370. IEEE (2012)

17. Wu, Y.P., Guo, J.J., Zhang, X.J.: A linear DBSCAN algorithm based on LSH. In: 2007 International Conference on Machine Learning and Cybernetics, vol. 5, pp. 2608–2614. IEEE (2007)

18. Yu, C., Ooi, B.C., Tan, K.L., Jagadish, H.: Indexing the distance: an efficient method to KNN processing. In VLDB, vol. 1, pp. 421–430 (2001)

19. Liu, Z., Wang, C., Zou, P., Zheng, W., Wang, J.: iPoc: a polar coordinate based indexing method for nearest neighbor search in high dimensional space. In: Chen, L., Tang, C., Yang, J., Gao, Y. (eds.) WAIM 2010. LNCS, vol. 6184, pp. 345–356. Springer, Heidelberg (2010). https://doi.org/10.1007/978-3-642-14246-8_34

20. Gionis, A., Indyk, P., Motwani, R., et al.: Similarity search in high dimensions via hashing. In: VLDB, vol. 99, pp. 518–529 (1999)

21. Kong, W., Li, W.J.: Double-bit quantization for hashing. In: AAAI, vol. 1, p. 5 (2012)

22. Youn, J., Shim, J., Lee, S.G.: Efficient data stream clustering with sliding windows based on locality-sensitive hashing. IEEE Access **6**, 63757–63776 (2018)

23. Kanj, S., Brüls, T., Gazut, S.: Shared nearest neighbor clustering in a locality sensitive hashing framework. J. Comput. Biol. **25**(2), 236–250 (2018)

Representing Unequal Data Series in Vector Space with Its Application in Bank Customer Clustering

Shohreh Tabatabayi Seifi$^{(\boxtimes)}$ and Ahmad Ali Ekhveh

Business Intelligence Group, Informatics Services Corporation, Tehran, Iran
sh_tabatabayi@isc.co.ir, Ekhveh@isd.co.ir

Abstract. Data Series are one of the major types of data producing in different domains. Performing data analysis on this domain currently have two obstacles. The big volume of data and also the large dimensionality and unequal size. In this research we have employed two famous methods from NLP algorithms named Word2Vec and Doc2Vec to represent and compress data series at the same time. These two approaches are based on Feed Forward Neural Networks. We have used a private bank dataset with more than 11 million transactions and converted it to data series. Then we employed two representation methods to convert each time series to a fix size $[1 \times 100]$ vector. The first approach ignores the order in data series and gains more than 96% compression ratio (30:1). The second method preserves order of data to some extent and gets about 24% compression ratio (1.35:1). The other advantage of proposed method is the fix size vector representation which makes the comparison of two data series with different length easily possible. The k-means clustering algorithm is performed on Bank Customer Data Series to show a usage of proposed data series representation.

Keywords: Data series representation · Bank transaction data base · Neural networks · Word2vec · Word embedding · Clustering of bank customers

1 Introduction

The volume of data producing in different fields are enormous nowadays and data series are one the pervasive types of this data e.g. stock market data, meteorology data series, customer online shopping list and bank customer transactions. One of the ways to conquer the size of this large volume data is to compress them to a smaller size somehow the properties of original series are preserved. Another issue in data series is the different size of each sample which makes the comparison task very difficult. In many data mining algorithms we need to compare the similarity/distance of two samples in order to do the relative task therefore unequal size of series will be a problematic issue. In this research we employed the famous Word2Vec and Doc2Vec vector representation method [1, 2] in a field of bank transaction data to solve the above problems. Word2Vec is a pioneer method in a larger context which is called Word Embedding. Generally speaking Word Embedding is a Neural Network based data representation method which is developed explicitly for Natural Language Processing

© Springer Nature Switzerland AG 2019
L. Grandinetti et al. (Eds.): TopHPC 2019, CCIS 891, pp. 315–330, 2019.
https://doi.org/10.1007/978-3-030-33495-6_24

tasks. Each word will be embedded in a fix size vector according to its distribution in a very large text corpus. Doc2Vec is an upgraded version of Word2Vec method which embeds a paragraph (or document) into a fix size vector.

In the current research we want to represent bank customers data series somehow we can do datamining tasks such as clustering, classification, regression and etc. The task we pilot here is k-means clustering on the sequence of transactional activity of each customer. This approach is relatively new owing the fact that most customer classification tasks are performed on customer's individual information such as occupation, income, education and etc. Each customer has a data series consisting of his/her transactions during a specific duration of time. It is obvious that different customers' data series have unequal length. On the other hand the size of this data base is getting large very fast which makes another obstacle for data mining processes. The proposed method contributes in solving both problems hence the vector representation of data series is of fix size even though the original series had different length. Moreover the database will be up to 30 times smaller in Word2Vec and 1.35 times smaller in Doc2Vec which is a good compression scale. This representation method can be used in a lot of different data Mining algorithms as an input data.

The rest of the paper is organized as follows: first we have an overview of representation methods for time series data. Afterward we briefly review the different similarity measurements which are employed in the field of time series clustering algorithms. Then at the end of literature review part we go over various clustering algorithms used on time series data. Next chapter will explain Word2Vec and Doc2Vec which are our basic algorithms in the current work.

Section 4 explains proposed representation methods WV-DSR[1] and DV-DSR[2]. Section 5 elaborates the bank transaction data base that we extracted from the transaction tables of a private bank. The coding we used to convert this data series and make a new artificial language is explained afterward.

Section 6 explains the implementation details and experimental results. We give some evaluation analysis of the obtained results in Sect. 7. In the two last chapters we encapsulate the advantages and disadvantages of our work in real applications and shed light on the road we plan to pursue.

2 Literature Review

2.1 Data Series Representation Methods

Data series and time series representation methods has been investigated a lot in the past two decades. The most effort has been put on time series representation most of the time in order to reduce the dimensionality of the original signal. In general there are two main branches for data series representation: Data Adaptive and Non Data Adaptive [3, 5, 6]. Figure 1 depicts the diagram of different representation methods. Our representation method is in Data Adaptive fields in this diagram.

---

[1] Word2Vec Data Series Representation.

[2] Doc2Vec Data Series Representation.

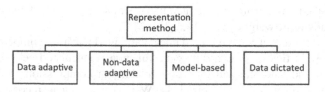

Fig. 1. Diagram of different representation methods of time/data series

In [4] a representation method has been proposed for the continuous series. First the input series is coded in string of symbols then a new similarity measure is applied on this sequence. The most important advantage of this representation is its dimension reduction power.

Article [25] encodes the input signal as follows: if the value of signal is greater than zero it gives back 1 and otherwise it gives back 0 therefore any input time series will be binary coded. They show that the time complexity of similarity computation drops significantly with this representation method while the accuracy of similarity measurement stays over a minimum threshold. The similarity measure which is employed in this paper is Dynamic Time Warping (DTW).

Paper [26] converts a time series to series of intervals with fix value. Then it suggests a new similarity method to compare these series.

In [7] Adaptive Piecewise Constant Approximation (APCA) is introduced. In this approach the main focus was on the minimization of the reconstruction error. Afterward they suggest two new similarity functions which are suitable to use on this representation method.

Non adaptive data transformation are suitable for equal size data series. Some of methods in this field are wavelet transformation [8], Discrete Wavelet Transform (DWT), Piecewise aggregate approximation [5] and random mappings [9].

2.2 Similarity/Distance Calculation Methods in Data Series Clustering

Clustering is data categorization algorithm without having any information about gold categories. In this operation the goal is to put the most similar samples in one cluster and to make clusters the most dissimilar that is possible. Therefore the concept of similarity or distance between samples plays the main role in this task. There exists plenty of different similarity/distance measurement method in this area the most famous of which is Euclidean distance. Most of clustering algorithms in the domain of Data series have employed this distance to perform their objective task. However there are some drawbacks for Euclidean distance. First it is very vulnerable toward noise and second, it cannot find similar series which are just skewed or expanded in time axes. We generally expect that this kind of series be assumed similar. Some papers put forward new distance measurement which are based on Euclidean distance but upgraded to levels which can be able to handle samples with different size [10].

Comparing two series is a challenging task. Most of early works in the field have used dimensionality reduction techniques which use feature extraction of data series. This approach has several advantages. Noise is handled very neatly in feature extraction methods and from the computation complexity point of view these algorithms are really light weight [4, 5, 11, 12].

Several researchers use Dynamic Time Warping (DTW) to compare time series [13–16, 22, 23]. In this method a difference function is calculated between main time series and the candidate one and then by stretching or compression of the candidate sample this function will be minimized. DTW is very strong in recognizing concept drift in data streams and can fit on new pattern very quickly.

The main assumption in DTW is that even though two series have different size but they can match together. This assumption is rather strong because in some cases only part of two time series may match and other parts could be different. Another approach is to calculate the maximum length common sub string of two series as the similarity measure [17].

The family of Edit Distance Measures are another similarity/distance class which is used in literature. Minimum Edit Distance (MED) and levenshtein distance are two famous measurement in this family. MED calculates the number of insert, delete and shift operations which are needed to transform the first string in to the second one. This distance measurement is very popular in domains like bioinformatics and natural language processing [18, 19].

In [20] a similarity method has been proposed for comparison of two short time series. It is very sensitive to scalability but can detect time changes which are dependent on each other but different in absolute values.

Paper [18] combines two common methods L1-norm and edit distance and gives a new measurement called Edit Distance with Penalty (EDP). The L1-norm is based on normalization of all data based on mean and variance of the whole data. In L1-norm it is presumed that two samples has the same length and time although this assumption is not required in edit distance however, EDP has the advantage of both methods.

Researchers in [21] convert the problem of maximum substring matching into the problem of finding shortest path in a directed acyclic graph and show that this method outperform DTW.

Edit distances mainly are applied on discrete data series but [19] invent a method to apply edit distance on real data. Its main objective is to stabilize the edit distance calculation to noise in data. They show that this measurement is more robust that previous distance measurements.

2.3 Data Series Clustering

Data series clustering has these main objectives: Recognition of outliers or new data samples, recognition of dynamic changes in sequences, hidden pattern recognition and prediction and suggestion based on clusters. In these researches the focus is either on the samples with high frequency of appearance or sample which are very rare and distinct. [24] shows that in data series clustering we can use the common clustering algorithms and the bottleneck is distance/similarity measurement method.

Hierarchical clustering performs the categorization task bottom-up. At the beginning it is assumed that each sample consists a cluster. In each step those clusters which have the most similarity are merged together to form a bigger cluster. The top-down hierarchical clustering algorithm is very similar. At First it takes all data as one big cluster and in each step one cluster will be divided into smaller ones [27].

K-means algorithm is the most famous clustering algorithm. The number of cluster, K, should be predefined in this algorithm. It starts with K random samples as clusters' centers and in each training epoch tries to partition all sample in K clusters somehow each sample drops in a cluster with the nearest center. Papers [12] use k-means clustering on time series data but each employs different similarity measure. There are some upgrade of k-means algorithm [28].

The prototype of each cluster in standard k-means algorithm is the mean of each cluster but defining the prototype in data series with different length is problematic [29].

K-medoid algorithm is similar to k-means with small difference in defining center of the clusters. In k-medoid the nearest sample to the mean of the cluster which is called medoid is considered as cluster center. But this algorithm costs more in time complexity with respect to k-means. [30] show that the stability of this algorithm toward noise and outlier samples is more.

Fuzzy C-means is the same as k-means with the difference that membership in each cluster is Fuzzy and not deterministic. Each sample can be the member of each cluster with some membership value. [31] uses this method to cluster speech time series and use these clusters for speaker identification.

Other clustering algorithms are either not tested on data series or used in few cases. From other methods we can mention [32] who use Hidden Markov Model (HMM) and spectral analysis to do the clustering task.

3 Vector Representation with Word2Vec and Doc2Vec

Using word embedding is a major trend in Natural Language Processing field nowadays. However according to our knowledge it is not used in other fields so much. We have found this work which uses word embedding to represent tabular data [33] in order to use as input to a deep convolutional neural network. The goal of their task is prediction of taxi destination based on its trajectory start points.

One of the main problem in NLP domain is the problem of word representation in vector space. Traditionally the one-hot representation were the dominate method for converting a document into a vector representing it. This has some major drawbacks. First of all document representation with combination of one-hot vectors is a very sparse matrix. For example a document with 100 words and vocabulary of 10000 words will be represented as a 100*10000 matrix which has only 100 ones and all the other elements are zeros. Secondly, the similarity between words in one-hot representation is absolutely zero because all one-hot vectors are orthogonal and their Euclidean or cosine similarity will be zero.

3.1 Word Representation with Word2Vec

Mikolov and his colleagues in Google introduced word embedding representation for the first time [1, 2]. Their method uses Feed Forward Neural Network with one hidden layer to produce word embedding. Their specific method is called Wrod2Vec. There are several other embedding emerges after that [34, 35]. The main idea is to take into account the context in which a word appears as a reference for its meaning. Similar words will appear in similar context. Although this assumption is not true in every situation (for instance one can argue that opposite words appear in the same context) but it gives very promising results in the corresponding field.

Figure 2 shows the architecture of the Feed Forward Neural Network which is used in Word2Vec. The task is to predict the missing word according to its surrounding words. The input layer is one-hot representation of each word and has V nodes. The hidden layer has H neurons with soft-max activation function and output layer has V nodes. Figure 2 depicts this network.

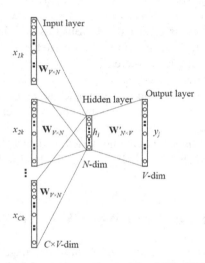

Fig. 2. The feed forward neural network of Word2Vec algorithm

The input of this network is a window of C surrounding words of center word y. The output to predict is the current word y. This model is called Continuous Bag of Words (CBW)[3]. The dimension of each one-hot representation of word x_{ik} is the size of vocabulary (V). The size of hidden layer (N) is up to user but in original paper it is set to 300 neurons. Output layer determines which word is the most probable word to be in the input context therefore its shape is V (size of vocabulary).

After training the network on the input corpus the weights of hidden layer which is a V * N matrix will be embedding of words of the vocabulary. For each word the corresponding 1 * N vector in the embedding matrix will be its vector representation.

---

[3] There is another version of Word2Vec in which the input is just w(t) and the goal is to predict n surrounding words. That version is called Skipped Gram.

3.2   Paragraph Representation with Doc2Vec

Doc2Vec is an extension of Word2Vec which can handle the problem of representing a paragraph in one fixed size vector representation [1].

In the updated version the index of each paragraph is feed to the network alongside the context words. Figure 3 shows the new architecture of the network [2].

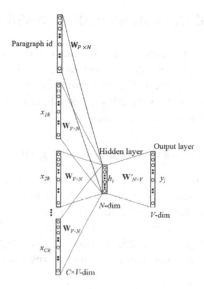

Fig. 3. The architecture of neural network which produces Doc2Vec embedding

Paragraph id in the input will be a one-hot vector and the dimension of this vector is number of paragraphs (P). Therefore after concatenating this paragraph vector with previous word vectors the dimension of input will be P + V. After training the network each paragraph has its own embedding which is a 1 × N vector. The advantage of this method is that paragraphs with different length can get a same size vector. Each two paragraph with similar context will have similar embedding. The drawback of this algorithm is the large dimension of input which reduce the compression power of using embedding enormously.

3.3   Time Complexity of Word2Vec and Doc2Vec

Writers of the original Word2Vec paper have explained the time complexity of this method in details. Here we just overview time complexity very roughly. Two terms in time complexity are number of training epochs (E) and number of words in the training set (T). Number of training epochs determines how many times should the algorithm goes over training set again to upgrade network weights. These two parameters are independent of network architecture we use. The main complexity factor comes from the Word2Vec architecture. The order of complexity of Word2Vec is as follows:

$$O = E \times T \times (C \times N + N \times \log_2 V) \qquad (1)$$

C is the size of input window, N is the size of hidden layer (size of embedding vector) and V is the size of vocabulary of training set. In Doc2Vec algorithm all the parameters are the same except that the vocabulary size will replace with (V + P) where P is the number of paragraphs in the training set.

4   Representation of Data Series with Embedding

We proposed two vector representations for data series in our work based on Word2Vec and Doc2Vec. First, we encode the data series into the symbols according to some transformation rules. The obtained strings are our paragraphs in a new language. The coding process can be very arbitrary and here we have used a set of simple rules which are explained in the next chapter.

4.1   Word2Vec Data Series Representation: WV-DSR

In this approach first we build the Word2Vec vectors of each word in our new language. After that we compute the average of all words in a string and consider the calculated vector as data series representation. This method is very simple but it ignores the order of words in the sequence which is not desirable. We keep this approach because the compression power of this method is very high and if the results is satisfactory in some applications it can be a good representation choice.

4.2   Doc2Vec Data Series Representation: DV-DSR

Our main approach is to use paragraph embedding as vector representation of data series. We build a Doc2Vec model on data series and for each input series, we get the vector representation. We expect that this model preserve order of words to some extent because during training phase local context will determine the weights of word embedding. On the other hand we know that all series with different length will embed in a fix size vector therefore it is obvious that we will lose some data. The important thing is that the similarity/distance between different series stays the same.

5   Representation of Private Bank Transactions Data Base

To evaluate our representation method on data series in the field of banking data we produce a relatively large database of transactional data. This data extracted from transactions of a private bank in December 2018. The whole data is almost 11 million transactions.

5.1   Preprocessing Step

First, we transform transactional data to data series. For each customer we make a series of transaction. Each element in series consists of two things: (Amount, Channel).

Amount can be positive for deposit and negative for debit transactions[4]. Channel shows the infrastructure on which the transaction is taken place e.g. Internet, POS, Mobile, ATM and etc.

Each customer who has less than 10 transactions or more than 500 transactions has been removed in order to get rid of sparsity.

5.2 Encoding Rules

To convert the input series into a symbolic series we have employed the following rules:

1- Negative Amount starts with N and positive amount starts with P
2- The most valuable digit in Amount determines the next character from the following dictionary:

$$\{1 : A, 2 : B, 3 : C, \ldots, 9 : I\}$$

3- The number of digits in Amount determines the next character from the following dictionary:

$$\{1 : A, 2 : B, \ldots, 16 : P\}$$

4- The channel of transaction has been kept unchanged.

According the above transformation rules (-124200000, POS) will be converted to NAGPOS.

5.3 Handling of Time

In our dataset time is removed and the only time information is the order of transactions. In this research we focused on similarities in sequence of actions of a customer not in the interval of his/her actions. It is obvious that this method basically has the ability to be extended to cover time or any other information. Table 1 represents the characteristics of the final data base.

Table 1. Characteristics of bank transactions data series

Number of customers	456550
Average length of series	25
Length of the longest series	500
Length of the shortest series	10
The number of all words in database	11402050
The number of words in vocabulary	6363

---

[4] We do not take into account non-financial transaction.

5.4   Training Word2Vec and Doc2Vec

After preparing the desired data series we train Word2Vec and Doc2Vec models. The details of these models and their architecture represents in Table 2.

Table 2.   Characteristics of neural networks for training Word2Vec and Doc2Vec

The embedding vector size in Word2Vec	100
The embedding vector size in Doc2Vec	100
Number of epochs in Word2Vec	300
Number of epochs in Doc2Vec	300
Training time for Word2Vec	16 min
Training time for Doc2Vec	1 h 36 min
Size of Word2Vec model on disk	8 Mb
Size of Doc2Vec model on disk	190 Mb
Size of original database in symbolic representation	255 Mb

6   Experiments and Results

6.1   Plotting the Embedding of Transactions on 2D

There are 6363 different words in our database where each represents an interval in amount along with a transition channel. We can reduce the 100-dimension vectors to 2D vectors with PCA. Figure 4 shows the obtained diagram of word embedding. From

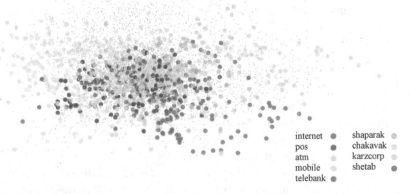

Fig. 4.   This figure shows the 2D-PCA transformation of words in our Word2Vec model. Small grey dots are those transactions with channels refer to a bank clerk in a branch. Bigger and colorful dots show channels out of bank branches like ATM, POS, INTERNET and etc. It is vivid that transactions with same channel types are closer to each other than transactions with different channels. Some of these channels is listed in the legend. (Color figure online)

this diagram we can see which types of transactions are more similar according to their similarity of contexts. In another words, when distance between two words are small it means that they have occurred in the similar contexts (surrounding words).

6.2   The Similarity of Each Paragraph to Itself

One of the most intuitive features of a similarity metric is that each sample should be one hundred percent similar to itself. Owing the fact that in DV-DSR we compress original paragraphs to vectors with fix size, we will lose some data. We pick a sample from training set and then get the most similar sample to this input sample with Doc2Vec model. After doing that on all training samples the percentage of samples which were most similar to themselves was 0.82%.

There is no need to do this evaluation in WV-DSR because in its paragraph representation each paragraph is most similar to itself.

6.3   Nearest Neighbor Similarity Length Analysis

To have another sense of comparison we calculate the nearest neighbors of 100 samples with two different representation methods. The length distribution of the test samples is similar to length distribution of training set. The similarity measure we use is Euclidean distance. Then we compare the series length of original sample, its nearest neighbor with WV-DSR and its nearest neighbor with DV-DSR. Table 3 shows the result of this comparison.

Table 3. The average length of nearest neighbor samples by each method

Average length of 100 samples	Average length of DV-DSR nearest neighbor	Average length of WV-DSR nearest neighbor
25.56	18.51	26.48

6.4   K-means Clustering with WV-DSR and DV-DSR

We have used these data representations for clustering task with k-means algorithm. In the first model we give the average of word vectors in a paragraph (data series) as input samples and in the next model we give paragraph embedding vectors as input samples. We have employed the elbow method[5] to define the best value for number of cluster K. Roughly speaking in K = 7 we have a very minor superiority in accuracy therefore we choose K = 7 as number of clusters. Figures 5 and 6 show the corresponding results with the clustering result in 2D and 3D dimensions PCA projection..

---

[5] In elbow method the k-means clustering is done on different k values, when the average of minimum distance error seize to reduce as the amount of k increases the optimum value of k is obtained.

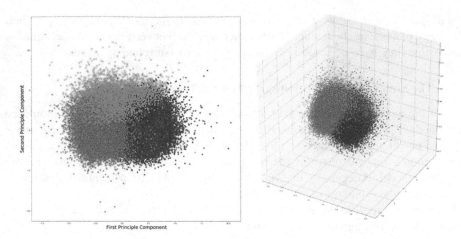

Fig. 5. 2D and 3D PCA transformation of k-means with k = 7 with DV-DSR representation method

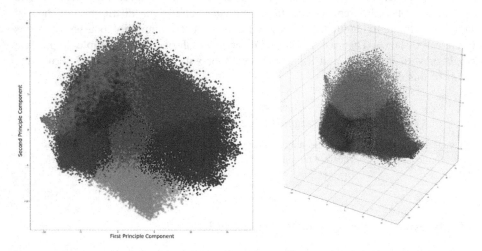

Fig. 6. The 2D and 3D PCA transformation of k-means with k = 7 with WV-DSR representation method

One of the difference between WV-DSR and DV-DSR is the difference between cluster center length. WV-DSR k-means tends to have long samples as its center of clusters whereas DV-DSR have short series as its centers. The average length of k-cluster centers in WV-DSR k-means is 46 and the corresponding average in DV-DSR clusters is 13.

Although the PCA transformation does not give a precise sense of the quality of clustering but it can be seen from the figures that WV-DSR has more discriminative power than DV-DSR. However this is not a concrete conclusion because we should investigate whether WV-DSR preserves the similarity relation in the original data well enough or not.

7 Evaluation of Clustering Task

Unfortunately there is no method to compute the absolute accuracy of clustering algorithm because there is no label for gold classes. Basically we use clustering in the situations where we do not know the true label of samples. However there are some investigations we can do to compare the quality of clusters with respect to themselves (not real data labels).

We have used two of those approaches. The first one determines how dense our clusters are. It calculates the percentages of samples in each cluster where the distance of sample is lower than mean distance of the cluster. We call it the density degree of a cluster.

The other approach tries to give us a sense of reliability of clusters by means of membership degree. For sample s with cluster label c the membership degree $M(s,c)$ is calculated as follows:

$$M(s, c) = 1 - \frac{Euclidean\ distance(s, c)}{\frac{\sum_{c' = all\ clusters\ except\ c} Euclidean\ distance(s, c')}{number\ of\ clusters - 1}} \tag{2}$$

The average of membership degrees of all samples can be a good measure of reliability. Table 4 shows the density degree and membership degree of our clustering models. It can be seen although the average density of both representation methods are almost similar but the reliability of clustering based on DV-DSR representation is very low.

Table 4. Two measurements to compare our clustering models.

Representation method	Average density degree	Average membership degree
WV-DSR	0.59	0.509
DV-DSR	0.54	0.085

The above experiments show that WV-DSR performs better than DV-DSR. It is somehow counter intuitive owing the fact that in WV-DSR we lose a lot of original data. The reason that WV-DSR gives better clustering results can be due to the k-means algorithm. Maybe we should try another algorithm too and then compare the results. Another very important issue which we should focus on later is to prepare a solid test set in which the true similarity of data series are determined by human expert. We can use it as a reference to check whether WV-DSR is still superior to DV-DSR or not.

8 Application of Bank Transaction Clustering

The proposed work can be used in several fields in banking sector such as Anti Money Laundering (AML), Fraud Detection, customer behavior modeling, Recommendation Systems and etc. Most of these tasks are based on either anomaly or similarity detection both of which can be done with the proposed methods. The representations themselves can be used as input to other datamining and machine learning algorithms too.

9 Conclusion and Future Works

This work has two main objectives: Data adaptive representation of different length time series and data compression. The idea was to employ two famous methods in NLP domain on data series database with special focus on bank customer transactions series. The empirical results show that first the representation methods which encode each series in a fix size vector can be an ease for comparing different series. They also have some compression power specially when we drop the order of input data series.

There are several things we have planned to do in future. First of all a better comparison should be done between proposed methods and other methods. Secondly we want to enlarge the vocabulary size by adding more feature and investigate the impact of this change on our data.

Another important thing to do is to use different similarity measurements and clustering algorithms to find the most appropriate one for Bank transaction data.

References

1. Mikolov, T., Chen, K., Corrado, G., Dean, J.: Efficient estimation of word representations in vector space. arXiv preprint arXiv:1301.3781 (2013)
2. Le, Q., Mikolov, T.: Distributed representations of sentences and documents. In: International Conference on Machine Learning, pp. 1188–1196 (2014)
3. Aghabozorgi, S., Shirkhorshidi, A.S., Wah, T.Y.: Time-series clustering–A decade review. Inf. Syst. **53**, 16–38 (2015)
4. Lin, J., Keogh, E., Lonardi, S., Chiu, B.: A symbolic representation of time series, with implications for streaming algorithms. In: Proceedings of the 8th ACM SIGMOD Workshop on Research Issues in Data Mining and Knowledge Discovery, pp. 2–11. ACM (2003)
5. Wilson, S.J.: Data representation for time series data mining: time domain approaches. Wiley Interdiscip. Rev.: Comput. Stat. **9**(1), e1392 (2017)
6. Wang, X., Mueen, A., Ding, H., Trajcevski, G., Scheuermann, P., Keogh, E.: Experimental comparison of representation methods and distance measures for time series data. Data Min. Knowl. Discov. **26**(2), 275–309 (2013)
7. Keogh, E.J., Pazzani, M.J.: An enhanced representation of time series which allows fast and accurate classification, clustering and relevance feedback. In: Kdd, vol. 98, no. 1, pp. 239–243 (1998)
8. Chan, K.P., Fu, W.C.: Efficient time series matching by wavelets. In: ICDE, p. 126. IEEE (1999)

9. Bingham, E., Mannila, H.: Random projection in dimensionality reduction: applications to image and text data. In: Proceedings of the Seventh ACM SIGKDD International Conference on Knowledge Discovery and Data Mining, pp. 245–250. ACM (2001)
10. Faloutsos, C., Ranganathan, M., Manolopoulos, Y.: Fast subsequence matching in time-series, vol. 23, no. 2, pp. 419–429. ACM (1994)
11. Ratanamahatana, C.A., Keogh, E.: Multimedia retrieval using time series representation and relevance feedback. In: Fox, E.A., Neuhold, E.J., Premsmit, P., Wuwongse, V. (eds.) ICADL 2005. LNCS, vol. 3815, pp. 400–405. Springer, Heidelberg (2005). https://doi.org/10.1007/11599517_48
12. Ratanamahatana, C., Keogh, E., Bagnall, Anthony J., Lonardi, S.: A novel bit level time series representation with implication of similarity search and clustering. In: Ho, T.B., Cheung, D., Liu, H. (eds.) PAKDD 2005. LNCS (LNAI), vol. 3518, pp. 771–777. Springer, Heidelberg (2005). https://doi.org/10.1007/11430919_90
13. Izakian, H., Pedrycz, W., Jamal, I.: Fuzzy clustering of time series data using dynamic time warping distance. Eng. Appl. Artif. Intell. **39**, 235–244 (2015)
14. Truong, C.D., Anh, D.T.: A novel clustering-based method for time series motif discovery under time warping measure. Int. J. Data Sci. Anal. **4**(2), 113–126 (2017)
15. Chu, S., Keogh, E., Hart, D., Pazzani, M.: Iterative deepening dynamic time warping for time series. In: Proceedings of the 2002 SIAM International Conference on Data Mining, pp. 195–212. Society for Industrial and Applied Mathematics (2002)
16. Seto, S., Zhang, W., Zhou, Y.: Multivariate time series classification using dynamic time warping template selection for human activity recognition. arXiv preprint arXiv:1512.06747 (2015)
17. Ozkan, I., Turksen, I.B.: Fuzzy longest common subsequence matching with FCM using R. arXiv preprint arXiv:1508.03671 (2015)
18. Chen, L., Ng, R.: On the marriage of lp-norms and edit distance. In: Proceedings of the Thirtieth International Conference on Very Large Data Bases, vol. 30, pp. 792–803. VLDB Endowment (2004)
19. Chen, L., Özsu, M.T., Oria, V.: Robust and fast similarity search for moving object trajectories. In: Proceedings of the 2005 ACM SIGMOD International Conference on Management of Data, pp. 491–502. ACM (2005)
20. Möller-Levet, C.S., Klawonn, F., Cho, K.-H., Wolkenhauer, O.: Fuzzy clustering of short time-series and unevenly distributed sampling points. In: R. Berthold, M., Lenz, H.-J., Bradley, E., Kruse, R., Borgelt, C. (eds.) IDA 2003. LNCS, vol. 2810, pp. 330–340. Springer, Heidelberg (2003). https://doi.org/10.1007/978-3-540-45231-7_31
21. Latecki, L.J., Megalooikonomou, V., Wang, Q., Lakaemper, R., Ratanamahatana, C.A., Keogh, E.: Elastic partial matching of time series. In: Jorge, A.M., Torgo, L., Brazdil, P., Camacho, R., Gama, J. (eds.) PKDD 2005. LNCS (LNAI), vol. 3721, pp. 577–584. Springer, Heidelberg (2005). https://doi.org/10.1007/11564126_60
22. Yi, B.K., Faloutsos, C.: Fast time sequence indexing for arbitrary Lp norms. In: VLDB, vol. 385, no. 394, p. 99 (2000)
23. Vlachos, M., Kollios, G., Gunopulos, D.: Discovering similar multidimensional trajectories. In: 18th International Conference on Data Engineering, Proceedings, pp. 673–684. IEEE (2002)
24. Liao, T.W.: Clustering of time series data—a survey. Pattern Recognit. **38**(11), 1857–1874 (2005)
25. Keogh, E.J., Pazzani, M.J.: A simple dimensionality reduction technique for fast similarity search in large time series databases. In: Terano, T., Liu, H., Chen, A.L.P. (eds.) PAKDD 2000. LNCS (LNAI), vol. 1805, pp. 122–133. Springer, Heidelberg (2000). https://doi.org/10.1007/3-540-45571-X_14

26. Keogh, E., Chakrabarti, K., Pazzani, M., Mehrotra, S.: Locally adaptive dimensionality reduction for indexing large time series databases. ACM Sigmod Rec. **30**(2), 151–162 (2001)
27. Łuczak, M.: Hierarchical clustering of time series data with parametric derivative dynamic time warping. Expert Syst. Appl. **62**, 116–130 (2016)
28. Paparrizos, J., Gravano, L.: k-shape: efficient and accurate clustering of time series. In: Proceedings of the 2015 ACM SIGMOD International Conference on Management of Data, pp. 1855–1870. ACM (2015)
29. Niennattrakul, V., Ratanamahatana, C.A.: Inaccuracies of shape averaging method using dynamic time warping for time series data. In: Shi, Y., van Albada, G.D., Dongarra, J., Sloot, P.M.A. (eds.) ICCS 2007. LNCS, vol. 4487, pp. 513–520. Springer, Heidelberg (2007). https://doi.org/10.1007/978-3-540-72584-8_68
30. Sheng, W., Liu, X.: A genetic k-medoids clustering algorithm. J. Heuristics **12**(6), 447–466 (2006)
31. Tran, D., Wagner, M.: Fuzzy C-Means clustering-based speaker verification. In: Pal, N.R., Sugeno, M. (eds.) AFSS 2002. LNCS (LNAI), vol. 2275, pp. 318–324. Springer, Heidelberg (2002). https://doi.org/10.1007/3-540-45631-7_42
32. Yin, J., Yang, Q.: Integrating hidden Markov models and spectral analysis for sensory time series clustering. In: Fifth IEEE International Conference on Data Mining, p. 8. IEEE (2005)
33. De Brébisson, A., Simon, É., Auvolat, A., Vincent, P., Bengio, Y.: Artificial neural networks applied to taxi destination prediction. arXiv preprint arXiv:1508.00021 (2015)
34. Busta, M., Neumann, L., Matas, J.: FASText: efficient unconstrained scene text detector. In: Proceedings of the IEEE International Conference on Computer Vision, pp. 1206–1214 (2015)
35. Pennington, J., Socher, R., Manning, C.: Glove: global vectors for word representation. In: Proceedings of the 2014 Conference on Empirical Methods in Natural Language Processing (EMNLP), pp. 1532–1543 (2014)

Multi-modal Emotion Analysis for Chatbots

Gijoo Yang[1], Jeonggeun Jin[1(✉)], Dongho Kim[2(✉)],
and Hae-Jong Joo[3(✉)]

[1] Department of Information and Communication Engineering,
Dongguk University, Seoul, Korea
gjyang@dgu.edu, vv0926@hanmail.net
[2] Convergence Software Institute, Dongguk University, Seoul, Korea
dongho.kim@dgu.edu
[3] Department of Computer Science and Engineering, Dongguk University,
Seoul, Korea
hjjoo@dgu.edu

Abstract. Developing chatbots that can recognize the emotions of users is a challenging problem of artificial intelligence. In order to build such a system, we need to define the emotion taxonomy to cover human-like feelings. Consequently, we need to prepare a large scale training data by using the defined emotion taxonomy. In this paper, we investigate methods of representing emotions and applying them in a deep neural network model that classifies the user's emotion into many dimensions. We also take into account auditory signals of spoken language in addition to contextual information for classifying the emotions of users. Furthermore, we tackle the compositional negation of utterances which may cause misinterpretation of the emotion in the opposite direction. Our experiment shows that our model improves the performance of baseline models significantly.

Keywords: Emotion analysis · Audio analysis · Chatbot · Recursive neural network

1 Introduction

To represent emotions, many approaches have used conceptual networks such as SenticNet (Cambria et al. 2014) and AffectNet (Liu and Singh 2004). In this type of work, the knowledge base is composed of assertions about concepts and emotions. Concepts or emotions of utterances can be identified through a matching process although the process is complicated because one concept or emotion can be related to hundreds of assertions. Despite complicated process, a training dataset can be developed using this method. There are other attempts to create a dataset for emotion analysis that use predefined emotion taxonomy. With predefined emotion taxonomy, it is easier to develop a large scale dataset for emotion analysis than using complex conceptual knowledge base. In this paper, we investigate the use of predefined emotion taxonomy applied in a deep neural network for our emotional chatbot which is designed to help patients of hospitals make reservations to see appropriate doctors.

© Springer Nature Switzerland AG 2019
L. Grandinetti et al. (Eds.): TopHPC 2019, CCIS 891, pp. 331–338, 2019.
https://doi.org/10.1007/978-3-030-33495-6_25

It should be noted that the negation in phrases or utterances can change the emotion into the contrary direction which causes problems in deciding the emotion properly and consequently devising appropriate responses. To alleviate this problem, we use a recursive neural network to capture the compositional negation of phrases.

In spoken language, the task of emotion recognition is a challenge because different individuals express their emotions differently (Anagnostopoulos et al. 2015). Although our work is focused on classifying emotions based on text content in the framework of dialogue handling, sound information is also used for emotion detection because the user's request to our chatbot can be either through online or by phone. With the success of deep neural networks in the last decade, many improvements have been observed in the area of computer vision and speech recognition. Many studies have shown that modeling the inherent structure in the speech signal with these networks is superior to the traditional methods (Hinton et al. 2012).

We employ a deep learning method that uses the representation of the input signal from the raw data. In the past, low-level descriptor (LLD) and/or high-level statistical function (HSF) have been used as features to deep neural networks and shown that its performance can be improved. However, recent researches rely on end-to-end learning where features are learned automatically since such hand-crafted features are hard to devise and update. We investigate the impact of using speech signal with text-based dialogue system in an end-to-end manner. We use Convolutional Neural Network (CNN) to extract the features of audio signals.

2   Related Work

Recurrent Neural Network (RNN) is widely used for sequential data processing, especially speech and language processing. But the problem that RNN suffers most is vanishing gradient descent problem. Long-distance dependency cannot be retained in vanilla RNN because information more than 7 steps apart is hardly transferred. To handle this problem Long Short Term Memory RNN was proposed and it has been widely used using gates of input, output, and forget (Hochreiter and Schmidhuber 1997). However, LSTM is complicated and computationally expensive. To alleviate this problem, Gated Recurrent Unit (GRU) has been proposed and used in many areas including language processing and speech recognition (Cho et al. 2014; Ravanelli et al. 2017). In this research we use GRU to encode the utterance.

A conceptual knowledge base including commonsense knowledge can be useful where reasoning through commonsense concepts is necessary. Although human seems to do this kind of knowledge analysis and gain supreme intelligence effectively (Cambria and Hussain 2015), building a large scale knowledge base using such networks, however, is not easy because transformation of concepts is computationally expensive and the amount of transformation is tremendous. Even though one such knowledge base can be built, obtaining a large scale dataset for emotion analysis for emotional chatbot from such knowledge base requires another complex computation for similarity comparison. Another reason that it is not preferred to obtain a large scale dataset through this kind of work is that it does not guarantee desirable preciseness because conceptual networks are manually prepared (usually for decades) and incorrect information is accidently or inherently incorporated.

There are other attempts for creating dataset for emotion analysis that use prede-fined emotion taxonomy defined by polarity words such as "negative", ..., "positive", or taxonomy defined by emotion terms such as "happiness", "anger", ..., "neutral" or taxonomy defined by non-emotion terms such as valence, personal strength, ..., social connection (Cochrane 2009). Various research on emotion analysis used simple cate-gories. Consequently existing emotional benchmark datasets contain three (or five) emotional categories from "negative" through "neutral" to "positive". Although such taxonomy is useful for classifying reviews or articles, it is not sufficient to describe real emotional situation for chatbots. It is necessary to define the emotion taxonomy to cover human-like feelings. For this purpose, the emotion taxonomy from Microsoft's cognitive service's emotion API which is designed to classify facial image into eight dimensions (happiness, surprise, anger, disgust, sadness, contempt, fear, neutral) is often used (Wu et al. 2017). In this type of work, seed words are selectively assigned to one of the emotion categories and are used to determine the emotion of utterances. Such methods are good for creating large scale dataset in that similarity check is less complicated than using conceptual networks. However, in this type of work, it is not easy to determine the emotion of utterances when contrary emotional words simulta-neously exist or negation of emotion words exists in the utterance.

Socher et al. (2013) proposed a neural network model to handle negation in a sentence. It uses compositionality in semantic vector spaces. Since analysis of com-positionality was held back by the lack of large and labeled compositionality dataset and models to capture the underlying phenomena, Stanford Sentiment Treebank corpus and a neural network model called Recursive Neural Tensor Network (RNTN) were developed. RNTN takes phrases (through word vectors and parse trees) as input and computes vectors for higher nodes in the tree using the same tensor-based composition function. In this paper, we adopted RNTN to analyze the user's utterance and obtain the emotional direction or emphasis point of the utterance. Since this model is RecursiVe Neural Network, we named our model as RVNN. The output of RVNN is fed to GRU to encode the underlying phenomena.

Recently, studies on neural network in many areas (e.g., Natural Language Pro-cessing, dialogue chatting, speech recognition, etc.) attempts to develop end-to-end architecture instead of the modular approach. This is because during optimization human apriori knowledge must be incorporated as little as possible. However, in the area of speech recognition, many approaches have used hand-crafted features such as Perceptual Linear Prediction (PLP) coefficients and Mel-Frequency Cepstral Coeffi-cients (MFCC). We use raw data as the representation of input signal. We examine the features extracted from deep neural network in an end-to-end manner.

3   Proposed Work

3.1   Training Data

Nowadays, various research on sentiment analysis have been carried out using Twitter data (Rosenthal et al. 2017). Thus, the emotion taxonomy forms with three emotional categories from "negative" to "positive" in many cases. However, it is insufficient to describe real emotional situation for chatbots. As with (Wu et al. 2017), we also adopted the emotion taxonomy from Microsoft's cognitive service's emotion API with customization. We modified it for our chatbot that is designed to deal with reservation request in hospitals. For this purpose, we changed eight dimension of emotions to five dimensions of happiness, surprise, anger, sadness, and neutral, that should be enough to deal with reservation request from possible patients.

In order to build large scale training dataset, we use Korean Twitter data. Since recent social media posts include emoticons, abbreviations, and informal words, we manually collected more than 100 widely used such expression as shown in Table 1. We also labeled each text data with five emotion category by examining the user's utterance and expression including emoticons and informal words. For developing dataset, we manually labeled more than 1,000 utterance for each emotion category to start. Started similarly with previous approaches, we built the seed word dictionary but expanded it by using WordNet's synset which is a synonym set. Then, we used the expanded dictionary to decide the emotion of utterance for the dataset. For simplicity and preciseness, we excluded expressions with multiple contrary emotional words. This is the first stage of dataset development.

Table 1. Emotions and emotional expressions

Emotions	Emotional Expressions
Happiness	ᄊ ^.^ :) ㅋㅋ ㅎㅎ
Surprise	O_O :O 헐; 헉;
Anger	>:O >:(>:-<
Sadness	:'(ㅠㅠ ㅠ·ㅠ ㅜㅜ
Neutral	o-o

In addition, we developed a classifier based on GRU which takes utterance as input and produces an output which decides the emotion category of an utterance based on its contextual information. We, first, fed our prepared utterances (manually labelled) and trained the classifier. We used this classifier as the secondary tool that decides the emotion category of utterances which were processed by the first stage processing.

To train our recursive neural network (RVNN), we manually labelled the conversation data collected in our hospital for 1 month which includes more than 1,000 utterances. Since there aren't enough negations in the collected utterances for training, we added some negated expressions that are frequently used in conversations.

For audio dataset, since there is no open Korean dataset for voice emotion recognition, we prepared our own dataset. For it, we analyzed voice data recorded in our hospital for 1 month and manually classified it into about 300 utterances for each emotion (happiness, surprise, anger, sadness, neutral). When there is not enough data for an emotion, we recorded several voice actors' utterances for a set of pre-selected utterances. For objectivity, 3 evaluators judge the labels of utterances respectively and the utterances with the labels with 2 or more evaluators agreed on were included in the dataset. As you expected, the portion of neutral emotion became the largest among all since during the judgement, uncertain emotion expressed in the utterances were often selected as neutral.

3.2 Model

We use Convolutional Neural Network (CNN) for extracting features from the speech signal. To reduce the influence of background noise, finite impulse response filters are used. Since we deal with audio signals, our model use 1-dimensional convolution function h defined as the following.

$$h(y) = \sum_{x=1}^{k} f(\mathbf{x}) \cdot g(y \cdot d - x + c)$$

where $d = stride$, $c = k - d + 1$ is an offset constant, g is input function and f is filter function. After convolution, we use max-over-time pooling function p defined as the following.

$$p(y) = max\, g(y \cdot d - x + c)$$

We use Gated Recurrent Unit (GRU) Recurrent Neural Network (RNN) for encoding utterances. The reason for using GRU instead of Long Short Term Memory (LSTM) is that GRU needs less computation. The output of CNN is fed to GRU to decide the emotion of the utterance and the dialogue system is able to select the appropriate response to the user's utterance.

Since user's utterance or expression can be negated or emphasized to contrary direction of emotion, we, first, examine the utterance by RVNN. The compositional negation that is captured as the output of RVNN is then fed to the utterance encoder which is in our case GRU. The final GRU is defined as the following.

Update gate:

$$z_t = \sigma(W^z x_t + U^z h_{t-1})$$

Reset gate:

$$r_t = \sigma(W^r x_t + U^r h_{t-1})$$

$$\tilde{h}_t = tanh(Wx_t + r_t \odot Uh_{t-1})$$
$$h_t = z_t \odot h_{t-1} + (1 - z_t) \odot \tilde{h}_t$$

where $\odot$ is a Hadamard product.

Let s be the output of CNN and n be the output of RVNN, then $h_0 = s \oplus n$. Where h_0 is the initial state of GRU and $\oplus$ is concatenation.

The architecture of our model is shown in Fig. 1.

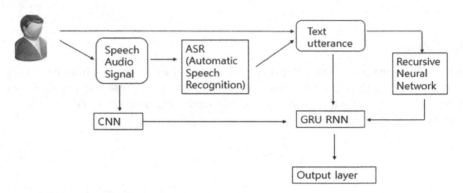

Fig. 1. Architecture of emotion recognition system

4 Experiments

To feed each word to the neural network, word embedding for each word is necessary. We used word2vec (Mikolov et al. 2013). We pre-trained more than 100,000 Korean words on a crawled news dataset (>1 GB). We used 200 dimensions for word2vec word embedding. For training our model, we use Adam optimizer with a learning rate of 0.001. Our implementation relied on Python, TensorFlow, and Keras. We initialized the recurrent parameter matrices as orthogonal matrices, and other parameters with Gaussian random distributions with a mean of zero and a standard deviation of 0.01. For the GRU RNN, hidden state dimension is set to 250. For the audio model, we fixed the input length as 30 s and the input shorter than 30 s are zero-padded. Since 98% of data is less than 30 s, most of the data is included. From the raw waveform of the audio data we extract 100 frames per second with the window (size = 25 ms) and draw 512-point FFT (Fast Fourier Transform) which is 257 dimension vector. For this audio data, we set the batch size as 25 samples. For regularization of the network, we used dropout with p = 0.7 for CNN. For comparison of models, we experiment GRU model with or without CNN and RVNN. In Table 2, comparison of these models is shown.

Table 2. Model accuracy comparison

Models	Accuracy (%)
GRU	80.1
GRU + CNN	87.2
GRU + RVNN	83.7
GRU + CNN + RVNN	89.1

5 Conclusion

We proposed an emotion detection model for our emotional chatbot for Korean Language. We used the audio signal as well as contextual information for emotion classification. We also used recursive neural network for handling compositional negation. Furthermore, we maintained our neural network differentiable so that it can be learned in end-to-end manner. Experimental results show that our emotion classification model, in particular, audio CNN module improves significantly. RVNN module also improves the base model but not as much as audio CNN module. This is because our dataset does not have many complex utterances with compositional negation. It would be interesting to see the success rate of GRU + RVNN model when it deals with data composed of compositional negation and emphasis only. We hope we can extend our work so that it can be applied to other languages. However, since emotion can be expressed differently in different languages and cultures, developing an emotion recognition system that can be generalized to various languages would not be easy.

Acknowledgment. This research was supported by the MSIT (Ministry of Science and ICT), Korea, under the National Program for Excellence in SW supervised by the IITP (Institute of Information & communication Technology Planning & evaluation)(2016-00017).

References

Anagnostopoulos, C.-N., Iliou, T., Giannoukos, I.: Features and classifiers for emotion recognition from speech: a survey from 2000 to 2011. Artif. Intell. Rev. **43**(2), 155–177 (2015)

Cambria, E., Hussain, A.: Sentic Computing. SC, vol. 1. Springer, Cham (2015). https://doi.org/10.1007/978-3-319-23654-4

Cambria, E., Olsher, D., Rajagopal, D.: SenticNet 3: a common and common-sense knowledge base for cognition-driven sentiment analysis. In: AAAI, Quebec City, pp. 1515–1521 (2014)

Cho, K., van Memenboer, B., Gulcehre, C., Bourgares, F., Schwenk, H., Bengio, Y.: Learning phrase representation using RNN encoder-decoder for statistical machine translation. In: EMNLP (2014)

Cochrane, T.: Eight dimensions for the emotions, Special issue: The Language of emotion – conceptual and cultural issues (2009)

Hinton, G., et al.: Deep neural networks for acoustic modeling in speech recognition: the shared views of four research groups. Signal Process. Mag. **29**(6), 82–97 (2012)

Hochreiter, S., Schmidhuber, J.: Long short-term memory. Neural Comput. **9**, 1735–1780 (1997)

Mikolov, T., Sutskever, I., Chen, K., Corrado, G.S., Dean, J.: Distributed representations of words and phrases and their compositionality. In: NIPS (2013)

Liu, H., Singh, P.: ConceptNet-a practical commonsense reasoning tool-kit. BT Technol. J. **22**(4), 211–226 (2004)

Ravanelli, M., Brakel, P., Omologo, M., Bengio, Y.: Improving speech recognition by revising gated recurrent units. In: Proceedings of Interspeech (2017)

Rosenthal, S., Farra, N., Nakov, P.: SemEval-2017 task 4: sentiment analysis in Twitter. In: Proceedings of the 11th International Workshop on Semantic Evaluation (SemEval-2017) (2017)

Socher, R., et al.: Recursive deep models for semantic compositionality over a sentiment treebank. In: EMNLP (2013)

Wu, X., Kikura, Y., Klyen, M., Chen, Z.: Sentiment analysis with eight dimensions for emotional chatbots. In: Natural Language Processing Conference (Japan) (2017)

A New Encoding Method for Graph Clustering Problem

Amir Hossein Farajpour Tabrizi[1]([✉]) and Habib Izadkhah[2]

[1] University College of Daneshvaran, Tabriz, Iran
ah.farajpour@gmail.com
[2] Department of Computer Science, University of Tabriz, Tabriz, Iran
izadkhah@tabrizu.ac.ir

Abstract. Clustering is used as an important technique to extract patterns from big data in various fields. Graph clustering as a subset of clustering has a lot of practical applications. Due to the NP-hardness of the graph clustering problem, many evolutionary algorithms, particularly the genetic algorithm have been presented. One of the most effective operators on the performance of the genetic algorithm is how to represent the solutions of a problem (i.e. encoding). The number of possible partitions of a graph is equal to Bell Number. In the literature, three encoding methods have been presented for graph clustering problem. The number of partitions that these encodings can generate is more than the Bell Number; which indicates that these methods generate a large number of same and iterative solutions which makes the speed of obtaining the solution unacceptable and leads to this fact that the good space search encounters a problem. To overcome this drawback, in this paper we present a new encoding method for graph clustering problem where the number of the generated solutions by this encoding is exactly equal to the Bell numbers. The initial results of our experiments represent that the quality of the obtained solutions by the new encoding is promising.

Keywords: Graph clustering · Pattern recognition · Genetic algorithm · Encoding

1 Introduction

Pattern recognition is one of the most important issues is the field of artificial intelligence and machine learning whose focus is on recognizing the patterns. The pattern recognition methods extract the given patterns from a data set and have many applications in speech recognition, face recognition, signature recognition or bioinformatics. A pattern can be either any object or a given data that is necessary to recognize and identify, and in fact the pattern recognition is to receive raw data and make decision based on their classification and its aim is to classify the pattern of the input data into a number of specific classes which are

© Springer Nature Switzerland AG 2019
L. Grandinetti et al. (Eds.): TopHPC 2019, CCIS 891, pp. 339–351, 2019.
https://doi.org/10.1007/978-3-030-33495-6_26

directly related to the clustering and classification problems. The major goal of clustering is to group a set of related objects into a cluster, so that the objects in a group (cluster) are more similar than other classes (clusters). Clustering can be classified into two data and graph clustering. It is worth to note that in most cases data can be represented as a graph, so that the datum represent nodes and the edges between the nodes show the datum links. Nevertheless, the subject of this paper is graph clustering, rather than data clustering. The graph clustering has a lot of applications such as community detection in web [1], community detection in the protein-protein interaction network [2], community detection in social networks [3], community detection in scientific writing networks [4], community detection in financial markets [5], extraction of software components with reusability and converting an application into small, manageable and understandable components [6]. The number of different clusters of n-vertex graph is equal to the possible number of partitions of an n-member set, Bell Number, denoted by B_n, and calculated by Eq. 1 [7].

$$B_n = \sum_{k=0}^{n-1} \binom{n-1}{k} B_k \tag{1}$$

where the number of partitions of a single-member set is 1, so the above equation can be calculated by assuming $B_1 = 1$.

The clustering problem, or generally, the set partitioning problem is an NP-hard problem [8], and therefore there is no applicable algorithm in the polynomial time for that. For this, nondeterministic and heuristic methods have been used to solve this problem. In these methods, a heuristic search is used to find a near-optimal solution and the search operations are directed by an objective function. In clustering, it is tried to place the close points within one cluster and the distant points are placed within the different clusters. Using evolutionary algorithms, particularly the genetic algorithm can have better results. Therefore, most researchers use heuristic based algorithms in order to reduce the time complexity where the genetic algorithm is considered as one of them. One of the major features of the genetic algorithm for a particular problem is encoding. Generally, a genetic algorithm cannot operate directly on the search space of a particular problem. Therefore, the search space should be mapped to a space on which the genetic algorithm could be operated. The way of solution representation and encoding has a significant influence on the quality of the obtained solutions by the genetic algorithm. The major problem of the given evolutionary methods to solve the clustering problem is how to encode the solutions by chromosomes. Palmer and Kershenbaum [9] have determined five general principles that are necessary for an accurate encoding. These principles are as follow:

1. An encoding must represent all the feasible solutions.
2. It must be able to represent only the feasible solutions.
3. All the feasible solutions must have an equal probability to be represented.
4. It must be able to perform the encoding by the minimum number of chromosomes.

5. Encoding must have a location. It means that the small changes in that location of the chromosome make small changes in the solution.

The principles 3 and 4 are also known as the Uniqueness of Mapping which means that there is a one-to-one correspondence between each chromosome and the solution space, namely each solution is the representative of one chromosome and each chromosome is only the representative of one solution.

In the literature, three different encoding methods have been presented which are useful in clustering a graph by the evolutionary algorithms. The space generated by these three encodings is much larger than the Bell Number which leads to this fact that the quality of the solutions generated by the algorithms applying these encodings is not acceptable. Therefore, the existing encoding methods for the graph clustering problem are in contrast at least with two principles 3 and 4, namely with the principle of Uniqueness of Mapping. Not following these two principles first results in algorithm's temporal and spatial overhead and second it leads the algorithm to converge to the more likely solutions.

This paper presents a new encoding method for graph clustering where its generated space is exactly equal to B_n. It means that the chromosomes set create a one-to-one mapping with the solutions set. The proposed encoding follows the principle of Uniqueness of Mapping. Therefore, exactly one chromosome is generated per possible solution. This feature improves the performance of the proposed algorithm over other methods. This way of encoding ensures the uniqueness of the chromosomes per possible solution. We have selected the software clustering problems to initial evaluation of the proposed encoding. The results of our comparison with a number of software clustering algorithms represent the superiority of the proposed encoding.

The remainder of this article is organized as follows. After presenting the introduction and problem statement in this section, we explain the related works in Sect. 2. The proposed method is presented in Sect. 3. In Sect. 4, we apply the proposed method on software clustering problem. Finally, in Sect. 4, the conclusion is given.

2 Related Works

One of the most important issues in solving the problems by graph clustering algorithms is the way of encoding or representing the problem's solutions as a chromosome which is the most difficult steps of solving a problem by the genetic algorithm. There are three encoding methods for graph clustering problem such as Vector-based encoding e.g., see [10, 11], Permutation-based encoding e.g., see [12] and Lucas-based encoding e.g., see [3]. In the following, we study each encoding.

Vector-Based Encoding. In this kind of encoding, each solution (chromosome) is as an array where the number of genes in each chromosome is equal to the number of graph nodes and the content of each gene is a cluster number in which its corresponding node is placed and its value is between one and the number of that graph's nodes. If the cluster number of a number of genes is same, it means that they are in the same cluster. For example, a sample graph is shown in Fig. 1. For example, to cluster this graph with four vertices, a chromosome can be shown as Fig. 2, where its clustering interpretation would mean that nodes A_1 and A_3 are in the same cluster (since they have the same cluster number, i.e. 2) and nodes A_2 and A_4 are also within the same cluster (as they have the same cluster number, i.e. 3). Three other possible chromosomes for this chromosome are shown in Fig. 3. As it can be seen, changing the numbers of these chromosome will not affect the actual solution of the clustering problem.

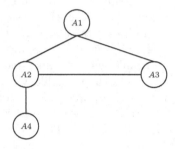

Fig. 1. An example of a graph

C1:   2 3 2 3

Fig. 2. A sample chromosome for a four-vertex graph by vector-based encoding

C2:   1 2 1 2

C3:   3 1 3 1

C4:   2 4 2 4

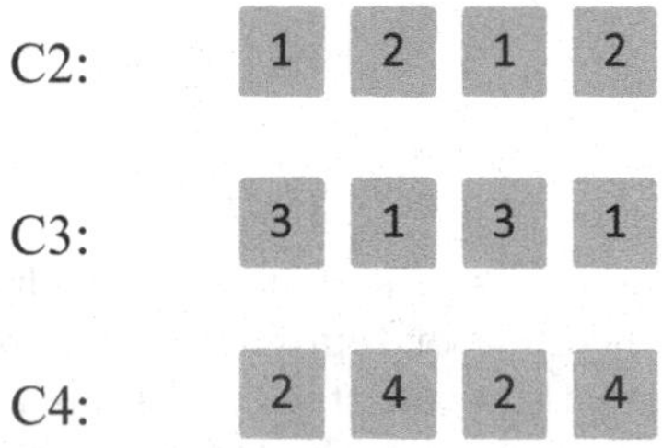

Fig. 3. Similar clustering for a four-vertex graph by vector-based encoding

It can easily be inferred that there are more chromosomes that encode this particular clustering. Therefore, it can be concluded that the space of the possible solutions produced by vector-based encoding is much larger than the real space. Let n denotes the number of nodes in a graph, the space generated to perform the search by this encoding is n^n. n^n is an upper bound for B_n, namely $\exists m, \forall n > m$: $B_n < n^n$. Also, the problem of this encoding is not only limited to the iterative solutions in different chromosomes. Another major drawback of this method is in the non-uniform distribution of these iterative states. It means that different solutions of the clustering may be generated by unequal number of chromosomes. For example, for graph Fig. 1, the clustering of Fig. 4 is generated by n! different chromosomes as non-iterative permutations of 1–4. While another clustering, shown in Fig. 5, is generated by only four chromosomes, c1: 1,1,1,1; c1: 2,2,2,2; c1: 3,3,3,3; and c1: 4,4,4,4.

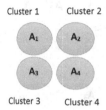

Fig. 4. A sample clustering for Fig. 1

Fig. 5. Another clustering for Fig. 1

This is in contrast with one of the basic principles of the evolutionary app-roach, namely the necessity of a completely random selection of chromosomes, and hence it leads to the convergence of the algorithm to those solutions that have more opportunity to be produced.

Permutation-based Encoding. The way of encoding in this method is that if the ith index of this chromosome has value c_i, and we have $i \leq v_i$, then a new cluster is formed and the ith node becomes a member of this new cluster. Otherwise, if $i > v_i$, the ith node will be assigned to the cluster in which the node c_i is placed.

It is obvious that the space generated by this encoding reduces more signif-icant than the vector-based encoding, since the total generated space is equal to the number of all n permutations, i.e., $n!$. However, the problem of itera-tive states also remains in this encoding. For example, all the chromosomes c1: 2,3,4,1; c2: 2,4,3,1; c3: 3,2,4,1; and c4: 4,2,3,1 are the same encoded clustering. Furthermore, the problem of the non-uniform distribution of the iterative states has also not been resolved. For example, although the above clustering has four generative chromosomes, the clustering is shown in Fig. 4 is generated only by one chromosome c5: 1,2,3,4. In general, $\exists m, \forall n > m : B_n < n!$.

Lucas-based Encoding. In this encoding, the solutions are represented in vector form, so that the indexes represent the nodes of a graph and their content represents the index of one of the adjacent nodes for the node. This encoding restricts the space of the solution, but, there are still duplicate encodings. For example, Fig. 6 shows three different chromosomes (encoding) for Fig. 1, so that the clustering achieved for them are the same.

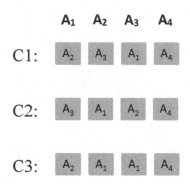

Fig. 6. Three different chromosomes with the same clustering

3   The Proposed Method

None of the existing evolutionary algorithms to solve the clustering problem have ever been able to present a method for one-to-one mapping of the solutions to the chromosomes. The number of the clustering generated by the given encoding methods is much more than B_n. This means that many distinct chromosomes may produce the same solutions. Another major problem is the lack of uniform distribution of the iterative chromosomes. It means that the number of the chromosomes that encode an iterative solution may be more than the number of chromosomes encoding the other solution. As a result, different solutions will not have the same probability of being appeared in the chromosomes population, which is incompatible with the basic principles of evolutionary algorithms.

The aim of this new encoding is to minimize the problem's state space and prevent iterative chromosomes. The state space of the proposed method is equal to the Bell Number which will have significant improvement over previous methods. Therefore, the reduction of the state space into the number of distinct solutions of the problem prevents the convergence of algorithm to the solutions with more generative chromosomes and high chance to be generated in the population of solutions.

3.1 Create a Set of Distinct Solutions

One can partition the set of all the chromosomes of the Vector-based encoding to the classes of chromosomes that generate the same solution. Then, a chromosome is selected from each group as the representative of that class and the given one-to-one mapping range is created by these chromosomes. In this way, each solution of the clustering problem is encoded by one chromosome. Other possible sequences are considered as invalid and will not be generated.

But which chromosomes do we choose? If we consider Vector-based encoding chromosomes as numerical values, then there could be an order relation for them as natural numbers. We simply contract that only a sequence with a lower numerical value will be produced from each group of chromosomes with the same solution. For example, among the four sequences studied earlier, only c2: 1,2,1,2 will be produced as a valid chromosome. Because the numerical value of 1212 is less than all the other sequences that encode the clustering of c2. As each of these valid sequences describes only one clustering state, it can be concluded that there is a one-to-one correspondence between two sets. Therefore, the number of these sequences is equal to the number of different clustering states, B_n.

In this method, we aim to prevent the generation of iterative solutions for the clustering method, so as the Vector-based encoding, we use n-member sequences of numbers from 1 to n to create chromosomes, but with this difference that here we use a simple rule to generate chromosomes and apply only a B_n-member subset of chromosomes and by this principle we control the validity of this chromosome, in a way that the value c, as a representative of each gene within a chromosome is only allowed to be placed in the ith gene of a chromosome when the value c-1 is placed at least once within one of the 1 to $i-1$ genes. According to the above contract, we generate each chromosome as gene by gene. In such a way that in each chromosome that encodes the clustering of M, the value of c is only allowed to be in the ith gene when the value of $c-1$ is placed at least once in one of the $1, 2, ..., i-1$ genes. As a result, we can be assured that the numerical value of the sequence to be generated is the minimum value among all the sequences that can encode M. In other words, the allowed range for the c_i value of the ith gene will be $1 \le c_i \le \max(c_1, c_2, ..., c_{i-1}) + 1$.

The main idea of this algorithm is that to construct n-length chromosomes, it is enough to attach the numbers 1-n at the end of all n-1 length chromosomes. For example, Fig. 7 represents the way of generating chromosomes with a length of four genes (n = 4) by the new method. In this figure, from the left, the value 1 in the column 1 shows that only value 1 can be placed in the first gene; column 2 represents that one of the values of 1 or 2 can be placed in the second gene, etc. The last column shows the whole possible clustering that can be generated for a graph with four nodes (15 clustering). The important point is that there is not any iterative clustering in these 15 clustering. For example, the encoding 1,1,1,2 is interpreted in this way that nodes 1, 2 and 3 are placed within one cluster and node 4 is placed within a cluster alone. Figure 8 shows the algorithm of this process. If this algorithm is called as chroGen (0,1,0, n), it will produce a valid chromosome to length n.

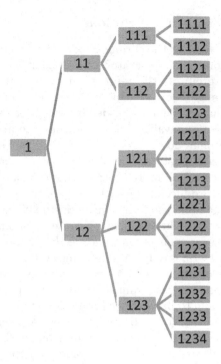

Fig. 7. Tree diagram of new chromosome's structure

```
Algorithm chroGen (lastPart, currentStep, currentMax, N)
begin:
        if currentStep = N + 1 then
                Return lastPart;
        else
                Let r be a random number in [1, currentMax+1];
                if r = currentMax + 1 then
                        currentMax ← r;
                Return chroGen(concat(lastPart, r), currentStep + 1, currentMax, N);
        endif
end.
```

Fig. 8. Creating a chromosome with new encoding

3.2   New Crossover and Mutation Operators

It is obvious that this method requires its own genetic operations, since the validity of a chromosome must be maintained until the end of the algorithm. For this reason, we provide some tools for validating or evaluating chromosomes by applying mutation and crossover operations. In order to complete the process of the proposed algorithm, we study the crossover and mutation steps, since after generating the initial valid chromosomes as the initial population by the given method, the newly generated chromosomes might not be valid according to the rule of the proposed algorithm due to applying crossover operation over the middle generation and finally applying the mutation operation over the middle generation, so in this case a method must be presented to its modification and correspondence with the main rule.

The correction principle in replacing the child's chromosome genes in our proposed method is that when after a crossover or mutation, an invalid chromosome such as V' is generated and a value such as n is placed incorrectly in a cell of the new chromosome like i, the values of the previous cells, 1 to i-1 must be scanned and the maximum value used in these cells is extracted as the maximum value. Then the value of n is replaced with max+1. This procedure can be continued until the last invalid gene in the new chromosome can be modified in order to transform the new chromosome into a valid chromosome.

Now, it is the turn of the mutation operation. At this stage, according to one of the mutation methods, we select a number of chromosomes for this operation. As we have observed at the crossover stage, some of these new chromosomes may lose their validation after performing mutation operations and changing the sequence of the mutated chromosomes' genes where to solve this problem, we can reuse the modification rule of the proposed method and replace its value by max+1 by scanning the first gene to the invalid mutated gene.

4   Case Study: An Applicable to Software Clustering Problem

As stated in the introduction, the graph clustering has different applications. For the initial evaluation of the proposed encoding, we apply it to the software clustering problem. In this problem, the relations between software's artifacts are represented as a graph, so that the nodes of this graph represent the software artifacts and the edges between the nodes represent the relationships between the artifacts. In a software system, an artifact can be anything such as classes, files, and functions. The software clustering aims to partition a software system from the source code into meaningful and understandable subsystems. In fact, this helps a software engineer to understand a software in the process of reverse engineering [13]. A number of search-based software clustering algorithms are EDA [14], E-CDGM [15], Bunch [16], DAGC [12], ECA [11], MCA [11], HC + Bunch [17], NAHC [10], SAHC [10], GA-SMCP [18], PSO [19], Harmony Search [20], and BCA [21].

Mitchell [16], in his dissertation, has presented a quality function called BasicMQ, which is used to direct the clustering process using evolutionary algorithms. The numbers generated by this function are in the range of -1 to $+1$. Let A_i indicates the number of internal relationships of cluster i and $E_{i,j}$ indicates the number of relationships between cluster i and j. The BasicMQ is defined as Eq. 2.

$$BasicMQ = \begin{cases} \frac{1}{k}\sum_{i=1}^{k} A_i - \frac{1}{\frac{k(k-1)}{2}}\sum_{i,j=1}^{k} E_{i,j} & k > 1 \\ A_1 & k = 1 \end{cases} \qquad (2)$$

where k is the number of clusters.

The goal of evolutionary algorithms is to maximize Eq. 2. In Table 1, the proposed algorithm, named BeEN, is compared with some search-based algorithms on eleven software systems in terms of BasicMQ. We selected the mean results for each algorithm over 20 independent runs. This table demonstrates that the results achieved with proposed encoding algorithm have higher or equal quality than those achieved with other algorithms for all software systems.

Table 1. Comparison of the proposed algorithm (BeEN) with the existing methods in terms of BasicMQ (MQ) and variance (Var)

Software system	Bunch		DAGC		ECA		NAHC		BeEN	
	MQ	Var	MQ	Var	MQ	Var	MQ	Var	MQ	Var
compiler	0.68	0.0006	0.58	0.004	0.32	0.0008	0.32	0.02	0.71	0.0006
boxer	0.54	0.0007	0.27	0.003	0.28	0.0008	0.27	0.009	0.61	0.0007
mtunis	0.62	0.0006	0.52	0.009	0.44	0.0005	0.53	0.06	0.78	0.0003
ispel	0.55	0.009	0.39	0.001	0.65	0.0009	0.45	0.01	0.55	0.0008
bison	0.73	0.0007	0.53	0.009	0.53	0.0007	0.53	0.008	0.73	0.0003
Cia	0.64	0.0006	0.58	0.009	0.60	0.0006	0.58	0.04	0.64	0.0006
ciald	0.61	0.008	0.58	0.003	0.68	0.0004	0.48	0.001	0.68	0.0001
nos	0.65	0.0008	0.59	0.005	0.63	0.0002	0.59	0.007	0.68	0.0002
rcs	0.71	0.0006	0.71	0.001	0.71	0.0009	0.45	0.009	0.71	0.0004
spdb	0.50	0.0003	0.41	0.0007	0.49	0.0003	0.34	0.09	0.58	0.0002
star	0.74	0.008	0.61	0.004	0.55	0.0007	0.69	0.01	0.78	0.0006

Mtunis is an operating system for academic purposes and since its design documentation (authoritative decomposition) is available we have used it to evaluate the proposed objective function reliability. The authoritative decompositions of "mtunis" as depicted in the design documentation, which consists of 20 artifacts, 57 edges. Table 2 shows the comparison of the BeEN with nine existing clustering algorithms on mtunis. This table shows that the proposed algorithm with new objective function is able to produce the clustering similar to clustering provided by a domain expert (the lower MoJo and edgeMojo represent more similarity between clustering produced by the algorithm with the one produced by an expert, while the larger F_m and MoJoFM indicate more similarity) [13]. The reason for this is that the proposed algorithm, with the guidance of the proposed objective function, is able to put utility artifacts into one cluster, which improves the quality of clustering achieved, while other objective functions can not identify utility artifacts.

Table 2. Comparison of BeEN with a number of state of the art genetic algorithms and other search based algorithms on mtunis

	MoJo	edgeMojo	F_m	MoJoFM
Bunch	6	7.47	57%	70%
DAGC	7	10.33	48%	60%
HC + Bunch	9	11.44	25%	70%
NAHC	5	13.14	53%	65%
SAHC	5	10.81	55%	70%
E-CDGM	5	10.81	55%	70%
EDA	5	7.47	57%	72%
MCA	5	7.41	57%	71%
ECA	5	7.41	57%	72%
BeEN	**4**	**4.66**	**62%**	**75%**

Table 3. Comparison of the number of clusters produced in the existing encoding methods and new method

Method	$N = 10$	$N = 20$
Bell Number	115,975	51,724,158,236,496 (14 digits)
Vector-based encoding	10,000,000,000	104,857,600,000,000,000,000,000,000 (27 digits)
Permutation-based encoding	3,628,800	243,290,200,817,664,000 (18 digits)
Proposed encoding	115,975	51,724,158,236,496 (14 digits)

As above mentioned in the introduction, the number of different partitions of a graph is equal to Bell Number. Table 3 shows the comparison of the total number of clustering produced by Vector-based encoding and Permutation-based encoding with proposed encoding on two graphs with 10 and 20 nodes. As it can be seen in this table, the number of clusters in Vector-based encoding and Permutation-based encoding is much more than the new method. In Vector-based encoding, the number of iterative clusters is also very high and although this problem has been slightly modified in the Permutation-based encoding method, there are still a number of iterative clusters; but in the new method, in addition to a significant reduction in the number of clusters, the iteration of clusters is completely prevented.

5   Conclusion

This paper presents an optimal encoding method to solve graph clustering problem where the search based methods can use it to perform clustering. The proposed encoding prevents the iteration of the similar solutions and guarantees the individual chromosomes where it can increase the speed of the algorithm's performance and the solution can be reached in more suitable time. The proposed encoding can be used for all graph clustering problems.

References

1. Mahdavi, M., Chehreghani, M.H., Abolhassani, H., Forsati, R.: Novel meta-heuristic algorithms for clustering web documents. Appl. Math. Comput. **201**(1–2), 441–451 (2008)
2. Pizzuti, C., Rombo, S.E.: Algorithms and tools for protein-protein interaction networks clustering, with a special focus on population-based stochastic methods. Bioinformatics **30**(10), 1343–1352 (2014)
3. Said, A., Abbasi, R.A., Maqbool, O., Daud, A., Aljohani, N.R.: CC-GA: a clustering coefficient based genetic algorithm for detecting communities in social networks. Appl. Soft Comput. **63**, 59–70 (2018)
4. Krapivin, M., Marchese, M., Casati, F.: Exploring and understanding scientific metrics in citation networks. In: Zhou, J. (ed.) Complex 2009. LNICST, vol. 5, pp. 1550–1563. Springer, Heidelberg (2009). https://doi.org/10.1007/978-3-642-02469-6_35
5. Tumminello, M., Lillo, F., Piilo, J., Mantegna, R.N.: Identification of clusters of investors from their real trading activity in a financial market. New J. Phys. **14**(1), 013041 (2012)
6. Beck, F., Diehl, S.: On the impact of software evolution on software clustering. Empir. Softw. Eng. **18**(5), 970–1004 (2013)
7. Duncan, B., Peele, R.: Bell and Stirling numbers for graphs. J. Integer Seq. **12**(09.7), 1 (2009)
8. Schulz, C.: Graph partitioning and graph clustering in theory and practice. Institute for Theoretical Informatics Karlsruhe Institute of Technology (KIT), pp. 24–187, 20 May 2016
9. Palmer, C.C., Kershenbaum, A.: Representing trees in genetic algorithms. IBM Thomas J. Watson Research Division (1994)
10. Mitchell, B.S., Mancoridis, S.: On the automatic modularization of software systems using the bunch tool. IEEE Trans. Softw. Eng. **32**(3), 193–208 (2006)
11. Praditwong, K., Harman, M., Yao, X.: Software module clustering as a multi-objective search problem. IEEE Trans. Softw. Eng. **37**(2), 264–282 (2011)
12. Parsa, S., Bushehrian, O.: A new encoding scheme and a framework to investigate genetic clustering algorithms. J. Res. Pract. Inf. Technol. **37**(1), 127 (2005)
13. Isazadeh, A., Izadkhah, H., Elgedawy, I.: Source Code Modularization: Theory and Techniques. Springer, Heidelberg (2017). https://doi.org/10.1007/978-3-319-63346-6
14. Tajgardan, M., Izadkhah, H., Lotfi, S.: Software systems clustering using estimation of distribution approach. J. Appl. Comput. Sci. Methods **8**(2), 99–113 (2016)
15. Izadkhah, H., Elgedawy, I., Isazadeh, A.: E-CDGM: an evolutionary call-dependency graph modularization approach for software systems. Cybern. Inf. Technol. **16**(3), 70–90 (2016)
16. Mitchell, B.S.: A heuristic search approach to solving the software clustering problem. Ph.D. theses. Drexel University (2002)
17. Mahdavi, K.: A clustering genetic algorithm for software modularisation with a multiple hill climbing approach. Diss, Brunel University (2005)

18. Huang, J., Liu, J.: A similarity-based modularization quality measure for software module clustering problems. Inf. Sci. **342**, 96–110 (2016)
19. Rajapati, A., Chhabra, J.K.: A particle swarm optimization-based heuristic for software module clustering problem. Arab. J. Sci. Eng. **43**(12), 7083–7094 (2018)
20. Chhabra, J.K.: Harmony search based remodularization for object-oriented software systems. Comput. Lang. Syst. Struct. **47**, 153–169 (2017)
21. Chhabra, J.K.: Many-objective artificial bee colony algorithm for large-scale software module clustering problem. Soft Comput. **22**(19), 6341–6361 (2018)

A Parallel and Improved Quadrivalent Quantum-Inspired Gravitational Search Algorithm in Optimal Design of WSNs

Mina Mirhosseini[1], Mahmood Fazlali[1(✉)], and Georgi Gaydadjiev[2,3]

[1] Department of Data and Computer Science, Faculty of Mathematical Sciences,
GC Shahid Beheshti University, Tehran, Iran
fazlali@sbu.ac.ir
[2] Department of Computing, Imperial College London, London, UK
[3] Maxeler Technologies Ltd, London, UK

Abstract. Wireless Sensor Networks (WSNs) are recently used in monitoring applications. One of the most important challenges in WSNs is determining the operational mode of sensors, decreasing the energy consumption while the connectivity requirements and the special properties are satisfied. This problem is an NP-hard one and is a time–consuming progress. In this study, an improved version of quadrivalent quantum-inspired gravitational search algorithm as a new metaheuristic, well suitable for quadrivalent problems is proposed using Not Q-Gate to optimize the performance of the WSN. Beside, to enhance the speed and the accuracy of the algorithm more, we used a parallelizing technique using Open-MP. Parallelizing this algorithm on mentioned problem is useful from four aspects; 1 - accelerate the speed of the algorithm, 2 - improving the quality of solutions by letting the increasing the population size, 3 - The possibility of using the algorithm, in larger-scale WSNs and 4 - power affectivity of the base station using multicore processors. To validate the performance of our proposed approach, a comparison between this approach and the previous methods is performed. Our experiments verified that the proposed method can effectively improve the performance more than 2.25 times and the speedup faster than 4 times on an 8-core CPU.

Keywords: Quadrivalent quantum-inspired gravitational search algorithm · Quantum computing · Parallel metaheuristic · Wireless sensor network

1 Introduction

Metaheuristics are approximate and population-based optimization methods that recently got so popular. They are considered as successful methods that provide acceptable near optimal solutions in a limited and reasonable time for solving NP-hard and complex scientific and engineering problems. An important challenge in NP-hard problems is computability; means there are no guarantees to solve these problems in a reasonable time or in a satisfactory style. Among various approaches to enhance solving NP-hard problems, Genetic Algorithms (GA), Particle Swarm Optimization (PSO), Simulated Annealing (SA), Ant Colony Optimization (ACO), Imperialist

© Springer Nature Switzerland AG 2019
L. Grandinetti et al. (Eds.): TopHPC 2019, CCIS 891, pp. 352–366, 2019.
https://doi.org/10.1007/978-3-030-33495-6_27

Competitive Algorithm (ICA) and Gravitational Search Algorithm (GSA) are some well-known and successful metaheuristics which are used frequently in different applications [1].

Although in comparison with the exact methods, metaheuristics can solve NP-hard problems in a reasonable time, they have relatively high execution time in some cases. Another problem associated with metaheuristic is that if the searching space is very wide, it may metaheuristics do not converge towards optimal or near-optimal solutions. The increasing population size of the algorithm can improve the quality of solutions in such cases, which it leads to increasing the execution time.

Parallelism of these algorithms can help to improve the precision of solutions and reduce the execution time. Recently, several parallel metaheuristics have been presented; parallel genetic algorithm [2], parallel PSO [3], parallel ant colony optimization (PACO) [4], parallel artificial bee colony (PABC) [5], parallel simulated annealing [6], parallel ICA [7] and parallel GSA [8] are some examples of that. Even some parallel metaheuristics achieve super-linear performance [9].

Gravitational search algorithm (GSA) is a new metaheuristic with fine exploration power. The main idea of this algorithm is the Newtonian laws of motion and gravity [10]. BGSA [11] is a binary version of this algorithm; an improved binary version called IBGSA [12] is presented to deal with binary optimization issues. The real-valued quantum version of GSA is suggested in [13] and its binary-valued quantum versions are presented in [14, 15]. Their exploitation and exploration powers are compared with their previous approaches. These comparisons indicate that BQIGSA [14] is the most effective algorithms of the binary optimization techniques. The performance of this algorithm tests on different applications like feature selection [16], unit commitment problem [17] and wireless sensor networks [18].

Wireless sensor networks (WSNs) have been recently applied in many monitoring applications like environmental monitoring, agriculture, medical care, target tracking and water networks. WSNs are generally composed of small sensor, which receive the data from the environment and send them to the base station to make some decisions.

Since the power source of sensors are batteries with low capacity and limitations in weight and cost, minimizing the energy consumption and consequently increasing the network lifespan are the main issues to design of WSNs. Another important concern is network connectivity requirements, which is highly depends on the used protocol of communication. Another challenge which is usually neglected in WSNs designing is considering the specific applications parameters. However, considering this type of parameters leads to be design of WSN more complex; it would certainly make a more practical and precise topology [19, 20].

A grid area of r by c length units is assumed in which sensors are deployed at all $r \times c$ junctions. Sensor nodes are identical and can get one of four possible operation modes such as cluster head (CH), sensor with either high or low signal range (HSR or LSR sensor respectively) or inactive (Off). The problem is optimal designing of wireless sensor networks by determining the operational mode of sensors while improving the energy consumption and satisfying the specific-application and connectivity necessities. This problem is known as an NP-hard one. Therefore, metaheuristics can be a good choice to solve this problem.

Ferentinos and Tsiligiridis [19] have adopted a binary genetic algorithm (BGA) to solve this problem. In this purpose, Hojjatoleslami et al. [20] have used the binary particle swarm optimization (BPSO). A modified version of binary quantum-inspired gravitational search algorithm has been proposed and applied to the current application in [18]. Since the mentioned problem is quadrivalent, a new version of quantum GSA suitable for quadrivalent encoded problems named Quadrivalent Quantum-Inspired Gravitational Search Algorithm (QQIGSA) has been suggested in [21]. This method gave fine achievements compared with used BGA [19] and BPSO [20].

In this paper, continued from the previous works, an improved version of QQIGSA is suggested by adding a Not Q-Gate to increase the diversity of the solutions. To deal with a wide searching space, a large initial population is needed to find acceptable quality of results. But for constraint of resources, we cannot meet this requirement. Also, for problems with complex computations, the execution time of the algorithm increases and may become not reasonable. Therefore, we use parallelism to accelerate the speed and improve the accuracy of the algorithm. Hence, a parallel version of the improved QQIGSA is suggested and implemented using Open Multi-Processing (Open-MP). It will be shown that the parallelism of the QQIGSA can lead to improvement in the quality of results in this problem.

Open-MP supports implementation of parallel programming multiprocessor architectures with shared memory. It has simple implementation compared to other parallel methods like CUDA applied on GPU. Recently, a lot of algorithms have been implemented in parallel by Open-MP and they achieved good speedup in addition to good accuracy equal to the accuracy of the sequential implementation [22].

The main contributions of this paper are as follows:

- An improved version of the QQIGSA using a Not Q-Gate is suggested.
- The suggested improved QQIGSA is parallelized using Open-MP library to speed up the exaction time and improve the results by increasing the population size of the algorithm. The algorithm is adapted to solve the problem of optimal design of WSNs with four operational modes of sensors.

The abilities and the speed of the improved and parallel QQIGSA in optimizing WSNs problem are compared with the state-of-the-art including, BPSO, BGA, QQIGSA and improved BQIGSA and the results are reported. The experimental results indicate the suggested approach prefers the other previous methods in terms of accuracy and speed as well as the energy consumption of the base station.

The remainder of this section is organized as follows. Section 2 explained the quadrivalent quantum-inspired gravitational search algorithm (QQIGSA). Section 3 introduced the suggested algorithm called improved and parallel Quadrivalent Quantum-Inspired Gravitational Search Algorithm. In Sect. 4, the proposed method is adopted in WSN to improve the performance of the network. Section 4 presents the experimental results and the paper concludes in Sect. 5.

2 Preliminary on Quadrivalent Quantum-Inspired Gravitational Search Algorithm (QQIGSA)

Quantum-inspired evolutionary algorithms (QIEAs) which recently attract a great interest of researchers are algorithms that have been made by merging quantum computing and evolutionary algorithms. One of these QIEAs is binary Quantum-inspired gravitational search algorithm (BQIGSA) [18] which is a very powerful algorithm in solving binary optimization problems. A quadrivalent version of this algorithm known as Quadrivalent Quantum-Inspired Gravitational Search Algorithm (QQIGSA) developed in [21] by a combination of quantum computing and GSA [11] to solve quadrivalent problems. Here, the procedure of QQIGSA is explained.

Qubit (Q-bit) is known as the smallest unit in quantum computing. In binary QIEAs, each Q-bit can get the state "0", "1", or a combination of both at a same time. It is called superposition in quantum computing. A Q-bit is shown as a pair of (α, β) which in $|\alpha|^2$ and $|\beta|^2$ are the probabilities of appearing Q-bit in the states of "0" and "1", respectively. Similarly, in QQIGSA, each Q-bit is a pair of (α, β) and can be observed in the states "0", "1", "2", "3" or a combination of them. $|\alpha|^2$ and $|\beta|^2$ are the probability of appearing the Q-bit in each state. In QQIGSA like BQIGSA [14], each object is represented by a vector including n Q-bits as Eq. (1).

$$q_i(t) = \left[q_i^1(t), q_i^2(t), \ldots, q_i^n(t)\right] = \left[\frac{\alpha_i^1(t)}{\beta_i^1(t)} \middle| \frac{\alpha_i^2(t)}{\beta_i^2(t)} \cdots \middle| \frac{\alpha_i^n(t)}{\beta_i^n(t)}\right],$$
$$i = 1, 2, \ldots, NP \tag{1}$$

where NP is population size and n is the dimension of search space. Each Q-bit, $q_i^d(t) = \left[\frac{\alpha_i^d(t)}{\beta_i^d(t)}\right]$, should satisfy the normalization Eq. (2).

$$\left|\alpha_i^d\right|^2 + \left|\beta_i^d\right|^2 = 1 \tag{2}$$

The initial swarm, $Q(t)$ composed of NP Q-bits which are generated in an n-dimensional searching space randomly. Then, the observing process is done on Q-bits to collapse the quantum state into one state by Eq. (3); that the solutions $SW(t) = \{X_1(t), X_2(t), \ldots, X_{NP}(t)\}$ are achieved, where $X_i(t) = (x_i^1(t), x_i^2(t), \ldots, x_i^n(t))$.

Afterward, the fitness evaluation is performed to quantify the fitness of each candidate solution $X_i(t)$. The set $SB(t) = \{B_1(t), B_2(t), \ldots, B_{NP}(t)\}$, in which $B_i(t) = (b_i^1(t), b_i^2(t), \ldots, b_i^n(t))$, shows the best found solutions during the iterations of the algorithm. If $X_i(t)$ be a better solution based on the fitness function, the $B_i(t) \in SB(t)$ is replaced by $X_i(t) \in SW$. It is noticeable in the first iteration, $SB(0) = SW(0)$. Then, $M_i(t)$ for $i = 1, 2, \ldots, NP$ are calculated regarding the fitness values of the solutions of $SB(t)$, using Eq. (4).

In QQIGSA, the value of movement toward "0", "1", "2" or "3" is determined by the angular velocity and applying the RQ-gate or quantum rotation gate on Q-bits. The angular velocity is computes as $\omega = \frac{v}{r}$ where v is the linear velocity and r is the radius of the circular system. It is clear if $r = 1$ then $\omega = v$. The angular velocity of Q-bits is computed by (5) and (6).

$$
\begin{aligned}
&\textbf{if} \quad rand[0,1] < (\alpha_i^d(t))^2 \quad \textbf{then} \\
&\quad \textbf{if} \quad \alpha_i^d(t) > 0 \quad \textbf{then} \\
&\qquad x_i^d(t) = 0 \\
&\quad \textbf{else} \\
&\qquad x_i^d(t) = 2 \\
&\textbf{else} \\
&\quad \textbf{if} \quad \beta_i^d(t) > 0 \quad \textbf{then} \\
&\qquad x_i^d(t) = 1
\end{aligned}
\tag{3}
$$

$$
M_i(t) = \frac{fit_i(t) - worst(t)}{\sum_{j=1}^{NP} \left(fit_j(t) - worst(t) \right)}
\tag{4}
$$

$$
a_i^d(t) = \sum_{j \in kbest} G(t) \frac{M_j(t)}{R_{ij}(t) + \varepsilon} \left(b_j^d(t) - x_i^d(t) \right)
\tag{5}
$$

$$
\omega_i^d(t+1) = rand_i \times \omega_i^d(t) + a_i^d(t)
\tag{6}
$$

where $a_i^d(t)$ is the acceleration of agent i in dimension d at iteration t. $M_j(t)$ denotes the mass value of agent j at iteration t. $kbest$ is the set of the first k solutions in $SB(t)$. This parameter is initialized to NP (the number of agents) and decreased linearly to 1 at the last iteration. $G(t)$ is the gravitational constant at iteration t defined as $G(t) = G_0(1 - 0.95\frac{t}{T})$; that is firstly set to G_0 and decreased by iterations. Here T is the total number of iterations. $x_i^d(t)$ is the value of the current solution $X_i(t) \in SW(t)$ in dimension d. Also, $b_j^d(t)$ is the one of the best found solutions, $B_j(t) \in SB(t)$ in dimension d. ε is a small number to avoid dividing into zero. $R_{ij}(t)$ is the distance between agents i and j in a quadrivalent space and is calculated by the normalized Hamming distance by Eq. (7).

$$
R_{ij}(t) = \frac{\sum_{l=1}^{n} \left| x_i^l(t) - x_j^l(t) \right|}{3 \times n}
\tag{7}
$$

In the polar system, the value of rotation is determined by $\Delta\theta = \omega \cdot \Delta t$. considering $\Delta t = 1$, we have $\Delta\theta = \omega$, which in depending on the difference between $b_j^d(t)$ and $x_j^d(t)$ and regarding Fig. 1, $\Delta\theta_i^d$ would be computed as Eq. (8). Updating each Q-bit is performed by Eq. (9) and t is increased by one. The procedure is redone while the stopping condition is satisfied. Figure 2 shows the procedure of the QQIGSA [21].

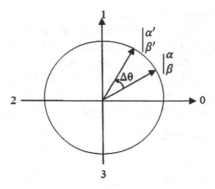

Fig. 1. Illustration of the rotational gate in polar plot [21]

---

1. **Begin**
2. $t = 0, SB(t) = \{\}$
3. Initializing $Q(t)$
4. **While** (not stopping condition) **do**
5. Making $SW(t)$ by observation of the states of $Q(t)$ by Eq. (3)
6. Evaluating $SW(t)$
7. Updating $SB(t)$
8. Updating $M_i(t)$, $i = 1, 2, \dots, N$ by Eq. (4).
9. Updating $Q(t)$ by modified RQ-gate to make $Q(t+1)$ by Eqs. (5, 6, 8, 9)
10. $t = t + 1$
11. **End while**

Fig. 2. The procedure of the QQIGSA [21]

$$\Delta\theta_i^d = \begin{cases} -\omega_i^d(t+1) & \text{if } \left| b_j^d(t) - x_i^d(t) \right| = 3 \\ \omega_i^d(t+1) & \text{if } \left| b_j^d(t) - x_i^d(t) \right| = 1 \\ 0 & \textit{otherwise} \end{cases} \qquad (8)$$

$$\begin{bmatrix} \alpha^d(t+1) \\ \beta^d(t+1) \end{bmatrix} = \begin{bmatrix} \cos(\Delta\theta^d) & -\sin(\Delta\theta^d) \\ \sin(\Delta\theta^d) & \cos(\Delta\theta^d) \end{bmatrix} \begin{bmatrix} \alpha^d(t) \\ \beta^d(t) \end{bmatrix} \quad d = 1, 2, \dots, n \qquad (9)$$

3   The Proposed Parallel and Improved QQIGSA

The performance of the QQIGSA can be improved by adding a new quantum gate called Not Q-Gate. The purpose of using Not Q-Gate in QQIGSA is like using mutation in genetic algorithm that is introducing and preserving the diversity. In fact, the Not Q-Gate prevents the population of the agents from becoming too similar to each other, and therefore avoids the algorithm to fall into local optimum.

The Not Q-Gate alters one or more agent in the population from its current state; that it may lead to change the solution entirely from the previous solution. Therefore, the improved QQIGSA can come to a better solution using a Not Q-Gate. This change can occur during the iterations of the algorithm by a user-defined probability. This probability should be adjusted low. If it is set too high, the search process will turn into a random search. A common way of implementing the Not Q-Gate involves selecting an agent randomly and change its α and β. For example, if agent i is randomly selected; this agent is changed from $q_i(t) = \begin{bmatrix} \alpha_i^1(t) & \alpha_i^2(t) & & \alpha_i^n(t) \\ \beta_i^1(t) & \beta_i^2(t) & \cdots & \beta_i^n(t) \end{bmatrix}$ to $q_i(t) = \begin{bmatrix} \beta_i^1(t) & \beta_i^2(t) & & \beta_i^n(t) \\ \alpha_i^1(t) & \alpha_i^2(t) & \cdots & \alpha_i^n(t) \end{bmatrix}$.

Although the experiment results show that the proposed improved QQIGSA can produce the best solutions in solving the optimal WSN design problem, in comparison with BGA [19], BPSO [20], QQIGSA [21] and improved BQIGSA [18], it is needed to increase the speed of the algorithm. Because when the algorithm executes on the network and the optimal operational mode of sensors were determined, the battery capacity of the sensors are updated and the algorithm should be rerun on the network with new battery capacity of the sensors in the next measuring cycle. Therefore, the exaction time of the algorithm is as important as the quality of the solution.

Parallelism is an efficient way to increase the speed of the algorithm. In the current problem, from experiments it was gotten that the part that can be parallelized and decreased the running time is re-evaluation of the fitness of agents that is the most time consuming (approximately 80% of total executing time). The proposed parallel QQIGSA performed by dividing the whole population into several sub-populations and evaluating the fitness value for each agent separately using Open-MP used on a multi-core processor. In each iteration, all agents are independent of each other and can be easily analyzed in parallel. With this approach, no communication is required between cores as individual fitness evaluations are independent of each other.

The pseudo code of improved and parallel QQIGSA is shown in Fig. 3. Compared to the sequential pseudo code of the improved QQIGSA, the pragma statement

---

1. **Begin**
2. $t = 0, SB(t) = \{\}$
3. Initializing $Q(t)$
4. **While** (not stopping condition) **do**
5. Making $SW(t)$ by observation of the states of $Q(t)$ by Eq. (3)
6. Pragma for each agent i // do in parallel by Open-MP
7. Evaluating $SW(t, i)$
8. Updating $SB(t, i)$
9. Updating $M_i(t)$, $i = 1, 2, \dots, N$ by Eq. (4).
10. Updating $Q(t)$ by RQ-gate to make $Q(t + 1)$ by Eqs. (5, 6, 8, 9)
11. Applying Not Q-Gate with probability p_N on $Q(t + 1)$
12. $t = t + 1$
13. **End while**

Fig. 3. The procedure of the proposed improved and parallel QQIGSA

considering shared and private variables is the only line of code that added before the for-loop to parallelize. This instruction leads to running the iterations of the for-loop in parallel. Parallelization of other section of the algorithms did not lead to speed up. The main reason is small for loops that their sequential time is not so much in comparison with the overhead of the parallelization.

Not only parallelism can increase the speed of the QQIGSA, but also, it can improve the quality of the solutions by increasing the size of population, that subsequently leads to achieve fitter solutions. Besides, it can help to apply the algorithm on larger problem. For example, in optimal design of WSN, since the execution of algorithm has been decreased by parallelism, the algorithm can run on larger networks in a logical time. In fact, in this case, parallelism helps to scalability of the problem.

3.1 Agent Representation

The WSN design involves determining sensors operation modes including cluster head (CH), inactive (Off), sensor with low signal range (LSR) and sensor with high signal range (HSR) such that the fitness function of Eq. (10) is maximized. In QQIGSA, each WSN topology or a candidate solution is presented as a string consisting of the sensor's operational mode. The length of this string is the same as the number of sensors. Each element of the string can get the values 3, 2, 1 and 0 which are respectively representative of CH, HSR, LSR and Off operational modes. Figure 4 shows a small scaled WSN and its agent representation as an instance.

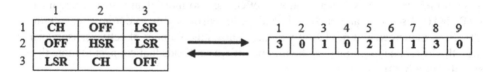

Fig. 4. An example of a small WSN topology and its agent representation [21]

3.2 Fitness Function[1]

To evaluate the effectiveness of the designed WSN in current problem, there are three groups of parameters. These parameters are energy-related parameters, application specific parameters and connectivity parameters. Considering these entire parameters makes a weighted multiple objectives function to evaluate the quality of each candidate solution as a WSN topology. All of these parameters should be minimized; they are combined into an objective function with their specific importance coefficients. The weighted fitness function which should be maximized is shown in Eq. (10).

$$f = 1/(\alpha_1.MRD + \alpha_2.SDE + \alpha_3.SCE + \alpha_4.SORE + \alpha_5.OE + \alpha_6.CE + \alpha_7.BCP) \tag{10}$$

---

[1] The main body of this subsection is adopted from [18, 21].

where, *MRD*, *SDE*, *SCE*, *SORE*, *OE*, *CE* and *BCP* are the main parameters and the coefficients $\alpha_1, \alpha_2, \ldots,$ *and* α_7 are the importance or the weight of them. The minimization forms of the main parameters are explained as bellows [18–21].

Mean Relative Deviation (MRD): One of the application-specific parameters is MRD which measures the uniformity of the active nodes. The mathematical definition of this parameter is shown in Eq. (11) where, m is the number of subareas, ρ_S denotes the area's special density and ρ_{Si} refers to the special density of the active nodes in subarea S_i.

$$MRD = \frac{\sum_{i=1}^{m} |\rho_{Si} - \rho_S|}{m.\rho_S} \tag{11}$$

Spatial Density Error (SDE): *SDE* is mathematically defined as Eq. (12) and considered as other specific-application parameters. This parameter is applies to penalize designs of WSN that do not meet the minimum amount of density of active points in the area. In Eq. (12), ρ_s denotes to the spatial density of the entire area and ρ_d represents the favorite special density.

$$SDE = \begin{cases} \frac{\rho_d - \rho_s}{\rho_d} & \text{if } \rho_s < \rho_d \\ 0 & \text{otherwise} \end{cases} \tag{12}$$

Sensors-per-Cluster-head Error (SCE): SCE as a connectivity parameter is applies to guarantee that each CH sensor does not connect to more than a specific number of LSR or HSR sensors; and each LSR and HSR sensor is able to communicate with its CH. SCE is defined as Eq. (13) in which, n_i shows the number of active sensors inside the i th cluster and *nfull* denotes the number of cluster heads that connect to more than a specific number of nodes.

$$SCE = \begin{cases} \frac{\sum_{i=1}^{nfull} n_i}{nfull} & \text{if } nfull > 0 \\ 0 & \text{otherwise} \end{cases} \tag{13}$$

Sensor-Out-of-Range Error (SORE): Another connectivity parameter is SORE which is defined as Eq. (14). In this equation, *nout* represents the number of LSRs and HSRs which cannot communicate with their CHs; and n is the number of sensors of WSN. In fact, this parameter that is related to the amount of signal ranges is used to ensure that each sensor can connect its cluster head.

$$SORE = \frac{nout}{n} \tag{14}$$

Operational Energy (OE): This parameter is an energy-related one and defined as the energy that sensors consume during a measuring cycle. Equation (15) computes this energy where, $x{:}y{:}z$ is respectively the proportion of CH, HSR and LSR energy consumption. Also, *nch*, *nhs* and *nls* is the number of CHs, HSRs and LSRs respectively.

$$OE = x.\frac{nch}{n} + y.\frac{nhs}{n} + z.\frac{nls}{n} \tag{15}$$

Communication Energy (CE): This energy-related parameter calculates the consumed energy for communication between LSRs or HSRs and their cluster heads. *CE* is computed by Eq. (16) in which, k and μ are constants, the number of clusters shown by c, the number of active nodes of cluster i shown by n_i, and the Euclidean distance between CH i and sensor j represented as d_{ji}.

$$CE = \sum_{i=1}^{c} \sum_{j=1}^{n_i} \mu.d_{ji}^{k} \tag{16}$$

Battery Capacity Penalty (BCP): This is another energy-related parameter computing by Eq. (18). In this relation, $BC_i^{[t]}$ is the battery capacity of sensor i at time slice t and depends on the operational mode of sensors. Equation (17) used to update battery capacity $(BC_i^{[t]})$ where $BRR_i^{[t-1]}$ is the battery reduction rate at time slice $t - 1$. $BC_i^{[t]}$ can get a value between 0 and 1; 0 expressing empty battery and 1 representing full battery capacity. In Eq. (18) $PF_i^{[t]}$ is the penalty factor depends on the operation mode of sensor i; also, *ngrid* is the total number of sensors [19–22].

$$BC_i^{[t]} - BC_i^{[t-1]} - BRR_i^{[t-1]} \tag{17}$$

$$BCP^{[t]} = \sum_{i=1}^{ngrid} PF_i^{[t]}.\left(\frac{1}{BC_i^{[t]}} - 1\right), \; t = 1, 2, \ldots \tag{18}$$

4   Experimental Results

To measure the performance and the speedup of the suggested approach, some experiments are performed on a WSN with 900 sensors placed in an area in the size of 30 by 30 square units. The aim is determining the sensors' operational mode to improve the energy consumption considering the specific application and connectivity requirements. The experiments are performed on a laptop with Intel® Core(TM) i7 CPU 1.80 GHz, RAM 8.00 GB, windows 10, 64 bit operating system and Dev C++ 5.11 compiler.

4.1   Parameter Setting

To conduct the experiments, there are some parameters to be set, including parameters related to fitness function and parameters related to the algorithms. Parameters related to fitness function are set as follows. The weighting coefficients, $\alpha_1, \alpha_2, \ldots, and \, \alpha_7$, show the importance of the fitness function parameters. Table 1 presents the value of coefficients. The other parameters are set as follows. In OE, x:y:z is assumed as 20:2:1 shows the rate of energy consumption. In the case of CE the constants μ and k are set to 1 and 3 respectively. In BCP, the penalty factor $PF_i^{[t]}$ uses the proportion 20:2:1 respectively for the CH, HSR and LSR operational modes as before. The battery reduction rates (BRR) for CH, HSR, LSR and Off modes are set to 0.2, 0.02, 0.01 and 0, respectively. In SCE, the number of sensors that can connect with a cluster head is set to be at most 15. In the case of SDE ρ_d is set to 0.2. In SORE, the signal range of HSR and LSR are respectively adjusted to 5 and 10 length units.

Table 1. The coefficients of the main parameters of the fitness function [18–21]

α_1	α_2	α_3	α_4	α_4	α_6	α_7
10^2	10^4	10^6	10^5	10	10^{-2}	10^{-2}

Parameter setting of the algorithms BGA, BPSO, improved BQIGSA, QQIGSA and proposed QQIGSA is performed as Table 2 based on the optimal values reported in the previous works. In all used methods, the population size is 20 and the number of iterations is 100.

Table 2. Parameter adjustment of the used metaheuristics

Algorithm	Parameter adjustment
BGA [19]	$p_c = 0.8$, $p_m = 0.005$
BPSO [20]	$c_1 = 2$, $c_2 = 2$, $w = 1$
Improved BQIGSA [18]	$G = G_0 \times (1 - 0.95 \times t/T)$, $G_0 = 3.78$ *kbest*: Linear function decreasing from N to 1
QQIGSA [21] and the Proposed QQIGSA	$G = G_0 \times (1 - 0.95 \times t/T)$, $G_0 = 0.125$ *kbest*: Linear function decreasing from N to 1

4.2   Comparative Study

Here, the performance of the suggested improved and parallel QQIGSA is compared with the previous method including BGA [19], BPSO [20], improved BQIGSA [18] and QQIGSA [21] applied on pre-described WSN. The near optimal operation mode of sensors of network is detected such a way that the fitness function of Eq. (10) is maximized. The reported results are the average of 10 independent runs of the algorithms through 100 iterations. Table 3 shows the comparison of the algorithm based on the mean and maximum value of fitness, the amount of parameters including, operational energy (OE), communication energy (CE), Mean Relative Deviation (MRD), Spatial Density Error (SDE), Sensor-Out-of-Range Error (SORE), Sensors-per-Cluster-head Error (SCE), Battery Capacity Penalty (BCP) and the number of sensor in each operational mode.

Table 3. Performance evaluation of the proposed QQIGSA with different number of population in comparison with the previous algorithms. The reported results are the average of 10 independent runs of the algorithms.

Parameter	BGA	BPSO	Improved BQIGSA	QQIGSA	Proposed QQIGSA with $NP = 20$	Proposed QQIGSA with $NP = 50$	Proposed QQIGSA with $NP = 100$	Proposed QQIGSA with $NP = 200$
Max fitness value	0.135	0.0158	0.0207	0.017	0.0214	0.0218	0.025	**0.0283**
Mean fitness value	0.0124	0.0152	0.0202	0.0162	0.0209	0.021	0.0243	**0.028**
SDE	0	0	0	0	0	0	0	0
OE	5.0397	4.597	3.0562	3.6017	5.302	5.196	4.98	**2.89**
CE	**1259.26**	1363.43	2252.93	2985.45	1176.22	1135.65	1192.43	2251.53
MRD	0.0166	0.0192	0.0147	0.0215	0.021	0.0186	0.0212	**0.0145**
SCE	0	0	0	0	0	0	0	0
SORE	0	0	0	0	0	0	0	0
BCP	1064.1	1099.1	554.86	1194.53	1052.26	1031.91	976.64	**523.679**
CH	192	171.6	105.6	113.8	209.4	204.5	193.3	**103.3**
HSR	231.2	235.4	233.6	316.4	173.1	169.1	188.1	**156.3**
LSR	233.4	234.6	231.4	332.8	246.2	248.8	244.3	**229.1**

Minimization of parameters and maximization of the fitness function are desired. It can be seen that the proposed QQIGSA with population size 20 has fine results compared with the BGA, BPSO, improved BQIGSA and QQIGSA with the same population size. By increasing the population size of the suggested method, as $NP = 50, 100, 200$, the quality of the results has been improved. The last column of Table 3 shows the preference of the proposed QQIGSA with population size of 200 in comparison with all others. It has the best results in terms of all parameters except CE, that it is due to the tradeoff between the OE and CE. The proposed algorithm tries to decline the operational energy consumption via decreasing the number of cluster heads that it subsequently leads to increase the communication energy.

Although increasing the population size leads to improve the solutions quality, it will lengthen the execution time of the algorithm. To handle this problem, the algorithm was parallelized. Speedup and parallel efficiency, respectively defined as Eqs. (19) and (20), are two important criteria to measure the parallelization quality. They indicate whether parallelism is useful or not for the given problem and algorithm. In these equations, $T(m)$ indicates the time taken to solve the problem with m processors.

$$Speedup = \frac{T(1)}{T(m)} \tag{19}$$

$$\textit{Parallel Efficiency} = \frac{T(m)}{m} \tag{20}$$

Tables 4 and 5 respectively show the speedup and parallel efficiency of the proposed algorithm by changing the number of cores from 2 to 8. The usefulness and efficiency of the parallel implementation of the proposed algorithm in optimal design of WSN are convincingly demonstrated in these tables. As it can be seen the speedup increases in average from 1.83 to 4.24 by changing the number of cores from 2 to 8. The speedup and efficiency values are the averaged of 10 independent runs of the algorithm.

Table 4. Speedup of the proposed parallel QQIGSA for different size of population

# of cores	2	3	4	5	6	7	8
$NP = 20$	1.84	2.41	3.06	3.33	3.46	3.92	3.97
$NP = 50$	1.8	2.45	3.11	3.24	3.68	4.04	4.35
$NP = 100$	1.84	2.35	2.71	2.89	3.60	4.01	4.28
$NP = 200$	1.86	2.44	3.14	3.36	3.60	4.04	4.38

Table 5. Parallel efficiency of the proposed parallel QQIGSA for different size of population

# of cores	2	3	4	5	6	7	8
$NP = 20$	0.92	0.80	0.76	0.67	0.58	0.56	0.50
$NP = 50$	0.9	0.82	0.78	0.65	0.61	0.58	0.54
$NP = 100$	0.92	0.78	0.68	0.58	0.60	0.57	0.53
$NP = 200$	0.93	0.81	0.79	0.67	0.60	0.57	0.54

The experimental results indicate parallelism of the algorithm helps to improve quality of solutions as well as the running time of the algorithm. Besides, it can help to improve the power efficiency. Since in most of WSNs, the algorithm runs on a sensor with limitation on energy, multi-core architecture can enhance the power efficiency. In these architectures, known as Chip Multiprocessors (CMPs), multiple low-power cores are inserted on a single chip. In fact, CMPs improve the throughput of the system by exploiting parallelism using multiple cores without significantly increasing the power of the processor. Moreover, parallelism let the algorithm applied and run on larger networks, containing more number of sensors, in a reasonable time.

5   Conclusion

In this research an improved version of Quadrivalent Quantum-Inspired Gravitational Search Algorithm (QQIGSA) using Not Q-Gate is presented to solve quadrivalent problems. To accelerate the speed of the algorithm, we use Open-MP library to parallelization. The improved and parallel QQIGSA is applied to a wireless sensor network to determine the operational mode of each sensor including cluster head, low signal range, high signal range and inactive so that energy-related and connectivity parameters as well as specific application requirements are optimized. Since this problem was NP-hard, using a robust metaheuristic and parallelism could be the best choice. The performance of the proposed method was compared to the state-of-the-art and the results show a satisfactory effectiveness of the proposed method based on both quality of the solutions and the speedup. The parallelism improves speedup over 4 times for 8 cores. Developing the parallel version of the algorithm using GPU and CUDA programming is one of our future works in this research.

References

1. Talbi, E.-G.: Metaheuristics: From Design to Implementation, 15th edn. Wiley, Chicago (2009)
2. Knysh, D.S., Kureichik, V.M.: Parallel genetic algorithms: a survey and problem state of the art. Comput. Syst. Sci. Int. **49**(4), 579–589 (2010)
3. Vanneschi, L., Codecasa, D., Mauri, G.: Comparative study a of four parallel and distributed PSO methods. New Gener. Comput. **29**, 129–161 (2011)
4. Pedemonte, M., Nesmachnow, S., Cancela, H.: A survey on parallel ant colony optimization. Appl. Soft Comput. **11**(8), 5181–5197 (2011)
5. Parpinelli, R.S., Benitez, C.M.V., Lopes, H.S.: Parallel approaches for the artificial bee colony algorithm. In: Panigrahi, B.K., Shi, Y., Lim, M.H. (eds.) Handbook of Swarm Intelligence. ALO, vol. 8, pp. 329–345. Springer, Heidelberg (2011). https://doi.org/10.1007/978-3-642-17390-5_14
6. Onbaşoğlu, E., Özdamar, L.: Parallel simulated annealing algorithms in global optimization. Global Optim. **19**(1), 27–50 (2001)
7. Majd, A., Sahebi, G., Daneshtalab, M., Plosila, J., Lotfi, Sh., Tenhunen, H.: Parallel imperialist competitive algorithms. Concurr. Comput. Pract. Exp. (2018). https://doi.org/10.1002/cpe.4393
8. Mahmoud, K.R., Hamad, S.: Parallel implementation of hybrid GSA-NM algorithm for adaptive beam-forming applications. Electromagnet. Res. **58**, 47–57 (2014)
9. Alba, E.: Parallel evolutionary algorithms can achieve super-linear performance. Inf. Process. Lett. **82**, 7–13 (2002)
10. Rashedi, E., Nezamabadi-pour, H., Saryazdi, S.: GSA: a gravitational search algorithm. Inf. Sci. **179**(13), 2232–2248 (2009)
11. Rashedi, E., Nezamabadi-pour, H., Saryazdi, S.: BGSA: binary gravitational search algorithm. Nat. Comput. **9**, 727–745 (2009)
12. Rashedi, E., Nezamabadi-pour, H.: Feature subset selection using improved binary gravitational search algorithm. Intell. Fuzzy Syst. **26**(3), 1211–1221 (2014)
13. Soleimanpour, M., Nezamabadi-pour, H., Farsangi, M.M.: A quantum inspired gravitational search algorithm for numerical function optimization. Inf. Sci. **267**(20), 83–100 (2014)
14. Nezamabadi-pour, H.: A quantum-inspired gravitational search algorithm for binary encoded optimization problems. Eng. Appl. Artif. Intell. **40**, 62–75 (2015)
15. Ibrahim, A.A., Mohamed, A., Shareef, H.: A novel quantum-inspired binary gravitational search algorithm in obtaining optimal power quality monitor placement. Appl. Sci. **12**(9), 822–830 (2012)

16. Barani, F., Mirhosseini, M., Nezamabadi-pour, H.: Application of binary quantum-inspired gravitational search algorithm in feature subset selection. Appl. Intell. **47**(2), 304–318 (2017)
17. Barani, F., Mirhosseini, M., Nezamabadi-pour, H., Farsangi, M.M.: Unit commitment by an improved binary quantum GSA. Appl. Soft Comput. **60**, 180–189 (2017)
18. Mirhosseini, M., Barani, F., Nezamabadi-pour, H.: Design optimization of wireless sensor networks in precision agriculture using improved BQIGSA. Sustain. Comput.: Inform. Syst. **16**, 38–47 (2017)
19. Ferentinos, K.P., Tsiligiridis, T.A.: Adaptive design optimization of wireless sensor networks using genetic algorithms. Comput. Netw. **51**, 1031–1051 (2007)
20. Hojjatoleslami, S., Aghazarian, V., Dehghan, M., Motlagh, N.Gh.: PSO based node placement optimization for wireless sensor networks. In: Proceedings of the IEEE Wireless Communication and Networking, pp. 12–17 (2011)
21. Mirhosseini, M., Barani, F., Nezamabadi-pour, H.: QQIGSA: a quadrivalent quantum-inspired GSA and its application in optimal adaptive design of wireless sensor networks. Netw. Comput. Appl. **78**, 231–241 (2017)
22. Caner, C., Dreo, J., Saveant, P., Vidal, V.: Parallel divide-and-evolve: experiments with Open-MP on a multicore machine. In: Proceedings of the 13th Annual Genetic and Evolutionary Computation Conference, GECCO 2011, Dublin, Ireland (2011)

Weakly Supervised Learning Technique for Solving Partial Differential Equations; Case Study of 1-D Reaction-Diffusion Equation

Behzad Zakeri[1]($\boxtimes$), Amin Karimi Monsefi[2], Sanaz Samsam[3], and Bahareh Karimi Monsefi[4]

[1] School of Mechanical Engineering, College of Engineering, University of Tehran, Tehran, Iran
Behzad.Zakeri@ut.ac.ir
[2] Faculty of Computer Science and Engineering, Shahid Beheshti University, Tehran, Iran
A_karimimonsefi@sbu.ac.ir
[3] Department of Aerospace Engineering, Sharif University of Technology, Tehran, Iran
samsam.sanaz@ae.sharif.ir
[4] Faculty of Mechanical Engineering, Islamic Azad University Central Tehran Branch, Tehran, Iran

Abstract. Deep learning as a valuable intelligence tool to deal with complicated problems plays a crucial role in the 21st century. The utility of deep learning in solving partial differential equations (PDEs) is an interesting application of AI, which has been considered in recent years. However, supervision of learning procedure needs to have considerable labeled data to train the network, and this method could not be a beneficial technique to deal with unknown PDEs which we do not have any labeled data. To tackle this issue, in this paper a new method will be presented to solve PDEs only by using boundary and initial condition. Weakly supervision as an efficient method can provide an ideal bed to tackle boundary and initial value problems. To have better judgment about this method we chose Reaction-Diffusion equation as a versatile equation in engineering and science to be solved as a case study. By using the weakly supervised method and the finite difference method reaction-diffusion equation have solved, and the results of these methods have been compared. It has been shown that the results of deep learning have high consistency with finite difference results, and weakly supervised learning can be introduced as an efficient method to solve different types of differential equations.

Keywords: Deep learning · Weakly supervised learning · PDEs · Dynamical systems

L. Grandinetti et al. (Eds.): TopHPC 2019, CCIS 891, pp. 367–377, 2019.
https://doi.org/10.1007/978-3-030-33495-6_28

1   Introduction

Machine learning can be mentioned as one of the most influential elements in the contemporary century [2]. This powerful tool plays a crucial role in large number of engineering purposes. For instance, smile recognition in cameras, intelligent assistants in mobile phones, and attack detection in the network were no longer possible, since machine learning made them possible [1,15,24]. Supervised learning as a subset of machine learning, needs a labeled database for learning procedure [10]. Although, providing a proper database considerably improve the performance of the learning process, but this database is not always available. To deal with this problem, weakly supervised learning has been introduced [14,18]. Typically, there are three types of weak supervision which are, incomplete supervision, inaccurate supervision and inexact supervision [30]. All of these methods are designed to make the learning process possible without the large labeled database. This feature of weakly supervised learning techniques makes it a comfortable bed for defining physical problems which we do not have sufficient data about. Among all of the techniques that we use for studying physical problems, one of the important ones is solving the partial differential equations.

Partial differential equations (PDEs) are one of the cornerstones in mathematical modelling. Most of the physical description of natural phenomena are being simulated by taking advantage of PDEs [13]. For example, fluid motion (Navier-Stokes equation), electromagnetic field (Gauss's law) and electrodynamics (Schrödinger equation) have been modelled using partial differential equations [3,5,22]. Solving these equations based on analytical methods in most cases is impossible, and numerical methods should be implemented to represent the approximate solution. Common numerical methods like finite element method (FEM) and finite volume method (FVM) have some inevitable problems, such as mesh dependency and long computational time, which motivate scientists to find an alternative to these methods. Considering recent advances in machine learning and specifically Deep learning, ML methods could be an efficient alternatives for conventional numerical methods for solving PDEs in near future [23].

One-dimensional Reaction-Diffusion equation is known as one of the versatile differential equations in science and engineering. This equation can be used for modelling several phenomena, such as Turbulent flows, diffusion of ions in a reactive medium and financial progress in competitive environment [9,12,26]. As a more sensible example, the case of sulfate attack to concrete is one of the famous cases that reaction-diffusion equation can successfully simulate. In this case, we are interested in finding the concentration of sulfate ions in a known time and position, and since sulfate and concrete react with each other we have to solve reaction-diffusion for finding the correct concentration distribution [6,20,27].

In this paper, we focus on the one-dimensional reaction-diffusion with Dirichlet boundary condition. We tend to encode the behavior of the equation into a loss function, in a way that deep learning algorithm can learn and generate correct solutions for any time frame without having any labeled data. For this purpose, a convolutional kernel has been designed which encode the constraints that must be satisfied in each time step and position, and this kernel was used to determine the loss function. By minimizing the loss function, the deep neural network learns to satisfy the constraints during solving the equation and learn the physics of the reaction-diffusion equation effectively. The reaction-diffusion equation with specified boundary conditions has been solved by weakly supervised method, also the numerical solution of this equation based on finite difference method (FDM) presented and the results of the deep learning algorithm have compared with the FDM method.

2   Related Work

This work is an interdisciplinary study between artificial intelligence and dynamical systems, and each one has been pursued by a large number of researchers. The former subject is studying by data scientists and AI developers. The main aim in weakly supervised learning is providing a general platform technique which learning algorithms will be able to learn with limited initial labeled data for training stage. The latter subject is searching to find an accurate method for solving partial differential equations. The utility of different numerical techniques for the approximation of the solution of PDEs is one of the important part of these researches.

Exchanging conventional numerical methods with alternative meshless techniques like machine learning in recent become increasingly popular. Especially in case of problems with the complex mathematical formulation machine learning schemes are replacing with classical models. Oquab et al. have used weakly supervised convolutional neural network for object classification in image processing to reduce the number of labeled input images [18]. This technique was a general concept, and has used in different applications, such as automatic classification, medical image analysis and solving differential equations [4,11,25]. Sharma et al. trained an encoder-decoder U-Net architecture, which was a fully convolutional neural network to solve the steady-state two-dimensional heat equation on a regular square. For this purpose, they used weakly supervised learning techniques in defining a proper convolutional kernel and loss function to train the network only by using the boundary conditions of the PDE rather than providing a large number of labeled data-sets [21]. Han et al. have introduced a new method for solving high-dimensional PDEs with the utilities of deep learning. They reformulate the PDEs in the form of backward stochastic differential equations (BSDEs) and then by using deep learning approximate the gradient of the solution. Although their method is accurate for dealing with high-dimensional cases, complexities of this method justify the searching for the comprehensive approach to tackle linear and low-dimensional PDEs [8,28].

On the other hand, conventional numerical methods such as FDM and FVM, have been widely developed to tackle different types of mathematical problems which representation of analytical solution for them is not available [17]. For instance, for the above example of the application of the reaction-diffusion equation in sulfate attack, Guo et al. have used the finite difference method to find the concentration distribution of sulfate ions in concrete [31]. Also, extended researches have conducted on the application of machine learning in different engineering fields, such as tackling with turbulent flows and control theory [7, 16, 29].

3   Physics

3.1   Reaction-Diffusion Equation

For solving 1-D reaction-diffusion equation, a simple line has been considered as domain with Dirichlet boundary condition at the ends of the line. By assigning the arbitrary constant to the diffusion coefficient, we can control the role of the material on the transport phenomena. Also, the reaction coefficient specified the effect of interaction between the diffusive substance and medium. In this simulation, a high concentration applied to the boundaries, and the aim is modelling the propagation of that substance among the domain. The boundary conditions are given by $C(0, t) = C(L, t) = C_0$, and we want to determine $C(x, t)$, the concentration field in arbitrary time.

The general form of the reaction-diffusion equation in one-dimensional space is shown in Eq. 1:

$$\frac{\partial C}{\partial t} = D \frac{\partial^2 C}{\partial x^2} - RC \tag{1}$$

Where $D, R > 0$ are the diffusion coefficient and reaction rate between specified material and domain respectively.

The analytical techniques for solving the reaction-diffusion equation are not as simple as pure diffusion. For this reason, we utilize numerical methods to obtain an accurate solution.

3.2   Finite Difference Method

The finite difference is a simple numerical method which is used to compute the accurate solution of the partial differential equations in regular domains. In this method governing equation and the domain both discretized, and the equations solved iteratively on the discrete domain. Considering the discretization of the domain 3 and time 4, the discretized form of reaction-diffusion equation for position m and time n would be:

$$\frac{C_m^{n+1} - C_m^n}{\Delta t} = D \frac{C_{m-1}^n - 2C_m^n + C_{m+1}^n}{\Delta x^2} - RC_m^n \tag{2}$$

$$\frac{x_L - x_0}{m} = \Delta x \tag{3}$$

$$\frac{t_\infty - t_0}{n} = \Delta t \tag{4}$$

Where C_m^n is the concentration in time n and position m, also indices 0, L and ∞ represent the initial step in time and position, end of the domain and last time step respectively.

With solving the Eq. 2 iteratively, the value of the C in each time and position converge to the real value.

4   Deep Learning Solver

The aim of this work is using the deep neural network to solve the reaction-diffusion equation with only using boundary and initial conditions, without knowing the numerical or analytical solution or even having any labeled data. For this purpose, the differential equation has been decoded into a *physical-informed loss function*. This technique helps us to find the solution of the PDE without using supervision in the form of data.

To import the initial and boundary conditions of the problem into the deep neural network, we used a $n \times m$ matrix which its columns and rows represent the positions and time steps respectively. All of the matrix elements for the input matrix are zero except the first and last columns which their values represent the boundary condition values (which in this study is C_0). Also, in this matrix each row demonstrates the concentration distribution in a specified time-frame.

4.1   Deep Learning Architecture

A fully convolutional encoder-decoder network in the form of U-Net architecture has been utilized in this study as Ronneberger *et al.* have used this architecture for biomedical image segmentation [19]. The main reason for choosing a fully convolutional architecture among other architectures is the flexibility of this structure to solve problems at multiple scales. The network contains several encoding convolutional and decoding pooling layers which save the input matrix size during the learning process. Finally, the output matrix represents the solution of the PDE in the discretized space-time domain. The schematic structure of the network is shown in Fig. 1:

As it is shown in Fig. 1, each encoding layer has been connected to the corresponding decoder layer using Fusion connection. The reason for the utility of fusion connections is to pass the boundary values of the input to the output layers, and by this technique, the network is not forced to memorize the structure of the input in its bottleneck layers. The number of layers in our architecture is arbitrary, and it is simply possible to add layers into the network as much as necessary.

4.2   Kernel

To make an intelligent network that can solve the equation in any time and position, it is necessary to define the governing rule in that equation in a simple way for the neural network. It is similar to the method that FDM use for solving the discretized equation. In fact, by discretization of a continues equation and transferring that equation into the algebraic form we can observe the governing rule for every point in space and time.

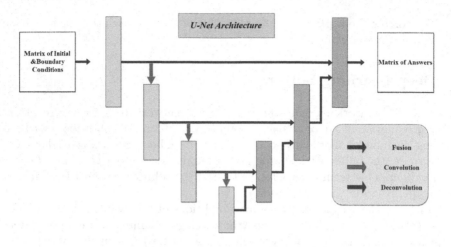

Fig. 1. Deep neural network diagram

By reforming the Eq. 2, we can find the state of an arbitrary point in the space-time domain based on its neighbours as shown in Eq. 5:

$$C_m^{n+1} = C_m^n + B\left(C_{m-1}^n - 2C_m^n + C_{m+1}^n\right) - RC_m^n\Delta t \qquad (5)$$

And B defined as follow:

$$B = \frac{D\Delta t}{\Delta x^2} \qquad (6)$$

For transferring the relation among variables into the neural network, Eq. 5 have been decoded into the 3×3 convolutional kernel as follow:

$$\begin{pmatrix} -a & -b & -c \\ 0 & 1 & 0 \\ 0 & 0 & 0 \end{pmatrix} \qquad (7)$$

Where:

$$a \to B \qquad c \to B \qquad b \to (1 - 2B - R\Delta t) \qquad (8)$$

Discussed Kernel has been convolved into the across the input matrix, and the output matrix after normalization was used to calculate the Loss function:

$$\sum_{i,j} (Conv2D(Kernel, Output)_{i,j})^2 \qquad (9)$$

By minimizing the Eq. 9, the deep neural network tries to make its' solution closer to the real values which can be found in Eq. 5 and changing in boundary and initial conditions train the network for solving any type of problems governed by reaction-diffusion physics.

5   Results

In this section, we will discuss the results of the deep learning solution, and compare the answers to finite difference results. In the presented study to have more realistic results, boundary conditions and all coefficients were chosen from the work of Zuo *et al.* on the sulfate attack to concrete [31]. As it was raised in the deep learning section, the output of the U–Net network for specific input is a matrix which its columns and rows represent the position and time, and the value of each element demonstrates the concentration in ith time and jth position. In Fig. 2, a sample output matrix of the deep learning solver with constant boundary condition in both sides of the domain is shown:

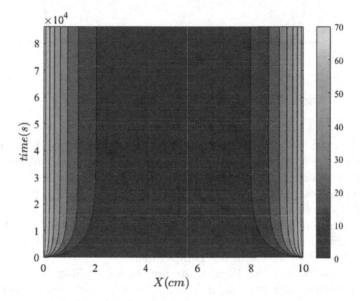

Fig. 2. Sample output matrix from neural network

Looking at Fig. 2 in more detail, the progress of concentration diffusion along the time axis is obviously visible. The value of all elements on the first and last column are same and equal to the boundary value. Also, it is clear that by passing the time, the gradient along time and position decrease, and the answer converge to the steady-state solution.

Figure 3 compares the deep learning solution with finite difference method results in terms of concentration distribution along domain in different times. With looking more precisely to Fig. 3, we observe that deep learning results have high consistency with FDM results. The only part that deep learning could not predict correctly was in the first and last part of the 1 s timeline. The reason for this disability of deep learning in these regions is the high gradient in these areas.

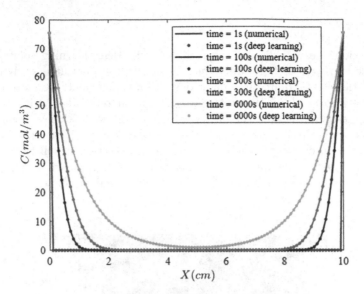

Fig. 3. 2D comparison of deep learning with FDM

In Fig. 4, solutions of the reaction-diffusion equation in three different sets of coefficients have been shown based on deep learning and numerical method. We have set the proportion of reaction and diffusion coefficients in a way that by solving the equation with these coefficients observe the physic of pure diffusion, pure reaction and reaction-diffusion.

For this purpose, a dimensionless coefficient has been defined which help us to calculate the correct proportion of reaction and diffusion coefficients to have all three state of the solution in our computation.

Damköhler number is an important dimensionless parameter in chemical engineering which clarifies the role of diffusion, reaction or simultaneous reaction-diffusion phenomena in transport phenomena and define as follow:

$$D_a = \frac{Rate\ of\ reaction}{Diffusion\ rate} \tag{10}$$

In our model Eq. 1, Damköhler number is defined as:

$$D_a = \frac{RL^2}{D} \tag{11}$$

This number represents the states of reaction-diffusion in different states where $D_a \cong 1$, $D_a \gg 1$, and $D_a \ll 1$ mean the physics of *Reaction-Diffusion*, *pure Reaction*, and *pure Diffusion* respectively.

To have a quantitative assessment of deep learning solution, we assumed one of the coefficients constant, and by changing the other coefficient MSE value has been computed, and the result of this analysis is reported in Table 1:

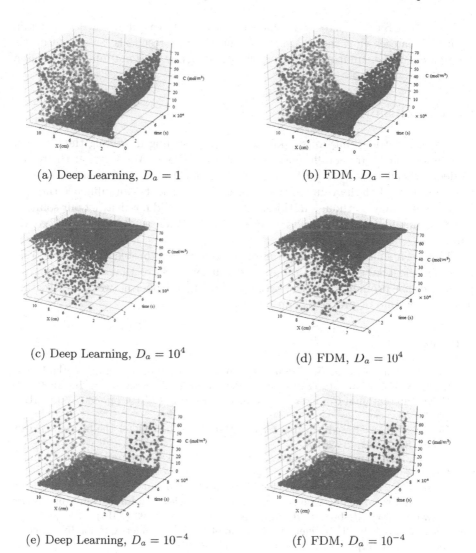

(a) Deep Learning, $D_a = 1$

(b) FDM, $D_a = 1$

(c) Deep Learning, $D_a = 10^4$

(d) FDM, $D_a = 10^4$

(e) Deep Learning, $D_a = 10^{-4}$

(f) FDM, $D_a = 10^{-4}$

Fig. 4. 3D comparison of deep learning with FDM

Table 1. Accuracy analyze based on changing coefficients

(a) $D = 2.7 \times 10^{-9}$

(b) $R = 2.25 \times 10^{-7}$

R coefficient	MSE value
2.25×10^{-2}	0.587
2.25×10^{-5}	0.495
2.25×10^{-7}	0.623
2.25×10^{-10}	0.341

D coefficient	MSE value
2.7×10^{-2}	0.305
2.7×10^{-7}	0.576
2.7×10^{-8}	0.588
2.7×10^{-10}	0.534

The Mean Square Error index has been utilized to calculate the deep learning error. We can see that the accuracy of deep learning results is dependent on the coefficients of the equation; however, this dependency does not influence the final results quality.

6   Conclusion

In this paper, the ability of weakly supervised learning in solving the transient one-dimensional partial differential equation has shown. We saw that the results of deep learning method perfectly were similar to the FDM results. Also, we have observed that the value of the equation's coefficients can influence the deep learning accuracy. Although in this study, this effect did not distract our solution, it could manipulate the results in critical cases such as stochastic differential equations.

This technique is a functional method to approach problems that we do not have sufficient labeled data about, and authors believe that by decoding the physics of the problems into the kernel and using a proper network architecture, a wide spectrum of problems can be solvable.

References

1. Bai, Y., Guo, L., Jin, L., Huang, Q.: A novel feature extraction method using pyramid histogram of orientation gradients for smile recognition. In: 2009 16th IEEE International Conference on Image Processing (ICIP), pp. 3305–3308. IEEE (2009)
2. Bell, S.: Project-based learning for the 21st century: skills for the future. Clearing House **83**(2), 39–43 (2010)
3. Chorin, A.J.: Numerical solution of the Navier-Stokes equations. Math. Comput. **22**(104), 745–762 (1968)
4. Ciompi, F., et al.: Automatic classification of pulmonary peri-fissural nodules in computed tomography using an ensemble of 2D views and a convolutional neural network out-of-the-box. Med. Image Anal. **26**(1), 195–202 (2015)
5. Feit, M., Fleck Jr., J., Steiger, A.: Solution of the Schrödinger equation by a spectral method. J. Comput. Phys. **47**(3), 412–433 (1982)
6. Gospodinov, P.N., Kazandjiev, R.F., Partalin, T.A., Mironova, M.K.: Diffusion of sulfate ions into cement stone regarding simultaneous chemical reactions and resulting effects. Cem. Concr. Res. **29**(10), 1591–1596 (1999)
7. Guo, X., Yan, W., Cui, R.: Integral reinforcement learning-based adaptive NN control for continuous-time nonlinear MIMO systems with unknown control directions. IEEE Trans. Syst. Man Cybern.: Syst. (2019). https://doi.org/10.1109/TSMC.2019.2897221
8. Han, J., Jentzen, A., Weinan, E.: Solving high-dimensional partial differential equations using deep learning. Proc. Natl. Acad. Sci. **115**(34), 8505–8510 (2018)
9. Kramers, H.A.: Brownian motion in a field of force and the diffusion model of chemical reactions. Physica **7**(4), 284–304 (1940)
10. LeCun, Y., Bengio, Y., Hinton, G.: Deep learning. Nature **521**(7553), 436 (2015)
11. Liu, P., Gan, J., Chakrabarty, R.K.: Variational autoencoding the Lagrangian trajectories of particles in a combustion system. arXiv preprint arXiv:1811.11896 (2018)

12. Majda, A.J., Souganidis, P.E.: Large scale front dynamics for turbulent reaction-diffusion equations with separated velocity scales. Nonlinearity **7**(1), 1 (1994)
13. Mattheij, R.M., Rienstra, S.W., ten Thije Boonkkamp, J.H.: Partial Differential Equations: Modeling, Analysis, Computation, vol. 10. SIAM, Philadelphia (2005)
14. Medlock, B., Briscoe, T.: Weakly supervised learning for hedge classification in scientific literature. In: Proceedings of the 45th Annual Meeting of the Association of Computational Linguistics, pp. 992–999 (2007)
15. Michalski, R.S., Carbonell, J.G., Mitchell, T.M.: Machine Learning: An Artificial Intelligence Approach. Springer, Heidelberg (2013)
16. Mohan, A.T., Gaitonde, D.V.: A deep learning based approach to reduced order modeling for turbulent flow control using LSTM neural networks. arXiv preprint arXiv:1804.09269 (2018)
17. Morton, K.W., Mayers, D.F.: Numerical Solution of Partial Differential Equations: An Introduction. Cambridge University Press, Cambridge (2005)
18. Oquab, M., Bottou, L., Laptev, I., Sivic, J.: Is object localization for free?-weakly-supervised learning with convolutional neural networks. In: Proceedings of the IEEE Conference on Computer Vision and Pattern Recognition, pp. 685–694 (2015)
19. Ronneberger, O., Fischer, P., Brox, T.: U-Net: convolutional networks for biomedical image segmentation. In: Navab, N., Hornegger, J., Wells, W.M., Frangi, A.F. (eds.) MICCAI 2015. LNCS, vol. 9351, pp. 234–241. Springer, Cham (2015). https://doi.org/10.1007/978-3-319-24574-4_28
20. Sarkar, S., Mahadevan, S., Meeussen, J., Van der Sloot, H., Kosson, D.: Numerical simulation of cementitious materials degradation under external sulfate attack. Cem. Concr. Compos. **32**(3), 241–252 (2010)
21. Sharma, R., Farimani, A.B., Gomes, J., Eastman, P., Pande, V.: Weakly-supervised deep learning of heat transport via physics informed loss. arXiv preprint arXiv:1807.11374 (2018)
22. Singh, C.: Student understanding of symmetry and Gauss's law of electricity. Am. J. Phys. **74**(10), 923–936 (2006)
23. Sirignano, J., Spiliopoulos, K.: DGM: a deep learning algorithm for solving partial differential equations. J. Comput. Phys. **375**, 1339–1364 (2018)
24. Sommer, R., Paxson, V.: Outside the closed world: on using machine learning for network intrusion detection. In: 2010 IEEE Symposium on Security and Privacy (SP), pp. 305–316. IEEE (2010)
25. Tajbakhsh, N., et al.: Convolutional neural networks for medical image analysis: full training or fine tuning? IEEE Trans. Med. Imaging **35**(5), 1299–1312 (2016)
26. Tang, L.H., Tian, G.S.: Reaction-diffusion-branching models of stock price fluctuations. Phys. A **264**(3–4), 543–550 (1999)
27. Tixier, R., Mobasher, B.: Modeling of damage in cement-based materials subjected to external sulfate attack. I: formulation. J. Mater. Civ. Eng. **15**(4), 305–313 (2003)
28. Weinan, E., Han, J., Jentzen, A.: Deep learning-based numerical methods for high-dimensional parabolic partial differential equations and backward stochastic differential equations. Commun. Math. Stat. **5**(4), 349–380 (2017)
29. Wu, J.L., Xiao, H., Paterson, E.: Physics-informed machine learning approach for augmenting turbulence models: a comprehensive framework. Phys. Rev. Fluids **3**(7), 074602 (2018)
30. Zhou, Z.H.: A brief introduction to weakly supervised learning. Natl. Sci. Rev. **5**(1), 44–53 (2017)
31. Zuo, X.B., Sun, W., Yu, C.: Numerical investigation on expansive volume strain in concrete subjected to sulfate attack. Constr. Build. Mater. **36**, 404–410 (2012)

A Parallel Hybrid Genetic Algorithm for Solving the Maximum Clique Problem

Mohammad Kazem Fallah, Vahid Salehi Keshvari, and Mahmood Fazlali[✉]

Department of Data and Computer Science, Faculty of Mathematical Sciences,
GC Shahid Beheshti University, Tehran, Iran
{mk_fallah,vs_keshvari,fazlali}@sbu.ac.ir
http://facultymembers.sbu.ac.ir/fazlali

Abstract. Finding maximum complete subgraph (maximum clique) from the input graph is an NP-Complete problem. When the graph size grows, the genetic algorithm is an appropriate candidate for solving this problem. However, due to the reduced computational complexity, the genetic algorithm can solve the problem, but it can still be time consuming to solve big problems. To tackle this weakness, we parallelize our hybrid genetic algorithm for solving the maximum clique problem. In this direction, we have parallelized producing, repairing and evaluation of chromosomes. Experimental results by using a set of benchmark instances from the DIMACS graphs indicate that the proposed meta heuristic algorithm in almost all cases, it finds optimal or near optimal answer. Also, the efficiency of the proposed parallelization is more than 3.44 times faster on an 8-core processor in comparison to the sequential implementation.

Keywords: Maximum clique problem · Genetic algorithm · Parallel algorithm

1 Introduction

Finding the maximum complete sub graph in a graph, called Maximum Clique Problem (MCP), has a wide range of usages in various fields, including social networks analysis [31,33], bio-informatics [20], coding theory [10], test planning [7], locating [9], scheduling [8] and signal transmission analysis [2]. Considering the importance and usefulness of the problem it seems that existence of an efficient method for solving the problem is really important and necessary.

MCP is an NP-complete problem [6,11,16]. Therefore, no algorithm with a polynomial time complexity has yet been found to solve it. Meta-heuristic approaches such as genetic algorithm have proposed solutions which can solve the problem in polynomial time. Therefore, up to now a large number of genetic algorithms have been provided for solving this problem [19,21,22,28,33,35,36].

In [36], the uniform crossover and inversion mutation as genetic algorithm operations and the repairing chromosome function have been used according to

© Springer Nature Switzerland AG 2019
L. Grandinetti et al. (Eds.): TopHPC 2019, CCIS 891, pp. 378–393, 2019.
https://doi.org/10.1007/978-3-030-33495-6_29

the vertex degrees. The repairing chromosome function has been used for converting chromosomes as a valid answer for the problem. This function consists of two main steps: extracting the clique and extending it to the maximal clique. In these two steps the vertex criteria for selecting the vertex, is the one with the greatest degree. The disadvantage of this type of selection is that it catches answers for some of local optimums. Based on an initial evaluation, the execution time can be reduced by using parallelization tools. However, issues such as the selection of parts to be run in parallel, the parallelization granularity, and the management of critical sections must be taken into account. The decision on these issues was parallelizing the proposed algorithm in the shared memory paradigm.

In this article we provide a parallel hybrid genetic algorithm for solving MCP that avoids local optimum with diversity in its selection. In this method the initial population is generated randomly. To convert the population of chromosomes into valid answers (each chromosome be a clique), a repair function is defined. This function consists of two main steps: extracting the clique and then developing it to a maximal clique. In these two steps, vertex selection criteria in each level unlike [36] is based on three parameters: randomize, vertex degree and core number. For the development of solutions we use random selection of the chromosome pairs for uniform cross over and reverse mutation operator. In the following, repairing and evaluation functions, are applied on child chromosomes. We use *Elitism* to transfer a number of fittest chromosomes to the next generation. In this way, the algorithm doesn't avoids some good answers that are obtained in the middle generations. In addition, we use parallelization libraries to reduce the execution time of the algorithm. In this method, production, evaluation and repairing operations are done sequentially for each chromosome, but these operations run in parallel for different chromosomes. In the same way, the cross over and the mutation operation on each chromosome pair and their children is done sequentially while these operations are performed in parallel on all chromosomes. Other operations such as random selection of parents and transferring children to the next generation are implemented in parallel whereas the measures taken to avoid repetitive selection of chromosomes at any stage of evolution.

The following sections of the article is organized as follows. Section 2 describes notations and previous related works. In Sect. 3 we describe the proposed parallel genetic algorithm which includes the problem encoding, the genetic algorithm operations and the parallelization strategies. Section 4 includes experimental results and related analysis. Finally, Sect. 5 is allocated to the conclusion.

2 Notations and Related Works

2.1 Notations

Definition 1. *Assume undirected graph $G = (V, E)$ consists of a set of vertices $V = \{v_1, v_2, \ldots, v_n\}$, and edges $E \subseteq V \times V$. A clique $C \subseteq V$ is a complete subgraph of G which means that $\forall v, v' \in C \Rightarrow \{v, v'\} \in E$.*

Definition 2. *Let $G = (V, E)$ is an undirected graph and $S \subseteq 2^V$ is the set of all cliques of G. Then, a clique $C \in S$ is maximal iff $\forall v \in V \Rightarrow \{v\} \cup C \notin S$.*

Definition 3. *Let $G = (V, E)$ is an undirected graph and $S \subseteq 2^V$ is the set of all maximal cliques of G. Then, a clique $C \in S$ is maximum if and only if $\forall C' \in S \Rightarrow |C'| \leq |C|$. We denote the size of maximum clique of G by $\omega(G)$.*

In other words, the number of vertices of a clique is called the size of the clique and each graph may contain a number of cliques with different sizes. The clique which cannot be expanded is called a maximal clique. This means by adding an additional vertex of the graph G to vertices of the maximal clique, it loses its feature of being a clique. The maximum clique is the largest maximal clique in graph G.

Example 1. In Fig. 1, graph G has 8 vertices and 11 edges. The set of vertices $\{v_1, v_2, v_4\}$ is one of the cliques in G. The set of vertices $\{v_4, v_5, v_7\}$ is a maximal clique and the set of vertices $\{v_1, v_2, v_4, v_5\}$ is the maximum clique. So the size of the maximum clique in this figure is 4. Let $\Gamma(v)$ denote the set of all adjacent vertices to $v \in V$. The size of $\Gamma(v)$ is the degree of v. The maximum degree of a graph G is denoted by $\Delta(G)$. As an example graph in Fig. 1, the degree of vertex v_1 is 3 and the maximum degree $\Delta(G)$ of the graph is 5.

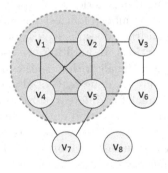

Fig. 1. Graph $G = (V, E)$ with the size of maximum clique 4.

Definition 4. *Let $G' = (V', E')$ be a subgraph of G where $V' \subseteq V$ and $E' \subseteq E$. The subgraph G' is k_CORE (or a core with size of k) if and only if for all $v \in V'$, the degree of vertex v (according to E') is greater than or equal to k. The greatest core number in graph G is denoted by $K(G)$.*

Definition 5. *The CORE number of each node is the largest K in the set of K_COREs containing that node.*

Proposition 1. *Let $G = (V, E)$ is an undirected graph. The order of time complexity for calculating CORE number of all nodes in V is $O(|E|)$ [3].*

Example 2. Figure 2 shows a graph G which contains cores with size of 0, 2 and 3. The core number of $V' = \{v_1, v_2, v_4, v_5\}$ is equal to 3 and $K(G) = 3$.

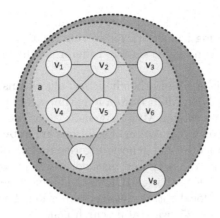

Fig. 2. Graph $G = (V, E)$ with $K(G) = 3$. The core size of each subgraph is (a) 3_CORE, (b) 2_CORE and (c) 0_CORE.

2.2   Related Works

Maximum Weight Clique (MWC) and Maximal Clique Enumeration (MCE) are NP-Complete problem like MCP [24,32]. Fazlali et al. [13] used a Branch and Bound (B&B) algorithm to solve MWC during efficient datapath merging in reconfigurable systems. B&B algorithms are a family of exact algorithms that facilitate searching large spaces by pruning a large part of them. Also, Li et al. [18] defined an incremental upper bound as well as some preprocessing techniques such as graph coloring and MaxSAT reasoning to reduce the search space for solving MCE. Although, B&B techniques tend to prune a large percentage of search space, they do not reduce the order of complexity and may not be efficient when the size of the input graph is large. An $O(|E|)$ time algorithm for MCE in large sparse undirected graphs is presented by [34]. To this end, they used a data structure called candidate map which holds maximal cliques as well as some non-maximal cliques. So, the candidate map is not appropriate for MCE in dense graphs. Lessley et al. [17] have presented a hashing-based MCE algorithm and performed data parallelism on shared memory, multicore architectures to achieve high performance in computations. Employing bit-parallel encoding to perform MCE for small and middle size graphs is presented in [26]. Also, they used some greedy strategies such as sorting vertices according to their degrees (in ascending order) to increase the speed of the algorithm. Their work was based on [29] and [23] that use B&B strategy and pivot selection method for MCE.

One of the methods for increasing the scalability of algorithms is the use of parallel processing techniques. Fazlali et al. [12] performed task parallelization to solve MWC. They used OpenMP library to perform divide and conquer task parallelism and accelerated solving MWC on multicore systems. San et al. [27] have splitted their heavy pre-processing steps on a 20-core system that enabled them to find the maximum clique in sparse graphs with more than three million vertices. Their algorithm was exact and based on the B&B strategy. Guo et al.

[15] have used Hadoop as distributed cloud computing platform and presented a parallel graph partitioning framework for solving MCP. They proposed a graph partition method based on degree sorting to improve the load balancing during the parallel processing.

Evolutionary algorithms and other heuristic algorithms [30] have presented different methods for solving MCP from a long time ago. One of the most popular provided methods for solving MCP is genetic algorithm. Unlike local search algorithms in this method a population of solutions are improved by evolutionary process. The first report on using genetic algorithms for solving MCP was presented in 1990. Following the proposed method, it was proved that pure genetic algorithm is not efficient for solving MCP [25]. Therefore genetic methods often use other techniques like local optimization combination. The first algorithm that was proposed after proof [25] was tabu search genetic algorithm. In this algorithm standard uniform crossover operator is used with the tabu search technique [14]. Generally, in recent years many algorithms were presented for solving MCP based on the genetic algorithm. The most important ones are HGA [21], GENE [22], EA/G [35], HSSGA [28], FGA [36] and [33]. In the following we will briefly refer to each one. HGA [21] algorithm is a simple heuristic algorithm based on genetics that was proposed by Marchiori. Following the algorithm MARCHIORI proposed GENE [22], where she used hybrid genetic with uniform crossover operator, inversion mutation and local search techniques. Hybrid evolutionary algorithm EA/G [35] was also proposed which used guided mutation operator And a combination of global statistical information and local information based on the results that has been obtained so far. For generating children the algorithm HSSGA [28] provided a combination of genetic algorithm with sequential random greedy approach. Harsh Bhasin in 2012 also provided a genetic algorithm for solving MCP in which he used single point crossover and single point mutation operators [5]. In 2013 Harsh Bhasin improved fitness function provided on his own method and used the roulette wheel as selection operator [4]. But he has not shown any results from his proposed algorithm.

FGA [36] presented in 2014 uses uniform crossover operator and inversion mutation. This algorithm consists of two main steps that the first step is to extract the complete subgraph and the next step is the development of the subgraph to maximal clique. In this algorithm the criteria for selecting vertices is greedy according to the vertex degree. In the article FGA [36] results have been compared with other algorithms in terms of time and accuracy. This article will focus on the latest proposed method FGA [36] which tries to improve this method in terms of time and accuracy of results.

3   The Proposed Parallel Genetic Algorithm for Solving MCP

In this section we present parallel genetic algorithm to attack MCP. The algorithm begins by generating a random population. For each member of population, repair operation is carried out. The merits of each member calculated by the fitness function which will be discussed later. The algorithm will improve the answers. For this purpose, paired chromosomes are randomly selected from the population. Next, the uniform crossover applies to pairs of chromosomes and the output is two children, therefore a four-member family is generated. Afterward, the inversion mutation operator is applied on the two children. Then the children of this family are sent to the repair function to become an answer for the problem. As the population size does not change and only members will change, two of the most qualified members in each family will be replaced with two other family members in the population. In this way, the population converges to the better population. This operation is done 100 times, in the end the chromosome with the most merit will be considered as the largest subgraph.

3.1   Encoding

Here, a chromosome is a subgraph $G' = (V', E')$ of the input graph $G = (E, V)$ where subgraphs are obtained by creating permutations on V, and having a subset $V' \subseteq V$, its corresponding subset $E' \subseteq E$ contains all the edges between the vertices V' in Set E. So, there is no need to store edges in chromosomes and we use binary coding. For this purpose, we assign a boolean array X of length n for each chromosome. The vertices of G are tagged by $v_1, v_2, \ldots, v_n$. If a vertex v_i exists in the chromosome, then $X[i] = 1$, otherwise $X[i] = 0$.

Example 3. All elements of the chromosome corresponding to the main graph are 1. Figure 3 shows two subgraphs and their binary encodings.

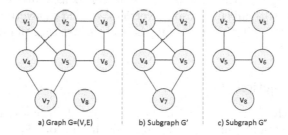

a) Graph G=(V,E)   b) Subgraph G'   c) Subgraph G"

Fig. 3. (a) The input graph $G = (V, E)$, (b) The subgraph corresponding to chromosome $X' = [11011010]$, (c) The subgraph corresponding to chromosome $X'' = [01101101]$.

3.2   Initial Population

As the first step, it is necessary to produce a set of chromosomes as the initial population. Members of this population have been produced randomly and its size has been obtained experimentally. The members of the initial population are generated randomly. In fact, after generating initial population, or after mutation or crossover which will be discussed in the following sections, there is no guarantee that each chromosome will be a valid answer for the problem (a maximal clique). So, it is required that when a chromosome is produced or altered, we repair it in order to transform it to maximal clique. We perform the repairing operation in two steps:

1. *Extracting clique from chromosome*
   Extracting clique from chromosomes is based on a greedy strategy. It should be considered that the order in which the vertices are processed affects the final results. So, the arrangement of vertices is very important. For extracting clique from a chromosome, we start with an initial vertex. Selecting the vertex with the maximum value of Γ, speeds up the convergence but, selecting a random initial vertex prevents the answer from getting trap in local optimality while the probability of the desired accuracy in a few numbers of execution is low. Our suggestion for avoiding both late convergence an local optimality is the combination of these two selecting methods with the core number. Therefor, we have three options for selecting this vertex: randomize, vertex degree or core number. We choose one of them by probability in each step. The next step is to randomly choose one of two following options for the selected vertex: removing the selected vertex from the chromosome or removing vertices that are not adjacent to this vertex. We start this method for each vertex and repeat it to choose other vertices.
2. *Developing the extracted clique to the maximal clique*
   Now that the chromosome has become a clique we try to expand it to a maximal clique. To do this we add other vertices to the clique in a way that it still stay a clique. So, at first we select the vertex as above and if this vertex was associated with all chromosome vertices, we add this vertex to the clique. We test all vertices in this way so in the end the existing clique converts to a maximal clique.

3.3   Fitness Function

In genetic algorithm, fitness function shows merits of chromosomes. As we are about to find the maximum clique we consider the size of clique represented by the chromosome as its criterion of merits. Since each chromosome represents a clique, the value of the fitness function is

$$F(X) = \sum_{i=1}^{n} X[i].$$

3.4   Uniform Crossover

In uniform crossover, two parents are copied into two children. Then, a random string of digits zero and one in length of the chromosome is created. Next, for each digit of the generated random string if it is one, there will be no change for its corresponding digits in children, but if it is zero, the values of the corresponding digits in children must be changed together. The probability of crossover is pc, as well.

Example 4. A sample uniform crossover is presented in Fig. 4. As can be seen, the values of zero at random string cause the corresponding values in the child chromosomes to be displaced together.

parent 1	1 0 0 1 0 1 0 1
parent 2	1 1 1 1 0 0 1 0
random string	1 0 1 0 1 1 0 0
child 1	1 1 0 1 0 1 1 0
child 2	1 0 1 1 0 0 0 1

Fig. 4. Generating two children from two parents by uniform crossover.

3.5 Inversion Mutation

After producing a chromosome by uniform crossover, it enters to the mutation cycle with probability pm. In inversion mutation, at first, two numbers between one and the length of chromosome is generated randomly. We also specify the maximum distance between the intervals. Then, we reverse the values corresponding to the generated interval. It means that one changes to zero and vice versa.

Example 5. Figure 5 shows a sample of inversion mutation, where the interval $[2, 4]$ is selected randomly.

before mutation	1 1 0 1 1 0 0 0
after mutation	1 0 1 0 1 0 0 0

Fig. 5. A sample of inversion mutation.

As mentioned before, the chromosomes produced by the uniform crossover and the inversion mutation operations may not be valid (maximal clique) and will be modified by the repairing operation.

3.6   Selection Strategy and Elitism

Selection must favor fitter candidate chromosomes over weaker candidates but beyond that there are no fixed rules. We use *roulette wheel selection* which is one of the most common selection technique, where the chances of selecting each chromosome as a parent has a direct relation with the value of its fitness.

Elitism is the simplest selection strategy that retains the fittest x% of the population. These fittest chromosomes are duplicated so that the population size is maintained. The elitism ensures that the quality of some best chromosomes does not come down. In other words, we pass on a certain percentage of the best chromosomes of each generation without change to the next generation, in order to finally get the best chromosome from all generations. The elitism is an easy selection strategy to implement but it can result in premature convergence as less fit chromosomes are ruthlessly culled without being given the opportunity to evolve into something better. Therefore, we only transfer 2 of the best chromosomes to the next generation.

3.7   Pseudo-code

Algorithm 1 is the pseudo code for the algorithm. The number of threads is determined to implement the program at the beginning of this pseudo-code. We made the number of threads equal to the number of CPU cores.

Block 1 shows commands related to parallel execution of generating initial population randomly. Each thread produces a chromosome, with regards to the N_t thread and as the dependency does not exist at this stage, production occurs $\frac{N_p}{N_t}$ times. N_p is the size of GA population. While in the serial execution, this operation is carried out N_p times.

In block 2 commands related to parallel execution of repairing each member of these populations and their evaluation is shown. Due to the association between repair and evaluation for each chromosome, this operation is performed in serial for each chromosome, but takes place in parallel for all chromosomes.

The elitism is done in block 3 by selecting two of the best chromosomes of the current generation and transferring them to the next generation.

In block 4 the random selection, the crossover and the inversion mutation operations are performed to produce new generation. These operations are performed in serial for selecting a pair of parents and producing a pair of children, but they are performed in parallel for all chromosomes. Also, there are some measures adopted in this implementation to prevent duplicate selection of chromosomes at every stage. To do so we use an array that is filled with 1 to n randomly and non-repetitively.

Block 5 involves moving to new generation which includes replacing the old population with the new population.

Algorithm 1. The proposed parallel GA.

Input: Undirected graph G, population size N_p, number of threads N_t

Output: Clique C as the maximum clique of G

0    $N_e \longleftarrow 2$ $pc \longleftarrow 0.9$ $pm \longleftarrow 0.01$

1   omp_set_num_threads(N_t);

2

3   // block 1 (generating initial population)

4   **#pragma omp parallel for shared** (C) **private** (X, i)

5   for $i = 1$ to N_p

6   Generate an individual X randomly

7        $C[i] \longleftarrow X$

8

9   for $iter = 1$ to 1000

10   // block 2

11   **#pragma omp parallel for shared** (C, F) **private** (i)

12   for $i = 1$ to N_p

13   Apply repairing method on $C[i]$ // modifying the individual

14            $F[i] \longleftarrow fitness(C[i])$ // calculating the fitness

15

16   // block 3 (elitism)

17        $(C'[1], \ldots, C'[N_e]) \longleftarrow$ the N_e best individuals of C

18

19   // block 4 (generating new population)

20   **#pragma omp parallel for shared** (C, C', F) **private** (i, p, c)

21   for $i = N_e$ to N_p steps 2

22            $(p_1, p_2) \longleftarrow select(C, F, 2)$

23            $(c_1, c_2) \longleftarrow uniform\_crossover(p_1, p_2, pc)$

24            $C'[i] \longleftarrow inversion\_mutation(c_1, pm)$

25            $C'[i+1] \longleftarrow inversion\_mutation(c_2, pm)$

26

27   // block 5 (replacing new population)

28   **#pragma omp parallel for shared** (C, F) **private** (i)

29   for $i = 1$ to N_p

30            $C[i] \longleftarrow C'[i]$

---

4   Experimental Results

In this section we introduce the tools and parameter settings for experiments. Then, we analyze the results according to result table and acceleration and performance diagrams.

4.1   Tools of Implementation and Parameters Settings

We implemented the proposed algorithm in C++ programming language and used Open MP for parallelism. Also, we employed common DIMACS maximum clique benchmark graphs [1] for evaluation. The algorithm is executed on a system with a *cori*7 processor, 4 GB RAM and Windows 7 operating system.

According to the experience that we achieved from setting parameters on the execution of the algorithm we conclude to produce a population with size of 100. Also, the evolution of chromosomes we have considered the possibility of crossover to 90%, and put mutation probability 1%. We put the termination condition to 100 iterations and due to the possibility of all parameters, for each sample we executed the algorithm 10 times and reported the average outputs.

4.2   Results and Discussion

Table 1 contains the results that the proposed algorithm's solution (the column K) is the same as the best solution ever found (the column ω) and Table 2 contains the results that the proposed algorithm's solution are near optimal. These two tables contain the same columns. We first describe these columns and then analyze the results. In the first column, the names of the graphs are arranged in alphabetical order, and the next three columns contain graph specs. The second and third columns contain the number of vertices ($|V|$) and the number of edges ($|E|$) of the graph, respectively. These two properties indicate the size of the problem and as they grow, the time needed to solve the problem increases. The fourth column specifies the density/sparsity of each graph where, if the value of this column is close to one, the corresponding graph is dense and if it is close to zero, then the graph is sparse. The last two columns represent the average execution time of the algorithm in seconds, so that T_1 corresponds to the execution time of the algorithm in sequential mode and T_8 corresponds to the execution time of the algorithm with 8 threads.

For example, the third sample in Table 1 is $c125.9$. It contains 125 vertices and 6963 edges which the number of edges in its complete graph is 7750 and its density is $\frac{6963}{7750} \simeq 0.898$. The best size of clique that has been achieved so far for this sample is 34 that is equal to the size of the maximum clique found by the proposed algorithm. Also, the sequential time obtained for this sample is 325 ms while its parallel execution time by 8 threads is 80 ms. The obtained speedup is $\frac{325}{80} = 4.0625$.

The results shown in Table 1 verify that the speedup for parallel execution time is more than 3.44 for all samples. The lowest speedup is for the second sample, $brock200\_4$ and the highest speedup is for the seventh sample,

$gen400\_p0.9\_75$, which are $\frac{0.413}{0.120} \geq 3.44$ and $\frac{1.641}{0.396} \geq 4.14$ respectively. However, in most cases, the speedup is about 4 which means that the parallel implementation of the algorithm (with 8 threads) is roughly four times faster than its serial implementation.

Table 1. The results of sequential and parallel implementation of the proposed algorithm on DIMACS benchmarks (in case $K = \omega$).

G	$|V|$	$|E|$	$Density$	ω	K	$T_1(s)$	$T_8(s)$
$brock200\_2$	200	9876	0.496	12	12	0.405	0.098
$brock200\_4$	200	13089	0.657	17	17	0.413	0.120
$C125.9$	125	6963	0.898	34	34	0.325	0.080
$DSJC1000\_5$	1000	499652	0.500	15	14	6.235	1.630
$DSJC500\_5$	500	125248	0.501	13	13	1.739	0.471
$gen200\_p0.9\_55$	200	17910	0.900	55	55	0.591	0.151
$gen400\_p0.9\_75$	400	71820	0.900	75	75	1.641	0.396
$hamming8 - 4$	256	20864	0.639	16	16	0.490	0.122
$keller4$	171	9435	0.649	11	11	0.309	0.083
$keller5$	776	225990	0.751	27	27	4.040	1.035
$p\_hat300 - 1$	300	10933	0.243	8	8	0.725	0.190
$p\_hat300 - 2$	300	21928	0.488	25	25	0.795	0.216
$p\_hat300 - 3$	300	33390	0.744	36	36	0.902	0.240
$p\_hat700 - 1$	700	60999	0.249	11	11	3.121	0.834
$p\_hat700 - 2$	700	121728	0.497	44	44	3.108	0.765
$p\_hat1500 - 2$	1500	568960	0.506	65	65	11.89	2.934

Comparison of columns ω and K of Table 2 shows that the maximum cliques found by proposed algorithm for the remaining 7 samples are near optimal. The result verify that the speedup for parallel execution time is more than 3.7 for these cases. The lowest speedup is for the first sample, $C4000.5$ and the highest speedup is for the seventh sample, $p\_hat1500 - 3$, which are $\frac{85.47}{22.879} \geq 3.73$ and $\frac{14.131}{3.315} \geq 4.26$ respectively. However, in almost all cases, the speedup is close to 4.

Figure 6 shows that by increasing the number of threads (with respect to the number of processing cores) the speedup grows. Also, different speedups for different samples are due to the different load balancing of threads, the source of which is the structure of the input graph.

Sample $kaller4$ in Fig. 6 shows that in some cases, using more threads may not result in maximum acceleration. In this case, the average of sequential execution time is 309 ms and the average of parallel execution time is 172, 128, 78 and

Table 2. The results of sequential and parallel implementation of the proposed algorithm on DIMACS benchmarks (in case $K \neq \omega$).

G	$|V|$	$|E|$	$Density$	ω	K	$T_1(s)$	$T_8(s)$
$C4000.5$	4000	4000268	0.500	18	17	85.47	22.879
$DSJC1000\_5$	1000	499652	0.500	15	14	6.235	1.630
$hamming10 - 4$	1024	434176	0.828	40	38	5.751	1.440
$keller6$	3361	4619898	0.818	59	52	64.168	16.185
$p\_hat700 - 3$	700	183010	0.748	62	61	3.550	0.875
$p\_hat1500 - 1$	1500	284923	0.253	12	11	13.521	3.550
$p\_hat1500 - 3$	1500	847244	0.753	94	91	14.131	3.315

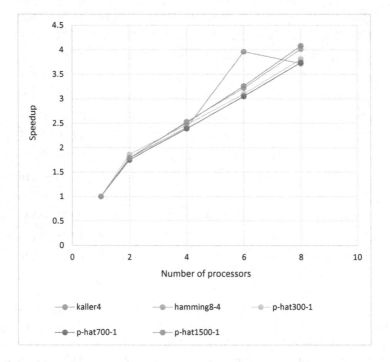

Fig. 6. Speedup of parallel versions using different thread numbers on a 8-core processor.

83 ms using two, four, six and eight threads where, the corresponding speedups are 1.49, 2.41, 3.96 and 3.72, respectively. Here, the parallel version of the algorithm with six threads (on a 8-core processor) leads to the most speedup. This is due to the trade-off between the acceleration of parallelism and related costs. The results show that when the size of the problem becomes larger, the use of more threads in parallelization results in more acceleration, and if the size of the

problem is not very large, the optimal number of threads may be less than the number of CPU cores.

The overhead of synchronization and communication of parallel threads affects the performance of CPU cores. Note that when the ideal usage of processor is 100%, it means that by adding one processor, run time should be reduced to half of the sequential time.

We calculate the performance of parallel processing under ideal conditions with different number of threads and show five samples in Fig. 7. For example the sequential run time is 100 s, the most ideal to run with two processors occurs if the run time is 50 s. For example, the performance of the processors in sample $hamming8 - 4$ in parallel execution with two, four, six and eight processors are 90%, 60%, 66% and 47%, respectively.

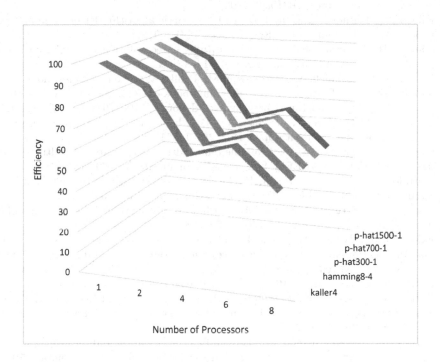

Fig. 7. The efficiency obtained using different number of CPU cores.

5 Conclusion

In this research we presented a parallel genetic algorithm for finding the maximum clique on multicore platforms. The main part of the proposed method is using task parallelization in genetic algorithm and we have employed OpenMP library to this end. In order to accelerate the search and prevent it from getting trapped in local optimums, a chromosome repairing technique is developed that uses core number, node degree and random selection in its greedy strategy. The proposed algorithm has been evaluated using DIMACS benchmarks and results indicates the efficiency of the parallelization.

References

1. DIMACS benchmark graphs. http://iridia.ulb.ac.be/~fmascia/maximum_clique/ DIMACS-benchmark. Accessed 03 Mar 2018
2. Balasundaram, B., Butenko, S., Hicks, I.V.: Clique relaxations in social network analysis: the max k-plex problem. Oper. Res. **59**(1), 133–142 (2011)
3. Batagelj, V., Zaversnik, M.: An O(m) algorithm for cores decomposition of networks. arXiv preprint cs/0310049 (2003)
4. Bhasin, H., Kumar, N., Munjal, D.: Hybrid genetic algorithm for maximum clique problem. Int. J. Appl. Innov. Eng. Manag. **2**(4) (2013)
5. Bhasin, H., Mahajan, R.: Genetic algorithms based solution to maximum clique problem. Int. J. Comput. Sci. Eng. **4**(8), 1443 (2012)
6. Bomze, I.M., Budinich, M., Pardalos, P.M., Pelillo, M.: The maximum clique problem. In: Du, D.Z., Pardalos, P.M. (eds.) Handbook of Combinatorial Optimization, pp. 1–74. Springer, Boston (1999). https://doi.org/10.1007/978-1-4757-3023-4_1
7. Brotcorne, L., Laporte, G., Semet, F.: Fast heuristics for large scale covering-location problems. Comput. Oper. Res. **29**(6), 651–665 (2002)
8. Chen, F., Zhai, H., Fang, Y.: Available bandwidth in multirate and multihop wireless ad hoc networks. IEEE J. Sel. Areas Commun. **28**(3), 299–307 (2010)
9. Dorndorf, U., Jaehn, F., Pesch, E.: Modelling robust flight-gate scheduling as a clique partitioning problem. Transp. Sci. **42**(3), 292–301 (2008)
10. Etzion, T., Ostergard, P.R.: Greedy and heuristic algorithms for codes and colorings. IEEE Trans. Inf. Theory **44**(1), 382–388 (1998)
11. Fakhfakh, F., Tounsi, M., Mosbah, M., Hadj Kacem, A.: Algorithms for finding maximal and maximum cliques: a survey. In: Abraham, A., Muhuri, P.K., Muda, A.K., Gandhi, N. (eds.) ISDA 2017. AISC, vol. 736, pp. 745–754. Springer, Cham (2018). https://doi.org/10.1007/978-3-319-76348-4_72
12. Fazlali, M., Fallah, M.K., Hosseinpour, N., Katanforoush, A.: Accelerating datapath merging by task parallelisation on multicore systems. Int. J. Parallel Emergent Distrib. Syst. **34**, 1–14 (2019)
13. Fazlali, M., Zakerolhosseini, A., Gaydadjiev, G.: Efficient datapath merging for the overhead reduction of run-time reconfigurable systems. J. Supercomput. **59**(2), 636–657 (2012)
14. Fleurent, C., Ferland, J.A.: Genetic and hybrid algorithms for graph coloring. Ann. Oper. Res. **63**(3), 437–461 (1996)
15. Guo, J., Zhang, S., Gao, X., Liu, X.: Parallel graph partitioning framework for solving the maximum clique problem using Hadoop. In: 2017 IEEE 2nd International Conference on Big Data Analysis (ICBDA), pp. 186–192. IEEE (2017)
16. Johnson, D.: Computers and Intractability-A Guide to the Theory of NP-Completeness. W. H. Freeman and Company, New York (1979)
17. Lessley, B., Perciano, T., Mathai, M., Childs, H., Bethel, E.W.: Maximal clique enumeration with data-parallel primitives. In: 2017 IEEE 7th Symposium on Large Data Analysis and Visualization (LDAV), pp. 16–25. IEEE (2017)

18. Li, C.M., Fang, Z., Jiang, H., Xu, K.: Incremental upper bound for the maximum clique problem. INFORMS J. Comput. **30**(1), 137–153 (2017)
19. Li, L., Zhang, K., Yang, S., He, J.: Parallel hybrid genetic algorithm for maximum clique problem on OpenCL. J. Comput. Theoret. Nanosci. **13**(6), 3595–3600 (2016)
20. Malod-Dognin, N., Andonov, R., Yanev, N.: Maximum cliques in protein structure comparison. In: Festa, P. (ed.) SEA 2010. LNCS, vol. 6049, pp. 106–117. Springer, Heidelberg (2010). https://doi.org/10.1007/978-3-642-13193-6_10
21. Marchiori, E.: A simple heuristic based genetic algorithm for the maximum clique problem. In: Symposium on Applied Computing: Proceedings of the 1998 ACM symposium on Applied Computing, vol. 27, pp. 366–373. Citeseer (1998)
22. Marchiori, E.: Genetic, iterated and multistart local search for the maximum clique problem. In: Cagnoni, S., Gottlieb, J., Hart, E., Middendorf, M., Raidl, G.R. (eds.) EvoWorkshops 2002. LNCS, vol. 2279, pp. 112–121. Springer, Heidelberg (2002). https://doi.org/10.1007/3-540-46004-7_12
23. Naudé, K.A.: Refined pivot selection for maximal clique enumeration in graphs. Theoret. Comput. Sci. **613**, 28–37 (2016)
24. Östergård, P.R.: A new algorithm for the maximum-weight clique problem. Nord. J. Comput. **8**(4), 424–436 (2001)
25. Park, K., Carter, B.: On the effectiveness of genetic search in combinatorial optimization. In: Proceedings of the 1995 ACM Symposium on Applied Computing, pp. 329–336. ACM (1995)
26. San Segundo, P., Artieda, J., Strash, D.: Efficiently enumerating all maximal cliques with bit-parallelism. Comput. Oper. Res. **92**, 37–46 (2018)
27. San Segundo, P., Lopez, A., Artieda, J., Pardalos, P.M.: A parallel maximum clique algorithm for large and massive sparse graphs. Optim. Lett. **11**(2), 343–358 (2017)
28. Singh, A., Gupta, A.K.: A hybrid heuristic for the maximum clique problem. J. Heuristics **12**(1–2), 5–22 (2006)
29. Tomita, F., Tanaka, A., Takahashi, H.: The worst-case time complexity for generating all maximal cliques and computational experiments. Theoret. Comput. Sci. **363**(1), 28–42 (2006)
30. Tomita, E., Yoshida, K., Hatta, T., Nagao, A., Ito, H., Wakatsuki, M.: A much faster branch-and-bound algorithm for finding a maximum clique. In: Zhu, D., Bereg, S. (eds.) FAW 2016. LNCS, vol. 9711, pp. 215–226. Springer, Cham (2016). https://doi.org/10.1007/978-3-319-39817-4_21
31. Wang, H., Alidaee, B., Glover, F., Kochenberger, G.: Solving group technology problems via clique partitioning. Int. J. Flex. Manuf. Syst. **18**(2), 77–97 (2006)
32. Wang, Z., et al.: Parallelizing maximal clique and k-plex enumeration over graph data. J. Parallel Distrib. Comput. **106**, 79–91 (2017)
33. Wen, X., et al.: A maximal clique based multiobjective evolutionary algorithm for overlapping community detection. IEEE Trans. Evol. Comput. **21**(3), 363–377 (2017)
34. Yu, T., Liu, M.: A linear time algorithm for maximal clique enumeration in large sparse graphs. Inf. Process. Lett. **125**, 35–40 (2017)
35. Zhang, Q., Sun, J., Tsang, E.: An evolutionary algorithm with guided mutation for the maximum clique problem. IEEE Trans. Evol. Comput. **9**(2), 192–200 (2005)
36. Zhang, S., Wang, J., Wu, Q., Zhan, J.: A fast genetic algorithm for solving the maximum clique problem. In: 2014 10th International Conference on Natural Computation (ICNC), pp. 764–768. IEEE (2014)

Using RNN to Predict Customer Behavior in High Volume Transactional Data

Hamed Mirashk[1][✉], Amir Albadvi[1], Mehrdad Kargari[1],
Mostafa Javide[2], Abdollah Eshghi[1], and Ghazaleh Shahidi[1]

[1] School of Engineering, Industrial Engineering Department,
Tarbiat Modares University, Tehran, Iran
hmirashk@modares.ac.ir
[2] Iran University of Science and Technology (IUST), Tehran, Iran

Abstract. Big data tools and techniques introduce new approaches based on distributed computing methods. When dealing with large data, one of these state-of-art approaches for analysing and predicting in the shortest possible time is the use of deep learning networks that provide real-time, accurate, and comprehensive analysis. This method has provided a new perspective to artificial intelligence with respect to increasing volume of data and complexity of real-world issues. The models used to predict customer behavior have mainly worked with limited features and dimensions. One of the applications of this method is to prevent customer churn, when predicting future behavior of customer transaction on point of sale (POS) devices.

In this paper, transactional data are analyzed using Recurrent Neural Networks (RNNs) that are capable of predicting high-dimensional data or time-series data, by which a model for predicting the behavior of POS holders are proposed. Hence, we used transaction data on POS devices to predict their future behavior using RNN algorithm. In order to validate the generated model, the real data of one of the private banks inside Iran has been used and the predicted results are about 87% accurate. The result is compared with previous methods which outperform other methods discussed later.

Keywords: Big data · Big data analysis · Deep learning · Deep neural network · Machine learning · Predictive analysis · Neural network

1 Introduction

With the improvement of artificial intelligence, new methods have been developed to empower computers to simulate human intelligence. One of these new methods is deep learning networks, which is a developed model of artificial neural networks. The history of deep networks can be traced back to 1943, when Warren McCulch et al. created a computer model based on neural networks of the human brain. They used a combination of algorithms and mathematics, which they called "threshold logic" to imitate the process of thinking (McCuluch and Pits 1943). Since then, deep learning has evolved steadily. The working of deep learning is quite similar to the working of animal neocortex (Yamins and DiCarlo 2016)

© Springer Nature Switzerland AG 2019
L. Grandinetti et al. (Eds.): TopHPC 2019, CCIS 891, pp. 394–405, 2019.
https://doi.org/10.1007/978-3-030-33495-6_30

In 1949, Donald Hebb corrected the binding strength of neurons of the neural network model with a simple rule, which is known today as the Hebbian rule (Shaw 1986). Eventually, in 1951, Marvin Minsky et al. succeeded in making the first computer in which the neural network was used (Saithibvongsa and Yu 2018). In 1956, a workshop hosted by IBM at Dartmouth University, in which McCarthy introduced the Artificial Intelligence. This introduction, which was the official birth of artificial intelligence, marked an important chapter in the evolution and development of this science (Mccarthy 1956). In 1965, Ladderberg developed the first knowledge-based expert system called DENDRAL at Stanford University (Lederberg 1987). The first unmanned car, a Mercedes-Benz van, equipped with cameras and sensors, was built at Bundeswehr University in Munich (Dickmanns and Zapp 1987). In 1988, Rollo Carpenter created a chat bot called Jabberwacky, which could talk to humans (Fryer and Carpenter 2006).

The next significant improvement of deep learning occurred when computers started processing information faster and the graphics-processing unit (GPU) was developed. Subsequently, neural networks began to compete with support vector machines (SVM). While the processing speed of the neural network was less than SVM, neural network provided better results (Supriya Pahwa 2016).

There have been many ideas about using deep learning in the past decades. But today, taking advantages of deep learning is highly welcomed due to some important reasons such as availability of data and computational scale. Today, people spend most of their time on using digital devices (laptops and mobile phones). These activities have produced a lot of data that can be used as a feed for the learning algorithm. It has been also only a few years now that computing power is available for learning a multi-layer neural network, which works with big data.

Figure 1 illustrates the performance of different learning algorithms in cases where data volume is increasing. In fact, if we move from old models to small neural networks, we will see better performance and, as the volume of our data is increasing, the use of old methods is discarded. Medium networks have the ability to process more data. But, when the volume of data increases dramatically, the volume of data cannot yet estimated and should be analyzed in larger dimensions. In this case deep multi-layer network is used, which has better performance compared with other methods (Ng 2018).

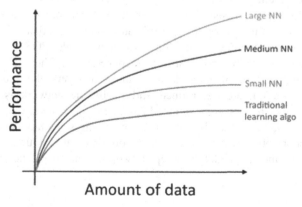

Fig. 1. Comparison of different machine learning algorithm performance (Ng 2018)

Deep learning has usage in various domains. One of the applications of deep networks is to predict customer behavior. Armando has done a research in the field of predicting customer behavior on online sales websites using deep learning (Armando Vieira Redzebra Analytics 2016). In this study, the final accuracy rate was upgraded from about 76% using random forest and logistic regression algorithms to about 89% using the deep learning algorithm.

In other aspects, Customer Relationship Management (CRM) is a comprehensive strategy for building, managing and strengthening loyal and long-lasting customer relationships. It has many applications in different fields such as telecommunications, banking, insurance, etc. Prediction of customer behaviour has a direct effect on customer churn prediction and prevention. This field of study in literature review is defined as "customer retention". The importance of this field is obvious because the cost of customer retention is greater than the cost of customer acquisition. Therefore, predicting customer status is a very critical issue in two aspects. One is to identify market growth and to plan for the future and the other one is to prevent customer churn, when significant changes have been identified in the customer's behavior.

Customer retention has many applications in different fields; one of the important eras of applications is bank industry to prevent customer migration or churn. Hence, prediction of customer behavior in bank area has been the subject of some recent researches. In one of the studies in this area, the importance of forecasting the future of banks status has been discussed. In the mentioned study, using the new approach of deep recursive networks, customer behavior is predicted based on their transactions to be used in fraud detection systems as a way to prevent customer churn (Zheng et al. 2013).

The rest of the paper is organized as follows: In Sect. 2, the methodology is described in detail in four parts. Next, in Sect. 3, the findings and results are represented. Section 4 concludes the paper and also limitations and future works are discussed in this section.

2 Methodology

In this section, different parts of the methodology are described. As previously mentioned, the purpose of this research is to predict the behaviour of bank customers in order to prevent future threats or risks. For this purpose, the real data of an Iranian bank have been used, which includes about 2.9 million point of sale (POS) transactions. This volume of data is acceptable for learning deep networks. Then the deep network structure including the type of network, the number of layers, the optimization algorithm, the Loss function are determined and finally the network is trained. After training the network, results are evaluated based on the test data and the network performance is validated. Generally, the problem-solving steps include pre-processing data (data collection, preparation), feature selection, removal of extra items, selecting a deep network type and its modeling, deep network learning, and network prediction using test data.

2.1 Preprocessing Data

In the deep learning concept, the method of collecting input data is so important. As mentioned above, in this work, the number of input data is about 2.9 million transactions, which are collected from about 35,847 POS devices. The used storage database is SQL-Server, and the data is linked from this database to the intended code using the JDBC connection. The general columns of the mentioned transactions are shown in Table 1.

Table 1. Columns associated with POS transactions

Column name	Type	Explanation
STerminalNo	int	The POS No.
BusinessWeekDate	int	Date of transaction occurrence
TERM_MODEL_CODE	int	Type of device code
Term_Grp_Code_LVL1	nvarchar(255)	Device model
TMA_Agency_Code	int	Agency code
Agency_State_Code	int	Code of agency's city
BAN_BANK_CODE	int	Bank code
BankZone	nvarchar(255)	Bank zone
BankZoneBranch	nvarchar(255)	Branch zone
Employer	int	Employment code
StuffCode	int	Personnel's code
TransactionCount	float	Number of transactions
TransactionAmount	float	Amount of transaction
TransactionCommission	float	Commission of transaction
ClusteringCode	int	Clustering code
COCU_ID	nvarchar(30)	Customer id
Contract_ID	nvarchar(30)	Contract id
Term_Grp_Code	int	Code of group of ATM
TCountRate	numeric(18, 2)	Rate of number of transactions
TAmountRate	numeric(18, 2)	Rate of transaction amount
Terminal_MAC	numeric(18, 2)	Moving average of benefit
Terminal_MAA	numeric(18, 2)	Moving average of cost
TargetTCountRate	int	Rate of number of target transactions
TargetTAmountRate	int	Rate of amount of target transactions
EffectiveEventCount	int	Number of effective events

2.2 Feature Selection and Removing Extra Items

In this step, the list of required features, which are inputs for deep network, is determined, and features that do not affect the prediction and learning of the network are eliminated. Then the data is retrieved from the database, and it is sorted based on the POS No. or terminal No. as well as the transaction occurrence time. Afterwards the additional fields such as the terminal No., the type of device code, the contract id, and the customer id are deleted from the data. The code is written in Python using the Pumpy, Pandads, Pyodbc, and Keras library.

In the next step, some categorical or non-numeric fields are converted to non-categorical ones so that they can be used in deep networks. For example, labelled or categorized data of business guild code should be defined as separate columns in Python data frame. Therefore, if a transaction is related to specific guild, in the corresponding column, number one will be written and in the other columns number zero will be displayed. An example is given in Table 2.

Table 2. Transforming categorical data to non-categorical data by column generation

Transaction	Guild 33	Guild 34	Guild 35	Guild 36	Guild 37	Guild 38	Guild 39
Trans01	0	0	0	0	1	0	0
Trans02	1	0	0	0	0	0	0
Trans03	0	1	0	0	0	0	0
Trans04	0	0	0	1	0	0	0
Trans05	0	0	0	0	0	1	0

Next, the process of eliminating business guild code and adding it to the column is done completely, and eventually its diagram is illustrated for the remaining fields. Due to the fact that the used data is a time series, a relation should be created between the current transaction data and the preceding transaction data, so they are put together in the same row. Creating conceptual relation between the current and the previous transaction during modelling is an important issue in the time-series concept. The deep network uses time series data to predict the behaviour of the customer in the future. Then, labels are passed to the deep learning network. These labels are used to train the network and the goal is to minimize the Loss Function or equivalent level of deviations. These labels are put into a new data frame. The result obtained from the network should be trained with minimum-square error, which is the main evaluation criterion of the network training.

At this stage, since the predictable labels of the network can be the number of transactions or transaction volume, accurate numerical prediction is not possible by logistic regression functions. Therefore, the selected labels of network output should include increasing or decreasing intervals related to the amount of increase or decrease in volume or number of transactions of next days, weeks or months. To this end, the clustering method is used to label various decreasing or increasing rates based on the train data. Therefore, the intervals of the rates labelling are based on train data, and then these labels are used to evaluate and train the deep network. Figure 2 represents a clustering output, which is obtained from prediction results.

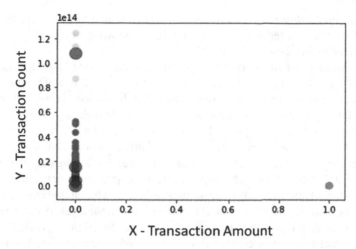

Fig. 2. Output (labels) resulted from transaction clustering

As shown in the Fig. 2, the labels result of clustering transactions are divided into four main clusters. Each of these clusters represents a degree of increase or decrease in transactions volume and frequency. Therefore, if the model is properly trained based on this type of labels, it will be expected that good predictions be obtained using the mentioned labels and test data. Table 3 shows the analysis of extracted labels for the prediction phase.

Table 3. Used labels for prediction in the deep network

Label	Description
First cluster (black colour)	Low number of transactions. Low transaction volume.
Second cluster (blue colour)	Medium number of transactions. Low transaction volume.
Third cluster (yellow colour)	High number of transactions. Low transaction volume.
Forth cluster (green colour)	Low number of transactions. High transaction volume.

2.3   Architecture and Modeling of RNN

In this section, the type of deep learning network and its architecture are explained. Considering that our data is a non-stationary time series, RNNs are used, which are capable of maintaining Long Short-Term Memory (LSTM). RNNs are designed for the purpose of remembering long sequences. Like a simple recursive network, RNN has its own repeatable mode, but its repeatable unit (cell) is a bit more complicated. These networks have the ability to receive input time series data and be trained recursively with them. Ultimately, after training, the mentioned networks are used to predict the future behaviour of the user.

In the architecture of this network, 80% of the data is used as train data and the remaining 20% is used as test data. The input of this network is a 3D tensor, its first dimension is the data input of the network, the second dimension is time step (one step) and, finally, the last dimension is the features of considered data.

As mentioned before, the Python language and Keras library have been used to design and run the deep network. The structure of this network is based on 50 LSTM layers, and in the next layer, a dense layer is considered for aggregating the results of the network that is called dense. The architecture of deep learning network using RNN and configuration of the network is presented in Fig. 3. This architecture consists of input layers for features, 50 multiple hidden layer and output layer, dense layer which aggregates all the result of output layer using soft-max function to predict the labels. Hence, the predicted result is estimated via sequential processes $(\hat{y})$. Then the estimated value compares against actual labels so that the best model is conducted and learning process is completed.

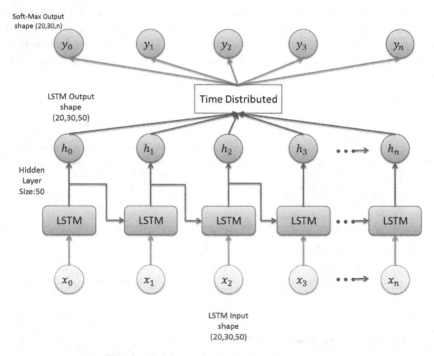

Fig. 3. Architecture of deep learning network

In this architecture, Adam Gradient Decent Algorithm, is used for minimizing the generalization error in the network and Root-Mean-Square Error (RMSE), is considered as Loss Function.

2.4   Deep Network Learning and Prediction

At this stage, the learning of the network takes place using input data and its features. In this process, the number of network epochs is 50 and the volume of training data in a batch is about 72. The "epoch" takes into account the number of times the learning network is run, so that all the data is not logged in entirely, and learning is done in each epoch using batch size in different iterations. After entering the above information, the test data can be placed on the learning model, so that based on this data and train data, the learning process is finished, and eventually the model is ready for the next sections.

As shown in Fig. 4, after training the network, we can also use the history of test and train data cost functions in order to check over-fitting or under-fitting in the network. When the number of epochs is about 50, this network is well-trained and there is no over-fitting or under-fitting in it. Under this threshold the learning becomes over-fitted and above the threshold it becomes under-fitted as the error rate increases.

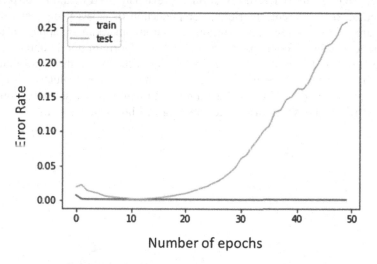

Fig. 4. History of cost functions of test and train data

After training the network, it should be evaluated thoroughly using test data to determine the prediction accuracy. By using the prediction method and considering test data as input, the final output is returned as the output of the prediction method. The output should be compared with the main test data labels.

3   Result

To determine the validity of the model, the final labels including growth rate or decrease in the volume and number of customer transactions are compared to the final output. If the prediction result has the correct coefficients, it can be claimed that the obtained result and the prediction model are valid and can be used to estimate and predict customer future behaviour. In this study, measures such as accuracy and mean-square error are used to evaluate network results. The results are represented in Table 4.

Table 4. Evaluation result of the prediction model

Evaluation measure	Evaluation result
Accuracy	0.8660124022466476
Mean squared error	0.25848191614379346

As can be seen, the prediction accuracy is about 86%, which is, in fact, a very good rate for deep learning networks. Meanwhile, the mean-square error is acceptable over the total numerical distance, and the coefficient of error is negligible. The confusion matrix also well illustrates how different predictions have occurred in the right clusters, with its remarkable results.

In order to compare our result with other previously used methods for prediction, we applied several different algorithms for prediction of labels using same data and same features. Table 5 shows the result obtained from other known, applied algorithms via our target dataset. It contains information about algorithm, the R library used, the given functions and parameters, and finally the accuracy of the algorithm which indicates how well they are predicted. All the methods are evaluated using Cross Validation method with 70% of data as train and 30% of data as test, except time-series method, which used 3 years of data for training and last year for test.

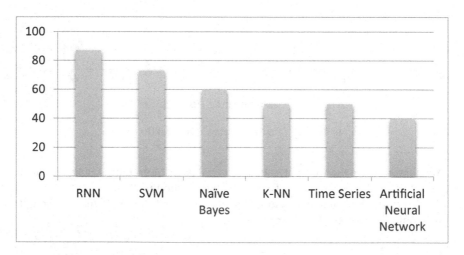

Fig. 5. Comparison of different algorithms based on accuracy measure

Table 5. Comparison of several traditional algorithms for prediction of customer behavior

	Algorithm	Used library	Function and configuration parameters	Result: Accuracy
1	SVM	library (e1071)	svm(group ~ ., data = trainset, method="eps-regression" , cost=1 ,kernel="polynomial" ,degree=1)	73%
2	Naïve Bayes	library (e1071)	selected_model = make-Learner("classif.naiveBayes") NB_mlr = train(selected_model, task) NB_mlr$learner.model	60%
3	K-NN	library (class)	cl <-train[,3] Knn_model <- knn(train, test, cl , k = 5)	50%
4	Time Series	library (forecast) Library (fUnitRoots) Library (data.table) library (tseries) library (timeSeries)	pacf(log(daily_dt_Train$Count)) pacf(diff(log(daily_dt_Train$Count))) fit <- arima(log(daily_dt_Train$Count), c(0, 1, 1),seasonal = list(order = c(0, 1, 1), period = 12))) pred_Train <- predict(fit, n.ahead = 156)	50%
5	Artificial Neural Network	Library (neuralnet) library (dplyr)	n <- names(train_) f <- as.formula(paste("cluster_result ~", paste(n[!n %in% "cluster_result"], collapse = " + "))) nn <- neuralnet(f,data=train_,hidden=c(6,4) , lifesign = "full", learningrate=0.90 , algorithm = "backprop", err.fct = "sse", act.fct = "logistic", linear.output = FALSE (	40%

As we can see from Table 5, the accuracy of SVM, which is about 70%, outperforms the other methods. However, our proposed method that predicts labels using RNN still has significant advantage in accuracy measures, which leads to 16%

improvement compared with the best possible outcome obtained from other algorithms. As Fig. 5 shows, we can conclude that the RNN method outperforms other traditional algorithms as dimensionality and volume of data increases, which was discussed before.

4   Conclusion

One of the most popular applications of deep networks is predicting customer behaviour, which could give us valuable information. Hence, identifying the pattern of customer behaviour and detecting anomalies, could be an important topic. In this research, using time series in recurrent neural networks, the behaviour of customer (owners of POS devices) has been modelled. Analysing the high volume of data is possible by using simple and fast deep networks.

The results of the user behaviour prediction model, which are obtained using standard evaluation measures, indicate that the model is comprehensive. By using this model, user behaviour can be predicted and if there is any anomaly in customer future behaviour, the system can alert in real time.

For future research, identifying differences between predicted behaviour and real behaviour, and also using concept drift or detecting anomalies to discover suspect cases, are suggested. Real-time customer analysis tools also can be used to identify customer behaviour and its real-time changes.

References

Ivakhnenko, A.G.: The group method of data handling-a rival of the method of stochastic approximation. Sov. Autom. Control **13**, 43–55 (1968)

Armando Vieira Redzebra Analytics: Predicting online user behaviour using deep learning algorithms. arxiv.org (2016)

Zheng, B., Thompson, K., Lam, S.S., Yoon, S.W.: Customers' behaviour prediction using artificial neural network. In: Preceding of the 2013 Industrial and Systems Engineering Research Conference (2013)

Dickmanns, E.D., Zapp, A.: A curvature-based scheme for improving road vehicle guidance by computer vision. In: Proceedings of SPIE, Mobile Robots I, vol. 0727 (1987)

Lederberg, J.: How DENDRAL was conceived and born (1987)

Fryer, L., Carpenter, R.: Bots as language learning tools. Lang. Learn. Technol. **10**(3), 8–14 (2006)

Mccarthy, J.: Measures of the value of information. Proc. Natl. Acad. Sci. U.S.A. **42**, 654 (1956)

Ng, A.: Machine learning yearning (2018). deeplearning.ai

Saithibvongsa, P., Yu, J.E.: Artificial intelligence in the computer-age threatens human beings and working conditions at workplaces. Electron. Sci. Technol. Appl. **1**(3) (2018)

Shaw, G.L.: Donald Hebb: the organization of behavior. In: Palm, G., Aertsen, A. (eds.) Brain Theory, pp. 231–233. Springer, Heidelberg (1986). https://doi.org/10.1007/978-3-642-70911-1_15

Supriya Pahwa, D.S.: Comparative study of support vector machine with artificial neural network using integer datasets. Int. J. Adv. Res. Comput. Sci. Softw. Eng. **200** (2016)

McCuluch, W.S., Pits, W.H.: A logical calculus of the idea immanent in nervous activity. Bull. Math. Biophys. **54**, 115–133 (1943)

Yamins, D.L., DiCarlo, J.: Using goal-driven deep learning models to understand sensory cortex. Nat. Neurosci. **19**, 356 (2016)

Performing Software Test Oracle Based on Deep Neural Network with Fuzzy Inference System

Amin Karimi Monsefi[1], Behzad Zakeri[2(✉)], Sanaz Samsam[3],
and Morteza Khashehchi[2]

[1] Shahid Beheshti University, Tehran, Iran
a_karimimonsefi@sbu.ac.ir
[2] University of Tehran, Tehran, Iran
{Behzad.Zakeri,m.khashehchi}@ut.ac.ir
[3] Sharif University of Technology, Tehran, Iran
samsam.sanaz@ae.sharif.ir

Abstract. One of the challenging issues in software designing is testing the product in different condition. Various software Oracles had suggested in the literature, and the aim of all of them is minimizing the time and cost of the testing process. Software test Oracles have designed to do this job automatically with as less as possible human contribution. In this work, a novel Oracle based on deep learning and fuzzy inference system introduced. For this purpose, by the utility of Takagi-Sugeno-Kang fuzzy inference, the output of software mapped to the fuzzy space, and the deep neural network has trained by this data. Finally, data has remapped to the primary form and used as the competitor stage input. To validate the performance of the Oracle, four different models have chosen to assess the Oracle enforcement, and after training the Oracle by the correct output of applications, source codes have changed manually, and the efficiency of the Oracle monitored. Several measures have been applied to evaluate the accuracy of the test Oracle, and it is observed that in most cases Oracle correctly could detect the correct and false results. Finally, designing Oracles requires several preliminaries and in this work we only focus on the architecture of the system.

Keywords: Software testing · Deep learning · Oracle problem · TSK fuzzy inference system

1 Introduction

Software testing is a vital part of software technology to test the quality and reliability of computer programs under various conditions. Since human-based software testing requires too much time and energy resources, complete testing of software in limited time is impossible. For this purpose, automatic testers have been suggested to decrease the cost of this process [3]. Oracles have widespread

© Springer Nature Switzerland AG 2019
L. Grandinetti et al. (Eds.): TopHPC 2019, CCIS 891, pp. 406–417, 2019.
https://doi.org/10.1007/978-3-030-33495-6_31

applications in testing different kinds of software, such as web search engines, embedded software and video games [5,6,10].

A test oracle is a mechanism to assess the performance of the software. It is a valid source to study the operation of the under-test software [25]. It also generates different test cases considering software specification, and evaluate the actual behaviour of software [16]. Test oracles are useful in forming the automated software testing platform.

To assess the correctness of the under-test software, after the result generation process, Oracle should compare the generated results by software with correct results. It is conventional to call produced results using under-test software as *actual outputs*, and the correct results which are utilized to verify actual output known as *expected outputs* [25]. To check the actual outputs, Oracle has to find suitable expected outputs; this process of finding proper expected outputs mostly known as oracle problem [1].

The function of an arbitrary test oracle to evaluate test results starts with the generation of expected outputs. The second step is executing the test cases. In the next stage, the initial domain should be mapped to the expected outputs, and corresponding actual outputs for each expected output should be specified. Finally, actual outputs compare with expected outputs to find out whether the behaviour of the software is accurate or not [20]. Although test oracles designed to tackle all of these stages, in most studies oracles are used to test the execute cases only.

The aim of test oracles is tackling the software testing process with as less as the human contribution. However, this process faces several challenges in each automation activities steps [18,20]. The preliminary challenge in software testing is automatic data generation, and Oracle should provide proper output results automatically. Another challenge is that Oracle should provide expected results of the corresponding software inputs, and this is impossible without correct automatic mapping between input domain and output domain. The last issue in Oracle function is making the decision about which actual outputs are acceptable and which results should be considered as a fault [19].

In this study, we focus on forming a precise oracle by taking advantage of deep learning and fuzzy logic. For this purpose, by using Takagi-Sugeno-Kang (TSK) structure, a sufficient number of software inputs and their corresponding expected outputs have been encoded to the fuzzy space [21,23]. By utility of these data, a deep neural network has trained. This network with cooperation with fuzzy encoder-decoder performs our target Oracle.

2   Backgrounds

In the year of 1965 fuzzy logic has been suggested by prof. Lotfi A. Zadeh to cover the inabilities of the classical logic [27]. In this framework instead of using 1's and 0's to evaluate the value of parameters, variables can have a value in the spectrum domain between 0 and 1 [26]. To assign a spectrum value to each variable, the Fuzzy Inference System (FIS) is using for mapping the value of data from binary space to fuzzy space. There are several FIS methods for mapping to fuzzy space, however, in action, only two FIS techniques are favourable which are TSK and Mamdani fuzzy models [13,17].

There are several differences between the TSK and Mamdani fuzzy structures. Computational cost is the main difference between these two structures. While Mamdani fuzzy model by computing the whole membership function, TSK model use simple formulas to computing the output. This feature of TSK structure makes it more useful FIS technique rather than the Mamdani model.

On the other hand, deep learning as a multilayer neural network with self-relying ability in feature extraction plays a crucial role in the prediction of systems' behaviour [11]. The range of deep learning applications covers a vast area of engineering and scientific topics, such as object detection, image classification and web search engines [4,7,8]. A deep neural network by the utility of several inputs and their corresponding output get trained and after the training procedure, can precisely predict the results of the system for arbitrary inputs.

Adaptive-network-based fuzzy inference system (ANFIS) is the most famous form of using the fuzzy logic in the neural networks. ANFIS is a kind of artificial neural network based on the TSK fuzzy inference system. This network has suggested by prof. Roger Jang in the early 1990s [9]. Since ANFIS uses the advantages of neural networks and fuzzy logic simultaneously, it has the capability of dealing with a wide range of problems like the prediction of nonlinear functions.

3   Methodology

In this study, with the integration of the advantages of deep learning and fuzzy inference system, it is tried to perform more efficient and accurate software test Oracle. Figure 1 shows the structure of the Oracle with containing two main stages which are Fuzzy encoder-decoder and deep neural network. The former stage by the utility of TSK structure maps the I/O of the under-test software to 0 to 1 and reverse it at the end of the Oracle structure. The latter stage uses these data to train and test of the deep neural network as the core of the Oracle. In the following context, each step will be discussed in detail.

3.1   Fuzzy Encoder-Decoder

Various fuzzy inference system has been suggested in the literature, however, most of the fuzzy reasoning can be classified into three main categories. Amongst these categories, Takagi and Sugeno fuzzy inference which uses a linear combination of input variables plus a constant term, is more functional than other inferences [12,22]. To have a better understanding of the TSK mechanism, firstly this inference is discussed for two input and the single output, and then the formula for multi-input-single-output (MISO) TSK model will be posed.

A conventional form of TSK fuzzy structure with two inputs and single output express as:

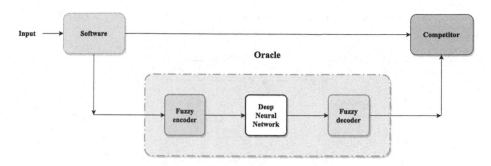

Fig. 1. Schematic Oracle structure

$$if\ x_1\ is\ A_i\ and\ x_2\ is\ B_i\ then\ y\ is\ y_i \tag{1}$$

Where in the Eq. 1, $A_i \in \{A^1, ..., A^{NA}\}$ and $B_i \in \{B^1, ..., B^{NB}\}$ represents the antecedent MF of the ith rule that belongs to the input variables x_1 and x_2 respectively. The sets $\{A^1, ..., A^{NA}\}$ and $B_i \in \{B^1, ..., B^{NB}\}$ are pre-defined antecedent MFs. The ith rule produces a partial output form which shown as:

$$y_i = f_i(x_1, x_2) \tag{2}$$

Where f_i are pre-defined functions in this study:

$$f_i(x_1, x_2) = r_i \tag{3}$$

And the $r_i = constant$, therefore characterizing a crisp consequent MF for the ith rule. Aggregation the partial outputs of each rule, the output is given by:

$$f = \frac{w_1 y_1 + w_2 y_2}{w_1 + w_2} \tag{4}$$

Where $w_i = AND(\mu_{Ai}(x_1), \mu_{Bi}(x_2))$ is the weight of the ith rule. The inference procedure has shown in the Fig. 2:

The mechanism of TSK inference system with two input and a single output for a better understanding of the operation of this model. However, it is noticeable that in this study all of the computations have conducted using MISO framework of TSK model. Since the schematic diagram of the MISO model and the governing equations are more complicated than the discussed model, in the following lines governing equation present briefly as:

$$R_i = if\ x_1\ is\ \widetilde{A_{11}}\ and(or)\ ...\ x_m\ is\ \widetilde{A_{1m}}\ then\ y = g(x_1, ...x_m) \tag{5}$$

Where m is the number of input variables, R is the fuzzy rule, $\widetilde{A_{ij}}$ is the fuzzy set corresponding to ith input variable for jth fuzzy rule and g_i is a function is defying as follow:

$$g(x_1, x_2, ..., x_m) = q + q_1 x_1 + ... + q_m x_m \tag{6}$$

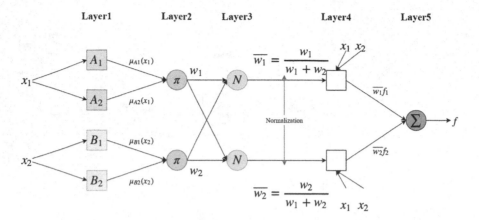

Fig. 2. ANFIS structure with two input variable

And finally, the fuzzy system can be described as follow:

$$y = \frac{\sum\limits_{i=1}^{R} g_j(.) T_{j=1}^{m_i} \mu_{ij}(x_j)}{\sum\limits_{i=1}^{R} T_{j=1}^{m_i} \mu_{ij}(x_j)} \tag{7}$$

Where μ_{ij} is membership function for the A_{ij} fuzzy set, and $m_i (1 < m_i < m)$ and T the number of inputs to the fuzzy inference and T-norm operator respectively.

After mapping the inputs by using Eqs. 5, 6 and 7 to fuzzy space, an n-dimensional hyperspace form by input variables and their output value conducted by FIS. To have a better understanding of the relationship between fuzzy inputs and outputs, a hyper-surface fit to these points in a way that contains as more as possible of points. Since digital computers and especially deep learning algorithm cannot accept a continues parameter as input, this hyperspace has discretized into several smaller hyper-surfaces. Each hyper-surface contains some of the fuzzy variables, and obviously, by increasing the number of this sub-hyper-surfaces, the precision increase and it is easier to define which point belongs to which element. Although increasing the number of elements in this stage increase the precision, since this sub-hyper-spaces were used as the learning input parameters in the deep learning module, it could be causing the over-fitting in the learning procedure. For this reason, it is significant to find optimum size of these elements which have been done by several experiments in this study. Sample model of hyper-surface and its discretization, and increasing the precision by increasing the number of elements in 3D fuzzy space is shown in Fig. 3.

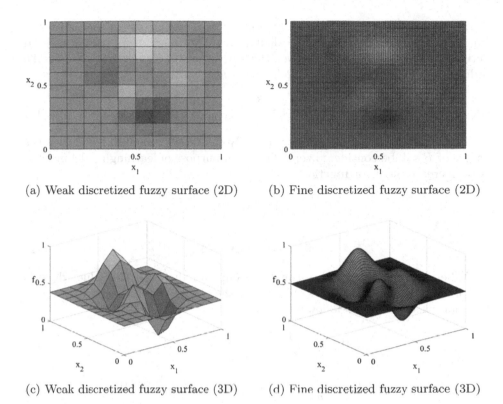

(a) Weak discretized fuzzy surface (2D)   (b) Fine discretized fuzzy surface (2D)

(c) Weak discretized fuzzy surface (3D)   (d) Fine discretized fuzzy surface (3D)

Fig. 3. Schismatic view of fuzzy space

3.2 Deep Learning

In this section, deep neural network as the core of the Oracle has described, and the governing equations of the deep learning algorithm have discussed.

The deep neural network architecture has divided into three main layers which are the input layer, hidden layer and output layer. Input layer receives the mapped data from the fuzzy inference and delivers them to the hidden layers. Hidden Layers by processing data try to fund the relation among data, and finally, the output layer reports the outputs of the user.

To study the governing equation for an arbitrary hidden layer, all the equations govern the lth layer have been driven as follow:

$$Z^{[l]} = W^{[l]} * A^{[l-1]} + B^{[l]} \tag{8}$$

$$A^{[l]} = g^{[l]}(Z^{[l]}) \tag{9}$$

Where in this network $W^{[l]}$ is the weight matrix which links the layer l to the layer $l-1$, and $w_{ij}^{[l]}$ is the weight between neuron i in layer l and neuron j in the layer $l-1$. Also $B^{[l]}$ is equal to bios vector for layer l.

In Eq. 9, $A^{[l]}$ is a the output matrix for layer l and $g^{[l]}$ is the activation function which employed in the calculations. The function $g^{[l]}$ is the utilized activation function in the $l'th$ layer the activation function which used in the hidden layer is function of *LeakyRelu* which are explained as follows:

$$\text{LeakyRelu}(x) = \begin{cases} x, & x > 0 \\ x * 0.001, & x \leq 0 \end{cases} \tag{10}$$

Firstly, the values of W will be randomly chosen between 0 to 1 and then the values of B will be considered zero. The main purpose of learning model method is to decrease the error function.

$$\min_{W,B} J(W, B) \tag{11}$$

$$J(W, B) = \frac{1}{m} \|Y' - Y\|_2^2 \tag{12}$$

Where Y' and Y are the estimated value and the expected value which is extracted from the case studies. To prevent the *Overfitting* in this part, the error function was changed as follows:

$$J(W, B) = \frac{1}{m} \|Y' - Y\|_2^2 + \frac{\lambda}{2m} \|W\|_2^2 \tag{13}$$

$$J(W, B) = \frac{1}{m} * (Y' - Y)^T (Y' - Y) + \frac{\lambda}{2m} W^T W \tag{14}$$

By conducting the discussed method, the deep neural network has been trained to predict the test cases results.

4   Results

In this part, it is tried to evaluate the performance of the discussed test oracle in action. Due to this, four different software has been used as our case studies which each one has specified features. By utility of these case studies, numerous test case were generated for training and testing the Oracle. After the training process, by using the generated test cases, the accuracy of the Oracle has assessed.

4.1   Case Studies

One of the significant issues in test Oracles is the ability to tackle with unknown source codes. For this purpose, four source codes have been chosen from www. codeforces.com. This web site is a host for competitive programming contests and contains more than 10 million source codes with various scopes.

Our case studies have written in Java and C++ languages, and the *I/Os* of these codes are in the numeric form. To the study of the complexity of these codes, two criteria were used. The first criterion is the number of the programs' code lines. The other criterion is *Cyclomatic Complexity (CC)* which is a standard programming measure based on the independent paths of the programming system. This criterion makes it possible to measure the complexity of arbitrary source code [24]. Table 1 summarizes all of the test cases features as follow:

Table 1. Specifications of test cases

#	Name	# code's line	# Input parameter	CC	Code's Language
1	Polyline	202	6	37	Java
2	Really Big Number	277	2	56	C++
3	Magic Number	368	4	31	Java
4	Karen & Neighborhood	256	2	29	C++

4.2 Oracle Evaluation

To train the Oracle for each case study 30K test case have generated. Producing of these data for each case study by considering the boundary of the case and utility of adaptive random data generation method has been done [2]. Also, 5K data with the same method have generated for testing purposes. For labelling these data, they considered as the input of the source code of each case study. Since the generation of a large amount of data is quite hard, it had forced us to generate the under-qualified number of data, and for filling this gap, we train the Oracle model with the same data iteratively with at least 500 iterations for each case study.

To analysis, the error value of the Oracle model for each case study, *mean square error (MSE)* formula has employed. The mathematical description of *MSE* become as follow:

$$MSE = \frac{\sum_{i=0}^{n} \left(y^i - y'^i\right)^2}{n} \tag{15}$$

Where n is the total number of data, and y and y' are expected and actual results respectively.

The following figures demonstrate the error value of each individual case study Fig. 4. Looking at the figures in more detail, in each case study by increasing the training iteration, the *MSE* value decreases considerably. It is also noticeable that the error value after a specified iteration converges to the certain value, and after that, by increasing the iteration the *MSE* value fluctuate around that value.

4.3 Assessment of Fault Detection

The aim of the Oracle is finding the existing errors in the under-test software. Meaning that for each test case certify that whether the output of the software is similar to the expected result (Oracle output) or not. To evaluate the function of

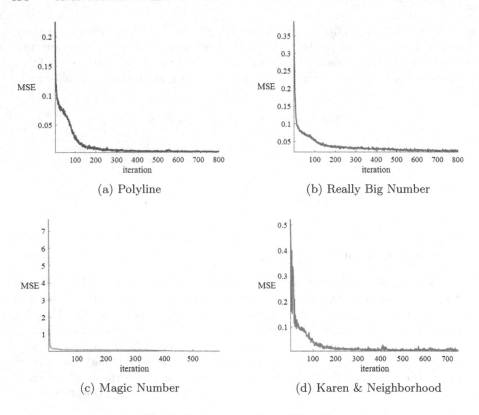

Fig. 4. Error per number of iteration

the Oracle, the software source codes were manipulated manually, and for each source code, one line of each code perturbed as follow:

As it is shown in the Table 2, by changing the source code of each program, expected results changes consequently. For instance, in the case of the Polyline program, 617 case of expected results out of 5K test data have been under the influence of our perturbation.

The main task of Oracle is distinguishing between changed and unchanged results. Due to this, the following parameters define as follow:

- True Positive (TP) is the number of test cases which their results have changed, and Oracle could find this change correctly.
- True Negative (TN) is equal to the number of test cases that did not change, and Oracle could figuring out this remain.
- False Positive (FP) define as the number of test cases that did not change, but Oracle knew them as changed cases.
- False Negative (FN) is the test cases that have changed, but Oracle defines them as unchanged results.

Regarding the fact that the value of the actual and expected results never be fully equal, By defining a threshold value, acceptability or unacceptability of

Table 2. Samples of the mutants

#	Original code	Mutated code	Error type	Affected output
1	$if(x_1 == x_2)$	$if(x_1 == x_3)$	Variable change	617
2	$if(temp \leq n)$	$if(temp < n)$	Operator change	811
3	++Digit	Digit++	Operator change	326
4	$if(L\_FB \geq L\_Cnt)$	$if(L\_FB > L\_Cnt)$	Operator change	64

the test case results have evaluated. The threshold in this study defined in the Eq. 16 as follows:

$$|y - y'| < threshold \tag{16}$$

The following table illustrates the values of TP, TN, FP, FN and threshold of the understudy software (Tables 3 and 4):

Table 3. The proposed approach evaluation results

#	Threshold	# TP	# TN	# FP	# FN
1	0.5	583	4263	71	83
2	200	781	4101	83	37
2	20	547	4293	104	56
3	0.5	311	4628	33	28
4	200	51	4929	13	7

Table 4. Effectiveness of purposed method

#	Threshold	P	TPR	FPR	TNR	FNR	ACC	FM
1	0.5	0.89	0.87	0.01	0.98	0.12	0.96	0.88
2	200	0.90	0.95	0.01	0.98	0.04	0.97	0.92
2	20	0.84	0.90	0.02	0.97	0.09	0.97	0.87
3	0.5	0.90	0.91	0.00	0.99	0.08	0.98	0.91
4	200	0.79	0.88	0.00	0.99	0.12	0.99	0.83

For evaluation of the Oracle's results and parameters, the favourable data science formulas have used [14,15], and define as follow:

- Precision: $P = \frac{TP}{TP+FP}$
- True Positive Ratio (Sensitivity or Recall): $TPR = \frac{TP}{TP+FN}$
- False Positive Ratio: $FPR = \frac{FP}{FP+TN}$
- True Negative Ratio (Specificity): $TNR = \frac{TN}{TN+FP}$

- False Positive Ratio: $FNR = \frac{FN}{TP+FN}$
- Accuracy: $ACC = \frac{TP+TN}{TP+TN+FP+FN}$
- F-Measure: $FM = \frac{2 \times P \times TPR}{P+TPR}$

5   Conclusion

In this paper, a new approach to designing software test Oracles by the utility of deep learning and fuzzy inference system has presented. To assess the performance of this architecture in action, a software test Oracle designed and tested with four different applications. The performance of this Oracle by importing the error data evaluated. For this purpose, firstly the Oracle trained by correct data, and then by manipulation in source codes of the applications, performance of the Oracle in finding the errors has been assessed. It is found that the Oracle has a high efficiency in detecting the error and also correct data.

The aim of all the proposed test Oracles is detecting the error with the best accuracy and the minimum human contribution in the process for all types of the software. Although the proposed test Oracle has acceptable accuracy in dealing with sample test software, this Oracle has a noticeable weakness. It is important to note that this Oracle can only be used for software that their output is numeric, and it cannot tackle with the string or graphical outputs.

References

1. Ammann, P., Offutt, J.: Introduction to Software Testing. Cambridge University Press, Cambridge (2016)
2. Chen, T.Y., Leung, H., Mak, I.K.: Adaptive random testing. In: Maher, M.J. (ed.) ASIAN 2004. LNCS, vol. 3321, pp. 320–329. Springer, Heidelberg (2004). https://doi.org/10.1007/978-3-540-30502-6_23
3. Desikan, S., Ramesh, G.: Software Testing: Principles and Practice. Pearson Education India, New Delhi (2006)
4. Erhan, D., Szegedy, C., Toshev, A., Anguelov, D.: Scalable object detection using deep neural networks. In: Proceedings of the IEEE Conference on Computer Vision and Pattern Recognition, pp. 2147–2154 (2014)
5. Fraser, G., Rojas, J.M.: Software testing. Handbook of Software Engineering, pp. 123–192. Springer, Cham (2019). https://doi.org/10.1007/978-3-030-00262-6_4
6. Gholami, F., Attar, N., Haghighi, H., Asl, M.V., Valueian, M., Mohamadyari, S.: A classifier-based test oracle for embedded software. In: 2018 Real-Time and Embedded Systems and Technologies (RTEST), pp. 104–111. IEEE (2018)
7. He, K., Zhang, X., Ren, S., Sun, J.: Deep residual learning for image recognition. In: Proceedings of the IEEE Conference on Computer Vision and Pattern Recognition, pp. 770–778 (2016)
8. Huang, P.S., He, X., Gao, J., Deng, L., Acero, A., Heck, L.: Learning deep structured semantic models for web search using clickthrough data. In: Proceedings of the 22nd ACM International Conference on Conference on Information & Knowledge Management, pp. 2333–2338. ACM (2013)
9. Jang, J.S.R., et al.: Fuzzy modeling using generalized neural networks and kalman filter algorithm. In: AAAI, vol. 91, pp. 762–767 (1991)

10. Khalilian, A., Mirzaeiyan, A., Vahidi-Asl, M., Haghighi, H.: Experiments with automatic software piracy detection utilising machine-learning classifiers for micro-signatures. J. Exp. Theor. Artif. Intell. **31**(2), 267–289 (2019)
11. LeCun, Y., Bengio, Y., Hinton, G.: Deep learning. Nature **521**(7553), 436 (2015)
12. Lee, C.C.: Fuzzy logic in control systems: fuzzy logic controller. I. IEEE Trans. Syst. Man Cybern. **20**(2), 404–418 (1990)
13. Mamdani, E.H., Assilian, S.: An experiment in linguistic synthesis with a fuzzy logic controller. Int. J. Man Mach. Stud. **7**(1), 1–13 (1975)
14. Perruchet, P., Peereman, R.: The exploitation of distributional information in syllable processing. J. Neurolinguistics **17**(2–3), 97–119 (2004)
15. Powers, D.M.: Evaluation: from precision, recall and F-measure to ROC, informedness, markedness and correlation (2011)
16. Ran, L., Dyreson, C., Andrews, A., Bryce, R., Mallery, C.: Building test cases and oracles to automate the testing of web database applications. Inf. Softw. Technol. **51**(2), 460–477 (2009)
17. Schnitman, L., Yoneyama, T.: Takagi-sugeno-kang fuzzy structures in dynamic system modeling. In: Proceedings of the IASTED International Conference on Control and Application (CA 2001), pp. 160–165 (2001)
18. Shahamiri, S.R., Kadir, W.M.N.W., Ibrahim, S.: An automated oracle approach to test decision-making structures. In: 2010 3rd IEEE International Conference on Computer Science and Information Technology (ICCSIT), vol. 5, pp. 30–34. IEEE (2010)
19. Shahamiri, S.R., Kadir, W.M.N.W., Ibrahim, S., Hashim, S.Z.M.: An automated framework for software test oracle. Inf. Softw. Technol. **53**(7), 774–788 (2011)
20. Shahamiri, S.R., Kadir, W.M.N.W., Mohd-Hashim, S.Z.: A comparative study on automated software test oracle methods. In: Fourth International Conference on Software Engineering Advances, ICSEA 2009, pp. 140–145. IEEE (2009)
21. Sugeno, M., Kang, G.: Structure identification of fuzzy model. Fuzzy Sets Syst. **28**(1), 15–33 (1988)
22. Takagi, T., Sugeno, M.: Derivation of fuzzy control rules from human operator's control actions. IFAC Proc. Vol. **16**(13), 55–60 (1983)
23. Takagi, T., Sugeno, M.: Fuzzy identification of systems and its applications to modeling and control. IEEE Trans. Syst. Man Cybern. **1**, 116–132 (1985)
24. Watson, A.H., Wallace, D.R., McCabe, T.J.: Structured testing: a testing methodology using the cyclomatic complexity metric, vol. 500. US Department of Commerce, Technology Administration National Institute of Standards and Technology (1996)
25. Whittaker, J.A.: What is software testing? and why is it so hard? IEEE Softw. **17**(1), 70–79 (2000)
26. Zadeh, L.A.: Outline of a new approach to the analysis of complex systems and decision processes. IEEE Trans. Syst. Man Cybern. **1**, 28–44 (1973)
27. Zadeh, L.A., et al.: Fuzzy sets. Inf. Control **8**(3), 338–353 (1965)

On a Novel Algorithm for Digital Resource Recommender Systems

Keivan Borna$^{(\boxtimes)}$ and Reza Ghanbari

Faculty of Mathematics and Computer Science, Department of Computer
Science, Kharazmi University, Tehran, Iran
borna@khu.ac.ir, reza9l@aut.ac.ir

Abstract. By increasing the use of users from online resource websites, we need a promising smart recommendation system for high-volume data which has been provided by high-speed service. A lot of research has been done to improve this trend in recent years. Undoubtedly, machine learning has played a main role and has created growing trend in proposer systems evolution. We can point to hybrid recommender systems and content based and collaborative filtering. Different data structures may be used in these methods and we intend to present a new method based on a tree data structure called ckd-tree. We compare this algorithm to different models with different data sets and we got that this algorithm provides a better result for a large system with massive data sets in comparison with other methods. This can be more valuable and also it would be an alternative to common methods like kd-tree.

Keywords: Data structures · Recommender systems · Binary tree · Cdk-tree

1 Introduction

Due to the fast growth of online digital resources, the use of recommender system has started to become necessary for ecommerce businesses. This paper gives an overview of how we can make the personalization process better, for customers and in particular, we recommend a k nearest neighbor method using ckd-tree data structure to improve the recommendation process. By using more precise and appropriate methods for each situation, we can help users to make their selection more accurately and deeply. In an online resource system, truly recommended objects, enhance the user's working experience. There are a lot of theories and methods that use content based, collaborative filtering and hybrid filtering [1]. Collaborative filtering is a mature method, that recommends the most related items to user, with the help of selected items by other similar users. Collaborative recommender systems have a lot of real world use cases. With an item-to-item 2D matrix of related items, facilitates the user experience and provides a faster access to later decisions [2]. These systems have been developed and used in many software companies. Youtube user recommender system, offers more pertinent videos to their users. This will allow users spend more time on their service with more interest, and also gives a sense of trust and a good experience to the users and they will welcome new recommendations with a sense of confidence. Amazon's recommendation engine for email marketing, has higher conversion rate than direct

© Springer Nature Switzerland AG 2019
L. Grandinetti et al. (Eds.): TopHPC 2019, CCIS 891, pp. 418–425, 2019.
https://doi.org/10.1007/978-3-030-33495-6_32

user searches on their website [3]. In this paper we present a collaborative filtering algorithm using ckd-tree data structure. ckd-tree is an optimal model of kd-tree. We are going to use ckd-tree for online digital resources. There are a lot of problems during developing recommender systems. In order to solve problems such as cold start or sparsity and get a better result from recommender system, we should use more adoptable addressing methods. Suppose we want to start an e-commerce company. At the beginning, we have no specific user number, no sold products. In case of using collaborative methods, we will have a sparse matrix, because we don't know users' taste and since our recommender system can't generate any valuable offer for sparse matrix, we will not have a solution for new users. As Rashid's [4] paper, we can have a more accurate collaborative matrix by asking users to rate products. It would be better to solve cold start problem by asking users to fill some information about their personal interests. Suppose a user has just been registered into an online system. It would be helpful for our recommender system to be asked him/her at the beginning of registration cycle, about books or any related stuff and after that we can create an item-to-user matrix for new registered users. There is another way to solve this problem. Case-Based reasoning is one of the historical based methods. It will recommends based on previously similar items [5] and after case definition we can estimate and figure out any valuable knowledge from resources. The discussions on this topic is not end here, and much has been discussed in many articles [6].

2 Preliminaries

Since we always intend to reduce the process calculation time and we have been trying to make existing processes faster and more accurately, the problem of reducing the recommendation time, that has long been discussed. This issue has particular importance for high volume data and we need to present a specific method to recommend users. In order to defining a new method, we use item-to-user matrix. The algorithms used for low-volume data are suitable for just those data size, and for problems with bulk data, specific methods need to be embedded and usually algorithms for low-volume data generate poor results for larger data which means they are slower than the others. Accordingly, to provide a better recommendation to users, we need to consider two factors, the speed and the process accuracy. Many studies have been carried out and many researchers have presented various models. In this paper we are going to present a fast method for high-volume data. Our results are based on the principles of other researchers and trying to improve existing methods. We are using an optimal tree data structure for search and faster compared to kd-tree. By using K nearest neighbor method, we plan to provide a recommender system for online resource systems. There are many articles on this subject, but we are going to provide a new way and to get this idea, we are using ckd-tree data structure. The k nearest neighbor method, calculates the distances (with specific metric) among requested points for a query set. We calculate the nearest values corresponding to k for each query set. It is common to use Manhattan, Euclidean and cosine similarity in k nearest neighbor and since Minkowski distance is a generalized form of Euclidean distance and Manhattan distance, we use this distance metric.

3   Existing Approaches and Their Limitations

According to the importance of recommender systems for data driven online resources, there are a lot of researches on this domain, mostly regression (formula 5) and collaborative methods for recommendation. The first part of the chapter presents the basic concepts and terminology of content-based recommender systems, a high-level architecture, and their main advantages and drawbacks. The Recommender Systems Handbook, provides a review of the state of the art of systems adopted in several application domains, telling the main approach of both classical and advanced techniques for indicating items and user profiles [7].

Amazon is a world-wide pioneer company in creating and using recommendation engines. They offer specific suggestions for their users to help them select a better option based on their taste and interest using cluster models and collaborative filtering. We can also refer to DieToRecs as another example. This company uses CBR model to recommend better tour and travel itineraries using three different models [8]. At the beginning, they use mentioned behavior, like asking questions from their users. After that, offering autocomplete items based on users search and finally they let their users to define a complete tour and recommend them based on their completed tours. Chef is another CBR, which creates a recipe for input values. The input values are user goals and it is an intelligent planner. Although if we have a bunch of recipes, we can recommend the similar one but it will learn from user inputs to recommend a better option in next usage. Roumani and Skillicorn [9] have presented an effective way to find positive k nearest neighbor in recommender systems. It means, they have gone beyond the simple recommendation methods and offered a way to make better recommender systems. They presented a method using SVM as a way to decrease dimension. They have offered a news recommendation using context trees [10]. Also there are another few works [11–13], using Markov chains for their recommendation method. Some researchers have created an adoptive framework with the help of hybrid recommender systems [14]. There are a lot of examples that all of them are came to expand and develop precision of recommender systems in different fields like movie or social media recommendation [15]. All of these methods have tried to provide a better model to reach a bidder. But these are not efficient for online resources like libraries with high-volume data. The mentioned methods are less responsive and accurate for high-volume data and we will present a method with greater speed and accuracy, especially in bulk data.

4   Materials and Methods

In this section, we intend to discuss the requirements of the algorithm discussed and talk a bit about them. Certainly you've learned about trees and graphs, different models and their features. A kd-tree data structure, is a binary search tree for arranging objects in k-dimension spaces. It is a useful data structure for nearest neighbor searches. It will be helpful to define specific k value according to the subject. A kd-tree, contains multiple levels and they are divided along a specific dimension, using a pillar hyperplane, with children based on root dimension. One of the most efficient ways of creating kd-tree is partition method, repeatedly spotting the root node with median object until the last trees, every object with smaller value will be on the left side and with larger value on the right side [16]. The algorithm is described in Maneewong-vatana and Mount 1999 [17]. The main idea behind this algorithm is each of whose nodes and objects in this tree indicates an axis-aligned hyper-rectangle.

In kd-tree, each node, determines an axis. Our conclusion will be based on coordinates around a node. It will use sliding midpoint method for nodes value comparison (which one is greater or less than a specific value). This process is in tree construction step, which ensures uniformity in cells distribution. Although the query efficiency for high-dimensional nearest-neighbor is an open problem we can search with closest k neighbors, for a specific point. ckd-tree, is a fast subset of kd-tree, developed in C++ and Cython. Same as kd-tree, it is a binary tree and represents a hyper rectangle with axis aligned. It is more efficient for sorting nearest neighbor queries. Unlike the parent data structure (kd-tree), at the phase of tree construction, the trained tree, must be re-created for serialization purposes and also ckd-tree provides dynamic node allocation, with sliding midpoint rule. Although because of nodes distribution, memory errors and leaks are possible but ckd-tree guarantees to cover all node dimensions, which lead to efficient build process and setup time [18]. Now we explain our dataset.

Book-crossing is a dataset, containing book-ratings, has been collected by Cai Nicolas Ziegler [19], contains million ratings of at least 250,000 books. All these ratings have given by more than two hundred thousand users. It is available for research usage in multiple formats. We have used this data-set to test our algorithm. As kd-tree, usually brings down the algorithm order to $O(n\log n)$ and the leaf size of tree shows the number of objects [20], we use a matrix model to fit data, and because of we use Minkowski metric for distance calculation, we put zero value instead of not as a number items. Actually we want to predict user opinion with collaborative filtering and also we want to find the closest object, classifying the nearest and most popular group of objects. According to ckd-tree data structure and k nearest neighbor algorithm we have implemented a recommender system using Python programming language, with the help of Scipy Python package that leads us to generate more powerful functionality for k nearest neighbors. We have used cross-validation technique to estimate k value. The main idea behind this method is to split the data-set into a randomly parts. For a fixed value, we can apply a random value to get predictions and after that evaluating errors. We use regression method to get the original value of error and the most effective one is sum of squared and also for classification, we define accuracy, that is the percentage of truly classified objects. This process is then successfully applied to all possible choices of v. At the end of the v folds (cycles), the computed errors are averaged to yield a measure of the stability of the model (how well the model predicts query points). The above steps are then repeated for various k and the value achieving the lowest error (or the highest classification accuracy) is then selected as the optimal value for k (optimal in a cross-validation sense). Note that cross-validation is computationally expensive and it should let the algorithm run for some time especially when we have large samples. Alternatively, we can specify k and we should have an idea of which value k may take (i.e., from previous k nearest neighbor analyses, that we may have conducted on similar data). Now to make predictions with k nearest neighbor method, for measuring the distance between requested object and other points of nearest sample, we should use a specific metric.

These are some of the most popular metrics:

$$d(x \text{ and } p) = \sqrt{(x - p)2} \text{ Euclidean} \tag{1}$$

$$d(x \text{ and } p) = (x - p)2 \text{ Euclidean squared} \tag{2}$$

$$d(x \text{ and } p) = |(x - p)| \text{ City block} - \text{Manhattan} \tag{3}$$

$$d(x \text{ and } p) = |(x - p)| \text{ Chebyshev} \tag{4}$$

After finding an appropriate value for k, we can make prediction based on selected k value. This is how we can make predications in regression problems:

$$y = \frac{1}{k} \sum_{i=1}^{k} y_i \tag{5}$$

In classification problems, to find a label for requested point, we use voting method in k nearest neighbor, that objects in a labeled distance are similar and we can clarify the difference between selected labels. Suppose we define a set of weights named by W, now we can define a weighting formula based on distances:

$$W(x, q_i) = \frac{e^{-d(x,q_i)}}{\sum_{i=1}^{k} e^{-d(x,q_i)}} \tag{6}$$

And specifically for our problem, classification problems, the maximum of above equation (formula 6) is chosen for every class and also the error rate is as follows:

$$\sqrt{\frac{1}{k-1} \sum_{i=1}^{k} (y - y_i)^2} \tag{7}$$

According to (formulas 6, 7), we have created a model to classify related online digital resources with the help of k nearest neighbor algorithm. As we mentioned before, there are a lot of work has been done on this topic, but we have benchmarked new data structure, ckd-tree with k nearest neighbor to generate more accurate results for massive data of digital resource like online libraries. The sliding midpoint splitting method, was introduced for approximate nearest neighbor searching, in ckd-tree queries are answered based on priority and also in comparison with kd-tree, it is a fixed version in number of data points. The dimensions and point distributions are also have been changed and made better. The cluster model in ckd-tree is Gaussian distribution. It is designed points in a full-dimension clusters. In this data structure for different point distributions, we have a better result for query points. In construction times, because this model uses ambiguity splitting that generates more effective trees, we have a better query time. It is noteworthy, however, that this method is sometimes worse than usual kd-tree method.

5 Results

We can see that in most cases we have a better solution in comparison with other tree data-structures. We can measure that in overall, we have a 10% growth in performance (Figs. 1, 2 and 3).

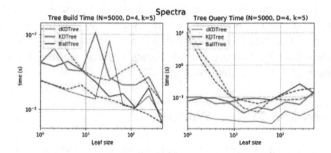

Fig. 1. Spectra model data

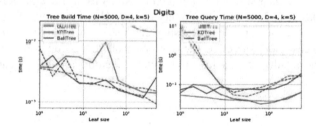

Fig. 2. Digits model data

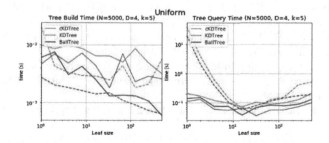

Fig. 3. Uniform model data

We should consider that, with larger leaf size, the build time will be decreased. Because we need fewer nodes to build. We can see that for small leaf sizes, the query is slow, this is because the algorithm needs to meet many nodes to complete the query and this will increase the query time, and this rule also applies to the high number. For very large leaf sizes, the query still is slow, because the algorithm should compute many distance values between pair nodes and this result motivates us to use a balanced leaf size value. Note that the ckd-tree shrinking model is different with other models. It means ckd-tree runs for each node and does not shrink their bounds and although we know that a difference in build time is not sensible for human but we can compare that the ckd-tree build time is up to 3× faster than the others. According to the figures, we know that ckd-tree is a bad choice for structural data. It is because of using midpoint rule in ckd-tree and can consider that it is really good for distributed data. Ckd-tree creates and propagates nodes which occupy all over the requested space, and most of these nodes are not useful and are empty. Because of this we may have a bit slower queries for this model of data distribution.

6 Conclusion and Future Works

We have mentioned that ckd-tree is less optimal for structured data, for these kind of data it's better to use other methods like usual kd-tree. Also there is a serialization disadvantage in ckd-tree, in allocating nodes with a dynamic manner. Actually usual kd-tree and ball tree data structures, have more flexibility in traverse, a lot of efficient metrics (formulas 1, 2, 3, 4) and other query options. Hence, it would be appreciated, if anyone can make a new model for using ckd-tree in highly structured data. This method also needs more effective metrics. With all this assumption, we know that according to the results, it is a fast and suitable method for the k nearest neighbor algorithm.

References

1. Nikhil, N., Srivastava, M.M.: Content based document recommender using deep learning. In: ICICI 2017, Coimbatore, India (2017)
2. Xue, F., He, X., Wang, X., Xu, J., Liu, K., Hong, R.: Deep item-based collaborative filtering for Top-N recommendation. TOIS **37**, 33 (2018)
3. Linden, G., Smith, B., York, J.: Amazon.com recommendation item-to-item collaborative filtering. IEEE Internet Comput. **7**, 76–80 (2003)
4. Rashid, A.M., et al.: Getting to know you: learning new user preferences in recommender systems. In: Proceedings of the 7th International Conference on Intelligent User Interfaces, pp. 127–134. ACM (2002)
5. Riesbeck, C.K., Schank, R.C.: Inside Case-Based Reasoning. Psychology Press, Routledge (2013)
6. Lund, S.S., Tandberg, Ø.: Design of a hybrid recommender system: a study of the cold-start user problem. Norwegian University of Science and Technology, May 2015
7. Lops, P., de Gemmis, M., Semeraro, G.: Content-based recommender systems: state of the art and trends. In: Ricci, F., Rokach, L., Shapira, B., Kantor, P.B. (eds.) Recommender

Systems Handbook, pp. 73–105. Springer, Boston, MA (2011). https://doi.org/10.1007/978-0-387-85820-3_3

8. Ricci, F., et al.: DieToRecs: a case-based travel advisory system. Destination recommendation systems: behavioural foundations and applications, pp. 227–239 (2006)

9. Roumani, A.M., Skillicorn, D.B.: Finding the positive nearest-neighbor in recommender systems. School of Computing, Queen's University (2007)

10. Garcin, F., Dimitrakakis, C., Faltings, B.: Personalized news recommendendation with context trees, Hong kong, China, 12–16 October 2013. ACM, New York (2013)

11. Kim, H.-N., Ji, A.-T., Ha, I., Jo, G.-S.: Collaborative filtering based on collaborative tagging for enhancing the quality of recommendation. Electron. Commer. Res. Appl. 9(1), 73–83 (2010)

12. Hinrichs, T.R., Kolodner, J.L.: The roles of adaptation in case-based design. In: AAAI, vol. 91, pp. 28–33 (1991)

13. Dey, A.K., Abowd, G.D., Salber, D.: A conceptual framework and a toolkit for supporting the rapid prototyping of context-aware applications. Hum.-Comput. Interact. 16(2), 97–166 (2001)

14. Sahal, R., Selim, S., Elkorany, A.: An adoptive framework for enhancing recommendation using hybrid techniques. Int. J. Database Theory Appl. 9(4), 107–118 (2016)

15. Chen, Q., Aickelin, U.: Movie recommendation systems using an artificial immune system. SSRN Electron. J. (2008). https://papers.ssrn.com/sol3/papers.cfm?abstract_id=2832022

16. Bentley, J.L.: Multidimensional binary search trees used for associative searching. Commun. ACM 18, 509–517 (1975). ACM student award. Paper: Second place

17. Maneewongvatana, S., Mount, D.M.: Analysis of approximate nearest neighbor searching with clustered point sets. In: Alenx 1999, Baltimore, MD, 15–16 January 1999. ACM E.1; F.2.2 (1999)

18. Maneewongvatana, S., Mount, D.M.: On the efficiency of nearest neighbor searching with data clustered in lower dimensions. In: Alexandrov, V.N., Dongarra, J.J., Juliano, B.A., Renner, R.S., Tan, C.J.K. (eds.) ICCS 2001. LNCS, vol. 2073, pp. 842–851. Springer, Heidelberg (2001). https://doi.org/10.1007/3-540-45545-0_96

19. http://www2.informatik.uni-freiburg.de/ ~ cziegler/BX

20. Brown, R.A.: Building a balanced k-d tree in O(kn log n) time. J. Comput. Graph. Tech. (JCGT) 4(1), 50–68 (2015)

Accelerating Decoding Step in Image Captioning on Smartphones

Behnam Samadi[1(✉)], Azadeh Mansouri[1],
and Ahmad Mahmoudi-Aznaveh[2]

[1] Department of Electrical and Computer Engineering, Faculty of Engineering,
Kharazmi University, Tehran, Iran
{std_b.samadi, a_mansouri}@khu.ac.ir
[2] Cyberspace Research Institute, Shahid Beheshti University, Tehran, Iran
a_mahmoudi@sbu.ac.ir

Abstract. In recent years, many efforts have been conducted to increase the accuracy of neural image captioning as one of the diverse applications of deep neural networks. Text-based image retrieval can be considered as one of the important applications of the image captioning. Moreover, improving the quality of life for visually impaired people is another crucial application of the image captioning. Accordingly, rapid and optimal implementations that can work effectively on mobile processors seems to be necessary. Despite the numerous image captioning approaches presented so far, few solutions are provided that consider the mobile computational capabilities. In this paper, we practically focused on the decoding step for the implementation of image captioning in android applications. Actually, iteration over variable lengths sequences can be performed using dynamic control flow. In other words, implementing such iterative algorithms using dynamic control flow may prevent unrolling the computation to a fixed maximum length. Using this facility will result in increased speed of the decoding routine in image captioning on smartphone devices. Experimental results on execution time validate the proposed approach.

Keywords: Image captioning · Mobile processors · Deep neural networks · Dynamic control-flow

1 Introduction

One of the most challenging areas in computer vision is image captioning. It refers to automatically generating properly formed textual descriptions in a natural language for a given image. This description has to determine the main objects in an image, their attributes and their relationships. It can help visually impaired people as an assistant or guide. For instance, it can be helpful when using webpages including images, or in daily life, by generating real-time responses about the surrounding environment. Considering large amounts of unstructured visual data, another important application of image captioning is natural language based image search.

Over the past few years, neural encoder-decoder based approaches in image captioning have gained a broad popularity [1, 2]. The main approach in such models, is to

© Springer Nature Switzerland AG 2019
L. Grandinetti et al. (Eds.): TopHPC 2019, CCIS 891, pp. 426–437, 2019.
https://doi.org/10.1007/978-3-030-33495-6_33

use CNN+RNN architectures. Firstly, the visual data is encoded into a compact representation (vision part). Then, this representation is decoded word by word into a textual description (language model part).

These models have been strengthened in approaches based on attention mechanism. Rather than having a static visual representation, attention based models are able to dynamically focus on various parts of the image during generating description inspired by the functionality of human visual system [3, 4]. Adding an attention mechanism to the architecture, improves the quality of caption generation; however, it requires a lot of trainable parameters. As a result, the model size and execution time increase.

Considering the applications of the image captioning, its optimized implementation on the smartphone devices is necessary. In spite of the progress made using deep neural architectures and attention based mechanisms, many of these architectures are not appropriate to be implemented on cell phones hardware due to the computational cost of these models. They need to be optimized to work effectively in the absence of high performance GPUs. To the best of our knowledge, "Camera2Caption" [5] is the only work which has deployed an android application for image captioning on smartphone devices. Their architecture is based on the "show and tell" model [2] which has been optimized to make it feasible for use on mobile devices.

In this paper, we focused on the decoding step of the "Camera2Caption". Due to the iterative nature of the decoder side, we employed dynamic control flow to prevent unrolling the decoding computations to a fixed maximum length. Stopping the decoding at the appropriate step leads to speedup without loss of accuracy.

In the following, a brief review of prominent image captioning methods is provided in Sect. 2. In Sect. 3, the architecture which is used in the android application is described. We proposed to employ the dynamic control flow for implementing the dataflow graph in Sect. 4. The experimental results which are demonstrated in Sect. 5 validate the effect of using dynamic control flow.

2 Related Works

Various approaches have been proposed to produce descriptions for images which can be divided into three main categories: Template-based methods, Retrieval-based methods and Neural image caption generators.

Earlier methods used some predefined sentence templates. They found suitable words for objects, relations and attributes. Then, this words are employed for inclusion in the template [6–8]. Another notable category is the retrieval-based methods. These approaches search a collection of images that are similar to the input image. Then, a novel caption is generated by using the corresponding captions of these images [9, 10]. In template-based techniques, the output caption is restricted to the few preset templates. Retrieval-based methods also produce captions which are strictly limited to the caption dataset.

In recent approaches, image captioning has been mainly considered in a similar way as utilized in machine translation algorithms. These approaches employ the neural encoder-decoder structures which are exploited in machine translation.

In fact, at the encoder side, instead of encoding an input sentence in the source language, the visual representations are extracted from the input image. At the decoder side, this representation is translated in order to generate the output caption.

In [11], the structure of the captioning model consists of three main parts: firstly, visual representation comes from a convolutional neural network which is pre-trained on a large-scale image classification dataset. This part can also be referred to as the vision part. Then, a language model part is presented using a recurrent neural network. Finally, a multimodal part connects the vision and language parts by a one-layer representation and the output of the model is the estimation of probability distribution of the next word, given the input image and the sequence of previous words.

This model was improved in [2] by exploiting a more powerful RNN (Long Short Term Memory). They also provide the visual part's output directly to the language model part. In [2], a static visual representation is extracted from entire input image only once at the beginning, and this is the all work of the visual part. In another scenario, in [3], by adding the visual attention mechanism, the model was able to devote more attention to the specific areas of the image, in order to generate each word of the output. Instead of extracting a feature vector from fully connected layers, they use lower convolutional layers whose output is a matrix of feature vectors corresponding to the different regions of the image. They introduced a 'soft' deterministic attention mechanism and a 'hard' stochastic attention mechanism under a common framework. In 'soft' attention mechanism which is trainable by standard backpropagation, weights are learned to calculate the probability distributions vector on different regions of the image, which also can be interpreted as the importance of the different regions for generating each word of the output. Conversely, in 'hard' attention mechanism, at each step, one region is selected for generating each word. Attention-based methods generate more accurate captions, but due to the huge number of trainable weights cannot be employed in real-time applications for low power systems.

According to our knowledge, the presented approach in [5] is the first attempt which addressed the practical challenges of image captioning on smartphone devices. Due to the computational complexity of image captioning and in accordance with mobile phone processing capabilities, they have created an optimized model for rapid performance on these low power devices. In the presented approach an android application is deployed which generates captions for input images from camera feed. They proposed an encoder-decoder model based on [2]. InceptionV4 [12] is utilized for feature extraction at the encoder side. Practically, the authors made use of a penultimate layer (*average pool layer*) for feature extraction. In order to generate the output caption, Long Short Term Memory cells are explored in RNN decoder.

In order to practically implement the mentioned method, the maximum caption's length is considered as 20 words. However, we investigated that this fixed value is not necessarily needful since in most of the cases the length does not proceed more than 14 words. This motivated us to implement the decoder side of the caption generating method in such a way that the time of decoding step is reduced by eliminating this overhead.

3 Architecture

The encoder-decoder structure is chosen for implementing the proposed approach similar to the Show and Tell method [2]. In fact, the optimized version of encoder-decoder structure, which is presented in [5], is employed for implementation of the proposed method.

In [5], exploiting pre-trained inceptionV4 for encoder side is demonstrated as a suitable choice in terms of both speed and accuracy. Furthermore, at decoder side, a neural architecture for sequence modeling is used to generate the output description. The purpose is to maximize the probability of generating the correct caption given the input image. This fact is formulated as Eq. (1):

$$\theta^* = \operatorname*{argmax}_{\theta} \sum_{(I,S)} \log p(S|I; \theta) \tag{1}$$

where θ are parameters of model, I is the input image and S is the correct sequence of words describing image. Considering that the length of caption (S) is N, by applying chain rule to model the joint probability over $S_0 \ldots S_N$ and by dropping the dependency over θ, the following statement is achieved (2):

$$\log(p(S|I)) = \sum_{t=1}^{N} \log p(S_t|I, S_0, \ldots, S_{t-1}) \tag{2}$$

$p(S_t|I, S_0, \ldots, S_{t-1})$ can be modeled by a recurrent neural network since recurrent neural networks are able to process variable-length sequences. In fact, the summarized information about what has been seen so far is stored in h_t, where h_t illustrates a fixed length hidden state or memory. Hidden state is then updated after each input is presented to the network through a recurrence function ($h_{t+1} = f(h_t, x_t)$).

As a result, the network behavior is dependent on the design of the f. For this recurrence function, long short term memory cell is introduced in order to deal with the two problems of vanishing and exploding gradient in recurrent neural networks [13]. To solve these mentioned problems and discover the long-term dependencies in the sequences, LSTM has a memory cell to remember values over arbitrary time intervals. The cell behavior is controlled by forget, input and output memory gates. The mentioned gates are referring to as f, i and o respectively. The input gate controls the extent to which a new state is added to the cell while the forget gate illustrates the extent to which the cell forgets its state. Moreover, the output gate expresses the rate of contribution of the value in the cell, to compute the output of the LSTM unit.

The gates interactions with cell state (c_t) can be modeled through following statements [2]:

$$
\begin{aligned}
i_t &= \sigma\big(W_{ix}\,x_t + W_{im}\,m_{t-1}\big) \\
f_t &= \sigma\big(W_{fx}\,x_t + W_{fm}\,m_{t-1}\big) \\
o_t &= \sigma\big(W_{ox}\,x_t + W_{om}\,m_{t-1}\big) \\
c_t &= f_t \odot c_{t-1} + i_t \odot h\big(W_{cx}\,x_t + W_{cm}\,m_{t-1}\big) \\
m_t &= o_t \odot c_t \\
P_{t+1} &= softmax\,(m_t)
\end{aligned}
\tag{3}
$$

where W matrices are parameters of affine transformation and $\sigma(.)$ is sigmoid and $h(.)$ is considered as hyperbolic tangent activation functions.

Each word is represented as a one hot vector with the dimension equal to the size of the word dictionary. Using word embedding vectors, the words are mapped into a lower dimensional space. In order to explain the overall functionality of the encoder-decoder structure, we denote image by I, the corresponding true caption in training set by $S = (S_0, \ldots, S_N)$, and finally the word embedding matrix by W_e.

In order to train the model, in the first step, the encoded visual features of the input image are fed into the recurrent structure as the following statement:

$$
(x_{-1} = encoder\,(I))
\tag{4}
$$

Then, the decoder takes the true caption's words one by one and finally the LSTM provides this probability distribution for the next word according to the image and previous words. These tasks can be formulated as follows:

$$
\begin{aligned}
x_t &= W_e\,S_t, \quad t \in \{0, \ldots, N-1\} \\
P_{t+1} &= LSTM(x_t), \quad t \in \{0, \ldots, N-1\}
\end{aligned}
\tag{5}
$$

Trainable parameters are optimized by minimizing sum of the log likelihood of the correct word at each step. Accordingly, the loss function should be minimized considering the following formula [2]:

$$
L(I,S) = -\sum_{t=1}^{N} \log P_t(S_t)
\tag{6}
$$

Two special symbols are added to the vocabulary, the Start, <S> and end token, </S> . Start symbol is added to the beginning and the end symbol is added to the end of all captions in the training set.

For inference, we have a trained probabilistic model and there exist multiple choices to generate a caption. In the sampling approach, the start sign is presented to the model as an initial input and according to the output probability distribution, one word is sampled and placed in the output caption. Then, the corresponding embedding is presented to the network as the next input and this process continues until sampling the end sign.

In beam search, there is a collection of size k, to keep the best k sentences up to time t. Among all the successors of each of these k sequences, k choices are selected as the best sequences of length t +1. Due to the computational complexity of this method, it is very time consuming to use on the smartphone, so such as [5] we use a greedy approach to generate captions. In each step, we choose the word with the maximum probability and present the corresponding embedding to the model as the next input and continue to reach the end symbol. Figure 1 shows an example of the process of generating a caption.

The mentioned encoder-decoder architecture is implemented using an open source software library developed by Google for numerical computations in mathematics and data science, TensorFlow [14]. There are special capabilities for implementation of machine learning algorithms and deep artificial neural networks provided in Tensorflow. Employing dataflow graphs to represent computations, gives Tensorflow the flexibility for deployment across a variety of platforms. In fact, Tensorflow maps nodes of the graph which represent mathematical operations across multiple devices in clusters, multi-core CPUs, general purpose GPUs and Tensor Processing Units (TPUs).

To make use of Tensorflow models on mobiles, firstly, the architecture is implemented in python language and trained on GPU. The trained weights are stored to tensorflow variables. By converting these variables to constants and serializing the graph, protobuf files (.pb) are prepared. Since there is no trainable parameter at pretrained convolutional encoder, feature extraction is performed once for the entire dataset and obtained feature vectors are used for decoder training. Thus, two separate probuf files are obtained. In inference stage, the goal is generating a caption for a single input image in the shortest time. Therefore, these protobuf files are merged into a single protobuf file. This final frozen graph is exported onto the smartphone device. In android application, the frozen graph gets the camera feed as an input and the generated captions are presented on screen.

4   Employing Dynamic Control-Flow

As stated above, it is desirable to have only one frozen graph which is utilized for the entire caption generation process for an input image. The length of the output caption is variable, and the LSTM-cell iterates until the end sign is reached.

In [5], the length of the output caption is considered up to a maximum 20 and truncated output caption's length to this maximum; Therefore, to generate a caption, 20 steps of decoding is required regardless of how long it is. The generated 20 length array is scrolled from left to right to reach the end sign ("</S>"). The words before the end symbol are placed in the output text and the rest, after this sign, are thrown away.

In order to extract the distribution of the caption's length, we analyzed the images of the MSCOCO test dataset. In this case, the caption length of each image is extracted and then the histogram of the obtained length is illustrated. Figure 2, shows the length distribution of captions which were generated using the presented architecture for MSCOCO test dataset that contains 40500 test images.

As it is shown in Fig. 2, the sentence length probability distribution is concentrated around the range 9 to 14. Therefore, in [5], in many cases time is wasted in producing a sequence of idle words. Thus, in the proposed approach, we tried to reduce the time of decoding step by eliminating this overhead.

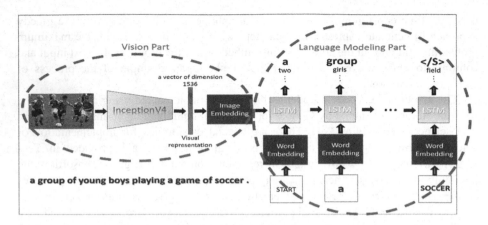

Fig. 1. Model architecture and an example of decoding routine.

Practically, conditional and iterative control flow can be implemented using the special facilities which are provided by Tensorflow [15]. Actually, recurrent neural networks are designed to model variable lengths sequences. In this case, the mentioned conditional (if statement) and iterative (while loop) programing constructs can provide special subgraphs which can be defined and executed dynamically for various number of times depending on the runtime values.

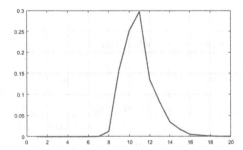

Fig. 2. Length probability distribution over 40500 responses of model to MSCOCO test dataset

We proposed using high-level control-flow constructs in order to provide a dataflow graph which includes a subgraph to check if the end sign is generated. The mentioned graph generally comprises both *condition* subgraph and a fragment to generate next word *body*. Tensorflow compiles these constructs to a dataflow graph that includes some special primitives in such a way that the body subgraph is executed till the condition fails.

Figure 3, illustrates the dataflow graph of the proposed dynamic dataflow architecture for the decoder side that corresponds to one of the loop variables. The illustrated dataflow comprises special primitives *Enter, Merge, Switch, NextIteration* and *Exit*. In order to run multiple iterations in parallel, for each loop variable, there exist a separate set of these primitives. Functionality of these primitives are described in [15].

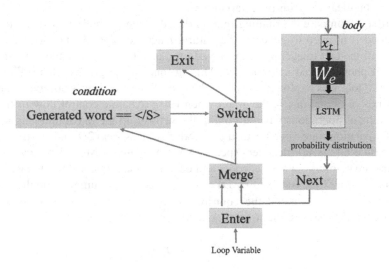

Fig. 3. Decoder dataflow for one loop variable

5   Experiments and Results

Practically, the main focus of the proposed method is related to analyzing the effect of employing dynamic control flow instead of unrolling computations to a fixed maximum length. Since the model architecture and the greedy behavior of the decoding routine are the same as the model presented in [5], the accuracy remains the same. In the following steps, we just present the execution time comparison.

In the first step, an experiment is conducted in order to evaluate the results of running time in both scenarios, "fixed caption's length" and the proposed "dynamic graphs approach". In this experiment, the results of running time are explored on a selected subset of MSCOCO dataset. In this case, for fair comparison, the average run-times are compared in two mentioned modes for the selected subset of images. Although the use of dynamic graphs may result in a small amount of run-time over-head, decreasing the number of decoding steps generally provides a reduced running time. In the following experiments, a mobile phone with a relatively weak processor was utilized with CPU configuration of Quad-core 1.2 GHz Cortex-A53[1].

Since the execution time in the dynamic graph mode is closely related to the output caption's length, a test subset including 30 images is selected from the MSCOCO test set in which the length distribution can be approximately modeled similar to the distribution illustrated in Fig. 2.

---

[1] Samsung Galaxy J5.

Practically, the performance comparison of the two mentioned scenarios is obtained by calculating the average value of 30 running-times in the decoder side. In addition to involving different test images, averaging helps to achieve more reliable results with respect to the measured run-time variations.

In order to compare the running-time in decoder side for both scenarios, the phone camera feed is utilized to generate the natural language captions. For simulating this platform, all the test images were displayed on a monitor; captured and then analyzed by the cell phone placed in front of it. During running the graphs, "tensorflow/tools/benchmark" tool was exploited to profile the execution time. It was observed that there are 36 kinds of operations in the graph when using dynamic control flow instructions and 22 kinds of operations when unrolling the network to the maximum length. Some kinds of operations are specific to the encoder such as *conv2D* and *maxpool*. Conversely, some others are decoder- specific such as *matmul*. Moreover, *add* and *mul* operations are common in both encoder and decoder sides. Practically, for each time of execution, decoder time (denoted by D) can be calculated by summing up the decoder-specific run-time $D^{specific}$ and the contribution of the decoding routine in common operations, D^{common}, using the following statement:

$$D_i = D_i^{specific} + D_i^{common} \tag{7}$$

Thus, by calculating the averages of decoder-specific and average contribution of decoder time in common operations, the average of whole decoding time in two scenarios is obtained with the following statement:

$$D^{avg} = D_{avg}^{specific} + D_{avg}^{common} \tag{8}$$

Since the encoding routine execution that considers the camera feed, is relatively straightforward; we generally employed the total and encoding execution time to achieve the average contribution of the decoder time in common operation run-times.

Accordingly, we performed an experiment using the frozen graph of the encoder part. This experiment was also performed on the mentioned subset of MSCOCO dataset including 30 test images. By analyzing common operations execution time as a random experiment, we can define three random variables: total elapsed time (T^{common}), encoder elapsed time (E^{common}) and finally decoder elapsed time (D^{common}). For each time of generating a caption, following relation is established:

$$D_i^{common} = T_i^{common} - E_i^{common} \tag{9}$$

Practically when N is large enough, we can get an approximation of D_{avg}^{common} using T_{avg}^{common} and E_{avg}^{common} for N repetitions of experiment. The relations of the mentioned variables are explained using following formula:

$$D_{Avg}^{common} \approx T_{Avg}^{common} - E_{Avg}^{common} \tag{10}$$

in which:

$$D_{Avg}^{common} = \sum_{i=1}^{N} D_i^{common} / N$$

$$T_{Avg}^{common} = \sum_{i=1}^{N} T_i^{common} / N \tag{11}$$

$$E_{Avg}^{common} = \sum_{i=1}^{N} E_i^{common} / N$$

Two right terms in (10) can be obtained by directly averaging the measured results. In this case, the D_{avg} for both approaches are calculated and reported in Table 1 by summing up $D_{avg}^{specific}$ and D_{avg}^{common} using (8). The two other reported parameters which are depicted in Table 1 are size of the frozen graph and memory usage of the model.

It should be noted that, the main focus of this paper is to devise dynamic control flow for decoding or language modeling part. In other words, we just changed the decoding side of graph and analyzed the decoding run-time for accelerating the sentence generation step. It is obvious that, the encoding step or vision part requires a lot of floating point computations since it includes several convolutional layers.

The model proposed in [5] achieved real-time performance on a mobile phone with CPU configuration of Octa-core (4 × 2.1 GHz Cortex-A53 & 4 × 1.7 GHz Cortex-A53)[2]. We employed this model on a weaker processor (Quad-core 1.2 GHz Cortex-A53 CPU[3]) to generate a caption. The decoding execution time and memory usage of static model [5] and the proposed dynamic decoding graphs are illustrated in Table 1. Furthermore, the whole model size related to both scenarios are expressed too. It is clearly shown that, the whole model size is slightly increased. Actually, the increment of the resultant value is very little in such a way that we can consider the same model size for both static and dynamic approaches. Conversely, the overall memory usage is slightly decreased which is related to the decrement of decoding steps of the recurrent neural network. More importantly, under these conditions the execution time illustrates about 31% speedup in decoding step.

Eventually, employing the proposed dataflow graph will result in substantially increasing the speed of the decoding routine with a small increase in the whole size of the model and a slight decrease in memory consumption.

Since the proposed method can considerably decrease the decoding execution time in comparison with the real-time static approach [5], the proposed model can be considered as a suitable candidate for real-time applications as well.

---

[2] Huawei Honor 6x.

[3] Samsung Galaxy J5.

Table 1. Decoding execution time, memory usage and model size of static and dynamic decoding graphs.

	Execution-time (ms)	Memory usage (MB)	Model size (KB)
Static graph [5]	711	110432	193867
Dynamic graph (proposed approach)	488	109596	193922

6   Conclusion

In this paper, a rapid implementation of the decoding side of image captioning on smart phone devices is presented. Practically, image captioning can be employed as a useful tool for improvement the quality of life for visually impaired people. By investigating the sentence length probability distribution, we tried to reduce the time of decoding step by eliminating the time overhead of computing the idle words. Intuitively, the recurrent neural networks are designed to model variable lengths sequences. Practically, the decoder uses its recurrent nature to loop over a fixed number of time steps which is considered as the maximum length of output caption. In the proposed method, using dynamic control flow may prevent unrolling the computation to a fixed maximum length. Using this facility will result in increased speed of the decoding routine in image captioning on smartphone devices. Experimental results on a subset of MSCOCO dataset validate our proposed approach.

References

1. Kiros, R., Salakhutdinov, R., Zemel, R.: Multimodal neural language models. In: ICML, pp. 595–603 (2014)
2. Vinyals, O., Toshev, A., Bengio, S., Erhan, D.: Show and tell: a neural image caption generator. In: Proceedings of IEEE Computer Society Conference on Computer Vision Pattern Recognition, 12–June, vol. 07, pp. 3156–3164 (2015)
3. Xu, K., et al.: Show, attend and tell: neural image caption generation with visual attention (2015)
4. Pedersoli, M., Lucas, T., Schmid, C., Verbeek, J.: Areas of attention for image captioning. arXiv (CVPR sub) (2016)
5. Mathur, P., Gill, A., Yadav, A., Mishra, A., Bansode, N.K.: Camera2Caption: a real-time image caption generator. ICCIDS 2017 – Proceedings of International Conference on Computational Intelligence Data Science, vol. 2018, no. 2015, pp. 1–6, (2018)
6. Farhadi, A., et al.: Every Picture Tells a Story: Generating Sentences from Images. In: Daniilidis, K., Maragos, P., Paragios, N. (eds.) ECCV 2010. LNCS, vol. 6314, pp. 15–29. Springer, Heidelberg (2010). https://doi.org/10.1007/978-3-642-15561-1_2
7. Kulkarni, G., et al.: Baby talk: understanding and generating simple image descriptions. IEEE Trans. Pattern Anal. Mach. Intell. 35(12), 2891–2903 (2013)
8. Elliott, D., Keller, F.: Image description using visual dependency representations. In: EMNLP, pp. 1292–1302, October 2013

9. Ordonez, V., Kulkarni, G., Berg, T.L.: Im2Text: describing images using 1 million captioned photographs. In: Shawe-Taylor, J., Zemel, R.S., Bartlett, P.L., Pereira, F., Weinberger, K.Q. (eds.) Advances in Neural Information Processing Systems 24. Curran Associates, Inc., pp. 1143–1151 (2011)
10. Gupta, A., Verma, Y., Jawahar, C.V., et al.: Choosing linguistics over vision to describe images. In: AAAI, p. 1 (2012)
11. Mao, J., Xu, W., Yang, Y., Wang, J., Yuille, A.L.: Explain images with multimodal recurrent neural networks, pp. 1–9 (2014)
12. Szegedy, C., Ioffe, S., Vanhoucke, V., Alemi, A.A.: Inception-v4, inception-resnet and the impact of residual connections on learning. In: AAAI, vol. 4, p. 12 (2017)
13. Hochreiter, S., Schmidhuber, J.: Long short-term memory. Neural Comput. 9(8), 1735–1780 (1997)
14. Abadi, M., et al.: TensorFlow : a system for large-scale machine learning (2016)
15. Yu, Y., et al.: Dynamic control flow in large-scale machine learning (2018)

Performance Issues and Quantum Computing

Scalable Performance Modeling and Evaluation of MapReduce Applications

Soroush Karimian-Aliabadi[1]([✉]), Danilo Ardagna[2], Reza Entezari-Maleki[3],
and Ali Movaghar[1]

[1] Computer Engineering Department, Sharif University of Technology, Tehran, Iran
skarimian@ce.sharif.ir, movaghar@sharif.ir
[2] Dipartimento di Elettronica Informazione e Bioingegneria,
Politecnico di Milano, Milan, Italy
danilo.ardagna@polimi.it
[3] School of Computer Engineering,
Iran University of Science and Technology (IUST), Tehran, Iran
entezari@iust.ac.ir

Abstract. Big Data frameworks are becoming complex systems which have to cope with the increasing rate and diversity of data production in nowadays applications. This implies an increase in number of the variables and parameters to set in the framework for it to perform well. Therefor an accurate performance model is necessary to evaluate the execution time before actually executing the application. Two main and prominent Big Data frameworks are Hadoop and Spark, for which multiple performance models have been proposed in literature. Unfortunately, these models lack enough scalability to compete with the increasing size and complexity of the frameworks and of the underlying infrastructures used in production environments. In this paper we propose a scalable Lumped SRN model to predict execution time of multi-stage MapReduce and Spark applications, and validate the model against experiments on TPC-DS benchmark using the CINECA Italian super computing center. Results show that the proposed model enables analysis for multiple simultaneous jobs with multiple users and stages for each job in reasonable time and predicts execution time of an application with an average error about 14.5%.

Keywords: Performance modeling · Scalable modeling · Stochastic reward nets · BigData frameworks · Map Reduce

1 Introduction

Huge amount of data is available in datacenters and is steadily, being produced in high velocity [1]. Data scientists need fast frameworks, specific algorithms, and even new programming paradigms to efficiently process this Big Data. The

© Springer Nature Switzerland AG 2019
L. Grandinetti et al. (Eds.): TopHPC 2019, CCIS 891, pp. 441–458, 2019.
https://doi.org/10.1007/978-3-030-33495-6_34

continuous struggle of data scientists with increasing size of data to be analyzed, led to handful of practical tools and methods. In 2008, Dean and Ghemawat proposed MapReduce (MR) paradigm [2] to process large amount of data on multiple node cluster to increase parallelism and therefor improved performance. The MR paradigm was not globally used until useful Hadoop framework [3] developed in 2011 by Apache. The Hadoop Distributed File System (HDFS) is a primary layer of the Hadoop ecosystem but not the only one. In 2013, Vavilapalli et al. [4], introduced YARN layer to the Hadoop cluster in order to specialize the resource management and make it dynamic rather than Hadoop's earlier static allocation scheme. With more complex dataflow in MR applications there was a need to cut down the complexity into multiple stages and thus Directed Acyclic Graphs (DAG) was chosen by Tez [5] developers to demonstrate the dataflow between stages of a complex application. Taking advantage of the memory's high speed and the Resilient Distributed Dataset (RDD) concept, Spark was created and became popular due to high speed and the ease of application development.

Tuning the framework and cluster parameters in order to reduce the execution time of a BigData application was a challenge from the earliest steps and a main part of this optimization process is to predict the execution time for a given set of parameters. But With each step in development of a more advanced framework for processing BigData, new set of parameters and complexity is created and execution time prediction made more and more challenging. A lot of works have been done in literature to simulate [6], model [7], or learn [8] the process, but their accuracy and scalability is only enough for simple runs with a single job running by one or more users and not for more complex applications with multiple multi-stage jobs running by number of users.

Among the approaches put into practice to predict execution time, analytical models play a prominent role. Simulation methods are time-consuming, and comprehensive simulators are rare and heavy to work with. Exploratory approaches like studying the history of past runs are also time-consuming and need a complete setup, while sometimes, predictions have to be made prior to the real setup. Learning traces is limited to a specified criteria and other performance measures need another whole learning process. Analytical methods on the other hand are fast to run, accurate enough, and give more insights on the process, hence number of performance measures can be studied using a single model. While deriving mathematical relations for a complex system like Spark framework is almost impossible, stochastic models are more feasible to work with. Fortunately, Petri Net (PN) and its stochastic derivations proved to be practical in literature and also have great tool support. Not all derivations of PN are analytically solvable, so the choice for the formalism is limited to few options which can be converted to Markov Models like Markov Reward Model (MRM). Stochastic Reward Net (SRN) is a formalism based on Stochastic Petri Nets (SPN) and is chosen as the formalism for building proposed models in this paper.

An inherent drawback of stochastic models like SRNs is the state space explosion problem due to increase in system variables and their cardinality. In the Spark framework which is usually running on top of the Hadoop cluster governed by YARN resource manager, multiple jobs are submitted simultaneously which have different execution DAGs. Number of Map/Reduce tasks in each stage is also different from other stages and jobs could be submitted by different users and thus dedicated priorities according to the YARN resource management policy. Modeling all of these variabilities in a monolithic model easily meets the state space explosion problem and is not feasible to analyze by regular hardwares.

To tackle this challenge, analytical models based on SRNs are proposed in this paper to accurately model the Hadoop and Spark framework running multiple jobs. We assume that MR Job is running on top of the Hadoop cluster and is governed by YARN resource manager with capacity scheduler. A lumping technique is also proposed to break down the complexity of the model and thus eliminate the state space explosion problem. Proposed models are validated against the real experiments of TPC-DS benchmark on CINECA supercomputer. Accuracy of proposed models in predicting execution time of Spark applications compared to results from experiments show an average error of 14.5% and the runtime of the analytic-numeric model solver is 15 s in average, which demonstrate the high scalability of the proposed SRN models.

The remaining parts of this paper are organized as follows. The Sect. 2 is dedicated to related proposals available in the literature, Sect. 3 presents the description of the features of the application frameworks. Our proposed SRN models for Hadoop MR and Spark applications in the single-class and multi-class forms are included in Sects. 4 and 5, respectively. In Sect. 6, we introduce The results obtained by the proposed models and their validation against the real systems. Finally, in Sect. 7, the paper concluded with some directions for future work.

2 Related Work

There are several researches on performance analysis of Big Data applications and tuning framework and cluster parameters. In this section, different methods for performing performance evaluation are being classified and reviewed one by one. Performing experiments and studying previous executions is a general way to reach insight on the performance of the framework. For example, in [9] MR job execution logs are used to assess performance measures and predict the future runs. Monitoring the execution of jobs in Hadoop, as proposed in [10], helps fine tuning the cluster and administrating the configuration parameters which are investigated thoroughly in [11].

A good predictor can learn from past executions and machine learning has a handful of tools and methods in this regard. From regression [12] to more sophisticated techniques like SVR [8] have been used in literature. Ernest [13], for example, is a tool set designed to predict Spark job execution time in large scale based on the behavior of the job on small samples of data. The sample set is produced using optimal experiment design and Non-negative Least Squares (NNLS) method to fit the model. Similar to Ernest, in [14] authors have explored sample representative mini dataset to train their model. Each stage of Spark application is first modeled using multiple polynomial regression and afterwards, Stage predictions are aggregated through the critical path of the execution DAG to estimate the whole job runtime. Combining analytical modeling power with machine learning methods has led to a operational system of MR job execution time estimator in [8]. Ataie et al. have combined queuing network model with SVR technique in [8] to further increase the accuracy and reduce the number of experiments to be performed for training the model.

Great efforts have been made to build a comprehensive simulator for Big Data frameworks like MR and Hadoop [6,15], and here, just few of them are mentioned. Ardagna et al. [6] have proposed DAGSim, a novel ad-hoc and fast discrete event simulator, to model the execution of complex DAGs, which can be used to predict Spark application runtime. Instead of building a Heavyweight simulator, others have designed simple graphical models which can be simulated. The approach, presented by Barbierato et al. [16], exploits Generalized Stochastic Petri Nets (GSPNs) alongside other formalisms such as process algebras and Markov chains to develop multi-formalism models and capture Hive queries performance behavior. More recently, Ruiz et al. [17] formalized the MR paradigm using Prioritized Timed Colored Petri Nets (PTCPNs). They validated the model and carried out a performance cost trade-off analysis. In [18], queuing network and Stochastic Well-formed Nets (SWN) simulation models have been proposed and validated for MR applications, considering YARN as resource manager. Requeno et al. [7] have proposed a UML profile for Apache Tez and transformed the stereotypes of the profile into Stochastic Petri Nets (SPNs).

Models solved analytically instead of simulation are the main focus of this work, specifically, models based on Markovian processes. Analytic model and formulation has a great value in getting to know a system, while the later, is difficult or some time impossible in complex systems like Big Data frameworks. Although finding a mathematic relation for execution time of a MR job is unachievable, Upper and lower bounds were analytically derived for MR job execution time in shared Hadoop clusters by authors in [19]. On the other hand, numerous works have used the more feasible approach of building analytical models. SPNs have been used by [20] for performance prediction of adaptive Big Data architecture. Mean Field Analysis was applied by authors in [20] to obtain average performance metrics. In order to cope with the Inevitable state space explosion problem, authors in [21] used Fluid Petri Nets to simplify the actual model. They proposed fluid models to predict the execution time of the MR and Spark application.

3   System Architecture and Application Structure

In order to better understand the behavior of proposed models, it's necessary to provide a background for the target system and application as well as the assumptions. The primary system of concern is Spark framework and Spark application. This choice is due to Spark's extensive use among enterprises and its popularity between data scientists [22]. However, modeling starts from simple Hadoop MR application, and then the applicability of models to Spark framework is discussed. Spark is able to run on top of the Hadoop cluster governed by YARN resource manager and since it is the usual deployment option [23], it is also the choice of our architecture in this paper. The outline of the system is depicted in Fig. 1.

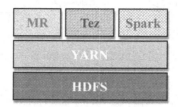

Fig. 1. The general schema of BigData frameworks

An MR job is consisted of three main phases: Map, Shuffle, and Reduce. Each of these phases include number of tasks that run in parallel on different cluster nodes. Map tasks perform computation on input data chunks and Shuffle tasks are responsible for gathering Map phase output to Reduce tasks. Reduce tasks perform aggregation on intermediate data and output the final result. Since Reduce and Shuffle tasks run on same thread, from hereafter, we consider a Reduce task an aggregation of corresponding Shuffle task and the succeeding Reduce task.

Spark programming model is similar to MR but extends it with a data-sharing in memory abstraction called Resilient Distributed Datasets (RDD). Every Spark application consists of a number of stages. A stage corresponds to an operation on RDDs and can be seen as Map, Shuffle, or Reduce phases in MR jobs. Each stage consists of multiple tasks running in parallel and distributed in the cluster. Stages are linked to each other in the form of a Directed Acyclic Graph (DAG) which demonstrates the flow of data between stages and also the execution order among them. The execution DAG specifies the parallelization degree and the critical path as well. A sample Spark application can be seen in Fig. 2.

Although, jobs were scheduled in earlier versions of the Hadoop framework by FIFO policy, better schemes are available today. Hadoop 2.x and Hadoop 3 let more complex schedulers (i.e. Capacity and Fair schedulers) to be plugged into the framework. A cluster is a resource pool in YARN, enabling dynamic allocation of resources (containers) to the ready tasks. We assume Capacity scheme for the YARN layer and this means that in the multi class environment jobs in each class run in FIFO manner and next job can only start if the last stage of the previous job has acquired all resources necessary to accomplish. YARN capacity scheme indicates a specified share of the resources for each class, so classes can race to acquire resources until they are under provisioned according

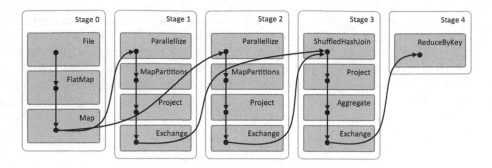

Fig. 2. A sample Spark application execution DAG

to their share. Once a class acquired all its share from the resources, it should leave available resources in the favor of other under provisioned classes.

In this paper, we consider target Hadoop clusters running on a set of homogeneous resources [24], including MR and Spark execution engines on top of the YARN Capacity scheduler [18]. This implies that the cluster capacity is partitioned into multiple queues and within a queue, multiple applications are scheduled in a FIFO manner. Multiple users can run the same query, which is submitted to a specific queue. Moreover, after obtaining results, end users can submit the same query again (possibly changing interactively some parameters) after a think time. In other words, a multi-class closed performance model is considered [25].

4   Single-Class Model

In this section, the proposed SRN model for a Spark application is presented. Formal definition of SRN formalism, its structure, and behavior are given in [26] which have been omitted herein for the sake of space limitation. SRN formalism is widely used in other areas of computer science such as Cloud Computing [27] and Computational Grids [28], for performance evaluation, and is proved to be practical. First, the proposed model is described in detail and afterwards, the validation and computation time of the model is discussed and lumping method is proposed to overcome the model complexity.

4.1   Proposed Model

The SRN model proposed for Spark application is shown in Fig. 3 and is modeling the execution of sample Spark application of Fig. 2. The model includes different stages, shared resources, think time and the scheduling mechanism. It should be noted that since stages in Spark are analogous to phases in Tez and MR frameworks, this model can be easily simplified for modeling Tez and MR applications as well. The detailed description of model behavior is as follows.

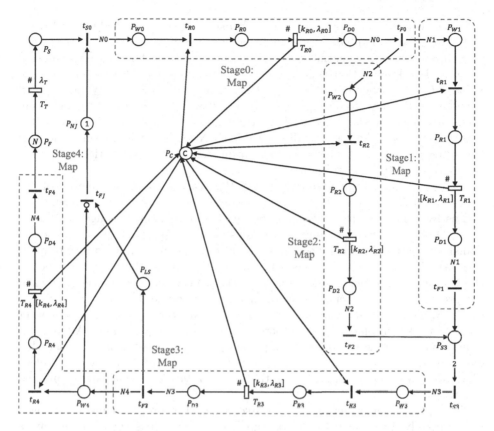

Fig. 3. The SRN model proposed for single class Spark application

In the initial state, there are N tokens in place P_F showing the jobs waiting to start execution. The marking dependent timed transition T_T models the think time of a waiting job. Upon completion of this transition, a token from place P_F is moved to place P_S with rate λ_T, which is the rate of the exponential distribution considered for transition T_T. Existence of a token in place P_S triggers instantaneous transition t_{S0} to start the job if the place P_{NJ} has a token to consume. The place P_{NJ} initially contains a token modeling the possibility of starting a waiting job according to the Capacity scheduler policy. At the start of a job, the instantaneous transition t_{S0} will produce N_0 tokens in place P_{W0}, each one representing a task in the first stage. Entities P_{Wi}, t_{Ri}, P_{Ri}, T_{Ri}, P_{Di}, N_i, and t_{Fi} where $0 \leq i \leq 4$ together simulate ith stage. Place P_{Wi} is starting point of stage i and shows tasks waiting for resource. Instantaneous transition t_{Ri} Allocates an available resource to a task, which removes one token from place P_C and one from P_{Wi}, and adds a token to place P_{Ri}. Place P_C is modeling the pool of containers, which is initially set to contain C tokens representing the total number of containers. The execution of a single task is modeled by the timed transition T_{Ri}, which returns the resource to the pool of available resources whenever a task is done. This transition is characterized by the Erlang distribution with shape k_{Ri} and a marking dependent rate λ_{Ri}. According to our experiments, the exponential distribution is not the case for the task execution time, and task execution time fits better with more general distributions like Erlang. On the other hand, for the SRN model to be analytically solvable, all timed activities have to be exponentially distributed [29]. Fortunately, an Erlang distribution can be simulated with a set of continuous exponential activities [30] helping us to use the analytically solvable SRN models, when some actions of the system follow Erlang distribution. Parameters of the distributions are being calculated from the experiment logs.

Once the number of tokens in place P_{Di} reaches the total number of the tasks, the ith stage is finished and the instantaneous transition t_{Fi} consumes N_i tokens from P_{Di}, where N_i denotes the number of tasks in ith stage, and starts $(i+1)$th stage with producing N_{i+1} tokens in place $P_{W(i+1)}$. The completion of transition t_{Fi} for the second to last stage, also results in adding a token to place P_{LS}, which indicates that a job is performing its last stage. Recalling from Sect. 3, the Capacity scheduler implies that the next job can start executing only when the previous job has received all of the necessary resources for completing its last stage. Similarly, in our model, instantaneous transition t_{FJ} enables, whenever there is a token in place P_{LS} and there is no token left in place P_{W4}. Afterwards, transition t_{FJ} removes a token from place P_{LS}, and puts a token into place P_{NJ} enabling instantaneous transition t_{S0} to start the next job. Notice that place P_{S3} and transition t_{S3} are assuring that both $stage_1$ and $stage_2$ are completed before $stage_3$ is started.

The performance measure to be assessed by the proposed model of Fig. 3 is the steady-state mean execution time of jobs, which is the average time a token needs to move from place P_S to place P_F. In order to compute the mean execution time, the reward shown in Eq. 1 is defined.

$$r = \frac{N}{throughput_{t_{F4}}} - \frac{1}{\lambda_T} \tag{1}$$

where $throughput_{t_{F4}}$ is the throughput of the instantaneous transition t_{F4} and can be calculated by Eq. 2.

$$throughput_{t_{F4}} = \mathbb{P}(\#P_{D4} = N_4 - 1) \cdot \lambda_{R4} \tag{2}$$

where $\mathbb{P}(\#P_{D4} = N_4 - 1)$ is the probability of being in a state where all but one tasks are finished in the last stage, so there are $N_4 - 1$ tokens in place P_{D4} and one token left to finish the entire job. This probability is multiplied by λ_{R4}, which is the rate of executing a task in the last stage.

4.2 Lumping Technique

Although the proposed model seems to conform with the Spark execution model, in real world, scalability remains a low point for this model. Technically, the model of Fig. 3 could easily grow in state space with increasing number of stages and face the state space explosion problem. The issue gets even worse in multi-class environments where multiple multi-stage Spark applications are running in parallel. Our experiments, also support this claim, so that, the SRN model of Fig. 3 takes more than 4 days to solve analytically. Details of experiment setup and tool set is described in Sect. 6. Therefore, a heuristic approach is introduced to decrease the complexity of the model by reducing the cardinality of the parameters. The basics of the heuristic is the fact that according to the experiments, tasks in a single stage run in waves. That is when the number of tasks is greater than number of cores, then at the beginning of the ith stage, all of the C cores are assigned to tasks and $N_i - C$ tasks are left. After a while, running tasks, eventually finish and release their acquired resources. With the assumption that the runtime of an individual task is almost similar to other tasks of the same stage, then according to Fig. 4, next C tasks will acquire C available resources. A group of C tasks, is called a wave.

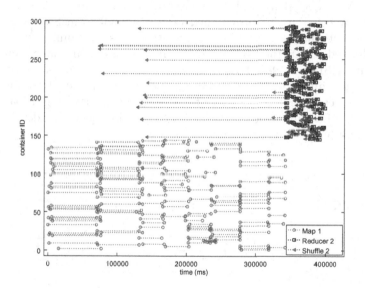

Fig. 4. The execution plot for a sample MR application

Modeling the execution of waves instead of individual tasks, reduces the cardinality of tasks and obviously increases the scalability of the model. This will be done by lumping tasks to waves and assume a single resource with $\lceil N_i/C \rceil$ waves. Although the assumption is not real and implies some error to the results, since tasks execution times are not exactly same, but experiments in following section shows that the error is still acceptable.

5   Multi-class Model

So far, the execution of a single class of application by multiple users is discussed and an analytic SRN model is devised to evaluate the execution time of such environment with a heuristic to make the model feasible to solve in reasonable time. In this section, the execution of multiple application classes is addressed, each running by multiple users and the SRN model of Fig. 3 is extended to evaluate the performance of a multi-class environment using the lumping technique discussed in Subsect. 4.2.

5.1   Proposed Model

According to the YARN scheduler, different classes of applications can be managed in multiple queues each has a share of the resources which is indicated by the framework operator. This means when a class of application is under provisioned according to its share, and at the same time is requesting for a resource then YARN will provide this class with more priority than other classes. In order to increase the utilization YARN also lets fully provisioned classes to acquire more resources than their share, only if remaining resources are free and not being demanded by other under provisioned classes. Recall from Sect. 3 that inside each queue different instances of the application class are running in FIFO manner. As claimed so far, the SRN model for multi-class environment can be built by replicating SRN model of Fig. 3 as many as the number of classes and let them race to acquire resources by the rules described earlier. Here, in order to be more concrete, assume 4 classes of MR applications running in parallel and the model for this sample environment is depicted in Fig. 5. The detail of model structure and behavior is as follows.

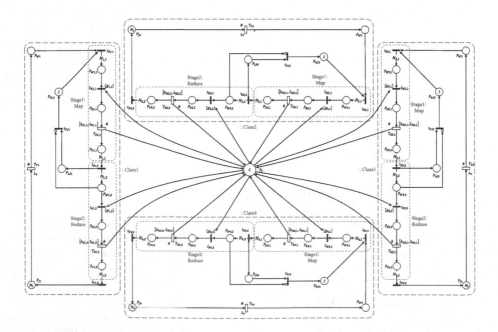

Fig. 5. The SRN model proposed for multi class MR application

For the sake of simplicity, MR applications are considered here, where as you can see in Fig. 5 there are just two Map and Reduce stages in each application. While the behavior of sub-models for each of the classes is similar to Fig. 3, the main difference is how instantaneous transitions $t_{Ri,j}$ enable. Instantaneous transition $t_{Ri,j}$ is responsible for acquiring resource in ith class, where $1 \leq i \leq 4$, and jth stage, where $1 \leq j \leq 2$ and enables according to the guard function $g_{i,j}$. The guard function $g_{i,j}$ is formulated as below.

$$g_{i,j} = \begin{cases} 1, & \text{if } (\#P_{Wi,j} > 0) \text{ and } (\#P_C > 0) \text{ and } (\\ & (\sum_{j=1}^{2} \#P_{Ri,j} < S_i \cdot C) \text{ or} \\ & \text{for each class } k \text{ and } k \neq i : \\ & (\sum_{j=1}^{2} \#P_{Wk,j} = 0 \text{ or } \sum_{j=1}^{2} \#P_{Rk,j} \geq S_k \cdot C) \\ &) \\ 0, & \text{otherwise} \end{cases} \tag{3}$$

where S_i denotes the share factor of ith class. The desired performance measure of the proposed model of Fig. 5 is similar to the performance measure devised for model of Fig. 3, that is steady-state mean execution time of each job in each class and is calculated by Eq. 4.

$$r_i = \frac{N_i}{throughput_{t_{Fi,2}}} - \frac{1}{\lambda_T} \tag{4}$$

where $throughput_{t_{Fi,2}}$ can be calculated by Eq. 5.

$$throughput_{t_{Fi,2}} = \mathbb{P}(\#P_{Fi,2} = N_{i,2} - 1) \cdot \lambda_{Ri,2} \tag{5}$$

5.2 Lumping Technique

The modal of Fig. 5 with multiple classes of applications, each running by multiple users has a huge state space which will be more unattainable if number of tasks in each stage increases. In Sect. 4 lumping technique proposed to scale down the number of tasks to number of waves. Here a similar technique is leveraged to make the model feasible to analyze. Assuming that c_i containers are assigned to each class i, c_{gcd} can be calculated as the gcd of c_i values. Afterwards, both the number of tasks and containers are divided by c_{gcd}, simulating a lumping technique. For example, if there are 240 total containers and each of 4 application classes have a share of 60 containers then a stage with 300 tasks will have approximately, 5 waves. This approximation is not realistic since sometimes the application will receive less or more resources than its share, however this assumption enables the model to be analytically tractable with an acceptable error.

6 Numeric-Analytic Results

The numeric-analytic solution to proposed SRN models is preferred to simulation results, since simulation methods can grow in runtime and usually lack in scalability as discussed in Sect. 1, therefore, we chose SPNP tool [35] and its steady-state iterative solver, in order to analyze our proposed models, considering that SPNP is the state of the art tool for analytic solutions of SRN models. In order to assess the accuracy of the numeric results obtained from the model we have conducted several experiments on real world platforms including public Clouds to private cluster and we defined the measure below as the error value.

$$\theta_{SRN} = |\frac{\tau_{SRN} - T}{T}| \tag{6}$$

where θ_{SRN} denotes the relative error between the SRN model and experiments, T is the execution time of a job on the real system under test which is measured from the experiment, and τ_{SRN} is the execution time received from the SRN model as the result of numeric solution. The experiment setup includes the PICO Big Data cluster available at CINECA [32] configured with number of cores ranging from 40 to 120, each of them assigned to a single container and the private cluster. Our IBM Power8 (P8) private cluster includes 4 VMs with 11 cores and 60 GB of RAM for each. Spark executors are configured to leverage 2 cores and 4 GB of RAM, while Spark drivers use 8 GB of RAM. The configurations vary in number of cores from 6 to 44 and number of executors from 3 to 22.

```
select avg(ws_quantity),
avg(ws_ext_sales_price),
avg(ws_ext_wholesale_cost),
sum(ws_ext_wholesale_cost)
from web_sales
where
(web_sales.ws_sales_price between
100.00 and 150.00) or
(web_sales.ws_net_profit between
100 and 200)
group by ws_web_page_sk
limit 100;
```

```
select inv_item_sk,
inv_warehouse_sk
from inventory
where
inv_quantity_on_hand > 10
group by inv_item_sk,
inv_warehouse_sk
having
sum(inv_quantity_on_hand) > 20
limit 100
;
```

(a) Q_1 (b) Q_2

```
select avg(ss_quantity),
  avg(ss_net_profit)
from store_sales
where
  ss_quantity > 10 and
  ss_net_profit > 0
group by ss_store_sk
having
  avg(ss_quantity) > 20
limit 100;
```

```
select cs_item_sk,
avg(cs_quantity) as aq
from catalog_sales
where cs_quantity > 2
group by cs_item_sk;
```

```
select
  inv_warehouse_sk,
  sum(inv_quantity)
from inventory
group by
  inv_warehouse_sk
having
  sum(inv_quantity) > 5
limit 100;
```

(c) Q_3 (d) Q_4 (e) Q_5

Fig. 6. MR queries (Q_3 to Q_5)

The dataset used for running the experiments was generated with the TPC-DS benchmark data generator [33], which is the industry standard for benchmarking data warehouses. Datasets are in the form of external tables for the Hive [34] queries and their size varies from 250 GB to 1 TB. Different queries are considered to be executed on datasets as MR or Spark applications. For the case of single-class scenario, both MR and Spark applications were chosen. Queries Q_1 and Q_2 are designed in Fig. 6 for which number of Map and Reduce tasks vary from 1 to 151 in different configurations. Queries Q_{26} and Q_{52} are selected from TPC-DS catalog for Spark applications and vary in number of stages from 4 to 8 each one having 1 to 1000 tasks. Finally, for the case of multi-class scenario MR queries were completed with queries Q_3 to Q_5 of Fig. 6 so different configurations of queries Q_1 to Q_5 can be executed in 4 queues. Number of users in each queue differs from 2 to 10 and number of tasks range from 1 to 600 in each Map or Reduce phase.

Table 1. Results obtained from the proposed SRN model for single class applications

Query	Users	Cores	Scale [GB]	T [ms]	τ_{SRN} [ms]	ϑ_{SRN} [%]
Q_1	1	60	250	80316	81285	1.21
Q_2	1	60	250	84551	86624	2.45
Q_1	3	20	250	1002160	1059403	5.71
Q_1	3	40	250	340319	380881	11.92
Q_2	3	20	250	95403	88982	6.73
Q_2	3	40	250	86023	76936	10.56
Q_1	5	20	250	1736949	1827978	5.24
Q_1	5	40	250	636694	688759	8.18
Q_2	5	20	250	145646	148453	1.93
Q_2	5	40	250	90674	106200	17.12
Q_{26}	1	24	250	178714	142446	20.29
Q_{26}	1	32	250	168041	116364	30.75
Q_{52}	1	24	250	181496	144862	20.18
Q_{52}	1	32	250	162232	121392	25.17
Q_{52}	1	48	750	279243	234573	16
Q_{52}	1	48	1000	359987	312014	13.33
Q_{26}	1	6	500	2532250	2720902	7.45
Q_{26}	1	8	500	2071159	2179066	5.21
Q_{26}	1	10	500	1778802	1878948	5.63

in order to estimate the mean execution time of tasks in stages which is necessary to solve our proposed SRN models a profiling step conducted which is a common idea as stated in different literatures [13, 14]. A pilot execution was designed for

Table 2. Results obtained from the proposed lumped SRN model for multi class MR applications

Configuration	Query	Users	T [ms]	τ_{SRN} [ms]	ϑ_{SRN} [%]
1	Q_2	5	118667	86094	27.45
	Q_5	5	120947	117429	2.91
2	Q_1	4	206938	235267	13.69
	Q_3	2	258220	275443	6.67
	Q_4	2	246750	250426	1.49
3	Q_2	10	252800	182187	27.93
	Q_5	10	246702	244860	0.75
4	Q_1	10	436212	555341	27.31
	Q_3	5	599848	710399	18.43
	Q_4	5	584583	633454	8.36
5	Q_1	5	264515	335193	26.72
	Q_2	5	998941	1163866	16.51
	Q_3	3	337716	356155	5.46
	Q_4	3	331327	338616	2.2
6	Q_1	5	363047	469673	29.37
	Q_2	5	1479480	1814878	22.67
	Q_3	10	1003998	1064639	6.04
	Q_4	10	1015219	1053188	3.74
7	Q_1	10	468085	601863	28.58
	Q_3	5	613296	658311	7.34
	Q_4	5	621132	675170	8.7
	Q_5	10	1060763	1317149	24.17
8	Q_1	10	452974	522958	15.45
	Q_2	10	1870229	2316278	23.85
	Q_3	5	613190	718536	17.18
	Q_4	5	587453	598438	1.87

each query with a minimum size cluster and mean execution times was measured through execution logs. Obtained values were then used for other cluster and dataset sizes. Task durations are measured as average values between 20 runs and fit better with Erlang distribution for Map tasks while for Reduce tasks exponential distribution fits good enough.

The results obtained from the experiments and the proposed SRN model are shown in Table 2 for single-class MR and Spark applications. The average error is 11.31% which offers the acceptable performance of lumping technique. Finally, the accuracy of the proposed model for multi-class MR applications is evaluated in Table 1 which shows an average error of 14.5%. Despite the increase in error, model runtime reduced from couple days to couple seconds and the improved scalability is the main contribution of lumping technique.

7 Conclusion and Future Works

In this paper we have discussed one of the challenges in BigData area. The fact that a BigData cluster which is equipped with the complete stack of frameworks and tools like Hadoop, YARN, Tez, or Spark, has a huge parameter set and tuning these values is not possible without a useful and accurate performance model which can predict the execution time of applications running in the cluster. Previous works in this field are mainly focused on simulation, learning, experiment, and log analysis which are time-consuming, costly, not accurate enough, and not general according to different performance measures. Therefore, analytical models were chosen, due to their low runtime, more general insight, and high accuracy.

New version of frameworks like Spark, support shared environment for simultaneous applications and users to run and make use of available resources. This feature will cause performance models to grow in state space and face state space explosion problem. While previous work on analytical model [18] considers multiple users and multiple stages, it is limited to single-class executions and suffers from state space explosion problem, our proposed model is scalable in the way that can predict execution time of applications in the presence of other simultaneous applications with different classes of jobs.

In this paper, analytical SRN models were proposed to evaluate most popular BigData frameworks Hadoop and Spark. Despite previous works [18] which have only considered single class executions, a lumping method is proposed to cope with the state space explosion problem and therefore, enable our model to evaluate the performance of multi-class executions. SRN models are then solved using numeric-analytic solver which outperforms other methods in low runtime and sufficient accuracy. Results from numeric-analytic solver are then compared to experiments on CINECA supercomputer considering TPC-DS benchmark workloads and the reported error is 11.3% and 14.5% in average for single-class and multi-class scenarios respectively, which is adequate to support capacity planning decisions and what-if analysis [25].

Future work will extend the models to support additional scenarios of interest like execution with faulty nodes, data placement, and speculative execution. Sensitivity analysis can also be derived in order to find the most effective parameters in execution time of applications. Effective parameters are those which will be optimized first to reach improved performance with the least effort.

Acknowledgment. The results of this work have been partially funded by the European DICE H2020 research project (grant agreement no. 644869).

References

1. Reinsel, D., Gantz, J., Rydning, J.: Data age 2025: the evolution of data to life-critical (2017). https://www.seagate.com/de/de/our-story/data-age-2025/. Accessed July 2018
2. Dean, J., Ghemawat, S.: MapReduce: simplified data processing on large clusters. Commun. ACM **51**(1), 107–113 (2008). https://doi.org/10.1145/1327452.1327492
3. Apache, Apache Hadoop. http://hadoop.apache.org/. Accessed July 2018
4. Vavilapalli, V.K., et al.: Apache hadoop yarn: yet another resource negotiator. In: Proceedings of the 4th Annual Symposium on Cloud Computing, SOCC 2013, pp. 1–16. ACM Press, Santa Clara (2013). https://doi.org/10.1145/2523616.2523633
5. Saha, B., Shah, H., Seth, S., Vijayaraghavan, G., Murthy, A., Curino, C.: Apache Tez: a unifying framework for modeling and building data processing applications. In: Proceedings of the 2015 ACM International Conference on Management of Data, SIGMOD 2015, pp. 1357–1369. ACM Press, Melbourne (2015). https://doi.org/10.1145/2723372.2742790
6. Ardagna, D., et al.: Performance prediction of cloud-based big data applications. In: Proceedings of the 2018 ACM/SPEC International Conference on Performance Engineering, ICPE 2018, pp. 192–199. ACM Press, Berlin (2018). https://doi.org/10.1145/3184407.3184420
7. Requeno, J.I., Gascón, I., Merseguer, J.: Towards the performance analysis of Apache Tez applications. In: Proceedings of the 2018 ACM/SPEC International Conference on Performance Engineering, ICPE 2018, pp. 147–152. ACM Press, Berlin (2018). https://doi.org/10.1145/3185768.3186284
8. Ataie, E., Gianniti, E., Ardagna, D., Movaghar, A.: A combined analytical modeling machine learning approach for performance prediction of MapReduce jobs in cloud environment. In: Proceedings of the 18th International Symposium on Symbolic and Numeric Algorithms for Scientific Computing, SYNASC 2016, pp. 431–439. IEEE, Timisoara (2016). https://doi.org/10.1109/SYNASC.2016.072
9. Zhang, Z., Cherkasova, L., Loo, B.T.: Benchmarking approach for designing a MapReduce performance model. In: Proceedings of the ACM/SPEC International Conference on International Conference on Performance Engineering, ICPE 2013, pp. 253–258. ACM Press, Prague (2013). https://doi.org/10.1145/2479871.2479906
10. Dai, J., Huang, J., Huang, S., Huang, B., Liu, Y.: HiTune: dataflow-based performance analysis for big data cloud. In: Proceedings of the USENIX Annual Technical Conference, pp. 87–100. USENIX Association, Portland (2011)
11. Jiang, D., Ooi, B.C., Shi, L., Wu, S.: The performance of MapReduce: an in-depth study. Proc. VLDB Endowment **3**(1–2), 472–483 (2010). https://doi.org/10.14778/1920841.1920903
12. Yigitbasi, N., Willke, T.L., Liao, G., Epema, D.: Towards machine learning-based auto-tuning of MapReduce. In: Proceedings of the IEEE 21st International Symposium on Modelling, Analysis and Simulation of Computer and Telecommunication Systems, pp. 11–20. IEEE, San Francisco (2013). https://doi.org/10.1109/MASCOTS.2013.9
13. Venkataraman, S., Yang, Z., Franklin, M., Recht, B., Stoica, I.: Ernest: efficient performance prediction for large-scale advanced analytics. In: Proceedings of the 13th Usenix Conference on Networked Systems Design and Implementation, pp. 363–378. USENIX Association, Santa Clara (2016)
14. Gibilisco, G.P., Li, M., Zhang, L., Ardagna, D.: Stage aware performance modeling of DAG based in memory analytic platforms. In: Proceedings of the 9th Interna-

tional Conference on Cloud Computing, CLOUD 2016, pp. 188–195. IEEE, San Francisco (2016). https://doi.org/10.1109/CLOUD.2016.0034

15. Liu, Y., Li, M., Alham, N.K., Hammoud, S.: HSim: a MapReduce simulator in enabling cloud computing. Future Gener. Comput. Syst. **29**(1), 300–308 (2013). https://doi.org/10.1016/j.future.2011.05.007

16. Gribaudo, M., Barbierato, E., Iacono, M.: Modeling apache hive based applications in big data architectures. In: Proceedings of the 7th International Conference on Performance Evaluation Methodologies and Tools, ValueTools 2013, pp. 30–38 ICST, Torino (2013). https://doi.org/10.4108/icst.valuetools.2013.254398

17. Ruiz, M.C., Calleja, J., Cazorla, D.: Petri nets formalization of Map/Reduce paradigm to optimise the performance-cost tradeoff. In: Proceedings of the 2015 IEEE Trustcom/BigDataSE/ISPA, Vol. 3, pp. 92–99. IEEE, Helsinki (2015). https://doi.org/10.1109/Trustcom.2015.617

18. Ardagna, D., Bernardi, S., Gianniti, E., Karimian Aliabadi, S., Perez-Palacin, D., Requeno, J.I.: Modeling performance of hadoop applications: a journey from queueing networks to stochastic well formed nets. In: Carretero, J., Garcia-Blas, J., Ko, R.K.L., Mueller, P., Nakano, K. (eds.) ICA3PP 2016. LNCS, vol. 10048, pp. 599–613. Springer, Cham (2016). https://doi.org/10.1007/978-3-319-49583-5_47

19. Malekimajd, M., Ardagna, D., Ciavotta, M., Rizzi, A.M., Passacantando, M.: Optimal map reduce job capacity allocation in cloud systems. ACM SIGMETRICS Perform. Eval. Rev. **42**(4), 51–61 (2015). https://doi.org/10.1145/2788402.2788410

20. Castiglione, A., Gribaudo, M., Iacono, M., Palmieri, F.: Exploiting mean field analysis to model performances of big data architectures. Future Gener. Comput. Syst. **37**, 203–211 (2014). https://doi.org/10.1016/j.future.2013.07.016

21. Gianniti, E., Rizzi, A.M., Barbierato, E., Gribaudo, M., Ardagna, D.: Fluid petri nets for the performance evaluation of MapReduce and spark applications. ACM SIGMETRICS Perform. Eval. Rev. **44**(4), 23–36 (2017). https://doi.org/10.1145/3092819.3092824

22. Spark, Apache Spark. http://spark.apache.org/. Accessed July 2018

23. Alipourfard, O., Harry Liu, H., Chen, J., Venkataraman, S., Yu, M., Zhang, M.: CherryPick: adaptively unearthing the best cloud configurations for big data analytics. In: Proceedings of the 14th USENIX Symposium on Networked Systems Design and Implementation (NSDI 2017), Boston, MA, USA, pp. 469–482 (2017)

24. Teng, F., Yu, L., Magoulès, F.: SimMapReduce: a simulator for modeling MapReduce framework. In: Proceedings of the Fifth FTRA International Conference on Multimedia and Ubiquitous Engineering, pp. 277–282. IEEE, Loutraki (2011). https://doi.org/10.1109/MUE.2011.56

25. Lazowska, E.D., Zahorjan, J., Graham, G.S., Sevcik, K.C.: Quantitative System Performance: Computer System Analysis Using Queueing Network Models, 1st edn. Prentice-Hall, Upper Saddle River (1984)

26. Ciardo, G., Trivedi, K.S.: A decomposition approach for stochastic reward net models. Perform. Eval. **18**(1), 37–59 (1993). https://doi.org/10.1016/0166-5316(93)90026-Q. https://www.sciencedirect.com/science/article/pii/0166531693 90026Q

27. Ataie, E., Entezari-Maleki, R., Rashidi, L., Trivedi, K.S., Ardagna, D., Movaghar, A.: Hierarchical stochastic models for performance, availability, and power consumption analysis of IaaS clouds. IEEE Trans. Cloud Comput. (to appear). https://doi.org/10.1109/TCC.2017.2760836

28. Entezari-Maleki, R., Trivedi, K.S., Movaghar, A.: Performability evaluation of grid environments using stochastic reward nets. IEEE Trans. Dependable Secure Comput. **12**(2), 204–216 (2015). https://doi.org/10.1109/TDSC.2014.2320741

29. Meyer, J.F., Movaghar, A., Sanders, W.H.: Stochastic activity networks: structure, behavior, and application. In: Proceedings of the International Workshop on Timed Petri Nets, Torino, Italy, pp. 106–115 (1985)
30. Reinecke, P., Bodrog, L., Danilkina, A.: Phase-type distributions. In: Wolter, K., Avritzer, A., Vieira, M., van Moorsel, A. (eds.) Resilience Assessment and Evaluation of Computing Systems, pp. 85–113. Springer, Berlin (2012). https://doi.org/10.1007/978-3-642-29032-9_5
31. Flexiant: Flexiant cloud management software & cloud orchestration. https://www.flexiant.com/. Accessed July 2018
32. Cineca: Cineca computing center. http://www.cineca.it/. Accessed July 2018
33. Poess, M., Smith, B., Kollar, L., Larson, P.: TPC-DS, taking decision support benchmarking to the next level. In: Proceedings of the 2002 ACM International Conference on Management of data, SIGMOD 2002, pp. 582–587. ACM Press, Madison (2002). https://doi.org/10.1145/564691.564759
34. Hive: Apache Hive. https://hive.apache.org/. Accessed July 2018
35. Hirel, C., Tuffin, B., Trivedi, K.S.: SPNP: stochastic petri nets. Version 6.0. In: Haverkort, B.R., Bohnenkamp, H.C., Smith, C.U. (eds.) TOOLS 2000. LNCS, vol. 1786, pp. 354–357. Springer, Heidelberg (2000). https://doi.org/10.1007/3-540-46429-8_30

Performance Improvement of Multimedia Kernels Using Data- and Thread- Level Parallelism on CPU Platform

Maryam Moradifar, Asadollah Shahbahrami(✉), Mina Nematpour, and Hossein Amiri

Department of Computer Engineering, Faculty of Engineering, University of Guilan, Rasht, Iran
shahbahrami@guilan.ac.ir

Abstract. Processor vendors have been expanding Single Instruction Multiple Data (SIMD) extensions in their General Purpose Processors (GPPs). These extensions have their own instruction set architecture and equipped with Special Purpose Instructions (SPIs) to exploit Data Level Parallelism (DLP). In addition, to these extensions, GPPs have been equipped with multicore technologies so that each processor has multicore to process program using exploiting Thread-level Parallelism (TLP). In order to exploit these technologies, SIMD and multicore, many parallel programming models such as Intrinsic Programming Model (IPM), and Compiler's Automatic Vectorization (CAV) and OpenMP have been proposed. Increasing performance using DLP depends on the number of data that can be processed in parallel using SIMD instructions. While performance improvement using TLP depends on the number of cores and program dependencies. In order to increase the performance of multimedia kernels, we exploit both DLP and TLP using parallel programming models such as IPM, CAVs, and OpenMP in this paper. Our experimental results show that the combination of DLP and TLP can improve performance significantly compared to each DLP and TLP individually. In addition, various compilers such as ICC, GCC, and LLVM are evaluated in terms of automatic vectorization. The obtained results show that ICC and GCC compilers have more ability to vectorize the kernels in comparison with LLVM compiler. In addition, despite the higher efficiency of IPM than CAVs, it is tedious and error-prone, and more attention is needed to develop and to extend auto-vectorization.

Keywords: Multimedia · Parallel processing · Single Instruction Multiple Data · Thread-level parallelism · Data-level parallelism

1 Introduction

Multimedia standards such as JPEG, JPEG2000, MPEG-2/4, and H26x are one of the most important applications available to all users. These applications have their own special features, including the ability to use fine and coarse-grained parallelism, considerable data reorganization, small loops, high memory bandwidth requirement, and small data types. These features cause some limitations for processing of multimedia

L. Grandinetti et al. (Eds.): TopHPC 2019, CCIS 891, pp. 459–467, 2019.
https://doi.org/10.1007/978-3-030-33495-6_35

kernels and applications in General-Purpose Processors (GPPs) [1]. In order to improve the performance of multimedia applications, GPPs vendors have extended their Instruction Set Architectures (ISAs). These ISA extensions use the subword level parallelism which processes multiple data using a single instruction that is called Single Instruction Multiple Data (SIMD) extension [2–4]. In addition to SIMD extensions which have been using for performance improvement using data-level parallelism, all processors have almost multiple cores that can process multiple threads in parallel using thread-level parallelism.

The purpose of this paper is to increase the performance of multimedia kernels using both data- and thread-level parallelisms. Data-level parallelism is based on the AVX2 SIMD technology, the Intrinsic Programming Model (IPM) and Compiler's Automatic Vectorization (CAV) and thread-level parallelism is based on OpenMP programming. The evaluation is based on an Intel's sixth-generation multi-core processor using ICC, GCC, and LLVM compilers. The results show that multi-threading improves performance, but with the combination of SIMD instructions and multi-threading capability, performance is significantly increased. In addition, our experimental results show that each compiler, ICC, GCC, and LLVM has different behaviors in dealing with each kernel. For example, ICC and GCC compilers can more efficiently vectorize kernels and use Special Purpose Instructions (SPI) compared to LLVM. Although the performance of the IPM is mostly higher than the CAV, due to problems in the IPM, time-consuming and error-prone, more attention is being paid for developing and extending the capabilities of compiler's automatic vectorization.

This paper is organized as follows. A brief introduction to SIMD technology and its programming models are given in Sect. 2. In Sect. 3 a few multimedia kernels that are used in the evaluation are briefly described. In Sect. 4 the evaluation environment and the implementation results are presented. Finally, the conclusion is presented in Sect. 5.

2   SIMD Technologies

Modern architectures employ a variety of parallel execution units such as multiple CPU cores for executing multiple threads and SIMD units for processing multiple data using a single induction simultaneously. In this section, SIMD technologies are briefly discussed.

In SIMD technology, multiple data are processed simultaneously by a single instruction. This technology is created against the older model, known as the Single Instruction Single Data (SISD). At the beginning of the SIMD development process, MultiMedia eXtension (MMX) technology was a powerful strategy for speeding up multimedia and communications applications. MMX technology performs SIMD calculations on integer data types of 8-, 16-, 32- and 64-bit packages. Data types were stored in 64-bit registers and could hold eight 8-bit packages, four 16-bit packages, two 32-bit packages, and a 64-bit number of integers on it [11]. After that Streaming SIMD Extensions (SSE), SSE2/SSE3/SSE4/SSSE3 for processing both fixed- and floating-point number with larger register width, 128-bit has been extended. Eventually, Advanced Vector eXtension (AVX) and AVX2 expanded vector registers to 256-bits and AVX-512 technology enlarged vector registers to 512-bit. In other words, these new technologies provide possibilities to process more multiple data using a single instruction compared to older SIMD extensions.

In order to use the capabilities of the SIMD ISAs, many SIMD programming models, intrinsic programming model, as an explicit, and compiler's automatic vectorization, as an implicit, have been proposed [4]. The IPM is used as the main reference for using the available SIMD technology on the processors. The set of instructions used in Intel processors is collected in various libraries, which include functions used in an extension and its prior extensions. The programmer is responsible to vectorize algorithms and uses suitable SIMD instructions and functions. It can be tedious and error-prone. In this paper, the x86intrin library is used for explicit SIMD programming and is referred to as the intrinsic or IPM [12]. On the other hand, in CAVs, compilers are used to translate codes written in high-level languages into parallel machine codes. Since each processor has specific code for that processor and the delay of instructions is different, the performance of a translated program into different machine codes can vary considerably. Therefore, the compiler used to create more efficient machine codes is an important factor in performance. Various optimization techniques are used in compilers. In the meantime, the optimization of the loops is important because of the time spent on the operation inside the loop.

3 Multimedia Kernels

Computer and Internet applications use multimedia concepts significantly. To this end, multimedia standards such as MPEG, JPEG2000, H.26x, MP3, etc. put challenges on both hardware architectures and software algorithms for executing different multimedia processing tasks in real-time, because each data in a multimedia environment needs different algorithms, processes, and techniques [5]. Some of the most frequently used kernels and applications in multimedia are summarized in Table 1, and a brief explanation is provided for each of them [6–8].

Table 1. Some of the most frequently used kernels in the multimedia domain.

Multimedia kernels	Applications	Loop numbers
Matrix operations	Image, video and audio processing three-dimensional rendering computer graphics	2–3
FIR filter	Different types of filters in image, video, and audio like sound filtering in AMR standard pattern recognition in digital signal processing	2–3
Motion estimation	Video compression standards such as MPEG and HEVC	5

Matrix operations are one of the most frequently used operations in the signal, image, video, and audio processing, 3D rendering, and computer graphics. Three examples of matrix-based operations are the addition, transposition, and multiplication of matrices, which each of them has their own algorithms. The matrix addition operation adds existing corresponding elements in the matrix and stores the result in the output. Matrix transposition operation is one of the infrastructure operations in many multimedia applications, engineering, and linear algebra. In the normal mode, the matrix transposition means to replace the index of the element and placing it in the corresponding place. Since data is stored the row in memory, the transposition algorithm can be considered that by reading the elements from the continuous memory addresses, it stores them in discontinuous memory addresses. This can be done in small matrices much faster than large matrices since the possibility of a cache miss for large matrices is more than small matrices.

Matrix multiplication is a computational and time intensive operation. Due to its importance and being time intensive, various algorithms have been introduced for matrix multiplication. One of them is the matrix multiplication in the other matrix transposition algorithm, which was developed to improve memory access and to meet the needs of the basic algorithm for reading from discontinuous memory addresses. Like this first, the second matrix is transposed, then the elements in the rows of both matrices are multiplied together and then sum of the result is stored in the matrix of the result.

The FIR filter is one of the most important algorithms in the field of multimedia processing, including voice, image, video, and pattern recognition. The calculations of this filter are based on the concept of convolution that multiplies the coefficients in the output values and stores sum of the results in an output element [9].

Motion estimation of two images is one of the main concepts in the field of video processing, which is applicable in video compression, video registration, image alignment and incremental conversion of images frame rate. One of the most widely used applications of motion estimation is in video compression which is based on the reduction of time redundancy in similar images. One of the most frequently used methods is to compare the blocks of the previous frame with the current frame and estimate the best motion vector for each block. The Sum of Absolute Differences (SAD) is the most widely used solution for this operation, which achieves the difference between the corresponding pixels in the frame, and by comparing it, obtains the least rate of output, which represents the best vector of motion [11]. Another block-based strategy is the Sum of Squared Differences (SSD) that can be used in the same way to estimate the motion. Considering the high volume of computations to find two blocks with more similarity, both strategies are very time intensive [10].

$$SAD(d_1, d_2) = \sum_{i=0}^{n1} \sum_{j=0}^{n2} |d_1[i,j] - d_2[i,j]| \tag{1}$$

$$SSD(d_1, d_2) = \sum_{i=0}^{n1} \sum_{j=0}^{n2} (d_1[i,j] - d_2[i,j])^2 \tag{2}$$

4 Performance Evaluation

4.1 Environment Setup

All mentioned kernels have been implemented using OpenMP, IPM, and CAV approaches in ICC, GCC, and LLVM compilers on an Intel multi-core processor. The specification of the evaluation environment is depicted in Table 2. To gain a performance boost, each algorithm has been executed many times and the lowest amount of pulse has been measured for evaluation. To evaluate the results, the parallel mode is compared with the scalar mode executed in the ICC compiler. It is important to note that the ICC compiler has a very good performance in the scalar mode than the other compilers, and the reason for choosing this compiler for the scalar mode is to compare the performance of the parallel mode with the best scalar mode.

Table 2. Platform specification.

CPU	Intel Corei7-6700HQ
Register width	Maximum 256 bits
Cache line size	64 Bytes
L1 Data Cache	32 KB, 8-way set associative, the fastest latency: 4 cycles, 2x32B load + 1x32B store
L2 Cache	256 KB, 4-way set associative, the fastest latency: 12 cycles
L3 Cache	Up to 2 MB per core, up to 16-ways, the fastest latency: 44 cycles
Operating system	Fedora 27, 64-bit
Compilers	ICC 18.0.1, GCC 7.2.1, and LLVM 5.0.1 at -O3 optimization level
Programming tools	Standard C, OpenMP, and x86 Intrinsics (x86intrin.h)
Based evaluation compiler	icc -O3 -xHOST -no-vec
Cycle count	_rdtsc();
Disable vectorizing	icc -O3 -xHOST -no-vec gcc -O3 -march=native -fno-tree-vectorize -fno-tree-slp-vectorize clang -O3 -march=native -fno-vectorize -fno-slp-vectorize

Table 3 depicts the abbreviations used to present the results. All implementations have been performed for integers, 8, 16, or 32-bit and some of them for floating-point numbers. When the data type is not mentioned, it means that the integer data type is used.

Table 3. Acronyms for implementations.

OMP-MLT	Multi-threading execution using OpenMP without vectorization.
IPM-MLT	Explicit vectorization used by combination the multi-threading and IPM using AVX2 SIMD instructions
CAV-MLT	Implicit vectorization used by combination the multi-threading and CAV using AVX2 SIMD instructions
ADD and ADD-float	The ADD is an implementation for the matrix addition with 32-bit integers, and the ADD-float is for floating-point numbers
TRA and TRA-float	The TRA is an implementation for the matrix transposition program with 32-bit integers and the TRA-float is for floating-point numbers
MUL and MUL-float	The MUL is an implementation for the matrix multiplication program in a transposed matrix with 32-bit integers and the MUL-float is for floating-point numbers
FIR	The implementation performed for the FIR filter program is based on an array of 32-bit integers
SAD	The implementation performed for the SAD kernel is based on a matrix of 8-bit integers
SSD	The implementation performed for the SSD kernel is based on a matrix of 16-bit integers

4.2   Evaluation of Implementation Results

Table 4 shows the speedups of parallel implementations using OpenMP without vectorization and combination of OpenMP with vectorization, either IPM or CAV using AVX and AVX2 technologies for some multimedia kernels for two image sizes over scalar implementations.

As the table shows, the first row of each implemented kernel shows the multi-threading performance of the OMP-MLT without vectorization compared to the scalar implementation. Given the existence of four cores, the maximum multi-threading performance for different kernels and the different size of the images is about 3.59, and the ICC and GCC compilers have a better performance than the LLVM compiler. On the other hand, for kernels that have a higher computing volume, such as multiplication, SAD, and SSD, there is not much difference in the performance of small and large matrices, but in kernels that are more memory-dependent and less computational volume, such as ADD, performance for smaller matrices is more than larger matrices. The use of multi-threading capabilities depends on the number of cores, the algorithm and the way of memory access, and the data type plays a fewer role in the performance. In addition, in the multi-threading mode, due to the presence of four physical cores on the processor, finally, we will have the theoretical 4x increase in efficiency, due to the cost of thread management in multi-threading mode this performance increase is less than 4x. Our processor also supports hyper-thread technology, and the processor at the operating system level is considered as an eight-core processor, but because of the fact that these kernels are independent of each other this technology does not influence the performance significantly.

Table 4. Speedup of OpenMP without vectorization and combination of OpenMP with vectorization, either IPM or CAV programming models for multiple multimedia kernels on multi-core processors for two image sizes 512×512 and 1024×1024 over scalar implementations.

Kernels	Implementations	512×512			1024×1024		
		ICC	GCC	LLVM	ICC	GCC	LLVM
ADD	OMP-MLT	3.42	3.54	0.97	1.22	1.21	1.10
	IPM-MLT	6.77	6.66	2.04	1.15	1.15	1.29
	CAV-MLT	6.59	6.84	1.99	1.16	1.18	1.29
ADD-float	OMP-MLT	3.38	2.86	0.96	1.31	1.32	1.13
	IPM-MLT	7.25	7.31	2.21	1.26	1.27	1.43
	CAV-MLT	7.16	7.39	2.13	1.26	1.26	1.42
TRA	OMP-MLT	1.80	1.80	0.46	2.42	2.47	0.74
	IPM-MLT	3.88	4.21	3.54	6.02	6.03	3.7
	CAV-MLT	1.79	1.8	0.45	2.46	2.47	0.74
TRA-float	OMP-MLT	1.76	1.76	0.44	2.53	2.54	0.80
	IPM-MLT	3.87	3.22	3.46	6.26	5.42	3.74
	CAV-MLT	1.75	1.75	0.44	2.53	2.56	0.78
MUL	OMP-MLT	**2.67**	**2.17**	**0.88**	**2.59**	**1.88**	**0.86**
	IPM-MLT	8.14	10.11	3.68	11.65	11.94	3.33
	CAV-MLT	13.24	13.51	3.36	13.86	12.53	3.34
MUL-float	OMP-MLT	2.47	1.88	0.28	2.42	1.83	0.27
	IPM-MLT	**9.21**	**10.18**	**2.84**	**12.04**	**12.22**	**2.56**
	CAV-MLT	**15.44**	**14.48**	**3.84**	**15.67**	**12.88**	**3.73**
FIR	OMP-MLT	3.53	3.58	1.02	3.53	**3.59**	1.02
	IPM-MLT	**21.6**	**21.9**	**6.49**	**8.24**	**8.35**	**5.7**
	CAV-MLT	3.52	16.66	1.01	3.53	7.98	1.01
SAD	OMP-MLT	**3.43**	**3.43**	**1.21**	**3.50**	**3.50**	**1.23**
	IPM-MLT	**71.63**	**76.4**	**22.44**	**65.86**	**66.48**	**19.38**
	CAV-MLT	**71.05**	**73.06**	**9.59**	**64.43**	**67.13**	**9.58**
SSD	OMP-MLT	2.94	2.59	1.05	3.01	2.65	1.05
	IPM-MLT	25.59	26.09	7.05	20.74	21.17	6.63
	CAV-MLT	12.33	11.96	4.17	13.16	13.3	4.14

With combining multi-threading and SIMD technology, parallelization can be achieved at two levels of thread and data. The second and third lines of each kernel implemented in the table above show the efficiency of the combination of multi-threading and IPM-MLT, and CAV-MLT of SIMD technology over scalar implementations. The results show that the combination of these two types of parallelism increases efficiency by about 76. Performance improvement of combination of multi-threading and SIMD technology that uses IPM is roughly more than performance improvement of combination of multi-threading and CAV programming, that this increase depends on the way those algorithms are mapped in the SIMD framework and the way that SIMD instructions are used by the programmer, that, of course, is a time consuming, tedious, and with error-prone. But in the CAV-MLT, the compiler is responsible for vectorization, and it has vectorized well in many kernels. For example, in the SAD kernel, automatic vectorization is performed using ICC and GCC compilers gains much more performance. The main reason for this is that special purpose SAD instruction that is available in SIMD technology has been used. That is why the performance has increased about 76, while improvement for multi-threading without SIMD technology is about 3.5. All compilers have the ability to vectorize kernels and use SIMD machine language instructions to generate codes. Even the ICC and GCC compilers use the Special Purpose Instructions (SPIs) in the processor architecture in SAD vectorization, while the LLVM compiler uses common instructions of SIMD technology to generate SAD code. That is why the performance of the ICC and GCC compilers is far more than LLVM.

Generally, different compilers show different behaviors in dealing vectorization for each kernel [13]. ICC vectorizes all kernels, however, for TRA and FIR kernels vectorization is not applied efficiently. GCC does not vectorize TRA kernel and scalar codes are multi-threaded. LLVM cannot utilize multi-threading capability of OpenMP efficiently and performance improvement of most implementation is gained through auto-vectorization [14]. The platform has four cores with various SIMD units inside each core. Therefore, multi-threading is expected to yield 4x speedups and vectorization speedups must be a factor of the size of (vector)/size of (data type) which for AVX2 and 32-bit integers 256/32 equals to 8x speedup is expected. Because of the following reasons, theoretical and practical performance improvements are different. First, the latency of memory instructions is more than computation instructions and for both scalar and vectorized implementations most of the time is consumed to bring data to the L1 data cache. As depicted in Table 2 vector elements are loaded from L1D cache with more latency than the scalar element. Second, threads need to be managed and it consumes time to create, manage and destroy. Third, there are more scalar ALUs than SIMD ALUs which executes the SIMD instructions in the platform. Fourth, front-end of the processor is a bottleneck for SIMD instructions, why, front-end can only deliver 4 micro-ops (μops) per cycle and most SIMD instructions have more than one μops thus scalar implementations are benefited by using ILP, more efficiently.

5   Conclusions

The purpose of this paper was to extract thread- and data-level parallelism of some multimedia kernels on today's general-purpose processors that have multiple cores and SIMD technology. Each core has the ability to process a thread, as well as it is possible to process multiple data using a single instruction that it provides the ability to extract thread and data-level parallelism. Multi-threading programming models, OpenMP, and SIMD such as Intrinsic Programming Model (IPM) for explicit vectorization and Compiler's Automatic Vectorization (CAV) for implicit vectorization by ICC, GCC, and LLVM compilers were used to improve the performance of multimedia kernels. Experimental results show that the combination of both thread- and data-level parallelism increase the functionality of the kernels considerably. The ICC and GCC compilers exhibit better performance than LLVM compiler in most implementations. Although the performance of the IPM is almost higher than CAV, but, because of difficulties of the IPM such as time-consuming, subjective and error-prone more attention is needed to develop and expand compiler's automatic vectorization.

References

1. Shahbahrami, A., Juurlink, B., Vassiliadis, S.: Matrix register file and extended subwords. In: Proceedings of the 2nd Conference on Computing Frontiers, vol. 5, no. 1, p. 171 (2005)
2. Shahbahrami, A.: Avoiding conversion and rearrangement overhead in SIMD architectures. Ph.D. dissertation, Computer Engineering Laboratory, Delft University of Technology, Delft, Netherlands (2008)
3. Intel Corporation: Intel® 64 and IA-32 Architectures Software Developer's Manual Volume 1: Basic Architecture, vol. 1, no. 253665-060US (2016)
4. Pohl, A., Cosenza, B., Mesa, M. A., Chi, C.C., Juurlink, B.: An evaluation of current SIMD programming models for C++. In: Proceedings of the 3rd Workshop on Programming Models for SIMD/Vector Processing - WPMVP 2016, pp. 1–8 (2016)
5. Shahbahrami, A., Juurlink, B., Vassiliadis, S.: A comparison between processor architectures for multimedia application. In: Proceedings of the 15th Annual Workshop on Circuits, Systems and Signal Processing 2004, pp. 138–152 (2004)
6. Choi, J., Dongarra, J.J., Walker, D.W.: Parallel matrix multiplication algorithms on distributed memory concurrent computers. Parallel Comput. 21(9), 1387–1405 (1995)
7. Zekri, A.S.: Enhancing the matrix transpose operation using Intel AVX instruction set extension. Int. J. Comput. Sci. Inf. Technol. 6(3), 67–78 (2014)
8. Chatterjee, S., Sen, S.: Cache-efficient matrix transposition. In: Proceedings Sixth International Symposium on High-Performance Computer Architecture, HPCA-6 (Cat. No. PR00550), pp. 195–205 (2000)
9. Kyo, S., Okazaki, S., Kuroda, I.: An extended C language and a SIMD compiler for efficient implementation of image filters on media extended micro-processors. In: Proceedings of Advanced Concepts for Intelligent Vision Systems, pp. 234–241 (2003)
10. Amiri, H., Shahbahrami, A.: High performance implementation of 2D convolution using Intel's advanced vector extensions. In: 2017 Artificial Intelligence and Signal Processing Conference, AISP, pp. 25–30 (2017)
11. Peleg, A., Weiser, U.: MMX technology extension to the Intel architecture. IEEE Micro 16 (4), 42–50 (1996)
12. Intel Corporation: Intel Intrinsics Guide, 29 January 2017. https://software.intel.com/sites/landingpage/IntrinsicsGuide
13. Amiri, H., Shahbahrami, A., Pohl, A., Juurlink, B.: Performance evaluation of implicit and explicit SIMDization. Microprocess. Microsyst. 63, 158–168 (2018)
14. Amiri, H., Shahbahrami, A.: High performance implementation of 2-D convolution using AVX2. In: 2017 19th International Symposium on Computer Architecture and Digital Systems (CADS), pp. 1–4. IEEE (2017)

A Parallel Algorithm for Eukaryotic Promoter Recognition

Seyyed Mohammad Shaheri Langroudi, Hamid Reza Hamidi[✉],
and Shokooh Kermanshahani

Computer Engineering Department,
Imam Khomeini International University, Qazvin, Iran
m.shaheri@edu.ikiu.ac.ir,
{hamidreza.hamidi,kermanshahani}@eng.ikiu.ac.ir

Abstract. One of the interesting problems in Bioinformatics is finding transcription start site in a gene. In fact, finding this site which separate promoter region from coding sequence, actually will end to promoter prediction. This leads to activate or inactivate some parts of gene which plays an important role after being translated to protein sequence. While traditional methods are reliable ways for promoter prediction, because of the large number of sequences and too much of information, it is not possible to study these sequences by those methods. Although some of these sequences have been already recognized and their information has been stored in big databases like NCBI, there are some sequences which their promoter regions have not been identified yet. This research aimed to design a parallel algorithm for one of the known promoter prediction algorithms, *Ohler*. We attempt to reduce the response time of *Ohler* algorithm, consequently increases the number of test samples, and improves the accuracy of the algorithm. The experimental results show that we have succeeded to achieve our purpose.

Keywords: Bioinformatics · Parallel algorithms · Promoter prediction · MPI

1 Introduction

Eukaryotic genes consist of two parts: coding sequence which contains information such as the locations of *exons*, and noncoding sequences called *intron* that intervening them. The second part which called promoter is located exactly before the coding sequence and facilitates binding enzymes to transcription factors [6,7]. Transcription is copying a particular segment of DNA into RNA by the enzyme RNA polymerase. Promoters are located in the vicinity of gene start sites and directly regulate gene expression [10,13]. Prediction of these regions is one of the problems that biologists faced to.

Like other areas of bioinformatics, the use of computational tools and computers to detect promoters is a great alternative for costly and time-consuming

© Springer Nature Switzerland AG 2019
L. Grandinetti et al. (Eds.): TopHPC 2019, CCIS 891, pp. 468–475, 2019.
https://doi.org/10.1007/978-3-030-33495-6_36

traditional methods [13]. However, computational identification of promoters and regulatory elements is also a very difficult task for several reasons. The first reason is that the promoters and regulatory elements are not clear enough and are highly diverse. Each gene usually has a unique combination of regulatory motifs that define its unique spatial expression. The second is that promoters and regulatory elements cannot be translated into protein sequences to increase the sensitivity for their detection. And finally, the elements of the promoters and regulatory sites have a very short length (6 to 8 nucleotides) and may be randomly seen in each part of the sequence [13].

1.1   Related Algorithms

One of the theoretical methods generally proposed for finding a, was the use of context-free grammar; but the practical use of this method is not possible because of the large ambiguity inherent in nucleotide sequences. For example, the markers **GT** and **AG** which are delimiters for *intron*, can be randomly seen anywhere of the sequence, even in the promoter region [7].

Some of the other algorithms are based on the presence of specific motifs in certain places of sequence and some of them use position weight matrix [5]. One of these algorithms is *Tata* which introduced by Bucher in [3,4]. The main problem with this algorithm was that the insertion, deletion or variable spacing of the promoter elements (in the promoter's scope) was not considered. In other words, this algorithm considered different parts of the sequence as independent parts [5]. These algorithms use a series of apparent features in a sequence, while each promoter region may have some of these features and have not the others.

Another method used in this field is using the number of special nucleotide motifs (with different lengths) appearance [5]. These algorithms have a database of known motifs that have already been proved as a sign for promoter region and they search input sequence for these motifs. The *PromoterInspector* algorithm was presented by Scherf in [12]. This algorithm detects the location of promoter in the input sequence just by matching parts of the input sequence with the motifs stored in the database, and of course in many cases can be problematic and actually cause a false positive recognition [5]. There are other algorithms in this category, such as *PromFD* [6] and *Mitra* [8].

Another category of algorithms, such as *Ohler*, uses Markov models in several ways [1,11]. These algorithms are more accurate than position weight matrices, because they did not convince by presence of a series of particular promoter motifs in the sequence. Another advantage of the methods of this category is that they also allow for a departure between adjacent nucleotides [5].

Experience has shown that designing applications which use advantage of powerful hardware completely is not easy and need to expand parallelization tools and parallel programs and algorithms [2]. This article tries to reduce the response time of *Ohler* algorithm, consequently increases the number of test samples, and improves the accuracy of the algorithm.

We describe the original algorithm briefly in the next section then presents parallel approach, and finally assets and analyzes the practical results.

2   Ohler Algorithm

This algorithm at the first stage, requires a series of training sequences (Initial sequences) that have been already identified as specific biological regions (named class) by absolute methods. These biological regions are *promoters*, *introns* and *coding* sequences. The algorithm needs a number of training sequences for each of these biological regions [11].

Divide the input sequence into 300bps windows
Do for all windows
      Divide the window into 10bps segments
      Do for all segments
            Calculate each class score
            Calculate score difference with background class
            // Specificity: number of false positives
            If (Specificity < suitable threshold)
                  Predict a Transcription Start Site (TSS)

Fig. 1. *Ohler* algorithm [11].

As shown in Fig. 1, for processing the input sequence, a window with a specified length (the original article proposed a 300 base pairs long [11]) is processed. Then the program sets out to process next window, and this procedure repeats until all input sequences would be analyzed and evaluated. Each time, once the window's location is specified, the window is divided into smaller segments (suggestion of *Ohler* is 10 base pairs) and the program calculates the score of each of these segments [11]. By adjusting suitable threshold, the sensitivity of algorithm can be changed. The higher value leads to the higher percentage of false positive predictions.

A few shortcomings in this approach are as following:

- If two (or more) promoter sites located in the same window, this algorithm will be able to detect one of them at most. This is not unexpected, since a whole 300 base pairs region is scored at once.
- Since the training sequences used in the original algorithm are human sequences, the percentage of true positive predictions for the non-human input sequences will be low. In fact, this algorithm highly depends on its training sequences; but this problem can be solved by changing the training database.
- The great difference in the length of the training sequences seems to influence the final results of the program, because program decides which segment belongs to which class based on the number of segments of windows in the input sequence.

The big advantage of this method is that deletion, replacement, or insertion nucleotides will not affect its recognition since this algorithm is not sensitive to the location of motifs. According to the *Ohler* article, due to the training sequences which was human sequences, behavior of the program on the Fickett data set, which many of them was non-human sequences, was quite good [9]. 12 of 24 promoter regions were correctly identified, and for seven of them the starting point was predicted with high accuracy [11].

Table 1. Dividing windows between tasks

Window 0	Window 1	Window 2	Window 3	Window 4	Window 5
P0	P1	P2	P0	P1	P2

3   Proposed Parallelization for Ohler Algorithm

In Fig. 1, the first loop is the part we parallelized. Windows will be distributed cyclic between them at the first phase. Table 1 shows how windows are distributed between tasks. Here, for example, the number of windows is 6 and the number of tasks is assumed 3.

The program needs 4 inputs:

- Input Sequence which user wants to process it.
- A file which contains *promoter* training sequences. If the training sequences belong to the same entity, the results will be more accurate because this algorithm relies on its training sequences severely.
- A file which contains *intron* training sequences.
- A file which contains *coding* training sequences.

In the first step, each task individually reads the files which contains training sequences and copies each of them into the corresponding array in order to accelerate the next operations. Then each task process its windows by dividing each window into the segments of 10 base pairs. The number of these segments is equal to *(window length)/(segments length)* for a window. According to the original article, we assume the windows length is 300 base pairs, thus the number of segments in each window will be 30. An array of 30 is considered for each segment in order to hold its results.

At the end, after each task processed all of its windows, it sends an array to the main task. This array contains the number of promoter segments for each window of the task.

Finally, in order to display the results, the main task stores all of them in an array with the size of the whole number of the windows of all tasks.

At last, the main process displays the promoter window numbers, the number of promoter segments that contains, and the corresponding starting point after receiving all results from all tasks.

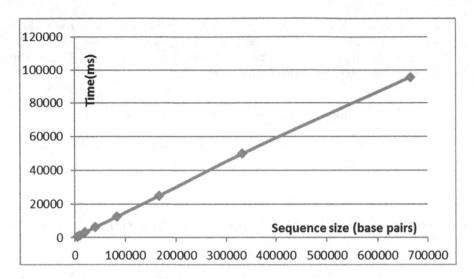

Fig. 2. The sequential program response time.

3.1   Time Complexity of Sequential Program

In this section, we present and analyze the results obtained from the sequential program[1]. Input and training sequences size are based on base pairs (bps).

To check the relationship between the runtime and input sequence size, we kept the training sequences constant, and we doubled the input sequence size for every experiment. The results of this experiment have been shown in Fig. 2. The *Ohler* program searches 10 motifs (with size of 1 to 10 which is the segment size) in the whole training sequences for every nucleotide. The program is implemented in such a way that it will continue its search for N nucleotide string in the training sequences only if $N-1$ previous nucleotide string has been founded in order to prevent time wasting.

According to the Fig. 2, we can say that time complexity of the algorithm is $O(n)$ (n is input sequence size). This was already predictable, since for each segment (10 nucleotides), program searches 10 times - with the size of 1, then 2 to 10 nucleotides - in the whole training sequences. The time complexity of the algorithm depends on the size of the training sequences too (for example if the total size of training sequences becomes double, the total runtime will also be doubled). If we assume that the total size of the three training sequences is t, then we can say that the time complexity of this algorithm is $O(nt)$.

---

[1] Programs have been tested on Intel core i7 processor in C language, and have been executed on windows 8.1 OS.

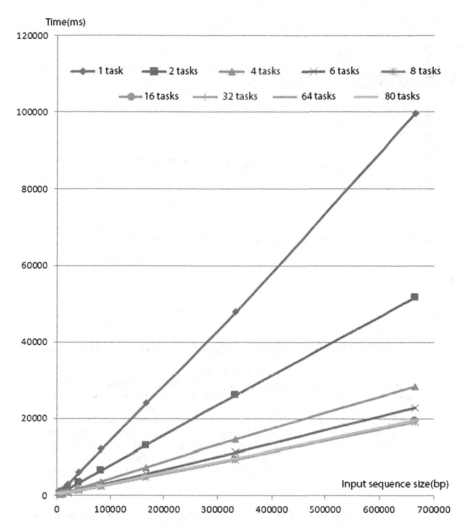

Fig. 3. The parallel program response time.

3.2 Time Complexity of Parallel Program

In this section, we analyze the proposed parallel program[2]. By assuming the number of processors p, the execution time of the parallel algorithm can be $O(nt/p)$ at best. The algorithm will get closer to this ideal time if it take less data transfer rate as much as possible and distribute tasks on processors fairly.

---

[2] Programs have been tested on Intel core i7 processor which has 4 physical cores and 8 threads; and the processor can support 8 processes in parallel at maximum. Programs designed and developed in C language, and we used Open MPI for parallelization. All programs have been executed on windows 8.1 OS.

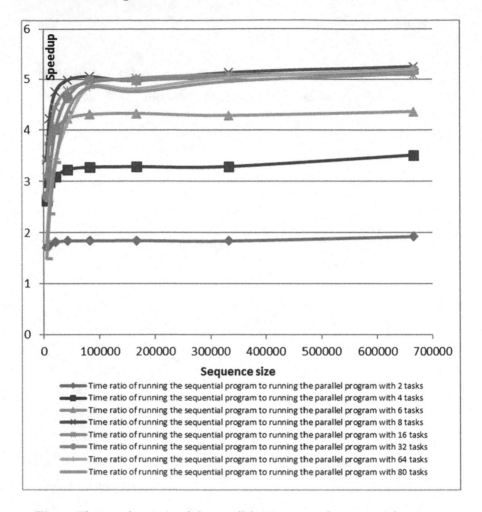

Fig. 4. The speedup ratio of the parallel program to the sequential program.

Figure 3 shows the results of practical testing and the comparison between the sequential program and the parallel program with different number of tasks. The worst execution time corresponds to the sequential execution, and all parallel runs respond in less time in compare to the sequential program. Programs have been lunch on a processor which can support 8 simultaneous tasks at most. Therefore, in Fig. 3, the algorithm runtime improved from 1 task to 8 tasks, and then the runtime of the algorithm decreased slightly.

4   Conclusion

Figure 4 shows the speedup ratio of the parallel program to the sequential program. There is no any part of graphs which places lower than 1 in the vertical axis. But each of the graphs from a point (according to the input sequence size) has reached to relative stability. In this diagram, it is obvious that usually by running the program with 8 tasks (which are the number of the processors), the best speed has been achieved.

If we focus on the Fig. 4, running the program with two tasks is mostly 2 times faster than the sequential program and it means that we get very close to the ideal time. If the number of tasks increases, the runtime of the program gets distant from optimal time span. One of the reasons might be that if the number of tasks increases, it takes more time to read input files and training sequences.

Acknowledgements. This research is done by companion of Agriculture Biotechnology Research Institute of Iran, Karaj. Thanks to all of the members of the Institute who helped during the project and especially thanks to Dr. Zahra Sadat Shobbar.

References

1. Audic, S., Claverie, J.: Detection of eukaryotic promoters using Markov transition matrices. Comput. Chem. **21**, 223–227 (1997)
2. Bischof, C., et al.: Parallel Computing: Architectures, Algorithms and Applications, vol. 38. John von Neumann Institute for Computing, Jülich (2007)
3. Bucher, P.: Weight matrix descriptions of four eukaryotic RNA polymerase II promoter elements derived from 502 unrelated promoter sequences. J. Mol. Biol. **212**, 563–578 (1990)
4. Bucher, P., Bryan, B.: Signal search analysis: a new method to localize and characterize functionally important DNA sequences. Nucleic Acids Res. **12**(1), 287–305 (1984)
5. Burden, S., Lin, Y.J., Zhang, R.: Improving promoter prediction for the NNPP2.2 algorithm: a case study using Escherichia coli DNA sequences. Bioinform. (Oxford J. Sci. Math.) **21**(5), 601–607 (2005)
6. Chen, Q.K., Hertz, G.Z., Stormo, G.D.: PromFD 1.0: a computer program that predicts eukaryotic pol II promoters using strings and IMD matrices. Comput. Appl. Biosci **13**, 29 35 (1997)
7. Cohen, J.: Bioinformatics - an introduction for computer scientists. ACM Comput. Surv. **36**(2), 122–158 (2004)
8. Eskin, E., Keich, U., Gelfand, M.S., Pevzner, P.A.: Genome-wide analysis of bacterial promoter regions. In: Pacific Symposium on Biocomputing, pp. 29–40 (2003)
9. Fickett, J.W., Hatzigeorgiou, A.G.: Eukaryotic promoter recognition. Genome Res. **7**, 861–878 (1997). Cold Spring Harbor Laboratory Press. ISSN 1054-9803/97
10. Huang, J.W.: Promoter prediction in DNA sequences, National Sun Yat-Sen University (2003)
11. Ohler, U., Harbeck, S., Niemann, H., Noth, E., Reese, M.: Interpolated Markov chains for eukaryotic promoter recognition. Bioinform. (Oxford J. Sci. Math.) **15**, 362–369 (1999)
12. Scherf, M., Klingenhoff, A., Werner, T.: Highly specific localization of promoter regions in large genomic sequences by PromoterInspector: a novel context analysis approach. J. Mol. Biol. **297**, 599–606 (2000)
13. Xiong, J.: Essential Bioinformatics. Cambridge University Press, Cambridge (2006)

A Novel Critical-Path Based Scheduling Algorithm for Stochastic Workflow in Distributed Computing Systems

Alemeh Matani[✉] and Asghar Darvishy

Department of Computer Engineering, Islamic Azad University,
South Tehran Branch, Tehran, Iran
A.matani91@gmail.com, A_Darvishy@azad.ac.ir

Abstract. The prominent characteristic of the workflow applications, i.e., complex dependencies among workflow tasks, have made workflow scheduling problem a challenging problem in any distributed computing paradigm such as cloud and grid computing. So far, various workflow scheduling strategies have been proposed whose usual assumption is that the parameters corresponding to workflow tasks, such as length and output size, are deterministic and known in advance. Whereas, due to uncertainties about loops and decision structures in task instructions, these parameters are never deterministic. Therefore, considering workflows as deterministic models, in order to prioritize interdependent tasks during mapping of tasks onto computational resources, will lead to an inefficient scheduling scheme. To cope with these uncertainties, we consider workflows as stochastic ones and model tasks parameters as normal random variables. But in order to simplify computation process, we approximate a stochastic workflow as several interval workflows. Eventually, we extend the traditional critical path algorithm and obtain more detailed ranking of the tasks. The simulation results show that thanks to this detailed ranking information, better decisions are made about assigning tasks to computational resources.

Keywords: Workflow scheduling · Stochastic workflow · Interval workflow · Criticality level

1 Introduction

The workflow is one of the more common application models in distributed computing paradigms such as cloud and grid computing. These applications are usually represented as a directed acyclic graph (DAG) including a set of ordered tasks that are linked by data dependencies. Scheduling is a vital issue in management of workflow execution which determines efficient mapping of workflow tasks onto distributed resources, so that precedence constraints on tasks are respected and some other requirements are satisfied as well. Numerous heuristic [1, 4] and meta-heuristic [5, 6] algorithms have been proposed aiming at addressing this prominent issue.

Critical path based algorithms [7–9] have been used widely for scheduling workflows. In all these algorithms, both the execution times and data transmission times of the workflow tasks are assumed to be constant and definitive. Although, because of

© Springer Nature Switzerland AG 2019
L. Grandinetti et al. (Eds.): TopHPC 2019, CCIS 891, pp. 476–489, 2019.
https://doi.org/10.1007/978-3-030-33495-6_37

uncertainties about loops and decision structures, these quantities would not be fixed and deterministic. In order to cope with these uncertainties, in this paper, task execution times and data transmission times are considered as normal random variables. In other words, we use a stochastic model of workflow that is more realistic than the classical workflow model which many of the previous works are based on. Then, we employ interval arithmetic to approximate the stochastic workflow as several interval workflows. Since transforming the stochastic non-determinism workflow to non-stochastic (interval) non-determinism one leads to information loss, in order to decrease this information loss we used three interval approximations instead of one. This has two benefits (1) Dealing with interval workflows is much easier than with stochastic ones and (2) as the results of our experiments reveal, the amount of information loss is fairly low.

The rest of this paper is structured as follows. The next section, introduces some related work briefly. In Sect. 3, we describe our problem along with some necessary definitions. Afterward, our proposed approach is presented in Sect. 4. Then, in Sect. 5 a comparison of the proposed approach with some existing approaches are given using a simulation study. At the end, Sect. 6 gives the conclusions and also discusses future work.

2   Related Works

An important goal in workflow scheduling strategies is to provide minimum execution time. However, numerous algorithms have been proposed to achieve an efficient scheduling solution. But large number of them aimed to schedule a deterministic workflow where the parameters relating to tasks are fixed and known in advance. Whereas, in practical environment, workflow tasks information is never deterministic. Therefore, these strategies may lead to low efficiency practically. Some of these heuristics are as follows. Heterogeneous Earliest Finish Time (HEFT) [10], the most popular list based heuristic for scheduling workflow applications which orders and schedules tasks based on their ranks. The rank of each task determines the length of the path from the task to the finishing task of the workflow. In addition, Critical-Path-on-a-processor (CPOP) [10], Min-min [11] and Max-Min [11] are amongst the other most important list-based heuristics which are designed to improve the performance of workflow scheduling. Further, in [12], two new mechanisms for ranking tasks in list-based workflow scheduling are developed based on novel criteria, as well as a method for mapping tasks into the computational resources.

Recently, some works [13, 14] have been proposed that deal with stochastic workflows and present a stochastic scheduling. Authors of [15] proposed an approach for scheduling stochastic DAGs which considers both the mean and standard deviations of the tasks execution and transmission times to calculate the earliest start times of the tasks statically.

In [16], a resource load based stochastic DAG scheduling method has been proposed where the distribution of the needed time for processing workflow tasks is estimated according to the workload levels of the resources. Also in [17], a randomized DAG scheduling heuristic is proposed based on HEFT which considers a stochastic modelling of task execution times and transmission times. In [18], a Monte-Carlo based approach has been presented where, in order to address uncertainty in task execution time, scheduling is built based on stochastic performance prediction.

Despite all these stochastic methods, the stochastic nature of the scheduling problem in our work arises due to uncertainty about workflow tasks parameters, i.e. tasks lengths and tasks output sizes, since in real world workflow applications, because of uncertainties about loops and decision structures, this information is never deterministic. Whereas, stochastic scheduling in previous works is only due to dynamic nature of resources behaviour.

3   Problem Statement and Definitions

In this section, we describe the problem considered in this study in more detail and give the preliminary definitions for scheduling workflows with non-deterministic information on the tasks.

3.1   Problem Statement

The previous workflow scheduling methods assume that both the length and output data sizes of the tasks are deterministic and fixed. Thus, task execution times and transmission times will be deterministic. Therefore, by using this deterministic information it is simple to determine not only if a task is critical or not, but also how much it can be delayed so that workflow execution time will not rise. But in the real word, due to available loops and decision structures in the instructions of the workflow tasks, this information cannot be deterministic.

In order to handle these uncertainties about workflow tasks, in the proposed method, we consider the length of the tasks (measured in million instructions (MI)) and size of their output data (measured in bits (b)) as normally distributed random variables with mean and standard deviation determined according to the historical data. The normal distribution is used in some previous works for modelling tasks sizes [13, 18]. Nevertheless, our reasons to choose a normal distribution for task parameters are as follows:

- Since each task is composed of a huge number of instructions and parts, whose execution times are almost independent, according to the Central limit theorem [19], normal random variables can be used to model the task sizes and their execution times.
- The normal distribution is an appropriate distribution for manipulating the stochastic values since the family of normal distributions is closed under various arithmetic operations unlike most other distribution families.

In other words, we use stochastic workflow instead of classical one used in existing works because using stochastic workflow is more realistic and help us to provide more convincing results. But since computation on normal random variables is complex, we use the interval approximates of the stochastic workflow to simplify the computation process. The aim of these approximations is to use an interval extension of the traditional critical path based algorithms and achieve more detailed ranking of tasks.

3.2 Workflow Models

Stochastic Workflow. A stochastic workflow is a triplet $(\mathcal{T}, \mathcal{D}, \mathcal{N})$ where $\mathcal{T} = \{t_1, \ldots, t_n\}$ is the set of workflow tasks, $\mathcal{D} \subseteq \mathcal{T} \times \mathcal{T}$ is the set of dependencies among tasks which forms a DAG and $\mathcal{N} : \mathcal{T} \to \mathcal{RV}$, with $\mathcal{RV}$ denoting the set of normal random variables, is a function that assigns to each tasks its size as a normal random variable.

Interval Workflow. An interval workflow is triplet $(\mathcal{T}, \mathcal{D}, \mathcal{J})$ where and are defined as stochastic workflow definition and $\mathcal{J} = \mathcal{T} \to \mathbb{IR} \geq 0$ is a function that assigns to each tasks its size as a non-negative interval number, assuming $\mathbb{IR}$ is the set of real intervals.

3.3 Resource Model

In our distributed computing environment, we consider a group of dynamic and heterogeneous computing resources with different processing capabilities and workload levels. These resources are assumed to be connected to each other through links with different transmission capacities. The processing capabilities of the resources and the transmission capacities of the links are respectively measured in MIPS (Million Instructions per Second) and Mbps (Megabits per second). In addition, we assume that each resource has processing elements, and resource policy is modelled as time sharing where resource workload is processed in parallel.

We also consider two kinds of workload for computational resources: (1) local workload which uses a portion of the processing capacity of each processing element and (2) the workload corresponding to other users requests that consists of n tasks executing on the resource.

In the proposed method, the available processing capability of a resource is calculated based on:

- n, the number of tasks executing on the resource,
- m, the number of processing elements,
- $\mathcal{PC}max$, the processing capacity of each processing element, and
- ℓ, the fraction of processor times consumed for processing the local workload.

Therefore, $\mathcal{PC}_{available}$, the available processing capability of the resource is estimated as follows:

$$\mathcal{PC}_{available} = \frac{\mathcal{PC}_{max} \times (1 - \ell)}{\lceil \frac{n}{m} \rceil} \tag{1}$$

Equation (1) simply says that $\mathcal{PC}_{available}$, the available processing capability of a resource, is computed by dividing $\mathcal{PC}_{max} \times (1 - \ell)$, the portion of the processing capacity of each processor which is not occupied by the local workload, by the average number of tasks executing on each processing element which is $\lceil \frac{n}{m} \rceil$.

4 Methodology

In this section, we describe our proposed scheduling algorithm named Multi Level Critical Path based scheduler (MLCP), whose basic idea is derived from critical path based algorithms. Unlike the traditional critical path based methods which classify tasks as critical and noncritical ones, in the proposed method we consider four classes of tasks according to their criticality levels from zero to three. Our approach consists of the following main steps:

- Generating stochastic workflow from the non-deterministic parameters of the tasks.
- Estimation of interval workflows according to stochastic workflow.
- Interval extension of the traditional critical path based algorithm.
- Calculation of the criticality level metric of all the workflow tasks.
- Scheduling workflow tasks onto appropriate resources.

In the following, we give more detailed description of our method steps.

4.1 Generating Stochastic Workflow from the Non-deterministic Parameters of Tasks

In most critical path based approaches, at first, the minimum execution time and transmission time of each workflow task is computed according to the length and output size of the task respectively, and then the critical path and lags of the tasks are determined by this deterministic information.

Here, we are facing with a more challenging issue because of uncertainty about the length of the tasks and their output size. In fact, since we consider these quantities as normal random variables, the execution time and data transmission time of the tasks will be non-deterministic and follow the normal distribution as well.

Assume μ_i^{length} and σ_i^{length} to be the mean and standard deviation for the length of the task t_i respectively (measured in MI). Therefore, the execution time of the task t_i is a normal random variable whose mean and standard deviation are calculated respectively by (2) and (3)

$$\mu_i^{et} = \frac{\mu_i^{length}}{\max_{j \,\in\, ResourceList}\{APC(R_j)\}} \tag{2}$$

$$\sigma_i^{et} = \frac{\sigma_i^{length}}{\left(\max_{j \,\in\, ResourceList}\{APC(R_j)\}\right)^2} \tag{3}$$

Where $APC(R_j)$ presents the available processing capability of the resource R_j.

In the same manner, the mean and standard deviation of the transmission time for task t_i can be calculated by (4) and (5), assuming that $\mu_i^{o\_size}$ and $\sigma_i^{o\_size}$ are respectively the mean and standard deviation of the output size of the task t_i, in terms of bits.

$$\mu_i^{dtt} = \frac{\mu_i^{o\_size}}{\max_{j \in ResourceList}\{BW(R_j)\}} \tag{4}$$

$$\sigma_i^{dtt} = \frac{\sigma_i^{o\_size}}{\left(\max_{j \in ResourceList}\{BW(R_j)\}\right)^2} \tag{5}$$

Where $BW(R_j)$ represents the transfer capacity i.e. Bandwidth of resource R_j.

In the stochastic workflow model, because of uncertainty about some information such as aforementioned parameters, critical path will not be deterministic. This means that we cannot be sure whether a workflow task is critical or not. Therefore, we need to define a new metric which indicates how likely a task is to be critical. Hence, we introduce a criticality level metric for each workflow task according to which tasks can be classified into several criticality levels. Afterwards, by taking advantage of this metric one can prioritize tasks for various purposes such as minimizing the execution time of the workflow, improving reliability and etc. Here, we focus on decreasing the execution time of the workflow.

In the following, since it is complex to work with stochastic workflows during computation of criticality level for the tasks, we approximate stochastic workflow as interval one to simplify the computation process. Furthermore, in order to decrease the information loss arisen due to this transformation we used three interval approximations instead of one.

4.2 Estimation of Interval Workflows According to Stochastic Workflow

Let $\mathcal{W}_S$ be a stochastic workflow containing n tasks, $T = \{t_1, \ldots, t_n\}$, where the execution time and data transmission time of the task t_i obey $N(\mu_i^{et}, \sigma_i^{et})$ and $N(\mu_i^{dtt}, \sigma_i^{dtt})$ respectively. Because of difficulty about computing the minimum and maximum of two normal random variables, we are better to approximate the interval workflow instead of stochastic one. In addition, since this transformation may leads to information loss, we used three interval approximations in order to decrease this information loss.

Assuming that $\mathcal{W}_I^{(k)}$ is the k-th interval workflow related to the stochastic workflow $\mathcal{W}_S$, the execution time of the task t_i in $\mathcal{W}_I^{(k)}$ is estimated as (6)

$$et_{(i,k)} = \left[\mu_i^{et} - k\sigma_i^{et}, \mu_i^{et} + k\sigma_i^{et}\right], \quad k = 0, 1, 2 \tag{6}$$

In the same manner, the k-th interval approximation of the transmission time of the task t_i is estimated as (7)

$$dtt_{(i,k)} = \left[\mu_i^{dtt} - k\sigma_i^{dtt}, \mu_i^{dtt} + k\sigma_i^{dtt}\right] \quad k = 0, 1, 2 \tag{7}$$

In general, the first approximation i.e. $\mathcal{W}_I^{(0)}$ is a degenerate workflow where the execution time of each task is a degenerate number. Figure 1 depicts the probability density function for the normal distribution. As can be observed, as k increases, it is less likely the corresponding interval occurs. Therefore, $\mathcal{W}_I^{(k)}$ represents a more pessimistic approximation of $\mathcal{W}_S$.

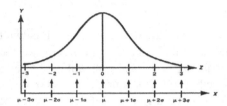

Fig. 1. Normal distribution probability density function

4.3   Interval Extension of Critical Path Based Algorithm

After obtaining three interval approximations of stochastic workflow $\mathcal{W}_S$, we need to perform an interval extension of the traditional critical path based algorithm on these interval workflows to obtain the criticality level metric of each workflow task. Before we describe the rest of the process, we first introduce some extended formulation of interval mathematics. Then, we explain how to compute the interval lags of the workflow tasks based on which the critical level of tasks are calculated. The formulations are as follows:

Assume $\mathbb{R}$ (respectively, $\mathbb{R}_{\geq 0}$) to be the set of real (respectively, non-negative real) numbers. Given the sets A and B, by A $\rightarrow$ B, we denote the set of total mappings from A to B. A (closed) interval denoted by $[a, b]$ is the set of real numbers given by $[a, b] = \{x \in \mathbb{R} \mid a \leq x \leq b\}$. The left and right endpoints of an interval X will be denoted by $\underline{X}$ and $\overline{X}$, respectively. Therefore, we have $X = [\underline{X}, \overline{X}]$. By definition, the sum of two intervals X and Y is the interval $X + Y = [\underline{X} + \underline{Y}, \overline{X} + \overline{Y}]$. For subtraction of two intervals X and Y we have $X - Y = [\underline{X} - \overline{Y}, \overline{X} - \underline{Y}]$. But, in this study, we need to reformulate subtraction operator as an element-wise one which denoted by $X \ominus Y$. Therefore, we have $X \ominus Y = [\underline{X} - \underline{Y}, \overline{X} - \overline{Y}]$

Calculation of the Interval Lag of the Workflow Tasks. Assume that $et_i = \left[\underline{et_i}, \overline{et_i}\right]$ and $dtt_i = \left[\underline{dtt_i}, \overline{dtt_i}\right]$ to be interval numbers denoting the execution and data transmission time of the task t_i respectively. At first, we need to compute the Earliest Starting Time (EST) and Earliest Finishing Time (EFT) of t_i as (8) and (9). Note that we assume the workflow starts with the single starting task t_1 which have no predecessors.

$$EST(t_i) = \begin{cases} 0, & i = 1 \\ \max_{t_j \in pred(t_i)} \left(\dfrac{\left[\underline{est_j}, \overline{est_j}\right] + \left[\underline{et_j}, \overline{et_j}\right]}{+ \left[\underline{dtt_j}, \overline{dtt_j}\right]} \right), & otherwise \end{cases} \tag{8}$$

$$\left(EFT(t_i) = EST(t_i) + \left[\underline{et_i}, \overline{et_i}\right]\right. \tag{9}$$

Where $pred(t_i)$ represents the set of predecessors of t_i. We also assume that the workflow finishes with the single finishing task t_n which have no successors.

In the following, the Latest Start Time (LST) and Latest Finishing Time (LFT) of t_i are computed as follows:

$$LST(t_i) = \begin{cases} EST(t_i), & i = n \\ \min_{t_j \in succ(t_i)} \left(\dfrac{\left[\underline{lst_j}, \overline{lst_j}\right] \ominus \left[\underline{s_i}, \overline{s_i}\right]}{\ominus \left[\underline{dtt_i}, \overline{dtt_i}\right]} \right), & otherwise \end{cases} \tag{10}$$

$$LFT(t_i) = LST(t_i) + \left[\underline{s_i}, \overline{s_i}\right] \tag{11}$$

Where $succ(t_i)$ represents the set of successors of t_i and $\ominus$ denotes the element-wise interval subtraction operator. In case we have multiple entry tasks and exit tasks in a workflow application, for simplicity, we connect them to dummy starting and finishing tasks with zero execution times. In the following, we compute the lag of task t_i as follows:

$$LAG(t_i) = LST(t_i) \ominus EST(t_i) \tag{12}$$

4.4 Calculation of the Criticality Level Metric of the Workflow Tasks

The criticality level of each task indicates how certain we can be that the task is on the critical path. Tasks with lower (higher) criticality level are the least (most) likely to be on the critical path. In other words, the criticality level of the tasks indicates how much time the task can be postponed without postponing the finishing of the whole workflow.

The criticality level of a workflow task t_i, which is denoted as $\mathcal{L}(t_i)$, is calculated according to the interval lags of the task t_i in all interval workflows $\mathcal{W}_I^{(k)}, k = 0, 1, 2$. To accomplish this, First of all, the criticality level is set to initial value zero and then if $\mu_{et}(t_i)$ is greater than the upper bound of the interval lag, $\overline{lag^{(k)}}$, the criticality level is increased by one otherwise if $\mu_{et}(t_i)$ is greater than the lower bound of the lag interval, $\underline{lag^{(k)}}$, the criticality level is increased by 0.5. Thus, the criticality level of each task is a number from zero to three. The algorithm for computing the criticality level is shown in Fig. 2.

Algorithm 1.CriticalityLevel $(\mathcal{W}_S, t_i)$

Input: A stochastic workflow $\mathcal{W}_S$

Output: Criticality level of the task t_i in $\mathcal{W}_S$

1:   $\mathcal{L}(t_i) = 0;$

2:   for $k = 0$ to 2 do

3:        Compute the interval workflows $\mathcal{W}_I^{(k)}$

4:        $[\underline{lag^{(k)}}, \overline{lag^{(k)}}] =$ Compute the lag interval of t_i in $\mathcal{W}_I^{(k)}$

5:        If $\mu_{et}(t_i) > \underline{lag^{(k)}}$ then

6:   $\mathcal{L}(t_i) = \mathcal{L}(t_i) + 0.5$

7:        If $\mu_{et}(t_i) > \overline{lag^{(k)}}$ then

8:   $\mathcal{L}(t_i) = \mathcal{L}(t_i) + 0.5$

9:        return $\mathcal{L}(t_i)$

Fig. 2. The algorithm for computing the critical level of a task in a stochastic workflow

---

Algorithm 2. Scheduling Algorithm

Input: Stochastic workflow $\mathcal{W}_S$

On ready queue updates do:

1:   Let $\mathcal{W}_S$ be the remaining part of the stochastic workflow

2:   For $k = 0$ to 2 do

3:        Compute the interval workflows $\mathcal{W}_I^{(k)}$

4:   For each ready task t_i do

5:   Compute the criticality level of t_i using Algorithms 1

6:        submit task to appropriate resource

On task completions do:

7:   If t_i is the finishing task then

8:        mark the workflow as completed

9:   else

10: For each $t_j \in Succ(t_i)$ if all predecessors of t_j are marked as completed do

11:        Update the ready queue by adding t_j

Fig. 3. The scheduling algorithm process algorithms

4.5 Scheduling Workflow Tasks onto Appropriate Resources

After calculating the criticality level of all the workflow tasks, the phase of workflow scheduling is commenced. At the beginning, the starting task is added to ready queue. Then, whenever a task execution is completed, the ready queue is updated and the scheduling algorithm in Fig. 3 is executed. The scheduling algorithm orders and submits the newly added tasks according to their criticality levels with more critical ones first. When choosing an appropriate resource for mapping a task, we pick the one which is expected to complete that task earliest, with consideration of available processing capability of resources.

5 Performance Analysis

To assess the performance of the proposed method, we compared it with two of popular heuristics for workflow scheduling including HEFT [10] and DCP-G [8]. In addition, we conducted our study by means of GridSim toolkit [20]. In order to perform this comparison, we evaluated these scheduling methods on the basis of the makespan which is, by definition, the time required for executing workflow.

In this study, we use randomly generated task graphs in where dependency and number of parent tasks of a task is generated randomly. Further, for each generated workflow, we run 40 iteration of experiment in order that the results in all experiments will be stable. The size of the workflow is varied by considering different number of workflow tasks. In addition, the length and output data size of each task t_l in workflow obey normal distribution $N\left(\mu_i^{length}, \sigma_i^{length}\right)$ and $N\left(\mu_i^{o\_size}, \sigma_i^{o\_size}\right)$ respectively. The values of different parameters used in our simulation environment are listed in Table 1.

Table 1. Value of the parameters used for simulation

Simulation parameters	Value
Number of workflow tasks (n)	20, 50, 100, 200
Processing capacities of resources	4000–10000 (MIPS)
Bandwidths of resources	100–120 (Mbps)
Local load of resources	20–50%
μ_i^{length}	100000–500000 (MI)
σ_i^{length}	$\mu_i^{length}/4$
$\mu_i^{o\_size}$	100–500 (MB)
$\sigma_i^{o\_size}$	$\mu_i^{o\_size}/4$

We carried out our experiments in two different environments: (1) in a static environment where the background workload of the computational resources is fixed and (2) in a dynamic environment where the background workload of the computational resources fluctuates during time. In the following, we present our experiments and results in both environments in detail.

5.1   A Comparison of Execution Time in a Static Environment

To verify the efficiency of our proposed method, we firstly compare it with HETF and DCP-G in an environment where the background workload of the computational resources remains stable during runtime. The results are illustrated in Fig. 4. As we can observe, for different sizes of workflows, our proposed method achieves a makespan 4 to 9% shorter than the one computed by DCP-G and also 9 to 18% shorter than the one reported by HEFT. Therefore, it provides remarkable improvement over other considered algorithms. The main reason is that in the proposed method, the workflow tasks are classified into four classes of criticality, and ready tasks are scheduled in order of their criticality levels from more critical tasks to less critical ones. Whereas, in previous works tasks classified into only critical and noncritical ones.

5.2   A Comparison of Execution Time in a Dynamic Environment

In the following, in order to compare the performance of the considered scheduling algorithms comprehensively, we also perform our experiments in a dynamic environment where the workload of the computational resources fluctuates during runtime. Figure 5 manifests clearly that our proposed method gives better scheduling solution than HEFT and DCP-G, and it is much more obvious as the number of the tasks (N) increases. As Fig. 5 shows, our proposed method provides a 9 to 18% improvement over to the DCP-G and an 18 to 30% improvement over to the HEFT. The two main reasons that lead to this improvement are as follows: (1) the detailed ranking of the tasks that causes better decisions when mapping tasks to computational resources (more details described in previous experiment) and (2) the choice of appropriate computational resource for processing a task based on the available computing capabilities of the resources and their maximum computing capacities. Therefore, since tasks are most likely to be submitted to light-loaded resources rather than being waiting on heavy-loaded ones, we have a significant improvement in makespan of the proposed approach.

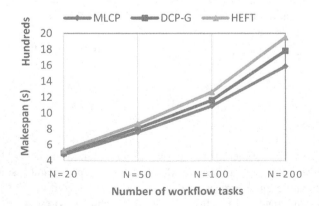

Fig. 4. A comparison of the MLCP, DCP and HEFT approaches in a Static Environment

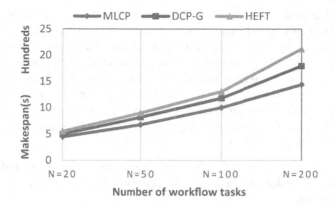

Fig. 5. A comparison of the MLCP, DCP and HEFT approaches in a Dynamic Environment

To sum up, in Fig. 6, the makespan of the proposed method in a static environment is compared against in a dynamic one, for different sizes of workflow graph. As it turns out, the makespan in dynamic environment is considerably lower. Because, in dynamic environment we consider the instantaneous workload levels of the resources and their fluctuations. Therefore, not only in a dynamic environment but also in a static one, our proposed strategy provides better solution as compared to both HEFT and DCP-G.

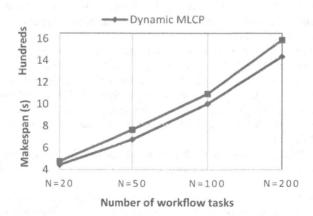

Fig. 6. A comparison of the makespan of the static and dynamic variants of the proposed method

6 Conclusions and Future Work

In this paper, we have tackled the problem of uncertainties about the parameters of workflow tasks and considered a stochastic model of workflow which is more realistic and lead to more convincing results. We have proposed a workflow scheduling method named MLCP, a multi-level critical path based method with different levels of criticality for tasks. This method first estimates interval approximates of stochastic workflow and then, injects them in an extension of the traditional critical path based algorithm and eventually, classifies task into four classes of criticality instead of classifying tasks into critical and noncritical ones as in previous works. The simulation of MLCP and other workflow scheduling methods have been performed in the GridSim simulator. The simulation results show that thanks to this detailed ranking information, better decisions are made about assigning tasks to computational resources in MLCP.

Our future work is aimed to:

- Propose a fault tolerant workflow scheduling with near-optimal task redundancy by making better decisions about redundancy type of workflow tasks according to their criticality levels.
- Design a decision making framework for choosing among different fault-tolerance solutions with respect to their makespan and cost and make a compromise between fault tolerance, execution time and cost.

References

1. Singh, V., Gupta, I., Jana, P.K.: A novel cost-efficient approach for deadline-constrained workflow scheduling by dynamic provisioning of resources. Future Gener. Comput. Syst. **79**, 95–110 (2018)
2. Jiang, J., Lin, Y., Xie, G., Fu, L., Yang, J.: Time and energy optimization algorithms for the static scheduling of multiple workflows in heterogeneous computing system. J. Grid Comput. **15**, 435–456 (2017)
3. Garg, R., Singh, A.K.: Adaptive workflow scheduling in grid computing based on dynamic resource availability. Eng. Sci. Technol. Int. J. **18**, 256–269 (2015)
4. Durillo, J.J., Nae, V., Prodan, R.: Multi-objective energy-efficient workflow scheduling using list-based heuristics. Future Gener. Comput. Syst. **36**, 221–236 (2014)
5. Casas, I., Taheri, J., Ranjan, R., Wang, L., Zomaya, A.Y.: GA-ETI: an enhanced genetic algorithm for the scheduling of scientific workflows in cloud environments. J. Comput. Sci. **26**, 318–331 (2016)
6. Rahman, M., Hassan, R., Ranjan, R., Buyya, R.: Adaptive workflow scheduling for dynamic grid and cloud computing environment. Concurr. Comput.: Pract. Exp. **25**, 1816–1842 (2013)
7. Singh, S., Dutta, M.: Critical path based scheduling algorithm for workflow applications in cloud computing. In: International Conference on Advances in Computing, Communication, & Automation (ICACCA) (Spring), pp. 1–6. IEEE (2016)
8. Rahman, M., Venugopal, S., Buyya, R.: A dynamic critical path algorithm for scheduling scientific workflow applications on global grids. In: IEEE International Conference on e-Science and Grid Computing, pp. 35–42 (2007)
9. Kwok, Y.-K., Ahmad, I.: Dynamic critical-path scheduling: an effective technique for allocating task graphs to multiprocessors. IEEE Trans. Parallel Distrib. Syst. **7**, 506–521 (1996)
10. Topcuoglu, H., Hariri, S., Wu, M.-Y.: Performance-effective and low-complexity task scheduling for heterogeneous computing. IEEE Trans. Parallel Distrib. Syst. **13**, 260–274 (2002)

11. Maheswaran, M., Ali, S., Siegal, H., Hensgen, D., Freund, R.F.: Dynamic matching and scheduling of a class of independent tasks onto heterogeneous computing systems. In: Proceedings of Eighth Heterogeneous Computing Workshop, HCW 1999, pp. 30–44 (1999)
12. Huang, K.-C., Tsai, Y.-L., Liu, H.-C.: Task ranking and allocation in list-based workflow scheduling on parallel computing platform. J. Supercomput. **71**, 217–240 (2015)
13. Zheng, W., Wang, C., Zhang, D.: A randomization approach for stochastic workflow scheduling in clouds. Sci. Program. **2016**, 13 (2016)
14. Zheng, W., Wang, C.: An experimental investigation into the approximation weight function of a stochastic list scheduling algorithm. In: 2015 International Conference on Cloud Computing and Big Data (CCBD), pp. 137–144 (2015)
15. Kamthe, A., Lee, S.-Y.: A stochastic approach to estimating earliest start times of nodes for scheduling DAGs on heterogeneous distributed computing systems. Cluster Comput. **14**, 377–395 (2011)
16. Dong, F., Luo, J., Song, A., Jin, J.: Resource load based stochastic DAGs scheduling mechanism for grid environment. In: 2010 12th IEEE International Conference on High Performance Computing and Communications (HPCC), pp. 197–204 (2010)
17. Zheng, W., Emmanuel, B., Wang, C.: A randomized heuristic for stochastic workflow scheduling on heterogeneous systems. In: 2015 Third International Conference on Advanced Cloud and Big Data, pp. 88–95 (2015)
18. Zheng, W., Sakellariou, R.: Stochastic DAG scheduling using a Monte Carlo approach. J. Parallel Distrib. Comput. **73**, 1673–1689 (2013)
19. Ross, S.M.: Introduction to Probability and Statistics for Engineers and Scientists. Academic Press, Cambridge (2014)
20. Buyya, R., Murshed, M.: GridSim: a toolkit for the modeling and simulation of distributed resource management and scheduling for grid computing. Concurr. Comput.: Pract. Exp. **14**, 1175–1220 (2002)

Toward Quantum Photonic Computers; Thinking May Not Be Realized by Digital Computers

Hassan Kaatuzian$^{(\boxtimes)}$ (iD)

Head of Photonics Research Laboratory (PRL), Electrical Engineering
Department, Amirkabir University of Technology, Tehran, Iran
hsnkato@aut.ac.ir

Abstract. Experts at the forefront of Artificial Intelligence (AI) have dreamed for more than half a century of autonomous thinking machines. But there's no hope that scientists can develop digital machines, capable of thinking process. They will never replace human mind in thinking. Simply we, ourselves are not logic machines. "I'm not a robot"! Human being have an intuitive intelligence cognition and consciousness that reasoning digital machines can not match. We should abandon some dogmas both in physics and computer technology. "Duality" in physics and "Digital Hardware" for High Performance Computing (HPC), both have failed when being applied in this area, no matter how many billions of dollars, world governments, invest in them. In this paper, at first, basic features for possible realization of thinking HPC machine using Quantum Optical Computer (QOC) versus classical digital computer will be discussed. Then we'll propose quantum optical processors work according to Quantum Photonic (QP) treatment which is based on corpuscular nature of light. Photons refraction phenomenon when travel through the interface between two different transparent solids will be simulated according to QP. Results, follow successfully from experimental measurements. If photons, as signal carriers in QOC be tracked by controlling Short Range Interatomic Forces (SRIF) in solids, then realization of QP computers (QPC), will be possible. It means QPC or QOC, will be accessible for HPC. Since QPC behaves in physics principle, more similarly to human brain in comparison with classical digital computers.

Keywords: Quantum Photonic Computers (QPC) · Thinking machine · Cognition · Consciousness · Artificial Intelligence (AI) · Bohm Theory · Hidden variables

1 Introduction

Every two years for the past several decades, computers have become twice as fast while their components have become twice as small. We're now at the age of Ultra Large Scale Integration (ULSI) electronic integrated circuits. Because of this explosive progress, today's machines are millions of times more powerful than their ancestors [1]. Advanced lithographic techniques can yield parts many hundred times smaller than even recently available [2]. So, computers are also to become much smaller. Then in

© Springer Nature Switzerland AG 2019
L. Grandinetti et al. (Eds.): TopHPC 2019, CCIS 891, pp. 490–503, 2019.
https://doi.org/10.1007/978-3-030-33495-6_38

the future, a new technology must replace or supplement what we now have. In spite of such a vast amounts of progress both in hardware techniques and software manipulations specifically in Artificial Intelligence (AI) domain, digital computers still can not think!

As, late Marvin Minsky, a well known AI professor at MIT and other AI entrepreneurs believed in talking about "intelligent systems", that will perform better than we can do in home, in class and at work.: Expert Systems. After nearly a quarter of a century, Hubert and Stuart Dreyfus challenged above idea and for the first time admitted why computers may never think like people [3]. Now after more than half a century and in spite of more huge progress still in computer manufacturing technology and software techniques, we're again should admit, digital computers may never think like human being. They are logic machines based on Boolean algebra. A logic physical hardware may not simulate "thinking" in brain. Since "thinking", does not necessarily works according to logic. We should have courage to admit the role of "wisdom" in this area, no matter we've never been able to define perfectly this notion, technically and no matter discussion at this domain seems to be very speculative!

It's more than several decades, pioneers in the world began investigating the physics of information processing circuits. Asking about how small can the components of circuits must be made in the course of computation? How much energy must be used? Because computers are physical devices, their basic operations are described by physics. But in very small scales, physical description must be given by Quantum theory, not Classic. Quantum computers should now be modeled and their differences with classical ones ought to be determined. In special, we should find out, whether quantum effects might be exploited to speed up computations or to perform calculations in novel ways. For example Hydrogen atoms could be used to store bits of information in a quantum computer [1, 4]. An atom in it's ground state can represent a "zero" and the same atom at a higher excited energy level can represent a "one". Using a laser pulse with enough energy, atomic bit can change it's state between "zero" and "one".

In Sect. 2, Quantum Optical Computer (QOC) versus digital computer for possible realization of thinking machine will be discussed. Quantum optical processors and how they work will be explained in Sect. 3. We'll not limit our theoretical viewpoints only on Quantum mechanics and Copenhagen interpretation of quantum world. Since they seem to be incomplete for explanation of qubits (quantum bits), theoretically [5, 6]. At this situation, Bohmian mechanics [7] and Hidden variables [8] must be taken into account. It's because we must know in theory (not in measurement), with exact certainty about photon's position and the time of it's interaction with qubits. This should be possible without violation of uncertainty in measuring process. Spintronics will also be mentioned in Sect. 3. Finally in Sect. 4, we'll introduce method of realization of Quantum Photonic computers (QPC) in continuation of QOC. Method of tracking light, at attosecond time scale theoretically, will be discussed at this section [9]. If we're able to control and track photons during their journeys between atomic layers, then QPC may be realized. We'll also have a conclusion section.

2   Quantum Optical Computer Versus Digital Machine

A conventional digital computing system, requires first of all a Central Processing Unit (CPU). It may consist of sequential (Von Neuman) or parallel (non-Von Neuman) architectures. All of the processing procedures on data signals for many different applications will be executed by clock pulses in CPU. For example, speech and image processing for pattern recognition or pattern generation are all in domain of CPU and it's coprocessors. Artificial Intelligence (AI) and developing advanced deduction systems are all in this domain too (see Fig. 1: Iranian first pedagogical robot PARS-1) [10].

Fig. 1. First Iranian pedagogic Robot called "**PARS-1B**", with AI Graph Search Algorithm to track and find barriers and targets. Developed by Author, in Sharif University of Technology, E.E. Dept, year 1986.

Basically, it's not still obvious whether we'll finally be able to make a thinking machine, but proposing a quantum CPU, will be a step more toward realization possibility of it. Cognition and consciousness happens in human brain. So, brain's physical shape, materials made from, methods of interconnections, all should be taken into account. Bio-photonic, bio-electric and bio-chemical methods of interconnections between nervous cells in human brain, work very differently in comparison with a logical CPU, which behaves intelligently, but not with cognition and consciousness [11–13].

Even Robot called "Sophia", revealed two years ago in United Nations (UN), can not think unless it's supervisor hints it. Although it's owners might claim it could think and will have consciousness soon, only for business considerations [14].

AI procedure in conventional computers is done symbolically, not physically. But nature consists of real matters (atoms). Where atoms interact physically with each other. They behave according to rules of quantum theory not symbolically. So, simulation of brain's behaviors for making an advanced machine, requires quantum CPU [15, 16]. It also requires optic and photonic interconnections [17], for development of fast image recognition systems based on Vander lugt optical filters and optical correlators for pattern recognition [18].

2.1 Memories in QOC

Memory in QOC should be able to save both magnitude and phase of input signals in 3-dimensional volume made from photo-refractive materials [19–22]. Data signals can be saved in 3D space domain in these materials, where we call them Holograms. Data may also be saved in spatial frequency domain, where we call them optical Vander lugt filters (see Fig. 2), phase only optical filters, etc. These kinds of memories both Holographic [20, 21] and Vander lugt [18], are able to store vast amounts of data, voice or images in physical volume of materials.

Fig. 2. First development and reconstruction of Spatial Optical Vander lugt filter for number "**5**". Developed for optical pattern recognition in machine vision, by A.H. Majedi and M.R. Chaharmir in PRL-E.E.Dept., Amirkabir University of Technology, year 1995 [18].

They're made from photo-refractive materials. It means applying suitable external electric field, for example in visible region, may alter material's optical properties like index of refraction. In turn, they cause a kind of optical storage for a limited time, from fractions of second up to several minutes, hours, days or even months. Inorganic materials like GaAs, BaTio3, LiNbO3, KNbO3 ..... crystals or organic crystals like 2-cyclooctylamino-5-nitropyridine doped with 7-7-8-8-tetracyanoquinodimethan [19, 22, 23]. Photo-refractive effect formulation for QOC-3D memories, has already been explained theoretically using Kukhtarev equations [22, 23]. Their applications are in high speed real time complex images correlations as in human face and alphanumeric symbols recognition. We've successfully developed hybrid optical computers for machine vision and recognition [18, 22]. Dynamic data storage and retrieving optical bit storage signals are main features necessary for developing real time QOC. It's already been studied that brain in pattern storage, behaves very similar to holograms [3].

Other alternative for temporarily data storage in QOC is called Slow Light optical bit memory. Quantum storage of light in atomic sodium vapor at very low temperatures (even micro-Kelvin) has already been reported experimentally [24, 26]. We've successfully explained this kind of storage theoretically according to Quantum Mechanics [25]. It's a kind of quantum entanglement of photons for example in Sodium or Robedium vapors at very low temperatures. QOC memory is also available in semiconductor solids [27, 28]. Slow light devices can also be used as optical delay lines. Some known optical properties like Stimulated Brillouin Scattering (SBS), Stimulated Raman Scattering (SRS), Electromagnetically Induced Transparency (EIT), Coherent Population Oscillation (CPO) may be applied in some nonlinear- dispersive optical devices to realize light speed reduction. We've already successfully simulated and analyzed such devices for designing new optical bit memories [29, 30].

2.2   Optical Keyboards and Display Devices as I/O in QOC

In QOC, man-machine interface must be more user friendly. Besides, computer size and it's hardware electronic components should be minimized or even be eliminated.

Limited reliability of a real keyboard causes to think about optical virtual keyboards illuminated by a laser beam as input device. As can be seen in Fig. 3a and b, Schema of a keyboard pattern can be projected virtually by laser on table, wall or even in the air. On the other hand, display devices are used as output. In QOC, 3D display devices are suitable choices [31, 32].

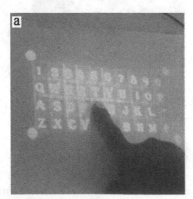

Fig. 3. a. A virtual optical keyboard illuminated by 550 nm Green laser. Developed by M. Farhadi, in PRL-E.E.Dept., AUT, year 2009. b. A virtual optical keyboard illuminated by Blue-Violet laser for Organ music instrument. Developed by M. Khalaj Babaie, in PRL-E.E.Dept., AUT, year 2017. (Color figure online)

3   Quantum Optical Processors

Old quantum Theory announced for the first time in 1905 after Hertz's and Planck's experimental and Einstein's theoretical works. Quantum mechanics developed since 1928 by N. Bohr and V. Heisenberg with non-causal Copenhagen interpretation of Quantum Theory. Finally in 50's decade of 20'th century, Bohmian mechanics developed by D. Bohm with intuitive- causal interpretation of quantum theory. For realization of QOC, we need to know quantum theory:

3.1   Quantum Mechanics

Quantum mechanics, for better or worse, predicts a number of counter intuitive effect that have been verified experimentally. To appreciate the weirdness of which quantum computers are capable, we need to accept a single strange fact called wave-particle duality [1, 15, 16]. In some weird quantum sense, an electron can sometimes be both here and there simultaneously. It's location will remain unknown, until some inter-action (such as photon bouncing off the electron) reveals it to be either here or there but not both. Two super imposed quantum waves behave like one wave. Atom is in a state equal to a superposition of a wave corresponding to zero and one, each having the same amplitudes. Such a quantum bit, is called qubit [5, 6]. In contrast a classical bit will always read either zero or one. While 3-bits equal 8 sequential states, 3-qubits equals 8 simultaneous states. It means exponential growth of processing speed. There are 5 things every quantum computer needs: (1) reset all qubits into zero(start). (2) readable at the end of calculation. (3) qubits last long enough during run program. (4) carry out two fundamental operations: NOT and EXOR and (5) ability to handle large amounts of qubits [5].

3.2   Bohmian Mechanics

Albert Einstein noted that Copenhagen interpretation of quantum mechanics, would violate all classical intuition about causality. In such a superposition, neither bit is in a definite state, yet if you measure one bit, there by putting it in a definite state, the other bit also enters into a definite state. In contrast with Copenhagen counter intuitive viewpoint, there's also another interpretation for quantum world that we know it as Bohmian mechanics [36, 37], based on viewpoints of late David Bohm [7, 9]. It's based on intuitive physics and introduces hidden variables. Because of universal ignorance of such hidden variables, we deduce superposition and duality. According to Bohm theory, there's a kind of what's called: "Straight forward conditionalization" during light-matter interaction [7, 8]. According to causality, it'll be possible to predict next position steps based on initial conditions. So, with certainty, we'll be able to trace information signals from starting point till to the end. Bohmian mechanics, has been deliberately ignored for most of the past six decades. It challenges the probabilistic, subjectivist picture of reality implicit in the standard formulation of quantum mechanics [7, 33].

In this theory, chance plays no role and every material object invariably does occupy some particular region of space. All the mathematical operations in quantum mechanics occurs in Hilbert space. But it's noticeable that quantum phenomena do not occur in Hilbert space. They occur in laboratories. In real space.

3.3 Spintronics

A growing band of experiments think they have seen the future of electronics, and it is "spin". This fundamental property of electrons and other subatomic particles underlies permanent magnetism, and is often regarded as a strange form of nano-world angular momentum. Paul Dirac for the first time postulated spin existence in 1920. Spin, like charge and mass is an intrinsic subatomic particles property. Bosons- like photon or pion- have integer spin. Fermions-like electron, proton, neutron- have half-integer spins. Spin can be used as memory cell or processor instead of electron. Huge amount of non-volatile magnetics RAM and ultra-fast spin microprocessors will be available. This technology which is called "Spintronics", requires special materials [34, 35]: Ferromagnetic metals (Cr, Co, Fe,...), magnetic semiconductors-(InCr, ...). Cryogenic temperature is needed. Or else spin scattering happens at room temperature. Moreover spin polarization (up-down) or (right-left) retains in microsecond orders. Spin field effect transistors have already been proposed in 90's decade [34]. Spintronics microprocessor is one of modern proposals for realization of quantum computers [35]. But some experiments based on electron's spin by researchers in physics, have demonstrated some kinds of violation of duality in quantum mechanics with Copenhagen interpretation of Niels Bohr [7]. Finally, Bohmian mechanics have found answer to this enigma.

4   Quantum Photonic Computer

Quantum Photonics (QP) revealed for the first time in 1994 [44] with the aim of finding intuitive description for atomic scale and attosecond time scale photon electron interactions. Annihilation and recreation mathematical operators in Hilbert space found intuitional meanings. If we want to realize Optical Photonic Computers (QPC), three dimensional space trajectory of photon when travelling inside a transparent material should be precisely tracked. For this reason, Short Range Interatomic Forces (SRIF), must be taken into account. For realization of QPC, duality will no more help us. Since QOC or QPC designer must know exactly the location of photon inside the matter. Moreover, straight forward conditionalization in Bohmian mechanics determines photon's trajectory in next following steps. We must know with certainty [33], where's the photon, and also the time of it's interaction with matter at different atomic layers.

4.1   Main Features of QP

This theory as it's called, is based on quantum theory. In addition, it delivers a "real", instead of "virtual" description of photon-electron interaction inside a transparent solid material. It not only emphasizes on mathematical point of views similar to quantum mechanics, but in addition, it also emphasizes on intuitive physics about what's really happening in space-time atomic world. In QP, we've four main postulates [9]: First, knowledge of real shape of molecules in 3D spatial coordinates. Second, knowledge of physical shape of materials lattice. Third, precise estimation of SRIF between molecules in a real material. Fourth, space-time analysis and simulation of electron-photon interaction and prediction of photon trajectory when travels inside the matter [38].

Moreover in QP, light itself is assumed to be a stream of billions of photons as particles. Each photon has quantized energy according to Planck's formula ($E = h \cdot v$). Photon has zero rest mass and potential energy. Carries momentum and has it's own electric and magnetic fields. It's electrical field has penetration depth on the order of it's wavelength and travels in vaccum at speed of 300,000 km/s. But in transparent solids experiences some kinds of retardation in every molecular layer of material [39–42].

4.2   Nano-Scopic Interpretation of Refraction Phenomenon in Q.P

Equation (1), shows the total time, which takes for photon to travel through a transparent media [9, 42]: (see also Fig. 4)

$$T = \frac{L}{C_0} + \sum_{i=1}^{N} \tau_{di} \tag{1}$$

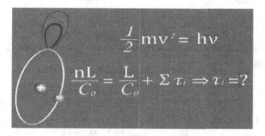

Fig. 4. During Photon-electron interaction, photon is Annihilated and it's energy as kinetic form will be transferred to electron for a short time. Then photon is recreated.

Where "L" is the material length. Travelling time of photon may be assumed as the sum of time that photon spends to pass through the intermolecular (inter-atomic) empty space $\left(\frac{L}{C_0}\right)$ and photon- matter interactions $\left(\sum_{i=1}^{N} \tau_{di}\right)$. The index of refraction is the ratio of C_0 (vacuum velocity of the light) over the average velocity of light in the medium for large values of N (number of interactions), so:

$$n = \frac{C_0}{C} = 1 + \frac{C_0}{L} \sum_{i=1}^{N} \tau_{di} \tag{2}$$

If now, we consider "τ_d" as the mean retardation time per interaction and "d", as the mean free pass between two successive interactions, we have:

$$n = 1 + \frac{C_0}{d}\tau_d \tag{3}$$

(where: $\tau_d = \frac{\sum\limits_{i=1}^{N}\tau_{di}}{N}$; and d $= \frac{L}{N}$).

During the interaction, as seen in Fig. 5, at first, the photon, annihilates and gives it's energy to the electron in the lowest energy level and perturbs it. Since the energy of annihilated photon is not sufficient to transfer the electron to a higher allowed energy state, the perturbed electron returns to it's initial orbit after a transit time, which we call (τ_p), ultimately the photon recreates.

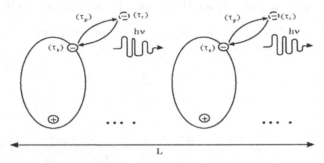

Fig. 5. Photon-Electron interaction with molecular layers of a transparent typical solid, based on Quantum Photonics and Bohm Theory.

Therefore, the retardation time (τ_d) for a more precise estimation, may be considered as the sum of photon annihilation time (τ_a), electron perturbation time (τ_p) and photon recreation time (τ_r) (see Fig. 5) and [42, 46]:

$$(\tau_d) = (\tau_a) + (\tau_p) + (\tau_r) \tag{4}$$

Generally these quantities can be considered as a function of wavelength but, for simplicity as a first- order approximation, we consider (τ_a) and (τ_r) as negligible constants on the order of zepto-seconds. Since according to Quantum Photonics assumptions, (τ_p) will be in the range of atto-seconds [9].

4.3   How to Obtain Hidden Variable: SRIF

When an atom is placed inside a crystal, the wave functions (or atomic orbits) of atoms are perturbed and altered because the neighboring atoms, exert electric field on the atomic electron, which results in the distortion of orbits and splitting of the energy levels. This electric field is known as crystal field. It's effect can be treated by Perturbation theory, a common approach in submicron scales related to quantum mechanics. In Perturbation theory, the potential in the presence of applied field becomes:

$$V = V_o(r) + V'(r) \tag{5}$$

where $V_o(r)$, is the atomic potential and $V'(r)$, is the potential due to the field. The details of this method may be found in books of quantum mechanics like [45].
the results are:

$$E_n \cong E_n^{(o)} + \langle n|V'|n \rangle - \sum_m' \frac{|\langle m|V'|n \rangle|^2}{E_m^{(o)} - E_n^{(o)}} \tag{6}$$

and:

$$\psi_n \approx \psi_n^{(o)} - \sum_m' \frac{|\langle m|V'|n \rangle|^2}{E_m^{(o)} - E_n^{(o)}} \tag{7}$$

Here, $E_n^{(o)}$ and $\psi_n^{(o)}$ are energy and wave function for an arbitrary level "n" in the absence of field. These results show the wave function of atoms inside the crystal in a stable state. So, if one atom goes farther or closer to other atoms in crystal structure, a restoring force makes it go back to it's stable state.

For simplicity, we assume that electrons of atoms in a typical lattice structure, form a uniform, negatively charged sphere surrounding the nucleus of atoms. Let F_o be the inter-atomic force between mother atom (which interacts with photon) and it's nearest neighbor at distance "a": constant lattice:

$$F_o = \frac{\text{Constant}}{a^r} \tag{8}$$

The interatomic force, when photon interacts with i'th layer from the interface surface, will be obtained as follows [9]: See also Figs. 6 and 7:

Fig. 6. Coulomb mother nucleus force and SRIF in a typical Orthorhombic or tetragonal crystal lattice.

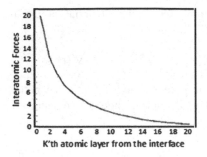

Fig. 7. Interatomic force in K'th layer from the interface for a typical crystal lattice. As can be seen, it decreases exponentially to zero when "K" becomes enough large. These results can be extended easily to amorphous transparent solids too.

$$F_{ii} = F_i + F_{i+1} + \ldots F_p + \ldots \tag{9}$$

4.4 Quantum Photonic Physical Model

When incident light strikes at the interface between two dielectric media and interacts with the surface layer electrons of a medium, it causes a small perturbation in electron orbits. It's been assumed that these electrons bear a transient stage at this moment, and their flight routes are determined by result of the columbic nucleus force (F_{cn}) and the inter-atomic force (F_i), and also depend on the angle of the incident light (θ_i).

We have shown that photon, when traveling through k'th layer from the interface region, deviates in an angle with the amount of ($\Delta\theta$) and we can write [9]:

$$\mathrm{tg}(\Delta\theta) = \frac{F_{ik} \cdot \sin(\theta_i)}{F_{cn} \cdot \cos(\alpha) + F_{sl}} \tag{10}$$

and:

$$\cos(\alpha) = \cos(\beta) \cdot \cos(\theta_i) \tag{11}$$

where F_{cn} is the columbic nucleus force, F_i is the inter-atomic force in the K'th layer from the interface region, and F_{sl} is the surface layer inter-atomic force.

Equation (10), is a nano-scopic refraction relation proposed by QP theory. It can predict the trajectory of photons, traveling through the first few atomic layers of the crystal- interface. When photons interact with atomic layers near the interface surface, the electrons of those atoms, bear rotational torque, because two main perpendicular forces exist: first, the inter-atomic force SRIF. Second: columbic nucleus force (Fq in Fig. 8). QP predicts photon's refracts gradually at the first few atomic layers from the interface surface (see Fig. 8).

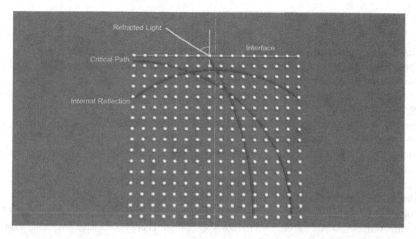

Fig. 8. According to predictions of QP, photon's flight route in first few atomic layers from the interface is not abruptly refracted as the case of macroscopic observation of Snell's law. But gradually refracts step by step and asymptotes to output angle value.

Now, we may propose here a real scenario: since the nature of SRIF is columbic, with applying external electric field for example, it'll be possible to change SRIF and control the rate of photon deviation inside the crystal. This is a kind of photon multiplexing or de-multiplexing through different atomic channels. Also in theory, it's also possible, when photon annihilates during interaction with bounded electron of an atom, it may recreate or may not: means a kind of photon atomic switch. Other photonic devices may also be realizable in atomic scales with switching times even on the order of attosecond. They all means realization of QP-CPU.

5 Conclusion

In QPC, photons assumed to be signal carriers. By controlling SRIF inside the material, computer is made from, we'll be able to track and interconnect optical signals even in attosecond time scales during single-photon interactions with electrons of matter.

This theoretical intuitive scenario, delivers a powerful tool for executing huge amount of mathematical calculations, but physically! This approach help us to claim that we're now one step more to realize QPC. Although using QP, we'll be able to track photons one by one inside brain, but it still does not mean we'll be able to realize thinking machine. Since we're not thinking machine, as some researchers claims [43]. We're not robots! However it seems emotional intelligence may be explained now in more details. AI will experience one more powerful step forward physically not symbolically. Human brain is a physical entity and is the container of thinking and intelligence processes. For understanding seriously about these processes we should abandon duality dogma in world of physics. Although duality works well in Politics and in Cinema, but is not a suitable tool to find out the truth!

References

1. Lioyd, S.: Quantum mechanical computers. Sci. Am. **273**, 44–50 (1995)
2. Kaatuzian, H.: Theory and Technology of Semiconductor Devices, 2nd edn. Amirkabir University Press, Tehran (2016). (In Persian)
3. Dreyfus, H., Dreyfus, S.: Why computers may never think like people. MIT Technol. Rev. **89**, 20–42 (1986)
4. Benioff, P.: Quantum mechanical models of turing machines that dissipate no energy. Phys. Rev. Lett. **48**(23), 1581–1585 (1982)
5. Stick, D., Sterk, J.D., Monroe, C.: The trap technique toward a chip-based quantum computer. IEEE Spectr. **44**, 30–37 (2007)
6. Vandersypen, L.: Dot-to-dot design. IEEE Spectr. **44**, 33–39 (2007)
7. Albert, D.Z.: Bohm's alternative to quantum mechanics. Sci. Am. **270**, 32–39 (1994)
8. Bohm, D.: A suggested interpretation of the quantum theory in terms of hidden variables, I, II. In: Wheeler, J.A., Zurek, W.H. (eds.) Quantum Theory and Measurement. Princeton University Press, Princeton (1983)
9. Kaatuzian, H.: Quantum Photonics, A Theory for Attosecond Optics. Amirkabir University Press, Tehran (2012)
10. Kaatuzian, H.: Pars-1 robot. Daneshmand Iran. Sci. Tech. Mag. **24**, 69–72 (1986) (1987)
11. Short, K.L.: Microprocessors and Programmed Logic. Prentice-Hall Inc., Upper Saddle River (1981)
12. Koch, C., Tononi, G.: Can machines be conscious? IEEE Spectr. **45**, 47–51 (2008)
13. Horgan, J.: The consciousness conundrum. IEEE Spectr. **45**, 28–34 (2008)
14. Goertzel, B., Mossbridge, J., Monroe, E., Hanson, D., Yu, G.: Loving AI: humanoid robots as agent of human consciousness expansion. arXiv: 1709.07791v1 [cs.AI], 22 September 2017
15. Shor, P.W.: Algorithms for quantum computation: discrete logarithms and factoring. In: Goldwasser, S. (ed.) 35th Annual Symposium on Foundations of Computer Science Proceedings. IEEE Computer Society Press, Washington, DC (1994)
16. Cirac, J.I., Zoller, P.: Quantum computations with cold trapped ions. Phys. Rev. Lett. **74**(20), 4091–4094 (1995)
17. Savage, N.: Linking with light. IEEE Spectr. **39**, 32–36 (2002)
18. Kaatuzian, H., Chaharmir, M., Majedi, A.H.: Design and development of optical spatial Vander Lugt filters. In: ICEE-97, 7–9 May, Tehran, Iran. Sharif University of Technology (1997)
19. Mok, F.H.: Angle-multiplex storage of 5000 holograms in lithium niobate. Opt. Lett. **18**(11), 915–917 (1993)
20. Hong, J.H., Mcmichael, I., Chang, T.Y., Christian, W., Paek, E.G.: Volume holographic memory systems: techniques and architecture. Opt. Eng. **34**(8), 2193–2203 (1995)
21. Psaltis, D.: Holographic memories. Sci. Am. **273**, 52–58 (1995)
22. Kaatuzian, H.: Photonics, vol. 2. 4th printing, Amirkabir University Press, Tehran (2018). (in Persian)
23. Yariv, A.: Quantum Electronics. Wiley, Hoboken (1989)
24. Phillips, D.F., Fleischhauer, A., Mair, A., Walsworth, R.L., Lukin, M.D.: Storage of light in atomic vapor. Phys. Rev. Lett. **86**, 783–786 (2001)
25. Kaatuzian, H., Rostami, A., Ajdarzadeh, O.A.: Analysis of quantum light memory in atomic systems. J. Opt. B: Quantum Semiclassical opt. **7**, 157–167 (2005)
26. Liu, C., Dutton, Z., Behroozi, C., Hau, L.V.: Observation of coherent optical information storage in an atomic medium using halted light pulses. Nature **409**, 490 (2001)

27. Kaatuzian, H., Shokri Kojori, H., Zandi, A., Ataei, M.: Analysis of quantum well size alteration effects on slow light device based on excitonic population oscillation. Opt. Quantum Electron. **45**(6), 947–959 (2013)
28. Kohandani, R., Kaatuzian, H.: Theoretical analysis of multiple quantum well slow light Devices under applied external fields using a fully analytical model in fractional dimensions. Quantum Electron. **45**(1), 89–94 (2015)
29. Abdolhosseini, S., Kohandani, R., Kaatuzian, H.: Analysis and investigation of temperature and hydrostatic pressure effects on optical characteristics of multiple quantum well slow light devices. Appl. Opt. **56**(26), 7331–7340 (2017)
30. Choupanzadeh, B., Kaatuzian, H., Kohandani, R.: Analysis of the influence of geometrical dimensions and external magnetic field on optical properties o InGaAs/GaAs quantum-dot slow light devices. Quantum Electron. **48**(6), 582–588 (2018)
31. Son, J.Y.: 3D displays. IEEE LEOS Newslett. **18**, 13–14 (2004)
32. Sullivan, A.: 3-deep. IEEE Spectr. **42**, 22–27 (2005)
33. Durr, D., Goldstein, S., Zanghi, N.: Quantum equilibrium and the origin of absolute uncertainty. J. Stat. Phys. **67**(5), 843–908 (1992)
34. Zorpette, G.: The quest for the spin transistor. IEEE Spectr. **38**, 30–35 (2001)
35. Svoboda, E.: Fresh spin on logic. MIT spintronics microprocessors. IEEE Spectr. **44**, 15 (2007)
36. Bohm, D.: Quantum Theory. Dover 1951, Reprint Prentice Hall (1989)
37. Bohm, D.: Causality and Chance in Modern Physics 1957, 1961, Reprinted by University of Pensylvania Press (1980)
38. Kaatuzian, H.: Quantum Photonics, an authentic concept for attosecond optics. J. Laser Opt. Photon. **4**(Suppl. 2), 65 (2017)
39. Kaatuzian, H., Wahedy Zarch, A.A., Ajdarzadeh Oskouci, A., Amjadi, A.: Simulation and estimation of normal dispersion phenomenon in an acentric organic crystal (NPP) by the quantum photonic approach. In: Modelling and simulation in materials science and engineering, vol. 15, no. 8, pp. 869–878. IOP Publisher, December 2007
40. Wahedy Zarch, A.A., Kaatuzian, H., Amjadi, A.: A semiclassical approach for the phase matching effect in the nonlinear optical phenomena. J. Opt.: Pure Appl. Opt. **10**(12), 125102 (2008)
41. Wahedy Zarch, A.A., Kaatuzian, H., Amjadi, A., Ajdarzadeh, O.A.: A semi-classical approach for electrooptic effect. Opt. Commun. **281**, 4033–4037 (2008)
42. Kaatuzian, H., Adibi, A.: Photonic interpretation of refractive index. In: SPIE Proceedings, vol. 2778, ICO-17, Taejon, Korea (1996)
43. Brooks, R.: I Rodney Brooks am a robot. IEEE Spectr. **45**, 63–67 (2008)
44. Kaatuzian, H.: Photonic interpretation of Refraction, Polarization, Dispersion… in submicron scales. Ph.D. dissertation in Electrical Engineering Department of Amirkabir University of Technology, July 1994
45. Fromhold, A.T.: Quantum Mechanics for Applied Physics and Engineering. Academic Press, Cambridge (1981)
46. Kaatuzian, H., Bazhdanzadeh, N., Ghohroodi Ghamsarim, B.: Microscopic analysis for normal dispersion based on quantum photonics treatment. In: CSIMTA 2004, Cherbourg, France (2004)

Quantum Advantage by Relational Queries About Equivalence Classes

Karl Svozil[(✉)] [ID]

Institute for Theoretical Physics, Vienna University of Technology,
Wiedner Hauptstrasse 8-10/136, 1040 Vienna, Austria
svozil@tuwien.ac.at
http://tph.tuwien.ac.at/~svozil

Abstract. Relational quantum queries are sometimes capable to effectively decide between collections of mutually exclusive elementary cases without completely resolving and determining those individual instances. Thereby the set of mutually exclusive elementary cases is effectively partitioned into equivalence classes pertinent to the respective query. In the second part of the paper, we review recent progress in theoretical certifications (relative to the assumptions made) of quantum value indeterminacy as a means to build quantum oracles for randomness.

Keywords: Quantum computation · Partitioning of cases · Quantum parallelism · Hidden subgroup problem · Quantum random number generators

1 Quantum (Dis-)advantages

Genuine quantum computations will be with us for a long time to come, because the miniaturization of electronic circuits is pushing the processor physics into the coherent superposition/complementarity/entanglement/value-indefinite regime (which has no sharp boundary just as the quantum–classical separation is fuzzy and means dependent). Moore's law, insofar as it relates to classical "paper and pencil" [45, p. 34] computation, has reached its effective bottom ceiling approximately ten to five years ago; this is due to exhaustion of minimization with respect to reasonably cooling, as well as by approaching the atomic scale. Most recent performance increases are due to parallelization (if possible).

Alas, this upcoming kind of "enforced" quantum domain computing, imagined by Manin, Feynman, and others, still poses conceptual, theoretical and technological challenges. Indeed, contemporary quantum information theory appears to be far from being fully comprehended, worked out and mature. It is based on quantum mechanics, a theory whose semantics has been notoriously debated almost from its inception, while its syntax – its formalism, and, in particular, the rules of deriving predictions – are highly successful, accepted and relied upon. Depending on temperament and metaphysical inclination, its proponents admit that nobody understands quantum mechanics [13,21], maintain that there is

© Springer Nature Switzerland AG 2019
L. Grandinetti et al. (Eds.): TopHPC 2019, CCIS 891, pp. 504–512, 2019.
https://doi.org/10.1007/978-3-030-33495-6_39

no issue whatsoever [18,22], one should not bother too much [10,14] about its meaning and foundations, and rather shut up and calculate [30,31].

By transitivity or rather a reduction, quantum information theory inherits quantum mechanics' apparent lack of consensus, as well as a certain degree of cognitive dissonance between applying the formalism while suffering from an absence of conceptual clarity [33], Strong hopes, claims and promises [1–3,16,17,41] of quantum "supremacy" [46] are accompanied by the pertinent question of what exactly, if at all, could make quantum information and computation outperform classical physical resources. Surely many nonclassical quantum features present themselves as being useful or decisive in this respect; among them complementarity, coherence (aka parallelism), entanglement, or value indeterminacy (aka contextuality). But if and how exactly those features will contribute or enable future algorithmic advances still remains to be seen.

The situation is aggravated by the fact that, although the quantum formalism amounts to linear algebra and functional analysis, some of its most important theorems are merely superficially absorbed by the community at large: take, for example, Gleason's theorem [23], and extensions thereof [8,36]. Another example is Shor's factoring algorithm [35, Chapter 5] whose presentations often suffer from the fact that its full comprehension requires a nonsuperficial understanding of number theory, analysis, as well as quantum mechanics; a condition seldom encountered in a single (wo)man. Moreover, often one is confronted with confusing opinions: for instance, the claim that quantum computation is universal with respect to either unitary transformations or first-order predicate calculus is sometimes confused with full Turing universality. And the plethora of algorithms collected into a quantum algorithm zoo [25] is compounded by the quest of exactly why and how quantum algorithms may outperform classical ones.

Quantum advantages may be enumerated in four principal groups, reflecting potential non-classical quantum features:

– quantum parallelism – aka *coherent superposition* of classically mutually exclusive bit states, associated with their simultaneous co-representation;
– quantum collectivism – aka entanglement (involving possibly nonlocal correlations) in a multi-particle situation: information is encoded only in *relational properties* among particles; individual particles have no definite property;
– quantum probabilities are vector-based (orthogonal projection operators), resulting in non-classical expectation values rendering different (from classical value assignments) predictions;
– quantum complementarity: in general quantized systems forbid measurements of certain pairs of observables with arbitrary precision: "you cannot eat a piece of the quantum cake & have another one too;"
– quantum value indefiniteness: there cannot exist classical (true/false) value assignments on certain collections of (intertwining) quantum observables.

In what follows the first and the last feature – parallelism and value indefiniteness – will be discussed in more detail.

2   Suitable Partitioning of Cases

One quantum feature called "quantum parallelism," which is often presented as a possible quantum resource not available classically, is the capacity of n quantum bits to encode 2^n classically mutually exclusive distinct classical bit states at once, that is, simultaneously: $|\Psi\rangle = \sum_{i=0}^{2^n-1} \psi_i |i\rangle$, where the index i runs through all 2^n possible combinations of n classically mutually exclusive bit states $\{0, 1\}$, $|i\rangle$ are elements of an orthonormal basis in 2^n-dimensional Hilbert space, and ψ_i represent probability amplitudes whose absolute squares sum up to 1.

Quantum parallelism, often presented rather mystically, may formally come about rather trivially: the alleged simultaneous quantum co-existence of classically mutually exclusive states is like pretending that a vector in the plane may simultaneously point in both directions of the plane [17]; a sort of confusion between a vector and its components. This seemingly absurd co-representability of contradicting classical states was the motivation for Schrödinger's cat paradox [37]. Note also that, in order to maintain coherence throughout a quantum computation, a *de facto* exponential overhead of "physical stuff" might be required. This could well compensate or even outweigh the advantage; that is, the exponential simultaneous co-representability of (coherent superpositions of) classical mutually exclusive cases of a computation.

The state $|\Psi\rangle$ "carrying" all these classical cases in parallel is not directly accessible or "readable" by any physical operational means. And yet, it can be argued that its simultaneous representation of classically exclusive cases can be put to practical use indirectly if certain criteria are met:

- first of all, there needs to be a quantum physical realizable grouping or partitioning of the classical cases, associated with a particular query of interest; and
- second, this aforementioned query needs to be realizable by a quantum observable.

In that way, one may attain knowledge of a particular feature one is interested in; but, unlike classical computation, (all) other features remain totally unspecified and unknown. There is no "free quantum lunch" here, as a total specification of all observables would require the same amount of quantum queries as with classical resources. And yet, through coherent superposition (aka interference) one might be able to "scramble" or re-encode the signal in such a way that some features can be read off of it very efficiently – indeed, with an exponential (in the number of bits) advantage over classical computations which lack this form of rescrambling and re-encoding (through coherent superpositions). However, it remains to be seen whether, say, classical analog computation with waveforms, can produce similar advantages.

For the sake of a demonstration, the Deutsch algorithm [32, Chapter 2] serves as a Rosetta stone of sorts for a better understanding of the formalism and respective machinery at work in such cases. It is based on the four possible binary functions $f_0, \ldots, f_3$ of a single bit $x \in \{0, 1\}$: the two constant functions

$f_0(x) = 1 - f_3(x) = 0$, as well as the two nonconstant functions: the identity $f_1(x) = x$ and the not $f_3(x) = (x+1)$ mod 2, respectively. Suppose that one is presented with a black box including in- and output interfaces, realizing one of these classical functional cases, but it is unknown which one. Suppose further that one is only interested in their parity; that is, whether or not the encoded black box function is a constant function of the argument. Thereby, with respect to the corresponding equivalence relation of being "(not) constant in the arguement" the set of functions $\{f_0, \ldots, f_3\}$ is partitioned into $\{\{f_0, f_3\}, \{f_1, f_2\}\}$.

A different way of looking at this relational encoding is in terms of zero-knowledge proofs: thereby nature is in the role of an agent which is queried about a property/proposition, and issues a correct answer without disclosing all the details and the fine structure of the way this result is obtained.

Classically the only way of figuring this ("constant or not") out is to input the two bit-state cases, corresponding to two separate queries. If the black box admits quantum states, then the Deutsch algorithm presents a way to obtain the answer ("constant or not") directly in one query. In order to do this one has to perform three successive steps [40, 44]:

- first one needs to scramble the classical bits into a coherent superposition of the two classical bit states. This can be done by a Hadamard transformation, or a quantum Fourier transformation;
- second, one has to transform the coherent superposition according to the binary function which is encoded in the box. This has to be done while maintaining reversibility; that is, by taking "enough" auxiliary bits to maintain bijectivity/permutation; even if the encoding function is many-to-one (eg, constant).
- third, one needs to unscramble this resulting state to produce a classical output signal which indicates the result of the query. As all involved transformations need to be unitary and thus reversible the latter task can again be achieved by an (inverse) Hadamard transformation, or an (inverse) quantum Fourier transformation.

This structural pattern repeats itself in many quantum algorithms suggested so far. It can be subsumed into the three- or rather fivefold framework: "prepare a classical state; then spread (the classical state into a coherent superposition of classical states) — transform (according to some functional form pertinent to the problem or query considered) — fold (into partitions of classical states which can be accessed via quantum queries and yield classical signals); then detect that classical signal."

Besides the (classical) pre- and post-processing of the data, Shor's algorithm [35, Chapter 5] has a very similar structure in its quantum (order-finding) core: It creates a superposition of classically mutually exclusive states i *via* a generalized Hadamard transformation. It then processes this coherent superposition of all i by computing x^i mod n, for some (externally given) x and n, the number to be factored. And it finally "folds back" the expanded, processed state by applying an inverse quantum Fourier transform, which then (with high probability) conveniently yields a piece of classical information (in one register) about the period or order; that is, the least positive integer k such that $x^k = 1(\text{mod } n)$ holds. As far as Shor's factoring algorithm is concerned, everything else is computed classically.

Partitioning of states may be related to the hidden subgroup problem [35, Section 5.4.3]: thereby, a function maps from some group to a finite set and is promised to be constant on cosets of the hidden subgroup. If those cosets are identified with the transformations "filtering" and "singling out" [15, 38–40] the elements of a partition of states associated with the particular problem, finding the hidden subgroup may yield an effective way of solving this problem (encoded by the state partition).

Whether or not this strategy to find "quantum oracles" corresponding to arbitrary partitions of classical cases is quantum feasible remains to be seen. There appears to be an *ad hoc* counterexample, as there is no speedup for generalized parity [20]; at least with the means considered.

3   Quantum Oracles for Random Numbers

Let me, for the sake of presenting another quantum resource mentioned in the beginning, contemplate one example for which, relative to the assumptions made, quantum "computation" outperforms classical recursion theory: the generation of (allegedly) irreducibly indeterministic numbers; or sequences thereof [7]. A recent extension of the Kochen-Specker theorem [4, 6, 8] allowing partial value assignments suggests the following algorithm: Suppose one prepares a quantized system capable of three or more mutually exclusive outcomes, formalized by Hilbert spaces of dimension three and higher, in an arbitrary pure state. Then, relative to certain reasonable assumptions (for value assignments and noncontextuality), this system cannot be in any defined, determined property in any other direction of Hilbert state not collinear or orthogonal to the vector associated with the prepared state [24, 36]: the associated classical truth assignment cannot be a total function. The proof by contradiction is constructive and involves a configuration of intertwining quantum contexts (aka orthonormal bases). Figure 1 depicts a particular configuration of quantum observables, as well as a particular one of their faithful orthogonal representations [28] in which the prepared and measured states are an angle $\arccos \langle \mathbf{a}|\mathbf{b}\rangle = \arccos\left[(1,0,0)\frac{1}{2}\left(\sqrt{2},1,1\right)^{\mathsf{T}}\right] = \frac{\pi}{4}$ apart [8, Table 1].

Whenever one approaches quantum indeterminacy from the empirical, inductive side, one has to recognize that, without *a priori* assumptions, formal proofs of (in)computability, and more so algorithmic incompressibility (aka randomness [29]) are blocked by reduction to the halting problems and similar [43]. The best one can do is to run tests, such as Borel normality and other criteria, on finite sequences of random number generators [5, 12] which turn out to be consistent with the aforementioned value indefiniteness and quantum indeterminacy.

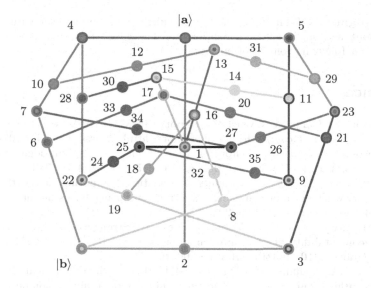

Fig. 1. Greechie orthogonality diagram of a logic [8, Fig. 2, p. 102201-8] realizable in $\mathbb{R}^3$ with the true–implies–value indefiniteness (neither true nor false) property on the atoms $|\mathbf{a}\rangle$ and $|\mathbf{b}\rangle$, respectively. The 8 classical value assignments require $|\mathbf{a}\rangle$ to be false. Therefore, if one prepares the quantized system in state $|\mathbf{a}\rangle$, the second state $|\mathbf{b}\rangle$ cannot have any consistent classical value assignment – it must be value indeterminate/indefinite.

4   Afterthoughts on Assumptions

Let me, as a substitute for a final discussion, mention a *caveat*: as all results and certifications hold true relative to the assumptions made, different assumptions and axioms may change the perceptual framework and results entirely. One might, for instance, disapprove of the physical existence of states and observables beyond a single vector or context [9,42]. Thereby, the problem of measuring other contexts would be relegated to the general measurement problem of coherent superpositions [27]. In this case, as von Neumann, Wigner and Everett have pointed out, by "nesting" the measurement objects and the measurement apparatus in larger and larger systems [19], the assumption of the universal validity of the quantum state evolution would result in mere epistemic randomness; very much like the randomness encountered in, and the second law of [34], classical statistical physics. From that perspective, quantum randomness might turn out to be valid "for all practical purposes" [10] through interaction with a huge number of (uncontrollable) degrees of freedom in the environment of a quantized system in a coherent state, "squeezing" out this coherence very much like a balloon losing gas [11].

Acknowledgments. I kindly acknowledge enlightening discussions with Cristian Calude about many of the subjects mentioned. All misconceptions and errors are mine. I declare that I have no conflict of interest.

References

1. Aaronson, S.: Happy new year! My response to M. I. Dyakonov (1999). http://www.scottaaronson.com/writings/bignumbers.html. Accessed 16 Mar 2017
2. Aaronson, S.: Quantum Computing Since Democritus. Cambridge University Press, New York (2013). https://doi.org/10.1017/CBO9780511979309
3. Abbott, A.A., Calude, C.S.: Limits of quantum computing: a sceptic's view. http://www.quantumforquants.org/quantum-computing/limits-of-quantum-computing/. Accessed 19 June 2016
4. Abbott, A.A., Calude, C.S., Conder, J., Svozil, K.: Strong Kochen-Specker theorem and incomputability of quantum randomness. Phys. Rev. A **86**, 062109 (2012). https://doi.org/10.1103/PhysRevA.86.062109
5. Abbott, A.A., Calude, C.S., Dinneen, M.J., Huang, N.: Experimentally probing the algorithmic randomness and incomputability of quantum randomness. Physica Scripta **94**(4), 045103 (2019). https://doi.org/10.1088/1402-4896/aaf36a
6. Abbott, A.A., Calude, C.S., Svozil, K.: Value-indefinite observables are almost everywhere. Phys. Rev. A **89**, 032109 (2014). https://doi.org/10.1103/PhysRevA.89.032109
7. Abbott, A.A., Calude, C.S., Svozil, K.: On the unpredictability of individual quantum measurement outcomes. In: Beklemishev, L.D., Blass, A., Dershowitz, N., Finkbeiner, B., Schulte, W. (eds.) Fields of Logic and Computation II. LNCS, vol. 9300, pp. 69–86. Springer, Cham (2015). https://doi.org/10.1007/978-3-319-23534-9_4
8. Abbott, A.A., Calude, C.S., Svozil, K.: A variant of the Kochen-Specker theorem localising value indefiniteness. J. Math. Phys. **56**(10), 102201 (2015). https://doi.org/10.1063/1.4931658
9. Auffèves, A., Grangier, P.: Extracontextuality and extravalence in quantum mechanics. Philos. Trans. R. So. A: Math. Phys. Eng. Sci. **376**(2123), 20170311 (2018). https://doi.org/10.1098/rsta.2017.0311
10. Bell, J.S.: Against 'measurement'. Phys. World **3**, 33–41 (1990). https://doi.org/10.1088/2058-7058/3/8/26
11. Bengtsson, I., Zyczkowski, K.: Geometry of quantum states - addendum (2018). http://chaos.if.uj.edu.pl/~karol/decoh18.pdf. Accessed 24 Mar 2019
12. Calude, C.S., Dinneen, M.J., Dumitrescu, M., Svozil, K.: Experimental evidence of quantum randomness incomputability. Phys. Rev. A **82**(2), 022102 (2010). https://doi.org/10.1103/PhysRevA.82.022102
13. Clauser, J.: Early history of Bell's theorem. In: Bertlmann, R., Zeilinger, A. (eds.) Quantum (Un)speakables: From Bell to Quantum Information, pp. 61–96. Springer, Berlin (2002). https://doi.org/10.1007/978-3-662-05032-3_6
14. Dirac, P.A.M.: The Principles of Quantum Mechanics, 4th edn. Oxford University Press, Oxford (1930, 1958)
15. Donath, N., Svozil, K.: Finding a state among a complete set of orthogonal ones. Phys. Rev. A **65**, 044302 (2002). https://doi.org/10.1103/PhysRevA.65.044302
16. Dyakonov, M.I.: State of the Art and Prospects for Quantum Computing, chap. 20, pp. 266–285. Wiley (2013). https://doi.org/10.1002/9781118678107.ch20

17. Dyakonov, M.I.: When will we have a quantum computer? (2019). https://arxiv.
org/abs/1903.10760, talk at the conference "Future trends in microelectronics",
Sardinia (2018). To be published in a special issue of Solid State Electronics
18. Englert, B.G.: On quantum theory. Eur. Phys. J. D **67**(11), 1–16 (2013). https://
doi.org/10.1140/epjd/e2013-40486-5
19. Everett III, H.: The Everett Interpretation of Quantum Mechanics: Collected
Works 1955–1980 with Commentary. Princeton University Press, Princeton (2012).
http://press.princeton.edu/titles/9770.html
20. Farhi, E., Goldstone, J., Gutmann, S., Sipser, M.: Limit on the speed of quantum
computation in determining parity. Phys. Rev. Lett. **81**, 5442–5444 (1998). https://
doi.org/10.1103/PhysRevLett.81.5442
21. Feynman, R.P.: The Character of Physical Law. MIT Press, Cambridge (1965)
22. Fuchs, C.A., Peres, A.: Quantum theory needs no 'interpretation'. Phys. Today
53(4), 70–71 (2000). https://doi.org/10.1063/1.883004, further discussions of and
reactions to the article can be found in the September issue of Physics Today, 53,
11–14 (2000)
23. Gleason, A.M.: Measures on the closed subspaces of a Hilbert space. J. Math. Mech.
(now Indiana Univ. Math. J.) **6**(4), 885–893 (1957). https://doi.org/10.1512/iumj.
1957.6.56050
24. Hrushovski, E., Pitowsky, I.: Generalizations of Kochen and Specker's theorem and
the effectiveness of Gleason's theorem. Stud. Hist. Philos. Sci. Part B: Stud. Hist.
Philos. Mod. Phys. **35**(2), 177–194 (2004). https://doi.org/10.1016/j.shpsb.2003.
10.002
25. Jordan, S.: Quantum Algorithm Zoo (2011–2019). http://quantumalgorithmzoo.
org/. Accessed 26 Mar 2019
26. London, F., Bauer, E.: La theorie de l'observation en mécanique quantique; No. 775
of Actualités scientifiques et industrielles: Exposés de physique générale, publiés
sous la direction de Paul Langevin. Hermann, Paris (1939). English translation
in [27]
27. London, F., Bauer, E.: The theory of observation in quantum mechanics. In: Quantum Theory and Measurement, pp. 217–259. Princeton University Press, Princeton
(1983). Consolidated Translation of French Original [26]
28. Lovász, L.: On the Shannon capacity of a graph. IEEE Trans. Inf. Theory **25**(1),
1–7 (1979). https://doi.org/10.1109/TIT.1979.1055985
29. Martin-Löf, P.: On the notion of randomness. In: Kino, A., Myhill, J., Vesley,
R.E. (eds.) Intuitionism and Proof Theory, p. 73. North-Holland, Amsterdam and
London (1970)
30. Mermin, D.N.: Could Feynman have said this? Phys. Today **57**, 10–11 (1989).
https://doi.org/10.1063/1.1768652
31. Mermin, D.N.: What's wrong with this pillow? Phys. Today **42**, 9–11 (1989).
https://doi.org/10.1063/1.2810963
32. Mermin, D.N.: Quantum Computer Science. Cambridge University Press, Cambridge (2007). https://doi.org/10.1017/CBO9780511813870
33. Mermin, D.N.: Making better sense of quantum mechanics (2019). https://arxiv.
org/abs/1809.01639
34. Myrvold, W.C.: Statistical mechanics and thermodynamics: a Maxwellian view.
Stud. Hist. Philos. Sci. Part B: Stud. Hist. Philos. Mod. Phys. **42**(4), 237–243
(2011). https://doi.org/10.1016/j.shpsb.2011.07.001
35. Nielsen, M.A., Chuang, I.L.: Quantum Computation and Quantum Information. Cambridge University Press, Cambridge (2010). https://doi.org/10.1017/
CBO9780511976667, 10th Anniversary Edition

36. Pitowsky, I.: Infinite and finite Gleason's theorems and the logic of indeterminacy. J. Math. Phys. **39**(1), 218–228 (1998). https://doi.org/10.1063/1.532334
37. Schrödinger, E.: Die gegenwärtige Situation in der Quantenmechanik. Naturwissenschaften **23**, 807–812, 823–828, 844–849 (1935). https://doi.org/10.1007/BF01491891. https://doi.org/10.1007/BF01491914. https://doi.org/10.1007/BF01491987
38. Svozil, K.: Quantum information in base n defined by state partitions. Phys. Rev. A **66**, 044306 (2002). https://doi.org/10.1103/PhysRevA.66.044306
39. Svozil, K.: Quantum information via state partitions and the context translation principle. J. Mod. Opt. **51**, 811–819 (2004). https://doi.org/10.1080/09500340410001664179
40. Svozil, K.: Characterization of quantum computable decision problems by state discrimination. In: Adenier, G., Khrennikov, A., Nieuwenhuizen, T.M. (eds.) Quantum Theory: Reconsideration of Foundations–3, vol. 810, pp. 271–279. American Institute of Physics (2006). https://doi.org/10.1063/1.2158729
41. Svozil, K.: Quantum hocus-pocus. Ethics Sci. Environ. Politics (ESEP) **16**(1), 25–30 (2016). https://doi.org/10.3354/esep00171
42. Svozil, K.: New forms of quantum value indefiniteness suggest that incompatible views on contexts are epistemic. Entropy **20**(6), 406(22) (2018). https://doi.org/10.3390/e20060406
43. Svozil, K.: Physical (A)Causality. Determinism, Randomness and Uncaused Events. FTP, vol. 192. Springer, Cham (2018). https://doi.org/10.1007/978-3-319-70815-7
44. Svozil, K., Tkadlec, J.: On the solution of trivalent decision problems by quantum state identification. Natural Computing (2009, in print). https://doi.org/10.1007/s11047-009-9112-5
45. Turing, A.M.: Intelligent machinery. In: Evans, C.R., Robertson, A.D.J. (eds.) Cybernetics. Key Papers, pp. 27–52. Butterworths, London (1968)
46. Wiesner, K.: The careless use of language in quantum information. https://arxiv.org/abs/1705.06768

Author Index

Printed in the United States
By Bookmasters